Springer Proceedings in Energy

The series Springer Proceedings in Energy covers a broad range of multidisciplinary subjects in those research fields closely related to present and future forms of energy as a resource for human societies. Typically based on material presented at conferences, workshops and similar scientific meetings, volumes published in this series will constitute comprehensive state-of-the-art references on energy-related science and technology studies. The subjects of these conferences will fall typically within these broad categories:

- Energy Efficiency
- Fossil Fuels
- Nuclear Energy
- Policy, Economics, Management & Transport
- Renewable and Green Energy
- Systems, Storage and Harvesting
- Materials for Energy

eBook Volumes in the Springer Proceedings in Energy will be available online in the world's most extensive eBook collection, as part of the Springer Energy eBook Collection. Please send your proposals/inquiry to Dr. Loyola DSilva, Senior Publishing Editor, Springer (loyola.dsilva@springer.com)

More information about this series at http://www.springer.com/series/13370

Paolo Bertoldi
Editors

Energy Efficiency in Domestic Appliances and Lighting

Proceedings of the 10th International Conference (EEDAL'19)

Springer

Editor
Paolo Bertoldi
European Commission Joint Research Centre
Institute for Energy, Transport and Climate
Ispra, Varese, Italy

ISSN 2352-2534 ISSN 2352-2542 (electronic)
Springer Proceedings in Energy
ISBN 978-3-030-79123-0 ISBN 978-3-030-79124-7 (eBook)
https://doi.org/10.1007/978-3-030-79124-7

This Springer imprint is published by the registered company Springer Nature Switzerland AG
The registered company address is: Gewerbestrasse 11, 6330 Cham, Switzerland

Contents

Energy Savings 2021: New European Energy Label and Ecodesign Requirements for Washing Machines

Gundula Czyzewski

Abstract The new EU Regulations on ecodesign and energy label for washing machines are due to be published. These two measures will create a momentum for energy savings under EU policies on energy efficiency and mitigate climatic change. Based on the framework for energy labeling the new classification will be again on an A–G scale like in the first energy label that entered into force in the 1990s. CENELEC TC 59X/WG 01-06 has already started its work for the next edition of the European standard EN 60456. Well-known performance parameters like washing performance and spinning performance will remain relevant criteria, rinsing performance is one of the new elements. A measurement procedure for the laundry temperature is under development. The review of EN 60456 includes adjusting the measurements for these parameters to be in line with the usage of washing machines by testing full, half and quarter loads at new programme cycle "eco 40-60" as required by the new regulations. A balance between test effort, repeatability and reproducibility is part of the draft of a new test scheme.

1 Introduction

The first European Energy Label for household washing machines was introduced in 1995. The green-yellow-red bars are well known and these 7 bars have stayed the same since the first introduction of the energy label. Many non-European countries have copied this successful visual element for their regional schemes.

Energy labelling and ecodesign requirements are the two key policy instruments for energy savings of household appliances in Europe. The European energy label has provided consumers for many years with standardized information on energy consumption with an A to G classification, linked to relevant performance criteria of major household appliances. For washing machines the label included water consumption as second resource related information. Performance related criteria like

G. Czyzewski (✉)
BSH Home Appliances, Munich, Germany
e-mail: Gundula.Czyzewski@BSHG.com

P. Bertoldi (ed.), *Energy Efficiency in Domestic Appliances and Lighting*, Springer Proceedings in Energy, https://doi.org/10.1007/978-3-030-79124-7_1

washing performance and spinning efficiency have been considered essential elements of the label since the beginning. A redesigned label with top classes A+, A++ and A+++ replaced the first European energy label in 2011. This second energy label was combined with eco-design measures.

The European Commission started in 2014 the second revision of the energy label and a revision of the first ecodesign measures for washing machines. The combined preparatory work for household washing machines and washer-dryers included market research, consumer studies and stakeholder consultations. The final report of the preparatory study was published in 2017 [1].

The draft documents for the next revision of policy measures for washing machines and washer-dryers became available early 2019 and the publication of the final regulations is expected for autumn 2019.

The draft documents are already available at European Commission's web pages.

Label—Delegated Act: COMMISSION DELEGATED REGULATION (EU) .../... of 11.3.2019 supplementing Regulation (EU) 2017/1369 of the European Parliament and of the Council with regard to energy labelling of household washing machines and household washer-dryers and repealing Commission Delegated Regulation (EU) No 1061/2010 and Commission Directive 96/60/EC [2].

Ecodesign: COMMISSION REGULATION (EU) .../... of XXX laying down ecodesign requirements for household washing machines and household washer-dryers pursuant to Directive 2009/125/EC of the European Parliament and of the Council, amending Commission Regulation (EC) No 1275/2008 and repealing Commission Regulation (EU) No 1015/2010 [3].

The two draft regulations (eco-design and label) are combined to cover both washing machines and washer dryers. Currently the washer dryer is still labelled according to the first energy label directive, so far there were no eco-design requirements for washer dryers defined. In future washer dryers have to follow in principle the same requirements and testing scenarios like washing machines, but there are additional requirements for washer dryer specifics like continuous wash & dry. The testing of washer dryers will be more complex than today. A joined work is currently running in standardization committee CENELEC TC59X WG01, WG01-06 and WG01-11 dealing with performance of laundry care appliances to prepare standards for both appliances in parallel. The target is to use as much as possible synergies in developing combined standardized testing schemes for washing machines and washer dryers.

This paper focuses on washing machines; requirements for washer dryers are mentioned only for reference and not explained in detail.

March 2021 is the final schedule: The third generation of energy labels in Europe linked to the second generation of ecodesign measures will become effective. The transition from the current energy label to the future label will already start in 2020.

Europeans know the Energy label and they keep it in mind when buying appliances. During the revision there was a lot of effort spent to understand consumer habits. A consumer study [4] revealed that consumers say they understand the label and they use the most energy efficient programmes, but it became obvious that this is not always the case. Consumers do not realize that the energy label refers to selected programmes but not to all programmes of washing machines. There were statements that their most energy saving programme lasts less than 2 h. This has not been the case anymore for many years. Time matters a lot to the consumers—therefore this was one of the relevant factors of this revision.

2 The Three Generations of European Energy Labels for Washing Machines

The European Energy Label displays the efficiency of electrical household appliances to consumers and focuses clearly on **energy efficiency**.

The first label introduced an A to G classification for energy ratings. The label classes were defined based on the measured energy consumption of a Cotton 60 °C programme at rated capacity and for classification the cycle energy consumption was calculated per kg. Class A for the most efficient machines required 0.19 kWh/kg.

Rated capacity stands for the maximum amount of dry cotton textiles according to the specification in the relevant standard EN60456 that can be loaded into a washing machine following a standardized loading scheme. In real life loading will vary a lot. Consumers will not always use the maximum loading capacity of a machine. This was considered in the second label scheme introduced in 2011.

Today household washing machines need to have an energy label indicating their energy efficiency on a scale from A+++ (most efficient) to D (least efficient). Three additional classes, A+, A++, and A+++ were added to the original classification scheme on top and eco-design measures have limited the range of washing machines in the European market to these 3 top classes since December 2013. This is not clearly visible at the point of sales, but market figures show that in 2018 more than 70% of washing machines are labelled A+++ or better.

The Energy Efficiency Index (EEI) was introduced to classify energy consumption from A+++ to D, it is the ratio of the annual energy consumption of a household washing machine to the standard annual energy consumption of a washing machine with the same capacity. The classification is based on the measurement of Cotton 40 °C at half load and Cotton 60 °C at rated capacity and at half load. The European measurement standard has defined details of testing, 3 test runs at Cotton 60 °C full load, 2 test runs each at Cotton 60 °C half load and Cotton 40 °C half load. The Energy Efficiency Index (EEI) refers to annual energy consumption, calculating a weighted average of the 7 measurements and including additionally low power modes for off mode and left on mode.

The next energy label will go back to the concept of the first label: Classes A–G will be used again, after rescaling. There should be no A class appliances when the new label is introduced. The Energy Efficiency Index as such will remain but new calculations are given in the draft regulation. The classification will be based on the measurement of only one programme but with measurements at quarter, half and full load to better mirror the user behaviour.

Different units are used to display the energy consumption figures below the coloured bars in the three generations of the European energy label: in the first edition it was kWh/cycle, in the second edition it is the annual energy consumption based on 220 washing cycles and starting 2021 it will be the weighted energy consumption for 100 washing cycles. Therefore it is not easy to compare the energy consumption between the different schemes. The first label encouraged manufacturers to optimize the energy consumption for the maximum load (rated capacity), with the second label a balance between full and half load was pushed and the next label aims at an overall improved energy consumption for small, medium and large loads.

The rating of the energy efficiency has always been linked to **washing performance**, so savings did not affect cleaning results in a negative way. The so-called washing efficiency index refers to a reference programme, which was established in 1995 with the first energy label. Washing performance was a key figure, even classified A to G on the label. After some years there were only class A machines regarding washing efficiency left on the market. Who would buy a machine that does not clean well? The second energy label did not display the washing efficiency classes anymore, an eco-design measure required a minimum washing efficiency index better than 1.03, the value of the former A class. This remains as a minimum requirement also for the upcoming eco-design measures linked to the third wave of regulations. Now the minimum washing performance must be reached at each load size tested and not only in average as today.

Water consumption is the second resource related factor being part of the label. But savings were not linked to rinsing effectiveness so far. This will now change, introducing a minimum rinsing effectiveness as an eco-design requirement.

Different units were used to display water consumption in the label: in the first edition, it was liters/cycle, in the second edition it is the annual water consumption based on 220 washing cycles and starting 2021 it will be again liters/cycle based on weighted water consumption of quarter, half and full load measurements. Therefore it is also not easy to compare the water consumption between the different schemes. The first label encouraged manufacturers to optimize the water consumption for the maximum load (rated capacity). With the second label an eco-design requirement was introduced to limit the water consumption at full load and that means that there was no real balance between full and half load. The next label aims at an overall improved water consumption for small, medium and large loads taking for the first time rinsing effectiveness into consideration.

The **spin-drying efficiency** classification was already introduced in 1995 and the A to G scheme has not been changed and will not change. Spin-drying efficiency is relevant when dryers are used, this needs to be communicated at the point of sales to help consumers taking the right decisions.

The **airborne acoustic noise emissions** were displayed on the first label but it was not mandatory for all European countries. It became a mandatory part of the label in 2011, with numbers expressed in expressed in dB(A) for washing and spinning. Only the noise emissions of the spinning phase will remain on display for next label. New is a A to D classification for noise.

One relevant new element of the next label is **programme duration**. The duration of the programme at rated capacity will be displayed in hours and minutes. This requirement is linked to ecodesign requirements limiting maximum programme duration.

The first energy label did not provide any information about programme duration Cotton 60 °C programmes at the end of the 90th had a duration of about 2 h maximum. That changed already with the second energy label, but programme duration is only revealed in the mandatory fiche, that should be available at the point of sales and in user manuals. Washing programmes today last many hours, machines with more than 6 h programme duration for full load are found in advertisements. That will change now, limiting the programme duration for a full load regardless of the capacity of a washing machine to a maximum of 4 h. The so-called time cap depends on rated capacity, details are explained in the chapter about Ecodesign reuirements.

3 A New Programme for Europeans: Eco 40-60

The programme offer of washing machine has changed over time. The most used programme is still the cotton programme. The first energy label had a focus on the Cotton 60 °C programme. Washing habits have changed too over time; lower washing temperatures and varying loads are consumers' reality. Studies show that most customers wash at 40 °C [1]. The current energy label, introduced in 2011, has already shifted to lower temperatures and considered half loads.

The next energy label will refer to a washing programme called 'eco 40-60', which is able to clean normally soiled cotton laundry declared to be washable at 40 °C or 60 °C together in the same cycle.

Today the label programmes are marked with an arrow. The challenge is now how to guide the consumer to use this new programme. Is a new care label required to help guiding consumers? (Fig. 1)

The name 'eco 40-60' shall be used exclusively for this programme. No other programme may have in its name the term 'eco'. The eco 40-60 programme shall be set as the default programme for automatic programme selection. The indications 'normal', 'daily', 'regular' and 'standard', and their translations in all EU official languages, shall not be used in programme names for household washing machines [3].

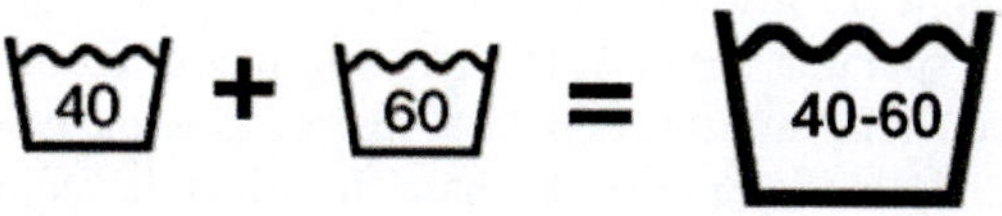

Fig. 1 New programme Eco 40-60 [5]

4 Ecodesign Requirements

The draft ecodesign regulation [3] includes well known measures but introduces also new requirements.

The first wave of ecodesign measures in 2011 combined specific and generic ecodesign requirements. Ecodesign requirements that will enter into force in 2021 will include programme requirements, energy efficiency and functional requirements. New requirements refer to programme duration, to laundry temperature and to new elements like reparability and spare part availability.

Specific requirements in 2011 defined minimum energy efficiency phasing out classes, maximum water consumption at full load only and functional requirements like a minimum washing efficiency index. These requirements are revised for 2021 adding new ones like minimum rinsing effectiveness and programme time limitations.

The EEI factor of the next regulation comes with a complex formula. The weighted average energy consumption is set in relation to the weighted standard cycle energy consumption. The weighted average energy consumption is based on testing the appliance in the new programme Eco 40-60 for a quarter, a half and a full load. The weighing factors depend on the rated capacity. The weighted standard cycle energy consumption is a quadratic formula linked to the rated capacity too.

With entry into force the EEI of household washing machines shall be lower than 105, that means class G of the new labelling scheme is already very narrow (G: EEI > 102). Three years later household washing machines with a rated capacity higher than 3 kg shall have an EEI of less than 91, that means class F and G are to be phased out. For better visualisation of relevant energy classes it is required that EEI classes that are no longer allowed will be shown in grey.

Generic requirements in 2011 defined the so-called standard cotton programmes and required that washing machines shall offer to end-users a cycle at 20 °C. The content of the booklet of instruction was regulated for aspects like indicative information on the programme time, remaining moisture content, and energy and water consumption for the main washing programmes, even though it was not clearly stated which are the main washing programmes.

Among the programmes requirements for 2021 is the introduction of the new programme Eco 40-60 and the refinement of the requirement for the 20 °C cycle, a washing cycle called '20 °C', which is able to clean lightly soiled cotton laundry, at a nominal temperature of 20 °C.

The mandatory content of the booklet of instructions includes now a lot of aspects. Required are e.g. an explanation about the eco 40-60 programme and that the most efficient programmes in terms of energy consumption are generally those that perform at lower temperatures and longer duration; information that the higher the spinning speed in the spinning phase, the higher the noise and the lower the remaining moisture content; instructions on how to find the model information stored in the product database. Information requirements include the temperature inside the laundry for several programmes.

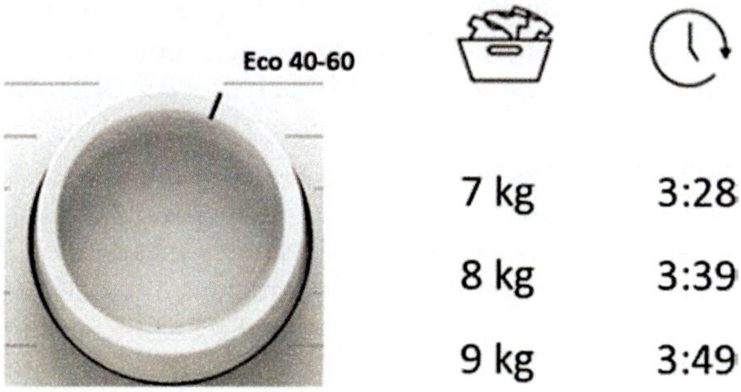

Fig. 2 Limitations of programme duration at rated capacity (full load)

As mentioned before time matters a lot, that's why time caps are introduced. Pictured is the maximum duration of a 7, a 8 and a 9 kg washing machine when fully loaded. This follows a formula and is linked to rated capacity, but the maximum accepted duration is 240 min, that means 4 h regardless of higher rated capacities. For half and quarter load a different formula applies that leads to shorter times, with an overall maximum of 3 h (Fig. 2).

There are revised eco-design requirements for low power modes. No later than 15 min after end of any activity a switch to standby or off will be mandatory. The new limits for low power modes are 0.50 W for off-mode and/or stand-by mode, if the stand-by mode includes the display of information or status, the power consumption shall not exceed 1.00 W, networked standby is limited to 2.00 W. The new regulation includes also limits for delay start, the power consumption shall not exceed 4.00 W and the mode should not be programmable by for more than 24 h.

For resource efficiency key topics are reparability, spare part availability and spare part delivery time. This new field of ecodesign requirements needs further considerations. The draft standardization request made available to CEN and CENELEC is under discussion now. Performance requirements are already under work and will be handled by CENELEC TC59X. For product specific resource efficiency topics new work items must be established soon.

5 CENELEC Work

CENELEC TC59X WG1 has already started its work to update the European standard EN60456 Clothes washing machines for household use—Methods for measuring the performance as a basis for a revised European Energy Label. This includes the development of a set of standard tests for new parameters like temperature inside laundry and a revised method for rinsing effectiveness.

A new test procedure for a combined test at 40 °C with full, half and quarter load is under development taking into account to facilitate washer dryer testing.

European standards are usually based on international IEC standards, defining test materials, test equipment, lab conditions and general procedures how to test. As time matters for the introduction of the next label it was decided to start preparing the next standard with measurement procedures for rinsing and temperature in advance as technical specifications (TS).

What is good rinsing? A questions asked for a long time and discussed among stakeholders of laundry business. How does consumers' observation link to something that can be measured? In some countries or regions it is mandatory to test for **rinsing effectiveness**, in others it is voluntary or checked in consumer tests. The IEC standard 60456 on washing machine performance so far includes the alkalinity method, but it is planned to add the new LAS method for the next edition.

CENELEC TC59X WG 01-08 has developed TS 50677. This Technical Specification was approved by CENELEC on 2018-12-31 and published in March 2019.

The new method specifies the evaluation of the rinsing effectiveness of household clothes washing machines, washer dryers and commercial washing machines based on the amount of residual linear alkylbenzene sulfonate surfactant (LAS), a key ingredient of the detergent, extracted from the unstained test swatches of the stain strips used in the washing performance test. Remaining LAS on textile surface is extracted by shaking the swatches in bottles with water. Samples from the water extract are investigated via ultraviolet (UV) light absorbance at the wavelength particular to LAS. A fixed linear relationship between LAS amount and quantity of detergent mixture is the basis for this evaluation. Using a concentration versus absorbance curve is part of this procedure. The absorbance values are converted into detergent concentrations, which together with the test solution mass data, yields detergent quantities. On the textiles, a linear relationship is not given, but it is nevertheless used to express the amount of LAS as determined by UV light absorbance measurements in terms of a detergent amount [6].

Washing temperatures have been subject of discussions for a long time. One needs to know: there is not "the washing temperature" of a washing programme. Usually a washing programme starts with filling cold water from the tap, flushing in detergent, wetting textiles. Later the pool of free water is heated in the lower part of the tub of a washing machine, where usually the heating element is located. The temperature inside the laundry starts at room temperature, is lowered by soaking up cold water and later being in contact with pool of heated water the temperature inside the laundry is raising slow but steady to a peak temperature until the heating element is switched off. What matters is the maximum temperature reached inside the textiles. The maximum temperature should not exceed the temperature given on the textile care label to avoid textile damages.

It is common today to reach lower temperatures than the programme name indicates. From consumer tests it is known that the temperatures reached inside the sump of a washing machine, close to the heating element, for a Cotton 60 °C programme vary a lot and do not reach 60 °C for most the machines in the market today. If the temperature in the sump of a washing machine is less than 60 °C the temperature inside the laundry is even less.

The current regulation on washing machines requires that the booklet of instructions provided by the manufacturer shall include an indication that the actual water temperature may differ from the declared cycle temperature [7].

This has caused a lot of discussion during the revision of the current energy label. There were opinions to specify a minimum temperature to be reached inside the laundry or to require the temperature reached to be the same like the programme cycle temperature given in the user interface. The discussion included also the question how long the temperature should be reached inside the laundry and how to check if there is an even temperature distribution inside the laundry. After all it will become mandatory to provide information about temperatures reached inside the laundry. User instructions shall include maximum temperature reached for minimum 5 min inside the laundry for a number of programmes but at least for the new eco 40-60 programme and for the mandatory 20 °C programme.

CENELEC TC59X WG 01-06 therefore currently develops a new TS, measurement and evaluation method to determine a representative temperature reached inside the load during the washing cycle.

The intention of this test specification is to define a measurement and evaluation method to determine a representative maximum temperature reached inside the base load during the washing cycle of a washing machine or washer dryer. The main idea is to measure a mean maximum temperature within the base load with three temperature sensors which are attached to the towels or pillowcases placed in different, representative locations inside the drum [8].

Testing washing machines becomes again more effort, but it is still possible to test a machine within 1 week. Ten test runs will be required to test a washing machine: three test runs with full load (rated capacity), four test runs with half load and three test runs with a quarter load. All tests combine measurements for cycle energy consumption and cycle water consumption with performance measurements of washing, rinsing and spinning and include the new measurement for laundry temperature.

The amendment to the European washing machine standard EN60456 was recently circulated for enquiry within CENELEC TC59X [9]. New elements are mentioned as follows:

The programme to be tested for the combined test series will be the new introduced ‘eco 40-60’ programme. This programme needs to be tested with default settings with the given temperature. Test loads are full, half and quarter of the rated capacity of the washing machine. A new quarter load is introduced and defined to be approximately a quarter of the rated capacity. The quarter load is treated as a separate load and not created by dividing full or half loads. The total number of test runs is 10, including 3 tests with full load, 4 tests with half load and 3 tests with quarter load. The time between two subsequent test runs within 1 day is reduced from 2 to 1 h.

The procedure to measure **low power modes** will be modified that required low power modes (delay start, off mode and standby mode) can be tested linked to any washing programme.

A new annex will define the testing procedure for **multi-drum washing machines** in simultaneous mode (multi-drum mode). A washing machine with more than one drum for the complete treatment of the textiles, where drums cannot be operated simultaneously is not regarded as having a multi drum mode. In this case each drum has to be tested separately.

The documents for the first enquiry for prEN60456 are already available. A round robin test is under preparation. Due to the fact that washer-dryers and washing machines are regulated together and testing is to be combined it was decided to run the next round robin test with washer dryers to be tested as washing machines and as washer dryer. This round robin test aims to check the clarity of the new standards for washing machines and washer dryers and to establish uncertainty figures. The accuracy of the measuring system in terms of repeatability and reproducibility is evaluated according to IEC 61923. Participating laboratories will be asked to carry out tests on at least one of the two machine types. A questionnaire will be used to check laboratory equipment and procedures.

6 Summary

The revision of the European energy label links energy and water savings to performance: washing and rinsing. The information requirements will provide transparency on the temperatures reached. Savings 2021 can be realized if the consumer follows the proposal to use the new programme Eco 40-60.

References

1. Boyano Larriba, A., Cordella, M., Espinosa Martinez, M., Villanueva Krzyzaniak, A., Graulich, K., Rüdinauer, I., Alborzi, F., Hook, I. and Stamminger, R., Ecodesign and Energy Label for household washing machines and washer dryers, EUR 28809 EN, Publications Office of the European Union, Luxembourg, 2017, ISBN 978-92-79-74183-8, https://doi.org/10.2760/029939, JRC109033
2. Public Register of European Parliament; Draft documents were published 2019-03-11 under the following document numbers: COM_ADL(2019)01804 and Annexes COM_ADL(2019)01804(ANN), Draft delegated acts (energy labelling of household washing machines and household washer-dryers) can be downloaded at: http://www.europarl.europa.eu/RegistreWeb/search/typedoc.htm?codeTypeDocu=ACTDEL&currentPage=7
3. European Commission. Draft documents, Comitology register, Can be downloaded at: http://ec.europa.eu/transparency/regcomitology/index.cfm?do=Search.Search&NewSearch=1. Use key word "washing machines" for your search; D06001702
4. Alborzi, Angelika Schmitz, and Rainer Stamminger (2017). Consumers' Comprehension of the EU Energy Label for Washing Machines. Tenside Surfactants Detergents: Vol. 54, No. 4, pp. 280-290.
5. Energy savings 2021: New label and EcoDesign requirements for washing machines, 49. International Detergency Conference 2019

6. Technical Specification CLC/TS 50677: Clothes washing machines and washer-dryers for household and similar use—Method for the determination of rinsing effectiveness by measurement of the surfactant content at textile materials—March 2019
7. COMMISSION REGULATION (EU) No 1015/2010 of 10 November 2010 implementing Directive 2009/125/EC of the European Parliament and of the Council with regard to ecodesign requirements for household washing machines, Official Journal of the European Union, L 293/21, 11.11.2010
8. Draft Technical Specification, Questionnaire on New Work Item Proposal, TC59X/Sec0723/NP, TC59X/Sec0724/Q, February 2019 and Enquiry Draft CLC/prTS 50707:2019 (69499)
9. Enquiry EN 60456:2016/prAA:2019

Towards Consumer-Relevant Product Testing

Gundula Czyzewski, Gerhard Fuchs, Alain Roux, Hans-Paul Siderius, Christoforos Spiliotopoulos, and Rainer Stamminger

Abstract Consumers and users of household appliances nowadays increasingly require information on the energy performance and other performance parameters of these products. The test methodologies from which this information is obtained subsequently need to have certain characteristics to be accurate and reliable. They need to be repeatable and reproducible so as to produce the same results regardless of test repetitions or different laboratories but also need to be representative of the conditions and user behavior observed when these appliances are used in real life. Furthermore, the cost of product testing also needs to be reasonable.

This poses a challenge of both a technical and political nature: How can test methods be repeatable and reproducible without compromising representativeness? Where can the trade-off be found amongst these characteristics in order for a test procedure to be consumer-relevant, in other words, appropriate for providing information to consumers and support regulatory provisions?

A general methodology to assess the consumer-relevance of test methods was presented at EEDAL 2017, offering a first step towards more systematic assessments

G. Czyzewski
BSH Home Appliances, Berlin, Germany
e-mail: Gundula.Czyzewski@bshg.com

G. Fuchs (✉)
BSH Home Appliances, Munich, Germany
e-mail: Gerhard.Fuchs@bshg.com

A. Roux
Brandt France, Orleans, France
e-mail: Alain.Roux@groupebrandt.com

H.-P. Siderius
Rijksdienst voor Ondernemend Nederland, Utrecht, the Netherlands
e-mail: hans-paul.siderius@rvo.nl

C. Spiliotopoulos
European Environmental Citizens' Organisation for Standardisation (ECOS), Brussels, Belgium

R. Stamminger
Institut für Landtechnik, Universität Bonn, Bonn, Germany
e-mail: stamminger@uni-bonn.de

P. Bertoldi (ed.), *Energy Efficiency in Domestic Appliances and Lighting*, Springer Proceedings in Energy, https://doi.org/10.1007/978-3-030-79124-7_2

of test methods. This paper now presents further findings from the implementation of the methodology at a product-specific level and at standardization working groups. It provides lessons learnt from its transposition to selected product categories, for example, vacuum cleaners, washing machines and refrigerators, and detailed insights about how specificities of such products (such as load definition and interruptions in operation) have an impact on the assessment.

1 Introduction

Testing products to the requirements defined in measurement standards is the usual way when assessing characteristics such as performance parameters, for example, energy consumption, water consumption or sound level. The results of such testing is ordinarily used to determine if the product meets legal requirements, such as whether the energy label classification is correct, or to benchmark the product against other products. Standards should ensure that tests are carried out in a uniform, standardized way so that results reflect product characteristics accurately and demonstrate the differences between products in a fair way in case several products are tested.

Such tests regularly take place within the context of the legislation of the European Union (EU), such as ecodesign and energy labelling, whereby the results are used to provide presumption of conformity and also inform consumers through, for example, product information and energy labels; this product testing should also be relevant to consumers.

It is explicitly stated in the new EU Regulation on labelling (EU) 2017/1369 that "harmonized standards shall simulate real-life usage …", [1].

Authors at the EEDAL 2017 conference, in Irvine, US, presented and discussed with the audience a novel assessment method for the representativeness of laboratory measurement methods with the aim of not compromising the testing accuracy in terms of reproducibility and repeatability and also considering the cost of the testing. The method was developed by CENELEC Technical Committee 59X Performance of household and similar electrical appliances, see [2, 3]. The term "consumer-relevant testing" was introduced at EEDAL 2017. This term is also widely used by experts and in literature. It is simply finding new compromises, which is difficult but possible when following the methodology closely.

2 Consumer-Relevant Product Testing

Consumer-relevant testing is still laboratory testing. Testing of products *in situ*, for example, a refrigerator in a kitchen, would not allow one to reach the repeatability nor reproducibility of the result desired.

Consumer-relevant product testing is that which provides results that correspond to results obtained when consumers use the product in practice. The general assumption is that if the situational conditions, input conditions and user behavior defined in the standard correspond to those found in practice (including extreme conditions of consumer usage) for a product with a given set of features, then the performance and energy consumption measured can be said to be 'consumer-relevant.' Consumer-relevant testing could be simply described as "bringing the home into the laboratory."

It is essential to find a balance between the following criteria, as described in detail in [2–4]:

Representativeness: The correspondence of the results from applying the test procedure to the results obtained in practice (with the end users);

Reproducibility: The consistency of results when the same product is retested under somewhat different conditions, for example, in another laboratory, but using the same test procedure;

Repeatability: The consistency of results, for example, regarding energy consumption or performance when the same product is retested under the same conditions, for example,. in the same laboratory by the same staff;

Costs: The costs of carrying out the test procedure.

It is important to note, that field data (if available, e.g. real energy consumptions of refrigerators measured in certain households) is an essential input for assessment of the correspondence with the laboratory test results.

3 Assessment Method

The method was developed by CLC/TC 59X/Working Group 22 Consumer-relevant Testing in 2016/2017. The aim is to develop measurement methods which reflect real life in a better way, for example, testing a vacuum cleaner with a (partly) filled receptacle instead of an empty one or testing a washing machine with different load sizes.

The work was supported by institutes, manufacturers, NGOs and stakeholders. Workshops have been carried out involving special products and aspects by the European industry association APPLiA and environmental NGOS working in standardization, represented by ECOS.

The method is described in [2]. It is a step-by-step approach.

Firstly, all the parameters (product features, situational conditions, input conditions and user behavior) have to be listed that influence product performance or energy consumption; analyze the interactions and then select the (most) relevant parameters. In addition to this, all the main performance aspects expected by consumers and how these are experienced by consumers are to be identified.

Secondly, the variations found in practice for the chosen parameters are to be listed and whether they are considered in the measurement standard should be checked.

Thirdly, after this check, how these expected performance aspects are measured according the present measurement standard has to be assessed.

Finally, the correspondence of the measurement standard to consumer practice is evaluated. This can be made either by theoretical analysis or based on real data.

As a result,

- missing performance aspects and parameters are identified and described;
- how the variability of the parameters can be taken into account is ascertained;
- and a clear understanding about the correspondence of the measurement of the performance with the performance experienced from the consumer side will be acquired.

4 Implementation and Application of the Method

CENELEC TC59X decided to implement the method on a broad scale at its working groups. Meanwhile, the method is applied and the results are turned into new measurement standards.

The method has been applied to some product categories, as shown below. It is also shown how the findings will be incorporated in new standards.

A pilot project on reporting the achievements was conducted by WG 22 in 2018. All 32 working groups at CLC/TC 59X have been asked to report about their present work on the consumer-relevant testing aspect. An annual report will be prepared starting in 2019. It should be made clear for the standardization committee members and experts and for all stakeholders outside the standardization committee what the annual achievements are.

Three working groups in the pilot project have reported that the assessment methodology has been applied and concluded: refrigeration, vacuum cleaning and washing machine. Furthermore, another three working groups have been reporting on the starting of the assessment of their measurement standards: ovens, hobs and the noise generated by vacuum cleaning. It was also announced that the assessments for some other standards will be started soon, for example, noise regarding hoods and washing machines.

Generic discussions are ongoing in seven standardization projects, and two other projects are just starting. Specific preparations are ongoing in 13 standardization projects, and another 9 have been started. This shows that there are many activities. However, many other assessments and concrete projects still need to be started.

The main achievements are briefly shown in the following chapters. Unfortunately, not all the details can be presented in this short paper.

5 Washing Machines and Washer-Dryers

There was a report on the upcoming EU regulations on labelling and ecodesign for washing machines at this EEDAL 2019 conference, see [5]. The related CENELEC measurement standards for those regulations are currently under preparation. The assessment method has been applied accordingly.

It has been identified that rinsing performance and actual washing temperature are important performance criteria relevant for consumers.

Water consumption is a very important performance criterion, as water is a valuable resource and scarce or expensive in many countries around Europe. Minimising the water consumption can be done easily by compromising the rinsing performance of the washing program. However, this is not a really good thing for people demanding a reasonable removal of detergent from the laundry.

Low temperature washing programs have been introduced in the past few years. This is a special part of the success story of energy efficiency around the world. However, consumers ask to be informed about the real temperature the wash load receives, for example, for hygienic considerations. This makes a temperature measurement method inside the load necessary.

Two technical specifications will supplement the European washing machine standard in the future: CLC/TS 50577:2019 on rinsing effectiveness by measurement of the surfactant content at textile materials and CLC/TS 50707:2020 on temperature measurement inside the textile load.

The future EU energy label be will be based on a cotton program 40 °C with ¼, ½ and full load. As there are three loads measured in the same program, it will be necessary to optimize a laundry appliance with some kind of automatic detection of load sizes for the full range of capacities in the future.

Publication of the new washing machine standards EN 60456:2016/A10 planned for 2020.

6 Vacuum Cleaners

An amendment to the EN 60312:2017 standard is under preparation. This will contain requirements for testing vacuum cleaners with a partly loaded receptacle. Regarding this, it should be noted that the standard has contained a test procedure for that for a long time. However, there was no respective data available regarding expanded uncertainty. Therefore, a respective Round Robin Test was concluded recently which allows one to determine the additional uncertainty due to tests with a partly loaded instead of an empty dust receptacle.

Furthermore, other items which also reflect consumer-relevant testing, such as debris pick-up or tests on market representative floors, are under preparation.

7 Refrigerators

A new refrigerator standard series is under preparation. The standards will support new EU regulations on energy labelling and ecodesign. At the same time, the IEC standards related are transferred to EN standards. The assessment method has been applied in the course of the development of the new EN standards. The most important achievement is testing at two ambient temperature levels instead of one. This allows a much broader scale of household conditions to be represented as out of this two tests all other conditions can be calculated. Discussions on door opening have led to a compensation of the effect in a reasonable way. Furthermore, the new standard includes requirements on anti-circumvention. (Note: EN 62552-1, -2 and -3 standards have been published in April 2020).

8 Dishwasher

Work on European dishwasher standard EN 60436 was started in 2017. The assessment method was under development in 2017. A more consumer-relevant load and market-relevant detergent have been integrated. The load is for example including plastic items which are widely used and behave differently in the dishwashing process. In this respect, it has been aligned with international Standard IEC 60436:2015. (Note: EN 60436:2020 has been published in March 2020.)

9 Ovens and Hobs

Work has been started recently on these appliance categories. The assessment method has already been applied.

A new brick method 2.0 is under preparation for the simulation of load. Furthermore, investigations on alternative loads have shown that artificial material represents real food reasonably well.

10 Noise Emission

The noise perception is depending on many factors. Also, acoustic measurements are very complex and there are many restrictions. This is mainly due to the fact that measurements have to be performed in test chambers. Appliances are often tested, for example, with empty load or clean water. It is, therefore, interesting to know what the effect on noise emission is when testing accordingly under different consumer-relevant conditions.

As an example, it has been investigated what effect in sound power emission level is to be expected when changing a washing machine program from 60 to 40 °C. It turned out that this effect is small or even negligible. A change in the measurement standard will be carried out in order to reflect consumer behavior conscientiously.

The same applies to the load of dishwashers, for which the load is also aligned. Artificial soiling in dishwashers is also introduced to prevent circumvention.

The standardization working group has started to apply the assessment method in order to integrate consumer practice in a more systematic way and in a broader sense.

11 Conclusion

A final conclusion cannot be made at the moment as the process is ongoing. For some appliance categories, there is already good progress as can be seen above. But, for other categories it will take some time as the related standardization activities need to be synchronized.

Very important is annual reporting about the activities which was started this year. This will allow more precise analysis in the future.

It was also recognized that consumer relevant testing is closely linked to the aspect of circumvention. This linkage needs to be explored more in detail.

References

1. REGULATION (EU) 2017/1369 OF THE EUROPEAN PARLIAMENT AND OF THE COUNCIL of 4 July 2017 setting a framework for energy labelling and repealing Directive 2010/30/EU, 2017
2. Gerhard Fuchs, Alain Roux, Hans-Paul Siderius, Christoforos Spiliotopoulos, and Rainer Stamminger (2017). A methodology to assess consumer relevant aspects for product testing. *Proceedings of EEDAL 2017 conference*
3. CENELEC TC59X (2018). *Internal Guide on Consumer Relevant Product Testing of Household and Similar Electrical Appliances.* TC59X/Sec0679B/INF
4. Christoforos Spiliotopoulos, Rainer Stamminger, and Hans-Paul Siderius (2019). Bringing the home in the lab: consumer-relevant testing for household electrical products. Energy Efficiency. https://doi.org/10.1007/s12053-018-9718-5
5. Gundula Czyzewski (2019). Energy savings 2021: New European Energy Label and Ecodesign requirements for washing machines. *EEDAL 2019 conference*, 2019

Shifting Consumers to Efficient Lighting: South Africa's Lighting Information Label

Theo Covary, Xolile Mabusela, and Toni Blumeris

Abstract Rapid technological improvements in residential lighting have yielded large electricity savings for consumers around the world. South Africa distributed over 70 million compact fluorescent lamps (CFL) to households from 2008, to stave off power outages during peak demand periods. The sheer volume and speed with which these lamps were introduced into the market, resulted in CFLs becoming the country's *de facto* symbol of energy efficiency (EE). Indeed, research based on 2018 data, has found that despite LEDs being as prevalent and costing the same, (often less than CFLs), South Africans continued to prefer the outdated technology by a ratio of over 2:1 (55% versus 24%). And this is notwithstanding that CFLs consume more electricity, have a shorter lifespan, and contain mercury content with very negative environmental consequences.

The Department of Mineral Resources and Energy's (DMRE) residential standards and labelling (S&L) program thus developed a strategy to correct this. Mandatory technology neutral technical specifications (whose first phase was introduced in 2020, with further expansion currently underway) would remove inefficient lamps. However, it was realized from the outset, that introducing legislation takes time. For the short term, an awareness campaign was devised to educate consumers at the point of sale. It recognized that lamps are an extremely low engagement commodity item, where the myriad of choices tends to result in consumers buying on a like for like basis. So, the campaign sought to develop a communication message to break the cycle. National focus groups and surveys were thus conducted, to develop an information label to influence purchasing decisions.

The paper describes the evolution of the design of the South African information residential lighting label, from inception to final version (before and after); detailing

T. Covary (✉)
Unlimited Energy, Johannesburg, South Africa
e-mail: theo@unlimitedenergy.co.za

X. Mabusela
Department of Mineral Resources and Energy, Pretoria, South Africa
e-mail: xolile.mabusela@energy.gov.za

T. Blumeris
Independent Consultant, Johannesburg, South Africa

P. Bertoldi (ed.), *Energy Efficiency in Domestic Appliances and Lighting*, Springer Proceedings in Energy, https://doi.org/10.1007/978-3-030-79124-7_3

consumer insights and decision-making criteria of household consumers from all income groups. In an effort to evaluate the effectiveness of the new information label, the authors contacted the original research participants to determine the impact, if any, that the campaign had on their purchasing decisions—the overwhelming majority of whom had been committed users of CFL technology, with a very limited understanding of LEDs.

1 Background and Context

Power shortages were first experienced regionally (Cape Town) in 2005 and then nationally in 2007. For the first time since the mid 1970s, the national vertically integrated utility, Eskom, was unable to meet demand. The supply shortages led to persistent national rolling blackouts, which have had a devastating effect on the economy. Fourteen years later, in 2021, Eskom is still not able to guarantee uninterrupted supply. The most severe supply shortages, i.e. rolling blackouts that can last for weeks at a time, occurred during three periods—with the first in 2007–2008, the second from 2014–2015 and the third from 2019 onwards. And although the power supply was relatively stable between these periods, the threat of the next period of outages always weighed heavily on citizens of the country, because they invariably return.

Government and Eskom's solution in the early part of the 21st century, was the construction of new generation plants (9.6 GW), which were commissioned in 2005. Such mega-projects take time however, and are thus a medium to long term solution. Therefore, the plan to address the immediate crisis, included upgrading and strengthening the existing, but aging, fleet; re-commissioning out of service power stations; and ramping up what was then a largely neglected Demand Side Management (DSM) Program.[1] The DSM program's primary objective was to manage demand during the morning and afternoon peaks, when the system was most vulnerable. Overall, total demand would decrease because of the various targeted measures it entailed but load shifting to reduce the usage of expensive diesel generation, was always the underpinning priority. Table 1 provides a timeline of key events during Eskom's DSM Program.

Eskom's CFL swap out program, where incandescent lamps of 100, 80 or 60 W were replaced with CFL's of 20, 16 or 11 W respectively, delivered sizable and immediate savings during peak periods. In 2012, Eskom reported that over 70% of the total electricity savings of the IDM program (2 164 MW peak reduction or 4 786 GWh [1]) as shown in Fig. 1 below, had been derived from the CFL initiative. This translated to more than seven million tons of GHG emissions mitigated over a decade [2]. Indeed, for the period 2005 to 2018, Eskom calculated that its IDM

[1] Eskom renamed the DSM Program in 2010 to Integrated Demand Management (IDM).

program delivered electricity savings of 4 521 MW. In this, CFLs' contribution for these years was was follows: 2016 69%; 2017 66%; 2018 95% [3].

In 2014/2015, Eskom suspended the IDM, (and with it the efficient lighting program), as a consequence of the following: Firstly, the Regulator (NERSA) only granted a portion of the annual electricity tariff increase requested by Eskom. The utility stated that it could no longer sustain an unfunded DSM program [4]. Secondly, additional supply became available (800 MW) [5] when the first unit of the new Medupi 4 800 MW coal power plant started producing. And the expectation was that the remaining units were to come online shortly thereafter. Simultaneously, electricity from the country's Renewable Energy Independent Power Producer Program (REIPPP) started feeding power into the grid. Thirdly, there was a genuine belief within Eskom that its power plant maintenance program had eliminated the servicing backlog, and that henceforth, the output of its existing fleet would increase and be more reliable. For the most part, the new (additional) power supplies and the stabilization of the existing plants, gave renewed hope to the country that the energy crisis was over.

Table 1 Timeline of Eskom's DSM program (2004–2015)

Year	DSM/IDM programs
2004	Regulator (NERSA) ratifies Eskom DSM approach
2005	Eskom undertakes first industrial EE project and CFL mass rollout
2007	Supply constraints cause blackouts—Eskom DSM responds with new CFL rollout and ESCO (Energy Service Company) projects for commercial lighting
2008	National supply crisis—DSM program intensified to alleviate constraints. Solar Water Heater rebate introduced
2012	IDM project expanded beyond single large industry interventions and lighting with Standard Product offering. CFL residential mass rollout extended to middle- and high-income households in the form of direct installations (swop out). Initiative included LED down lighters, electric water blankets, shower heads and timers
2014	CFL rollout ramped up and focus shifted to larger scale efficiency projects via ESCOs
2015	IDM program suspended due to lack of funding and additional generation capacity

Fig. 1 Electricity Savings from Eskom's IDM Program (Eskom 2012)

Eskom had distributed over 70 million CFLs when it ended the program in 2015 [6]. Without doubt, this initiative achieved almost immediate reduction in electricity consumption during peak periods. And this most certainly avoided, or reduced, the severity of electricity blackouts during periods of insufficient supply. Eskom's unconsidered exit strategy, however, led to sizable unintended consequences:

1. Nationally, CFLs came to symbolize EE lighting, thus precluding the uptake of the next generation of efficient lighting—Light Emitting Diode (LED) lamps;
2. The severity of the electricity crisis allowed Eskom to withdraw general service lamps from the DMRE's S&L program. It was argued that the adoption of mandatory national standards to regulate quality, performance and safety of lamps, were a priority and could not risk being delayed by the S&L initiative. Regulations VC 9091 and 8043 were duly promulgated in 2014 and 2015 for incandescent lighting and CFLs, which set quality performance standards for these lamps. Being excluded from the S&L program, and with Eskom soon exiting the scene, unfortunately then left general service lamps without a caretaker. Thus, in 2019 there were no mandatory standards to regulate LED lamps, while energy performance requirements were technology-specific and varied across lighting technologies;
3. Consumers, especially in the lower income groups, were no longer able to access free CFLs, and therefore tended to revert to incandescent lamps. A 2015/2016 study [7] found that retail stores serving these communities had effectively stopped stocking lighting products, due to low demand, because of free distribution. Moreover, evidence suggested that low income households were reverting to illegally imported incandescent lamps, which cost a fraction of CFLs ($0.30 versus $1.20). Indeed, The Regulator reported that over 2.5 million illegal incandescent lamps [8] were seized in 2018—while noting that they had not come across any such meaningful quantities since 2012. Additionally, during visits to low income areas to conduct focus groups for the lighting information label, researchers reported seeing large quantities of illegal incandescent lamps in informal stores;
4. Finally, in becoming price competitive, the majority of LEDs sold, had lower technical specifications and quality attributes. This always has the potential to compromise user experience; where a premium is paid for a product that does not meet expectations. Here, key challenges remained in quality, longevity, flicker, light degradation and low power factor. And poor user experiences are then likely to result in consumers reverting to lamps that they are not only familiar with, (CFLs, halogen or incandescent), but whose upfront cost also may be lower.

The South African S&L program, as an initiative of the DMRE, recognized the high risk of hard-gained electricity reductions during peak hours, being lost to inefficient lamps; and decided to act. The Department resolved, with the Regulator, to start the process of developing technology neutral technical specifications for general service lamps. By targeting performance, rather than a specific technology, all lamp types would in future need to comply to the minimum lumens/watt energy efficiency requirement. This approach makes the regulation non-discriminatory toward specific technologies and avoids the need to develop additional regulations should a new lamp technology enter the market. The second action taken, and the subject of this paper, was the development of a point-of-sale lighting information label, to be placed on retail shelves to guide and inspire consumers to: (1) Purchase a lamp that truly meets their needs; and (2) Include EE in the decision-making process.

2 Research Methodology

The overarching goal was to develop a lighting information label of lamps at point-of-sale, to influence consumers to think about EE and consider purchasing LED technology. Thus, it was necessary to comprehend consumers' perceptions and understanding of lamp technology, as well as their needs, so as to decipher what informed their choice of purchases and shopping destinations. It was also necessary to evaluate the various iterations of information label design, in terms of its level of comprehensibility and the impact of the intended message. As South Africa has a diverse population of varied ethnicities, languages, education and income levels, it was also required that the final recommended label be understood by all South Africans. As such, a dual or mixed method research approach—using both qualitative (exploratory) and quantitative (evaluative) methodologies—was recommended [9]. This combination provides flexibility, as well as a sample large enough to achieve a national reach and allow comparison.

The following research methodology was proposed, which consisted of three consecutive stages. This allowed prior learnings to be incorporated into each stage and to evolve the concept of the lighting information label accordingly:

1. Stage 1: Primary Qualitative Research—9 × 2 h focus groups
2. Stage 2: Primary Quantitative Research—10–12 min online survey
3. Stage 3: Follow-up Qualitative Research—3 × 2 h focus groups

2.1 Qualitative Research Methodology

The qualitative research sample reached 94 respondents against the following agreed recruitment criteria; and some interesting observations emerged:

- All purchasers of electricity and lamps for domestic use;
- Age—a spread per group:
  - 18–21 vs. 22–35 vs. 36–49 vs. 50–65 years;
- LSM (Living Standards Measure): Sampling for this research opted to use LSM, rather than household income, as a more stable measure in South Africa's current

economic depression. The high level of unemployment and retrenchment, particularly in low and middle-income groups, result in variable household income levels month- to- month. This was particularly evident in LSM 3–5 groups, where almost all respondents were unemployed, and in LSM 6–7 groups, where up to half had recently lost their jobs.

- LSM 3—new to electricity in the last few years:

  Contrary to available information on the originally identified areas, when qualitative recruiters arrived, they discovered that many had already been electrified for quite some time. This necessitated changing the location for Group 5 and Group 9.

- LSM 4–5—had electricity for 4–10 years;
- LSM 6–7—had electricity for as long as could be remembered;
- LSM 8–10+—had electricity for as long as could be remembered;
- LSM 1–2 were excluded because of no access to electricity in the home - and therefore having no need for, and less awareness of, the lamp purchasing process;

• Race—all races, either separate or mixed where appropriate;
• Gender—equal split of males and females per group:

  - Lower LSM groups were skewed towards females and younger males, as older males tend not to do the shopping. The few older males recruited, tended to live alone and had to do the shopping.
  - Higher LSM groups were balanced, male vs. female - having a higher prevalence of shared shopping/home responsibilities, as well as single parents of both genders, looking after home and children.

• Regions:

  - Six of the country's nine provinces were visited; making it a national research study - Gauteng; Western Cape; Kwa-Zulu Natal; Eastern Cape; Free State; Limpopo.

All group survey sessions were 2 h long and moderated by one of two specialist moderators. This allowed for continuity of insights and learning, as well as accommodating the various home languages across South Africa. Moderators followed a discussion guide that was designed to direct each interaction with a consistent flow. However, as the information label evolved, so did the discussion guide; of which there were 3 versions by the end of the research. Each discussion comprised 7 or 8 respondents, recruited against specific demographics (Ref. Fig. 3) and each was given a cash incentive and meal for participation. Discussions were audio recorded, from which transcripts were produced. These, alongside observations made in the field, formed the basis of a content and discourse analysis.

2.2 *Quantitative Research Methodology*

Respondents in the quantitative research online sample were screened, based on having purchased lamps for domestic use within the last year. The chart below demonstrates the demographics of the quantitative sample achieved.

For the quantitative stage, use was made of a single, cross-sectional, descriptive methodology. It utilized a structured online questionnaire, which was designed and informed by the findings and preferred language from the qualitative consumer research. Convenience sampling was used to contact participants through purchasing lists. As response rates are generally low for online surveys, efforts were made to improve the rate of response, by offering an incentive. This took the form of a lucky draw for three cash prizes of R1 000 ($70), R500 ($35) and R300 ($20) respectively. Invitations to participate were sent out to the list of potential respondents with email addresses, via the selected online survey program, Qualtrics. In line with the Protection of Personal Information (PoPI) Act, those to whom an invitation was sent, were given the option to opt out of the research. A total of 12,122 invitations were sent out; to which the successfully completed response rate was 2.1% (255 participants). The average completion time was 12 min, including screening and demographic questions.

Consistency checks were carried out on the data; and 254 usable questionnaires were obtained, upon which to run the analysis. This used descriptive statistics, frequency distribution and cross-tabulation, to provide insights into the data.

Results were analysed in terms of overall total, and by the main demographic breaks of age (four groups), gender (males and females) and gross monthly household income (four groups). As most of the responses came from Gauteng, base sizes were not large enough in other provinces to analyse the data by province. The results for all questions used a rating scale to analyse data, by excluding the 'don't know/cannot answer' responses. Figure 2 below shows details of the final sample achieved.

3 Final Research Sample and Reports Delivered

As the final information label needed to be understood by all South Africans, the overall research sample was designed to be nationally representative. The qualitative focus groups engaged more with lower LSMs 3–6; with only 2 groups for LSM 7–10 and the quantitative survey, focusing more on the latter. This was to account for the low reach of online methodologies to lower LSMs, where there was less frequency of smartphones and access to internet. The first 5 focus group discussions, represented LSM 3–10, to gauge the level of comprehension across the market, of the originally proposed information label. They also sought to determine the extent to which the label required upweighting of messaging elements. As the original information label performed so poorly across the market, it required a complete

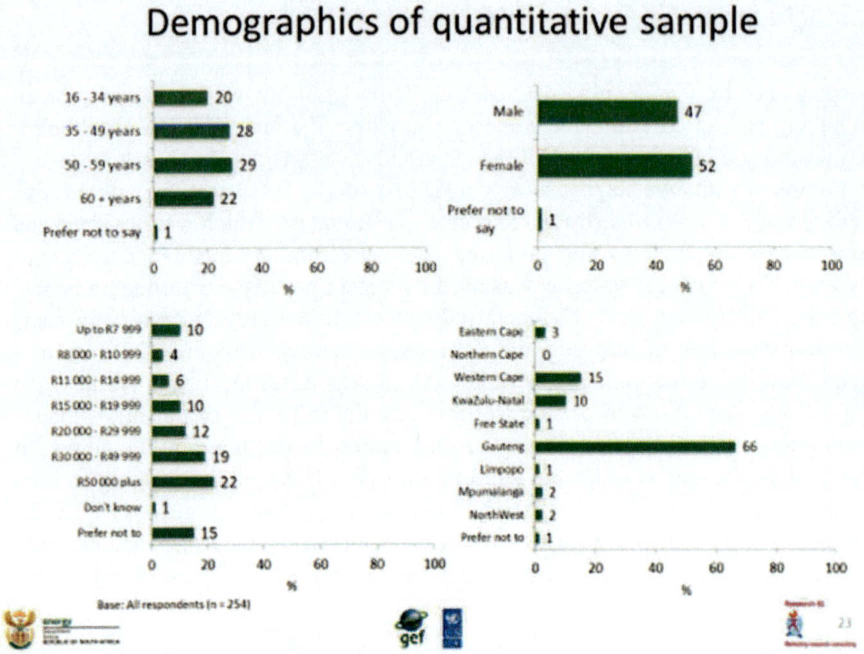

Fig. 2 Quantitative sample demographics

re-design. Thus, the qualitative sample for the remaining focus group discussions also needed adjustment, so that the second version of the information label could again be tested qualitatively across the entire market, to ascertain comprehension levels. Interim reports were delivered for Stage 1 and Stage 2, with a full report after Stage 3. Figure 3 demonstrates the final and revised sample for all three stages of research.

4 Follow Up Impact Interviews

As the lighting information label had not yet been rolled out to retail point-of-sale, the research team re-contacted a small, yet nationally representative sample of research participants, to establish the impact of the communication material in the lamp purchasing decision-making process. 5–10 min qualitative telephonic discussions were conducted with 14 research participants in June 2019.

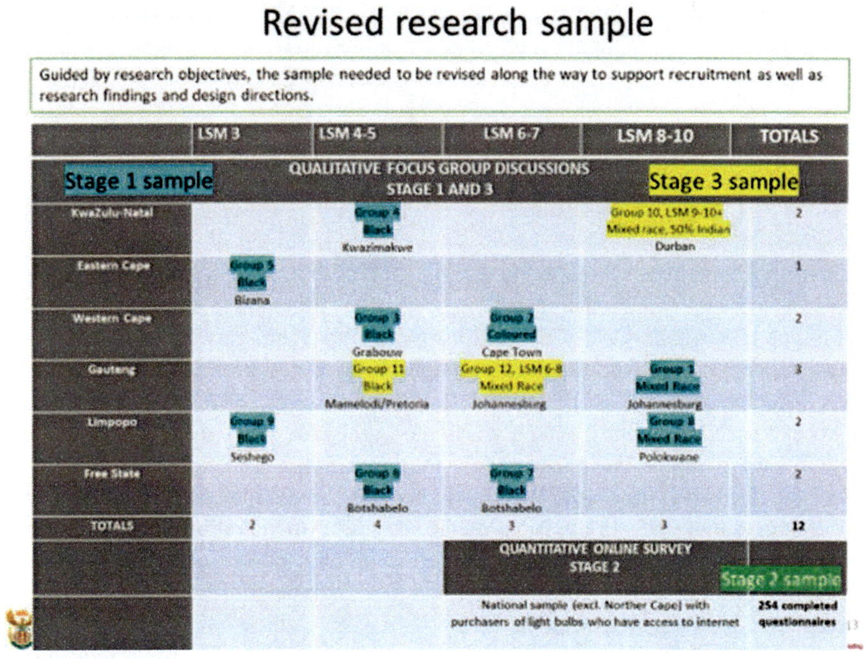

Fig. 3 Revised overall research sample

5 Key Findings of the Research and Design Journey

5.1 Growing Opportunity and Desire for Educational Energy Efficiency Campaigns

There is an almost unanimous sentiment amongst consumers that South Africa needs to use electricity more efficiently; with high awareness of the country's electricity shortage, and therefore, the increasing cost of the resource. However, it's a 'catch 22' scenario. High misuse exists, mainly through illegal connections and unnecessary wastage. Simultaneously, there is a growing need for bright outdoor security lighting at night, given the increasing crime rate. Consumers ensure they have sufficient units of electricity to power outdoor security lighting throughout the night; and will compromise indoor lighting or other electricity needs if necessary, to keep the outdoor lighting on.

Consequently, there is a strong ongoing opportunity and desire for educational EE campaigns. These can help to curb wastage of a dwindling, yet critical resource, and to reduce overall electricity costs in the home - particularly in the current economic downturn, where overall living costs continue to increase.

Further, in comparison to the foundation research conducted in 2011 [10] there is substantial growth in what are now regarded as energy saving practices, by residential consumers. Particularly pertinent to this research on lighting—although at a relatively low level—is some awareness and usage of LED, exterior sensor lights, solar lights and even one mention of using a wind turbine to power lights.

5.2 *Limited Knowledge of Efficient Lighting Technologies Despite Varied Needs and Access*

Needs for lighting are as varied as the levels of access to lighting technology options between urban and rural consumers. The level of awareness and knowledge, however, is not as variable. One of the key insights, is that the 'gap' in knowledge of lighting technologies, is not as wide as might be expected between low and high LSM consumers.

Those exposed to lighting and purchasing of lamps for many years, have formed their knowledge around incandescent technology and thus purchase based on wattage = brightness, i.e. that 100 W is brighter than 60 W which is brighter than 40 W. Whereas those new to electricity in the last couple of years, have been exposed to CFL first, where energy saving is a more dominant purchase driver. In some cases, albeit limited, LSM 3 consumers were found to know more about LEDs, than upper income LSM 8–10 consumers.

Ultimately, through the Eskom swop out campaign, the need to use a more EE technology took hold in the market; with the technology growing in popularity. Consumers stated that CFL is their preference, either in purchasing, or wanting to purchase, the technology [7, 11] (Ref. Fig. 4). But a significant barrier was the lack of understanding of how to purchase 'like brightness' in comparison to the 100 W/60 W incandescent. This resulted in the perception of 'dim' performance that prevents its use for outdoor security and high brightness areas, such as kitchen and bathroom. Additional barriers included perceived high cost and lack of in-store availability in rural and peri-urban areas. Consequently, these factors appeared to encourage continued use of incandescent and halogen lamps, particularly in middle to lower income homes, but not excluding some upper income consumers too.

Very few consumers seemed to know about halogen technology, even though this presents higher quality incandescent lamps, as the shapes of both are identical. The few who knew to differentiate between the technologies, opted for halogens in outdoor security lighting areas, as they deliver good levels of brightness, last longer and use less electricity.

The 100 W incandescent was the most popular for its brightness, and perceived by most to serve outside security lighting needs better than CFL. Then, 60 W and 40 W were seen as offering affordable levels of brightness throughout the home.

Perceptions of LED were the most polarized in the market; with accurate knowledge more limited to upper income, older homes, while also being dependent on prior engagement and purchases in the category. Across the market, LSM 3–10+,

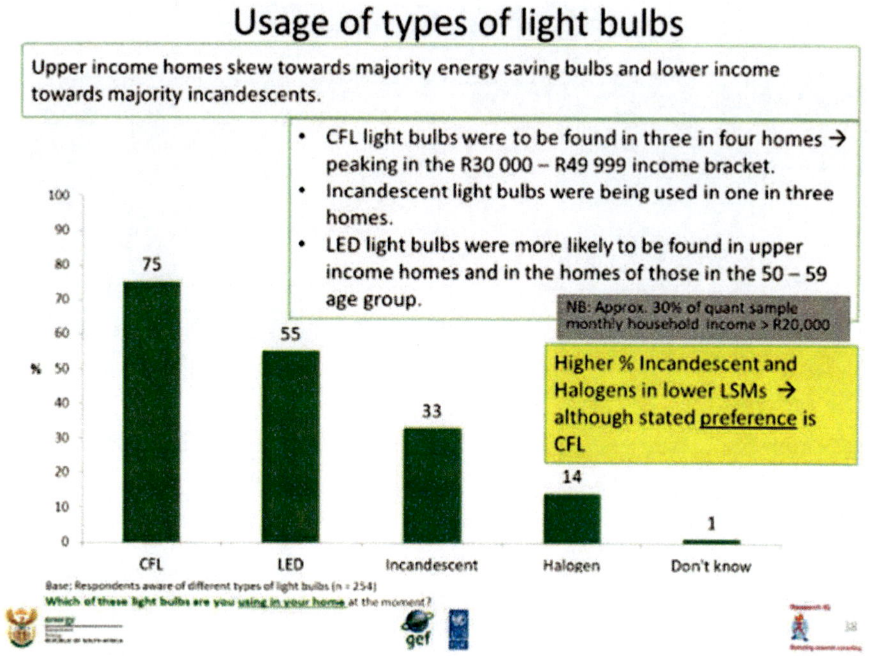

Fig. 4 Usage of General Service Lamps in South Africa

there were those who either knew very little or nothing about LED. But for the limited few who up-traded to LED, it was a preferable option to CFL in terms of value, overall life expectancy and EE. These consumers were also aware of 'bright' LED options to replace incandescent and halogen lamps, as well as experiencing LED as cheaper at point-of-sale than CFL. But, for the majority with little experience of LED, it was dismissed for its perceived expense and not delivering the same level of relative brightness of 100 W and 60 W incandescent equivalents.

Whilst General Service Lamps (GSL) constantly appear to be the most frequently shopped category in comparison to other appliances, it also seems to be the least understood, and therefore a very low engagement category. Most consumers did not understand the product information supplied on GSL packaging, such as light output levels, colour rendering, life expectancy and energy usage. This makes purchasing a new type of lamp very confusing. Consequently, most consumers continued to buy what is familiar, rather than learning about new and better options. Also, insightful assistance at small 'spaza' shops in rural and peri-urban areas, as well as in general retail grocers, was not on hand, as compared to major retailers and outdoor and speciality lighting stores in urban areas; most of whom are capable of making recommendations on what to purchase, if requested.

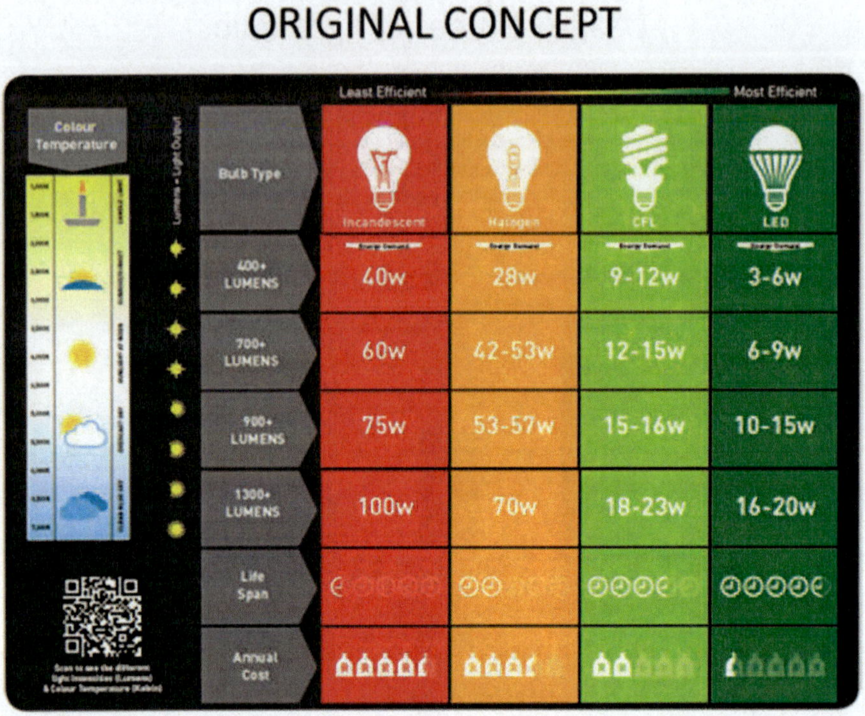

Fig. 5 Original lighting information concept

5.3 Reactions to the First Information Label Concepts

The first lighting information label concept (Fig. 4) exposed to Group 1, an LSM 8–10 pilot group, was completely rejected. Consumers were unable to understand the information pertaining to the value proposition of each lighting technology.

Between Group 1 and Group 2, the layout was revised (Fig. 5) to improve comprehension, but the content was mostly kept the same. Likewise, Group 2 to Group 5 found the material as complicated and confusing as the lamp package (box), where the information completely failed to engage new learning or interest.

The focus group participants unanimously agreed that the initial concepts were visually off-putting; with content that was too technical and a layout that was cluttered and difficult to understand. It did not tell a story to engage in the learning process. Initial reactions confirmed that the baseline knowledge of lighting technologies across the target market, up to and including some LSM 10+ respondents, was immature. However, through explanation of the material's content and intent, awareness and knowledge levels shifted. This indicated a desire to learn more about lighting, so as to make better EE lamp choices.

In such a low engagement category, an unsuccessful lighting information label that did not engage interest, clearly would not shift consideration away from familiar 'type' choices. For the majority of the target market, these were incandescent or halogen lamps, selected for price and brightness, or CFLs as the longer lasting option. In shifting such perceptions, important knowledge gaps that needed to be addressed, came to the fore:

1. High familiarity and frequent engagement with incandescent technology has established a mental conceptual framework that believes *brightness equates to the number of watts*. Therefore, the information label needed to educate on Lumens, for consumers to believe than a low watt LED lamp can deliver the same level of brightness as a 100 W/60 W incandescent equivalent.
2. Similarly, high wattage incandescent (and halogen) lamps are associated with lamp heat and the ability to 'warm a room'. Thus, the referencing of colour temperature as 'warm' vs. 'cool' white, appeared to result in a literal interpretation, which prevents many from understanding colour rendering as mood/tone. Therefore, the information label needed to also build both awareness and relevance of colour rendering, measured by Kelvins.

5.4 *Reactions to the Infographic Lighting Information Label Concepts*

These insights, together with those gleaned from perceptions and behaviors in the category from Group 1 to Group 5, led to the development of an infographic lighting information label (Fig. 6). The intention was to tell a straightforward story, by building from common and familiar knowledge (e.g. screw vs. 'pin') towards introducing

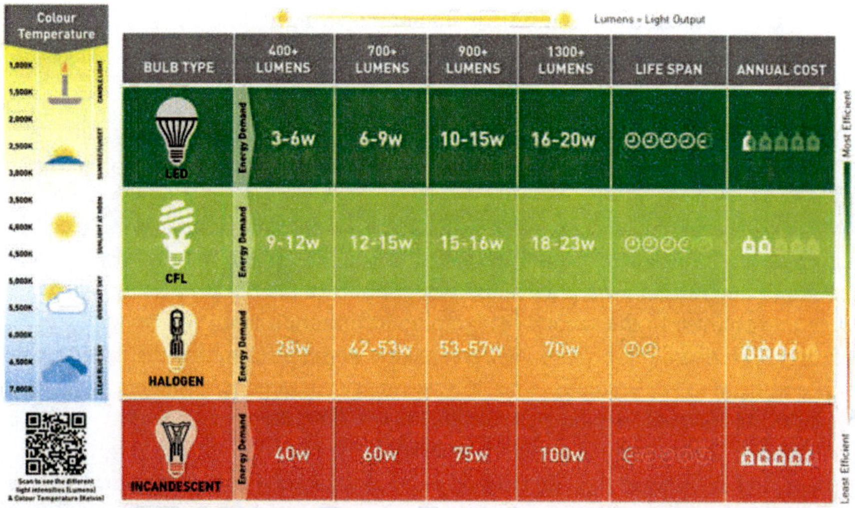

Fig. 6 Revised lighting information concept

new information and terminologies. And it was specifically geared to ease the journey of choosing a lamp that best meets consumers' needs, whilst also raising the influence of EE and performance. Additionally, the order of the instructions (Step 1, Step 2 etc.) dealt with a consumer's primary needs, before introducing new concepts. This again prioritized the purchasing of the right size and type of lamp (Step 1), then introduced the different technologies (Step 2), then focused on quality (Step 3) and so on.

The first iteration of the new infographic concept was tested in Group 6 to Group 9 (LSM 3 to 10) and was successful in spontaneously engaging attention and raising awareness of LED. In particular, it helped the following new awareness and knowledge emerge for many:

- There are different fitting sizes for lamps;
- New lighting technologies are available;
- Higher awareness of LEDs and that their value proposition is a better option.

The communication however was not sufficiently strong enough to:

- Shift the belief that brightness is dependent on watts rather than Lumens: Consequently, low watt (in comparison to equivalent incandescent) LED lamps would be interpreted as having very 'dim' lighting capability and would be overlooked as not satisfying lighting needs. Further, the perceived expense of high wattage LEDs would continue to deter engagement with the category, even if in reality, the mainstream target market does not need speciality lamps such as these.
- Drive further understanding of colour rendering: The need for tone/mood, created through lighting, was not a high priority or frequently mentioned need in the mainstream. It was only important for some LSM 10/10+, where there was low level understanding of how to determine this from packaging. As the concept needed to be legible on an A4 layout for research purposes, the recommendation was made to exclude further testing of colour rendering at this point, and to rather redesign and introduce it for the final production. It was more a priority to address the Watt/Lumen issue.

At this point in the research, it was clear that the lack of understanding around lumens was priority, which if unaddressed, would prevent widespread adoption of LED technology (Fig. 7).

Prior to Stage 2 of the research, the stimulus material was revised and the second iteration of the infographic (Fig. 8) tested in the quantitative online survey, which was followed by further in-depth discussion in Group 10 to Group 12 (LSM 4–10+).

Overall response to the second infographic information label was very positive (Ref: Fig. 9). Indeed, there was 91% agreement that the information displayed was useful and easy to understand; especially amongst younger respondents, females and those in the lowest income group.

Ultimately, the communication was particularly successful (Ref. Fig. 10) in increasing awareness and new understanding, which directly impacts a shift in purchase drivers around the following:

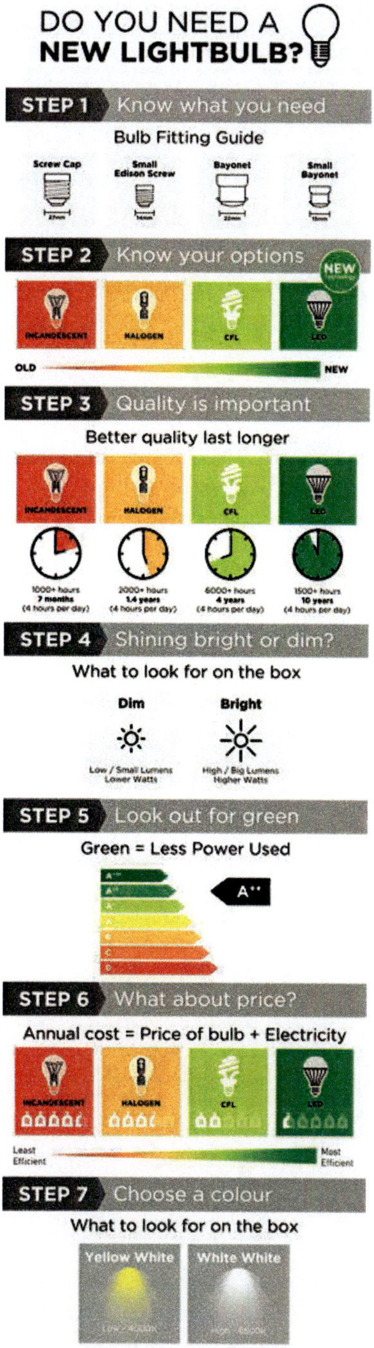

Fig. 7 First iteration of infographic information label

Fig. 8 Second iteration of infographic information label

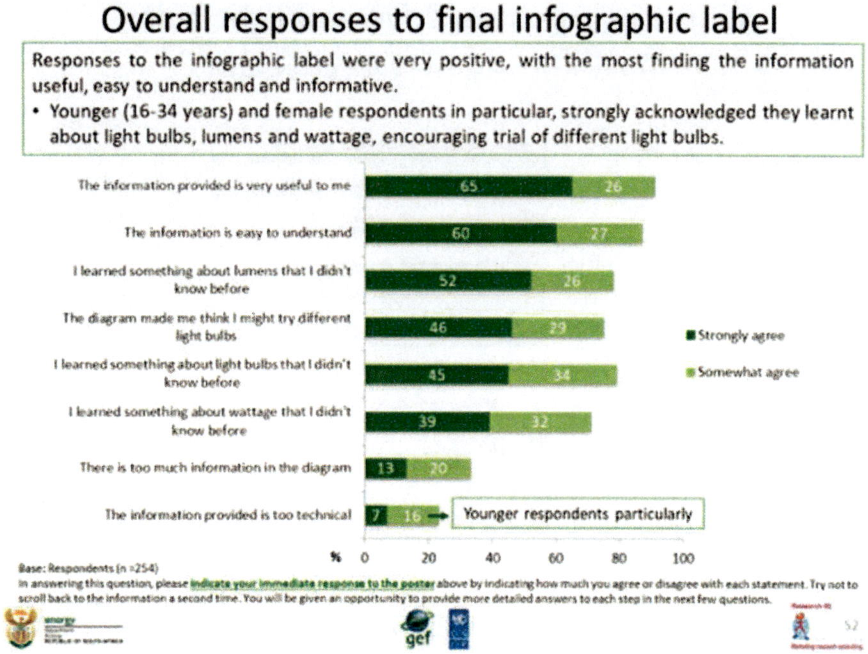

Fig. 9 Overall responses to final infographic lighting label

- The role of lumens as indicators of brightness, which increased fivefold from 11% to 55% post exposure; whilst the role of watts decreased by 19%. Some grasped the information quicker than others, depending on literacy levels, age and previous experience in the category, but repeat exposure to the material is shown to assist in building knowledge over time;
- New and innovative lighting technologies, which offer better quality, higher performing lamps that use less electricity (increase of 12%), and last longer (increase of 6%).

Even without knowing that LED pricing has reduced, consumers stated that their consideration of LED in the next purchase cycle will be stronger than CFL or incandescent; particularly, but not exclusively, amongst upper LSMs.

5.5 *Impact of Exposure to Communication 10 Months Later in June 2019*

The research team telephonically re-contacted 14 participants in the qualitative research, LSM 3–10+, of which 12 have either shifted purchase choice completely to LED, or a mix of LED and CFL, over CFL and incandescent. Some had started shopping for LED in different stores that stock them and shared their knowledge

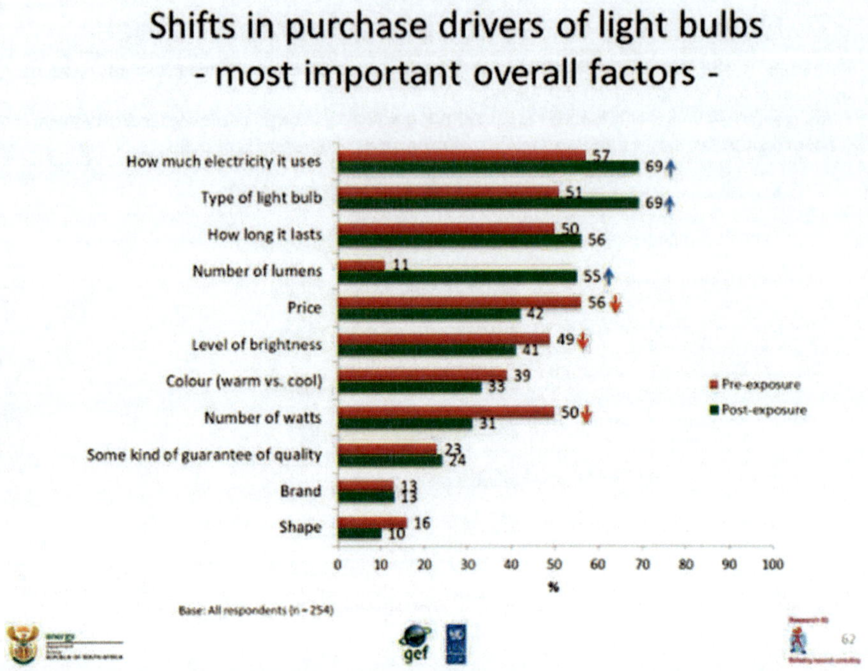

Fig. 10 Impact of exposure to infographic lighting label

with others. All noticed that the LED lasts longer; and some reported a reduction in electricity usage. However, many commented on the how the light 'starts dim' then gets brighter and that LED light is less harsh than that of incandescent lamps.

Interestingly, the two respondents who had not shifted purchase choice, were exposed to the original and revised concepts, not the infographic information label; but one of these was already purchasing CFL, and claimed that the high price of LED offers no extra value.

6 Conclusion

The graphic information label strongly drives the desire to trade up to LED, by outlining the LED value proposition and, targeting the performance of all technologies, rather than focusing on a particular one. However, the reality of South Africa's current economic situation, means that pricing of LEDs may still remain a barrier for some, who cannot afford the up-trade. Then, those purchasing cheaper LED lamps, are trading off cost saving benefit or performance against a lesser quality brightness. However, a highly iterative and collaborative process between research, design and strategy emerged, based on what was learnt throughout the research. Indeed, the success of the final information label is underpinned by the agility of this collaboration, which resulted in a highly impactful piece of communication, generated with maximum consumer benefit in mind.

References

1. Eskom (2012): Energy Efficiency and Renewable Energy—the way forward. Presentation by Andrew Etzinger Senior General Manager to the German Business Chamber 24 October 2012.
2. Eskom (2011): COP17 Fact Sheet: Efficient Lighting Program. Available at: www.eskom.co.za/OurCompany/SustainableDevelopment/ClimateChangeCOP17/Documents/Efficient_Lighting_Programme_involving_the_rollout_of_Compact_Fluorescent_Lights.pdf.
3. Eskom (2019). Written response to request for information from author. Email available on request.
4. Eskom (2016): *Effective Management of Energy Demand.* Eskom presentation to industry. Available at: www.eskom.co.za/sites/idm/Business/Documents/EskomAdvisoryService14042016.pdf
5. Online news website (9 October 2018): *Eskom announces Unit 2 of Medupi is operational.* Available at: www.fin24.com/Economy/eskom-announces-unit-2-of-medupi-is-operational-20181009.
6. Eskom (2018): *Eskom IDM—The way forward.* Presentation to energy planning workshop. Available at: www.savingenergy.org.za/wp-content/uploads/2018/11/Eskom-IDM-The-Way-Forward.pdf.
7. Harris, P., Baran, C. & Hazard, C. 2017. *Identify, Assess and Design a Market Based Economic Incentive for Energy Efficient Appliances in South Africa.* Study funded by the Danish Embassy for the Department of Energy.
8. Author discussion with NRCS Inspector, May 2019
9. Wisdom, J. (2013): *Mixed Methods: Integrating Quantitative and Qualitative Data Collection and Analysis While Studying Patient-Centered Medical Home Models.* George Washington University
10. Unlimited Energy (2012): *Energy Performance and Labelling Requirements for Specific Electrical Appliances and Equipment.* Project Amber by T. Blumeris (KLA)—independent research undertaken as an input into the final report. Available at: https://www.savingenergy.org.za/reports/
11. Covary, T., de la Rue du Can, S., Shuma-Iwisi, M. (2019): *A Review of the Residential Efficient Lighting Programme Rollout in South Africa.* Paper presented at ECEEE Summer Study 2019.

Little Appliances: Big Power Appetites

Vesa A. Lappalainen

Abstract Energy efficiency labelling of domestic appliances and lighting are discussed in the context of technical progress, consumer and market behaviour, and the law. The chief postulate is that many ubiquitous, small appliances escape energy labelling, although their peak power consumption vastly surpasses that of larger appliances—such as refrigerators and boilers—which must carry energy efficiency labels. Labels are mandatory in the European Union and some 100 jurisdictions worldwide.

The electric bread-toaster, hot-water kettle and hair-dryer can each consume up to 3000 W when used. Their retail prices can be €10. By comparison, large refrigerator-freezers may have a rating of just 100 W. The toaster, kettle and dryer.

- consume large quantities of electricity, but only for short periods, with repeated daily use common;
- represent mature technology, with first devices appearing in markets 100 years ago;
- rely on NiCr resistance filament or heating rod element, based on an invention patented in 1906.

For light sources, the light-emitting diode (LED), patented in 1966, created a paradigm change, but it has taken 50 years for LEDs to become mainstream. The reasons for LED's success are technical and economic rather than policy-related. For the toaster, kettle and dryer, no change of paradigm has occurred. The same is true for most other miscellaneous electric loads (MELs).

While ecodesign and energy labelling have brought good results for large base-load appliances, no progress is discernible for the usually unlabelled MELs. No new heating technologies, such as thick film, are tested in consumer markets. Rather than better energy efficiency, suppliers concentrate on design and colour, unhampered by any labelling obligations.

V. A. Lappalainen (✉)
University of Vaasa, Vaasa, Finland

P. Bertoldi (ed.), *Energy Efficiency in Domestic Appliances and Lighting*, Springer Proceedings in Energy, https://doi.org/10.1007/978-3-030-79124-7_4

1 Introduction

From the perspective of law and economics, the most obvious externality of MELs is their influence on power demand in distribution systems which must be designed and built to withstand increasing household loads, including peaks. This externality affects all home appliances. While base-load appliances—mostly white goods—are subject to mandatory ecodesign and energy labelling in the European Union (EU) and many other jurisdictions, toasters, kettles, dryers and many other MELs usually are not. There seems to be no explanation for this anomaly, except perhaps that white kitchen appliances are large and easily catch the eye, whereas MELs are rather inconspicuous.

This paper addresses the significance of technical development of lighting and certain home appliances, their influence on power demand, distribution and peak load. The toaster, kettle and dryer are used as references mainly because they all rely on a basic 100-year-old technology for heat generation, and also because they share an almost complete market penetration in many large developed consumer economies.

2 Domestic Appliances and Lighting

The driving force behind the electrification of national economies, which started in the 1880s, was to provide light, mechanical power, and, finally, also heat and cold derived from electric current. Electricity offered convenience and utility when compared to the earlier technologies based on solid and liquid fuels, such as firewood, coal and lamp oil. The key inventions and their inventors of this era are still household names in the industrialised world: the incandescent light bulb[1] and Thomas Alva Edison; the electric motor[2] and Nikola Tesla, Werner Siemens and George Westinghouse; and the nickel-chromium heat-resistant filament[3] and Albert L. Marsh. All of these inventions represented a breakthrough also because they established new business by means of making the commercialisation of the related basic patents feasible.

Since the electrification of societies is a process which started some 100–130 years ago, it is no coincidence that also most of the common electricity-driven domestic appliances currently available were first introduced at that time, like the electric

[1] U.S. Patent No. 223,898, granted on 27 January 1880 for "Electric lamp" to T.A. Edison. Many other patents were granted around this time to Edison and his competitors.

[2] U.S. Patent No. 381,968, granted to Nikola Tesla on 1 May 1888 for "Electro Magnetic Motor". Tesla, who worked for Westinghouse, and others were granted a large number of related and relevant patents at the same time.

[3] U.S. Patent No. 811,859, grated on 2 June 1906 to Albert Leroy Marsh for "Electric Resistance Element". It is admitted that Marsh may not be a household name.

vacuum-cleaner (1908),[4] clothes-washer (1910),[5] dishwasher (1886),[6] refrigerator (1915)[7] and the air-conditioner (1906).[8] All these rely on electric motors, heating elements, and cooling devices. Since the time when the original patents expired by the 1920s, technology has evolved only slowly. The notable exception is lighting.

Until 1937—for some 50 years—the incandescent light bulb faced no competition as the standard source of lighting. While incandescent bulbs were based on a well-tested technology, were relatively cheap and had a market penetration of *ca.* 100%, it was long understood that some 95% of the energy spent by the bulbs was wasted on by-products (heat, infrared light), not the primary product: light visible to the human eye.[9] Consequently, the fluorescent light tube (FLT) was developed with energy efficiency in mind, and after the Second World War the FLT gradually became the solution of choice for industrial, commercial and public indoor lighting.[10] The energy saving achieved by fluorescence is some 75% in comparison to incandescent bulbs, with energy efficacy that is *ca.* 3.4 times better.[11] For domestic purposes, a compact version of the fluorescent light (CFL) was developed in response to the 1973–1974 energy crisis by the American inventor Edward E. Hammer, but it was not patented, as Mr. Hammer's employer, the General Electric Company, chose not to commercialise the invention due to expected high investment cost for new manufacturing plant. Competitors seized the opportunity.

For the last two decades of the twentieth century, the CFL became increasingly popular as an energy-saving option for households. In 1966, however, the first light-emitting diode (LED) had been patented,[12] and gradually the LED has been successfully commercialised, a process which is ongoing at a rapid pace. The first LEDs were constrained by limited colour spectrum—initially, only red—low luminous flux, and high prices. For years, LEDs were limited to specialised purposes, such as number displays of expensive medical and technical instruments. During the second

---

[4] U.S. Patent No. 889,823, granted on 2 June 1908 to J. M. Spangler, of Canton, Ohio, for "Carpet Sweeper and Cleaner". Mr. Spangler sold his patent to W.H. Hoover in the same year.

[5] U.S. Patent No. 966,677, granted on 9 August 1910 to Alva J. Fisher for "Drive Mechanism for Washing Machines". There were many other innovators active in the washing machine business at the same time.

[6] U.S. Patent No. 355,139, granted on 28 December 1886 to Josephine Cochran for "Dish Washing Machine". This first dishwasher was driven by hand.

[7] U.S. Patent No. 1,126,605, granted on 26 January 1915 to F.W. Wolf for "Refrigerating Apparatus". This was, for practical purposes, the first electric refrigerator.

[8] U.S. Patent No. 808,897, granted on 1 January 1906 to Willis H. Carrier for "Apparatus for Treating Air". The Carrier Corporation is still in business.

[9] Gayral, Bruno. *LEDs for lighting: Basic physics and prospects for energy savings/Les LEDs pour l'éclairage: physique de base et perspectives pour les économies d'énergie.* Comptes Rendus Physique, Volume 18, Issues 7–8, September–October 2017, pp. 453–461.

[10] The source on fluorescent lights here is the Edison Tech Center in Schenectady, NY, USA, see http://edisontechcenter.org/Fluorescent.html#inventors.

[11] *Do.*

[12] U.S. Patent No. 3,293,513, application filed in 1962, granted on 20 December 1966, to James R. Biard and Gary E. Pittman for "Semiconductor Radiant Diode".

decade of the current millennium, the LED is fast replacing fluorescence in both domestic and non-domestic illumination, with energy savings of 85% or more, as compared with incandescent bulbs, and eight times the energy efficacy of such bulbs, at the time of writing.

The above discussion attempts to explain that the driver for more energy-efficient electric lamps is, above all, energy cost. The basic utility of all light sources, irrespective of technology, is visible light. The driver behind technical development has been the need to generate more light with less energy, *i.e.* less cost for consumed, metered electricity. The so-called life cycle assessment, which considers the entire life of a product, is a recent approach, as witnessed by *e.g.* the ISO 14040:2006 Standard, "Environmental management—Life cycle assessment—Principles and framework",[13] and concentrates on the total environmental impact of a product.

As regards energy efficiency labelling, which was first introduced in the EU in 1979, with the adoption of two **voluntary** directives,[14] it would be an exaggeration to claim that the domestic lighting paradigm change happened because of the directives and compulsory labelling. This is proven by the time when the basic patents were granted: 1937 for the FLT and 1966 for the LED, well before the 1973–1974 oil crisis. The invention of the CFL in 1975 was, however, a direct result of a perceived energy shortage, and the first EU energy labelling directives were adopted. The directives did not cover light sources of any kind, but rather eight other product groups: water heaters, ovens—both electric and "ovens using other sources of heat"—refrigerators and freezers, washing machines, television sets, dishwashers, tumbler dryers, and ironing machines.

For household lamps, the first EU labelling directive was adopted in 1998.[15] The currently applicable Regulation 874/2012[16] took effect partially on 1 September 2013 and finally on 1 March 2014, and it concerns not only the energy labelling of electrical lamps but also of *luminaires*, which are defined as—

> apparatus which distributes, filters or transforms the light transmitted from one or more lamps and which includes all the parts necessary for supporting, fixing and protecting the lamps and, where necessary, circuit auxiliaries together with the means for connecting them to the electric supply.[17]

A luminaire is essentially an all-in-one light source, a fixture that includes both an electric bulb (light source) and a "lamp" as used in daily parlance—"table lamp", "floor lamp", "chandelier", etc. Where a bulb becomes disposable after it has burnt

---

[13] https://www.iso.org/standard/37456.html. The standard was last updated in 2016.

[14] Council Directive 79/530/EEC of 14 May 1979 on the indication by labelling of the energy consumption of household appliances; and Council Directive 79/531/EEC of 14 May 1979 applying to electric ovens Directive 79/530/EEC on the indication by labelling of the energy consumption of household appliances.

[15] Commission Directive 98/11/EC of 27 January 1998 implementing Council Directive 92/75/EEC with regard to energy labelling of household lamps.

[16] Commission Delegated Regulation (EU) No 874/2012 of 12 July 2012 supplementing Directive 2010/30/EU with regard to energy labelling of electrical lamps and luminaires.

[17] Regulation 874/2012, Art. 3, Definition (26).

up its useful hours—never mind if that is 2000 h, as in the case of incandescent light bulbs, or 20,000 h, as in the case of many LEDs—the bulb only will be thrown away. In the case of a luminaire, also the entire apparatus may become disposable.

From the point of view of life-cycle energy efficiency, it is obvious that luminaires could lead to significant waste of material resources. For example, a floor lamp would have to be discarded because the light no longer burns and the replacement of the light-emitting part is impossible, at least for the ordinary consumer. The whole concept of a *luminaire* is contrary to some of the very basic tenets of frugality, and also of the EU Circular Economy Action Plan of 2015,[18] which establishes as energy efficiency policy goals "the **reparability, upgradability, durability, and recyclability** of products by developing product requirements relevant to the circular economy" (part 1.2, box on p. 4). When juxtaposing disposable luminaires and the circular economy goals, an exception can perhaps be made for *recyclability*, if that concept is interpreted to include *disposability*. Something that is durable and reparable usually is not readily disposable.

On 11 March 2019 the EU Commission issued a proposal for a new regulation[19] on the labelling of light sources and repealing Regulation 874/2012, just some 5 years after the current statute had been implemented. The proposed new regulation disposes of the *luminaire* entirely, with the following justification given in the explanatory memorandum (p. 5, part 3.1):

> Light sources as defined by this act are always in the scope, even when they are parts of 'containing products', such as luminaires, mirrors, fridges or shelves. However, the containing products themselves are not in the scope of this act (but they may be in the scope of other energy labelling acts). As a result, this act no longer requires an energy label for luminaires.

The short regulatory history of the luminaire as an object of energy efficiency regulation and labelling demonstrates that the EU lawmaker is in a race with technology, running after technical innovations. This race, however, concerns only light sources. As noted at the outset of this article, there has been little or no *de facto* progress in the energy efficiency of most household appliances during the past 100 years.

---

[18] Communication from the Commission, COM (2015) 614 final, "Closing the loop—An EU action plan for the Circular Economy". Brussels, 2.12.2015.

[19] European Commission. Proposal for Commission Delegated Regulation (EU) supplementing Regulation (EU) 2017/1369 of the European Parliament and of the Council with regard to energy labelling of light sources and repealing Commission Delegated Regulation (EU) No 874/2012. Brussels 11.3.2019, C (2019) 1805 final.

3 Domestic Appliances Other than Lighting

While just eight domestic product groups were covered by the EU mandatory labelling regime in 1979, the number of appliance groups currently regulated by EU law for energy efficiency labelling is 14. In early 2019, vacuum cleaners[20] were removed from the number of regulated appliances by court order.[21] The General Court of the EU gave the following grounds for its decision[22]:

> Since the Commission adopted a method for calculating the energy performance of vacuum cleaners based on an empty receptacle, the General Court holds that that method does not comply with the essential elements of the directive. The General Court finds, therefore, that the Commission disregarded an essential element of the directive and annuls the regulation, since the method for calculating energy performance is not an element which may be severed from the remainder of the regulation.

The *Dyson Ltd vs European Commission* case exposes at least four potentially weak points in the current energy labelling system, gradually and carefully built up in the EU during the past 30 years:

1. Since the energy labels are mandatory, they constitute a part of EU law, they are open to legal scrutiny and may consequently be challenged in the European courts of law.
2. As the dictates of EU energy law are based increasingly on technical language, mathematical formulae and engineering science, they are becoming less comprehensible to the energy-consuming general public and even to the judges, who after all are just lawyers, not energy engineers.
3. The current alphabet-based Energy Efficiency Class system relies on the transposition of a number of measurements into a single alphabet code, rather like the performance of pupils and students at schools and universities in many countries. The ABC system is easily exposed to suspicion, even when combined with evocative arrows and bright colours, or maybe especially so.
4. In the *Dyson Ltd vs European Commission* case, the General Court annulled an entire regulation, citing Dyson's argument that the Commission had disregarded an essential element of the directive,[23] which requires the method for calculating a vacuum cleaner's energy performance to reflect normal conditions of use. After

[20] Regulated on the basis of Commission Delegated Regulation (EU) No 665/2013 of 3 May 2013 supplementing Directive 2010/30/EU with regard to energy labelling of vacuum cleaners, effective as of 2 August 2013.

[21] Judgment of the General Court (Fifth Chamber) of 8 November 2018 in *Dyson Ltd vs European Commission*; http://curia.europa.eu/juris/documents.jsf?num=T-544/13#.

[22] General Court of the European Union. Press Release No 168/18. Luxembourg, 8 November 2018. https://curia.europa.eu/jcms/upload/docs/application/pdf/2018-11/cp180168en.pdf.

[23] Directive 2010/30/EU of 19 May 2010 on the indication by labelling and standard product information of the consumption of energy and other resources by energy-related products. The directive has since been repealed and replaced by Regulation (EU) 2017/1369 of 4 July 2017 setting a framework for energy labelling and repealing Directive 2010/30/EU.

5 years, the General Court finally accepted this argument, and, as a consequence, energy efficiency labels and stickers were diligently removed from the sides of vacuum cleaners and their packaging in retail stores and warehouses all over Europe. These were vacuum cleaners that had already been placed in the common market.

The question arises whether the same could not happen to other household appliances or, in an extreme case, the entire EU energy labelling regime?

With the exception of vacuum cleaners and light sources, the household electrical appliances subject to energy labelling can broadly be classified as "white goods", such as washing machines and refrigerators, or in some cases "brown goods", such as televisions. All share a large physical size. The plethora of small appliances available in shops—mixers, blenders, razors and toothbrushes—has escaped regulatory attention, as has a very large group of mostly spare-time-related devices usually referred to as "consumer electronics".

4 Toasters, Kettles and Dryers

Admittedly, most such devices consume very little electricity and **should** be exempted from energy labelling, as they fall within the *de minimis* rule, in the substantive sense: *de minimis non curat lex*, *i.e.* the law does not concern itself with trifles.[24] However, there is a large number of ubiquitous, physically small electric appliances which share high peak-load power consumption but escape labelling regimes for no obvious reason. These products include (with minimum and maximum wattages indicated):

Bread-toaster	600–2000 W
Hot-water kettle	1300–3000 W
Hand-held hair-dryer	1200–2400 W

By comparison, the rated wattage of a typical 300-L refrigerator-freezer (with 200 L and + 100 L of space, respectively), standard in modern households having two or more members, is 100–120 W, with annual consumption of ca. 300 kWh, in energy class A+ according to the currently effective EU labelling regulation for refrigerators and freezers.[25] As an electrical home appliance, the refrigerator stands at the other end of several continua vis-à-vis the toaster, kettle and dryer: physical

[24] Hojnik, Janja. "*De Minimis* Rule within the EU Internal Market Freedoms: Towards a More Mature and Legitimate Market?" European Journal of Legal Studies, Volume 6, Issue 1 (Spring/Summer 2013), pp. 25–45.

[25] Commission Delegated Regulation (EU) No 1060/2010 of 28 September 2010 supplementing Directive 2010/30/ EU with regard to energy labelling of household refrigerating appliances.

size (large—small), consumption pattern (constant—switch-on/switch-off) and power demand (base load—peak load).

At the same time, both the refrigerator and the three appliances discussed represent mature technology, having appeared in the consumer market *ca.* 100 years ago. As noted above, the first electric refrigerators were introduced to the consumer market in the United States in 1915, and the first automatic "pop-up" toaster was granted a patent in 1921. The invention was made by Charles Perkins Strite in 1919.[26] The device is technically indiscernible from most of today's basic toasters.

The toaster, kettle and hair-dryer are all based on intensive use of electric power conducted through resistance wire, an invention made by Albert L. Marsh, who was granted a patent for an "Electric Resistance Element" in 1906. The original NiCr alloy filament, or filaments very similar to it, is still used in toasters. All three devices are now heavily commoditised in mature consumer markets, so that retail prices can be below €10 or US$10 for the so-called entry-level models, while prices for other models range between 20–30 € and 200–300 €, as provided in Table 1.

The table reflects information collected by means of a desktop study made by the author in 2017 and 2019 and shows, *ceteris paribus*, that in just 2 years.

Table 1 Market Survey—Toasters, Kettles and Dryers

1. Electric Toasters		**7.7.2017**	**27.7.2019**
Number of models		67	82
Number of brands		16	19
Price Range, Euro	Min	23.90	16.95
	Max	359.00	269.00
Power Load, Watt	Min	600	600
	Max	1800	2000
Number of models		66	127
Number of brands		16	20
Price Range, Euro	Min	19.90	19.95
	Max	299.00	199.00
Power Load, Watt	Min	1300	1300
	Max	3000	3000
Number of models		30	53
Number of brands		10	10
Price Range, Euro	Min	9.90	9.99
	Max	249.00	399.00
Power Load, Watt	Min	1200	1200
	Max	2400	2400

A desktop study conducted by the author through the website of Gigantti Oy, which claims to be the largest retailer of home appliances and consumer electronics in Finland, with a market share of 25%. https://www.gigantti.fi

[26] U.S. Patent No. 1,394,450, granted on 18 October 1921 to Charles P. Strite for "Bread Toaster".

- the number of models has almost doubled;
- the minimum and maximum power ratings have not changed;
- the prices have not changed—admittedly, there is some fluctuation in the highest, "luxury" segment, but this hardly reflects any improvements in basic technology, much less energy efficiency.

The study proves that manufacturers and suppliers have made no effort to improve basic technology or energy efficiency of their products, but rather seek differentiation in their customers' eyes through branding and design. For example, one particular toaster model is now available in seven colours, including pink and mint green. Except for body colour, the units are identical. For hair-dryers, the desktop study proves that, in marketing, wattage is used as a proxy for efficiency—the more watts the device consumes, the more efficient (better) it is.

Why, then, have toasters, kettles or hair-dryers not become more energy efficient? Could the reason be that it is not possible to replace the current technical solutions based on resistance wire and coil made of NiCr or similar alloys? While heat-emitting diodes, or "HEDs", probably are and will remain the stuff of science fiction, a number of technical solutions exist for the purpose of improved electricity-derived heating, such as

- electric convection;
- electric induction;
- fat film resistance;
- halogen;
- infrared, including
- far infrared; and
- microwave.

Many devices, including household kitchen stoves and ranges, exist with heating elements based on infrared, halogen, microwave and electric induction technologies. Infrared saunas were all the rage in Finland some years ago. However, no toasters, kettles or hair-dryers utilising other than conventional resistance wires and coils seem to be available in consumer markets in the EU or elsewhere. The notable exception is the induction hob (cooktop), which is fast becoming standard in European kitchens and is claimed to be much more energy efficient than a conventional hob. These claims are difficult to verify, since hobs carry no energy labels—as all-in-one cooking ranges and separate ovens do. The very high nominal wattages, of 7000 W or more, of many hobs in the market make one suspect that energy efficiency has not been a top priority for the designers of these products.

Another exception is the so-called toaster oven, which in effect is small table-top oven used to prepare home fast food, such as pizza and grilled sandwiches, for small audiences. Infrared heating is used in many toaster ovens, sometimes combined with microwaves and traditional NiCr grill elements. Toaster ovens are not subject to EU energy labels.

Refrigerators and freezers are the household appliances that probably have benefited most from advances in energy efficiency. The single most important technology improvement has been the more efficient insulation of the big white boxes. No such change has been discernible for toasters, kettles or hair-dryers. The extra heat, paid for dearly in the electricity bill, is happily dispersed into the atmosphere. Indeed the most immediate technical concern of many manufacturers of small MELs seems to be the risk overheating the unit. Toasters may be used by children, who could burn their fingers. The little box should be cool on the outside.

That said, there has been no change in the heat insulation of the three devices discussed. The basic technical solutions are 100 years old, or more. There is not the slightest doubt about the virtues of insulation: it saves energy. Why not insulate a toaster? Because—it costs money, as it makes manufacturing more complicated, probably requiring not one or two but many new technical solutions. If there is no incentive to improve the energy efficiency of the device by adding insulation, why do it?

Fat film heaters have been commercially available since many years,[27] but manufacturers and their customers accustomed to appliances that offer cheap-and-cheerful heating utility are likely to resist innovations that may increase prices significantly. For example, water-heating thick film thermal elements are claimed by component suppliers to have thermal efficiency of 96%[28]—a big number. One supplier of thick film heating elements for kettles offers 10% improvement in energy efficiency[29]—a rather small number. A quick look at the raw materials used for thick film heat elements gives a hint of an explanation for lack of interest among kettle manufacturers: silver and palladium. Not iron, nickel and chromium, as in traditional heat resistance elements.

The heat elements inside each apparatus are as primitive as ever. If *e.g.* hair dryers based on presumably more expensive infrared, halogen or fat film technologies were developed, it would make more sense to reduce the waste of heat energy on the outer boundary of each piece of equipment.

The MELs discussed here address some rather basic daily needs in EU and other households in developed countries: crispy toast, curly hair, and hot tea. The related devices have become through and through commoditised. Consumers expect to be able to purchase their appliances at a low price. Energy efficiency currently plays no role, nor does the durability of the devices. Design and colour matter more, at least from the point of view of many manufacturers and retail sellers.

In this situation, introducing new mandatory energy efficiency standards, including ecodesign and labelling, could be the only way out to a more sustainable market

[27] European Commission. Directorate-General for Enterprise and Industry. *Preparatory study to establish the Ecodesign Working Plan 2015–2017 implementing Directive 2009/125/EC.* Draft Task 3 Report. 16 June 2014, p. 25; R.B. Tait, R. Humphries, J. Lorenz. *Thick film heater elements and temperature sensors in modern domestic appliances.* IEEE Transactions on Industry Applications. Volume 30, Issue 3, May/Jun 1994.

[28] For a supplier among many, see for example https://www.made-in-china.com/showroom/xxjieda/product-detaillbExWvzjXFAI/China-Replacing-Quartz-Heating-Tube-Thick-Film-Electric-Instant-Water-Heater-Element.html.

[29] http://www.ottercontrols.co.uk/kettletechnologies.html.

situation. If the basic utility of the MELs discussed here can be purchased for the equivalent of €10 each and, concurrently, the same utility can be sold to better-off consumers for many times that amount, up to several hundred euro, there definitely must be significant room for product and market development for energy-efficiency-conscious consumers.

At the outset, it may be assumed that all toasters, kettles and hair-dryers currently available in the EU market would fall into the lowest energy efficiency classes, F or G, like the incandescent light bulb. This is just a guess made by the author, as no ecodesign standards, directives or regulations exist for these devices at the time of writing. There is a long way to go.

From the point of view of law and economics, the most obvious, expensive externality of the toaster, kettle and dryer is their heavy influence on load demand in households. The problem becomes acute in situations where many base-load units, such as refrigerators and hot-water boilers, are operated simultaneously, and peak-load appliances, including one or several of the three appliances discussed in this article, are switched on: the main fuse of the house blows. The only secure way permanently to avoid this annoyance is to request the main fuse be upgraded. Your local power distribution company will be happy to comply—and charge a fee for this. This, again, makes it necessary eventually to strengthen the capacity of the distribution network, as an increasing number of customers demand more load. Heavier cables, more heavy-duty transformers, and other larger power supply items are installed.

Household electricity load is often divided into: (1) load necessary for core household functions and, (2) miscellaneous electricity loads (MELs).[30] The core functions include space and water heating, ventilation, air conditioning and refrigeration—and, of course, usual home cooking. Refrigeration of foodstuffs no doubt belongs to the most essential household functions in modern societies, where the refrigerator-freezer has a market penetration of almost 100%. Among MELs, the toaster, kettle and dryer also share a market penetration close to 100% and extremely high daily peak loads, yet for short periods each time.

To demonstrate the relative load patterns of the four appliances, some simple household arithmetic is employed, as follows:

1. A typical 300-L refrigerator-freezer of EU Energy Class A+ consumes some 300 kWh/a. This consumption is based is on 24 h/day cooling and freezing of foodstuffs inside. Most of the time, the double doors of the unit are closed, and the inside (LED) light is off. The fast-freeze function, to freeze e.g. raw meat, fresh berries, or large portions of cooked food, is used infrequently.
2. A typical two-slice bread-toaster has no EU energy class. A toaster is used two to three times a day. The typical duration of each time of use is 2 × 3 min. The daily use is 6 min, or 90 Wh, based on an average two-slice toaster with power

[30] Source: U.S. Department of Energy, Office of Energy and Renewable Energy. https://www.energy.gov/eere/buildings/articles/miscellaneous-electric-loads-what-are-they-and-why-should-you-care.

rating of 900 W. This equals to *ca.* 33 kWh/a. These estimates do not apply to larger, four-slice models, or even larger, mostly commercial units. It is assumed that just four pieces of toast are consumed daily in a dual-occupancy household using a single toaster, which is probably frugal. On the other hand, no weekly or annual variation is accounted for.

3. A typical 1.5- to 2-L hot-water kettle has no EU energy class. A kettle is used frequently, typically five or more times a day, but just for some minutes each time, until the water reaches 100 °C boiling temperature. The kettle is often overfilled because the users (usually tea-drinkers) typically consume 1–2 cups or mugs (1.5–4 dL) each time, yet fill more water in the kettle. Many users do not pour out the hot water, but let it remain in the kettle for the next use, which makes stochastic estimation necessary. Therefore, we assume that a 2-L, 2000-W kettle, is used just once a day to boil up its full content. Such daily use would be 6 min, or 200 Wh = 0.2 kWh. The consumption of a kettle equals to 73 kWh/a.
4. A typical hand-held blow-dryer has no EU energy class. Hair-dryers are used frequently, typically one or two times a day, for 6 min each time. A nominal wattage of 2000 W is common also for dryers. As use patterns vary significantly, stochastic estimation is necessary. We assume that a 2000-W dryer is used for 6 min just once a day, with no variation for heat (temperature) and blow speed (air flow). Such daily use would translate to 200 Wh = 0.2 kWh. The consumption of a hair-dryer equals to 73 kWh/a.
5. Consequently, the cumulative annual energy consumption of the toaster, kettle and dryer used in this simplified example is 33 + 73 + 73 ≈ 180 kWh, which is 60% of that of the refrigerator-freezer referred to above in Item 1. The peak load of the three appliances, however, is 4900 W, or some 40 times that of a standard refrigerator-freezer, which has a load of 100–120 W.

Understanding that electric power networks must be built to sustain both base load and peak loads, which are highly irregular, the networks have for decades been constructed with ever-increasing load reserves. This development has been slow but certain, as consumers are increasingly disinclined to accept the risk of a power cut-off every time they switch on another appliance. It is not a realistic proposition to think of a household with 40 refrigerators, but one with two blow-dryers, two kettles and indeed dozens of other sundry MELs would not be unimaginable.

Also situations where the three common peak-load appliances discussed in this paper are switched on simultaneously in just one household are common. Indeed such "morning rush" situations are probable in entire residential neighbourhoods. Distribution networks have been designed and built to withstand peaks, but this has come at a cost. In a mature, unbundled electricity market system, the cost of distribution is approximately 1/3 of the final cost of electricity for households.

Since security of supply is probably the first requirement made to electric power systems, both technically and from the point of view of energy and social policy, it is not likely that investment costs for the "last mile" will be reduced in the foreseeable future. On the contrary, the privatisation of electricity utilities, which is an ongoing process in the EU and many other regions worldwide, is based on at least the following postulates:

1. Sufficient investments in all links of the electricity supply chain must be secured.
2. Security of supply to end customers is paramount.
3. Since household customers, unlike industrial and institutional customers, normally do not have power reserves or alternative supply sources, such as back-up generators, households should be the highest priority to the energy sector regulators.
4. Securing the uninterrupted functionality of distribution networks, and the resultant investments, are likely to become a major guideline to regulatory decision-making.
5. The requirement to allow private-sector investors a reasonable return for their investments is a necessary constituent in the concept of power system unbundling, as is the understanding that distribution networks cannot, or should not, be subjected to free competition.

These considerations combined result in a conclusion that the cost of power distribution is likely to increase, or at least not decrease, despite all regulatory efforts. This tendency seems largely independent of any overall increase in distributed electricity loads. That said, energy consumption **is steadily increasing**, according to statistics published by the International Energy Agency (IEA), at a fast pace worldwide and also in most national economies.[31]

5 Conclusions

The energy policy intentions behind energy efficiency labelling and ecodesign regimes are no doubt respectable in the EU and in other jurisdictions. Households should be nudged into making better choices when selecting energy-using appliances. The underlying question, however, is "what is better?" Indeed this question is just another version of "what is good?" And what is good to whom?

The discussion about light sources shows that, after some 100 years, there has been an energy-saving paradigm change with the introduction of the LED. This change has not happened because of energy labelling or ecodesign but rather because of advancements in technology, which have been spurred by a need to save electric energy, and economic considerations in general. This paper has not discussed whether labelling regimes have possibly accelerated the paradigm change in domestic lighting by increasing consumer awareness of new technologies.

Within the EU—since 1979, when the first voluntary energy labelling directive was adopted—regulatory ambition has gradually grown to include 14 energy-using

[31] https://www.iea.org/newsroom/news/2019/march/global-energy-demand-rose-by-23-in-2018-its-fastest-pace-in-the-last-decade.html. NOTE: the original paper was written and presented before the coronavirus pandemic.

household products and product groups. The current mandatory system no longer applies to vacuum cleaners, which were removed from the roster of labelled appliances by order of the General Court of the EU in 2018. The *Dyson vs. Commission* case exposes the risks taken when making energy efficiency a legal obligation. On the other hand, voluntary labelling tends to be ineffective, as is witnessed by the proliferation of marks and labels in shops. The energy label competes not only with other, voluntary environmental and quality labels, but also with the manufacturer's trademark and, perhaps most importantly, the price tag.

The current EU energy labelling regime is targeted at "slow and bulky" home appliances, especially after vacuum cleaners have been removed from the list. Their other main feature of the currently regulated appliances is that they constitute base electric load in households—or, indeed, also in typical commercial buildings, providing illumination, heating and cooling, ventilation, refrigeration and basic cleanliness. Practically all MELs have escaped ecodesign and energy efficiency labelling regimes, although many MELs consume very large amounts of electricity, albeit for short periods of time. The three devices discussed in detail—the toaster, kettle and hair-dryer—have escaped energy labelling regulation for no good reason. Their market penetration nears 100% in most developed countries, where they are used daily in most households, and their combined annual load equals some 60% of that of a standard household refrigerator-freezer.

A desktop market survey conducted by the author reveals an increasing proliferation of models, including pink toasters, flaming red kettles and purple hair-dryers, but no reduction in energy use. For dryers, wattage is clearly used in marketing as a proxy for efficiency and quality: the more power, the better. Together with many other MELs, the toaster, kettle and dryer are likely to proliferate with increasing standards of living. Their history as consumer appliances in the EU and North America invites speculation on a ***"Two Hundreds Market Penetration Rule": from zero to hundred percent market penetration in one hundred years.***

With no ecodesign and energy efficiency labelling regimes in place, the toaster, kettle and dryer are likely inflict an unwanted additional demand on base and peak loads of distribution networks. This externality is not adequately reflected when "entry level" devices with heat-resistant filaments or rods are sold under unknown brands, without energy labels, at very low prices, and there is no connection with their energy performance and prices, be they high or low.

The discussion, above, of electric induction, infrared, halogens and microwaves as potentially energy-efficient methods for transforming electricity into usable heat in household appliances shows that there has been progress away from conventional solutions, which could also mean reduction in energy demand while the same energy service—*e.g.* a piece of warm toast—is obtained. With many brands and models of infrared toaster ovens available, it is utterly surprising that no toasters using infrared heat elements are in the market. If there are consumers who are prepared to pay almost 300 euro for a mint green "classic" household bread toaster relying on NiCr wires,[32] with no attempt to improve energy efficiency, there must be room in the market for more energy-efficient, actually green models—with better heat generation technologies, perhaps with insulation—given the significance of the warm-glow effect and the willingness of many consumers to be in the forefront of technical progress. Proving this postulate should merit some empirical research.

[32] A desktop revisit by the author at the website of Gigantti Oy, Finland. https://www.gigantti.fi.

What Can Market Surveillance Learn from Reducing Violent Crime in Chicago? It's Time that EU and Similarly Organised Market Surveillance Authorities Adopted a New Paradigm

Chris Evans

Abstract Talk of non-compliance has become part of the routine when discussing Ecodesign (MEPS) and Energy labelling success. Seemingly, paying little more than lip service to the EU's energy efficiency performance regulations is all too common amongst product suppliers.

Stakeholders describe this as a failure of market surveillance and call for more actions by those authorities. One solution to this in the EU has been to throw more market surveillance product testing at it. This delivers improvements but it does not cure the problem and it is extraordinarily expensive.

The time has come for EU market surveillance authorities to develop a different approach; time to stop treating non-compliance as a misdemeanour, time to recognise that non-compliance is a widespread epidemic, time to treat it as a disease just like any other epidemic.

A useful parallel with violent crime in Chicago can be drawn. Here, the authorities stopped throwing more and more police at it and treated the increasing levels of violent crime as a disease instead. That has now evolved into a hugely successful operational model (www.cureviolence.org) that could be adapted to "treat" non-compliance in the energy efficiency sector too.

At its core are the same three components that are used to reverse epidemic disease outbreaks. (1) Interrupting transmission of the disease. (2) Reducing the risk of the highest risk. (3) Changing community norms.

The paradigm shift required is to refocus market surveillance in the EU and in any other jurisdiction where products can enter the market based on little more than the equivalent of a suppliers' self-declaration. In these jurisdictions, it is time to shift away from an end-of-pipe market surveillance model to one focussed on:

C. Evans (✉)
S2E4 Ltd, Cork, Ireland
e-mail: chris.evans@s2e4.com

P. Bertoldi (ed.), *Energy Efficiency in Domestic Appliances and Lighting*, Springer Proceedings in Energy, https://doi.org/10.1007/978-3-030-79124-7_5

1. Inspections at points of import and places of manufacture;
2. Prioritisation of interventions based on risk analyses of non-compliance;
3. Exposing the anti-societal nature of non-compliance through training and support programmes provided to product suppliers and stockists.

The suggestions in this paper are based on reflections of 15 years' experience of developing market surveillance methodology across numerous national jurisdictions. Results of recent pan-EU market surveillance activities have shown levels of non-compliance continuing at much the same >50% levels for products as were detected in the pan-EU programme, ATLETE 1, 8 years previously. These, and other published results e.g. EEPLIANT 1 (http://www.eepliant.eu/images/Documents/WP1/EEPLIANT2014-Final-Report-12-12-2017.pdf) in 2020, support the core argument that non-compliance is a persistent and widespread disease requiring a new and effective form of treatment.

1 Preamble

Dr. Gary Slutkin was an epidemiologist with the World Health Organisation where he had spent much of his career fighting the spread of infectious diseases across the globe.

He had tackled the spread of Aids in Africa with some success but had become jaded by the death and misery he had been exposed to so took a break by returning to Chicago. On arriving there, instead of finding a tranquil safe society he was shaken to find that there were high levels of premature death there too. Not caused by Aids but, instead, by youth violence.

"I saw that all this violence was happening in America and I didn't even know, as I'd been away for so long, I thought America had no problems," he is reported to have said. "When I came here, I saw in the newspapers and TV that there were 14 year olds shooting 13 year olds in the head. Killing them. Just little kids shooting each other. What is this?"

Between 1994 and 1999, 4663 people were murdered in Chicago [2] By comparison, Los Angeles—which had a larger population—had seen fewer (3380) murders [3].

Puzzled, Dr. Slutkin began to study the problem. A review of the statistics showed similarities between the levels and patterns of violence in Chicago and the Aids epidemics he had been working on in Africa.

He surmised that the same techniques used in combating the spread of disease in Africa could be applied to reducing the spread of violence in the USA. He realised that like Aids, violent incidents were occurring in clusters at certain locations and at certain times. The violence appeared to be self-replicating, much in the way an infectious disease does. One violent incident would lead to another and then another, and so on. Finally, violence was increasing rapidly in a fashion very similar to an epidemic wave.

As an epidemiologist, he knew there were three features to look for in order to class a disease as contagious: clustering, self-replication and epidemic waves. Dr. Slutkin concluded Chicago was facing an epidemic disease just as bad as he had witnessed in Central Africa.

He decided to try to deal with the problem in the same way. A local university made funding available and Cure Violence [4] was set up—a project dedicated to using tried and tested public health strategies for tackling violent crime.

As with the fight against Aids, the starting point was to stop treating violence as "a problem with bad people". Instead, it was to be dealt with as a contagion that infected people. This meant focusing on stopping violence before it begun and mitigating it once it had.

The success of this approach to reducing such crime has been remarkable. Results have shown 41–73% reduction in shootings and killings [5] in areas where this public health-type approach is applied.

2 Introduction

Non-compliance with regulatory performance requirements amongst energy using products in the EU may not impose the same strains on society as violent crime, but its persistent and widespread nature could be described as an epidemic and has parallels with the behaviours of infectious disease.

Results from a recent pan-EU market surveillance programme, EEPLIANT 1, have shown levels of non-compliance at much the same >50% levels for domestic refrigeration appliances as were detected in the 2011 pan-EU programme, ATLETE 1. Additionally, suppliers' failure to correctly produce the legally required documentation also exceeded 50% across all product sectors examined. These indicative results and those from the EEPLIANT 1 project, where it was reported for LEDs *"Only 14% of the tested models were fully compliant..."* reinforces the core argument that non-compliance is a persistent and widespread problem requiring new and effective forms of treatment.

This paper explores the proposition that if the methods of disease control were applied to the problem of non-compliance, then it is likely that significant reductions in non-compliance could be achieved. And achieved at no increase in the overall current costs of operating market surveillance by Member State authorities in the EU.

It is time to stop chasing around the EU trying to spot non-compliant products. It is time to stop throwing ever more money at testing products taken from the marketplace and time to stop treating such non-compliance as a minor misdemeanour. Evermore testing may deliver improvements but these seemingly are short term and there is no evidence that they cure the problem. And testing products is extraordinarily expensive.

3 The State of Market Surveillance Today in the EU

The market surveillance authorities in the EU who are responsible for the enforcement of the Ecodesign (MEPS) and Energy Labelling Regulations have an unenviable task. The Ecodesign Directive calls up some 30 different regulations, one for each different product sector. Products range from the familiar domestic to much less familiar and technically more complex industrial units. The skills needed to fully interpret the technical documentation that can be demanded from suppliers of, say, a heat pump, are very different to those required to wrestle with the complexities of evaluating network standby performance. The cost of testing can be so high that some EU Member State authorities do no testing of their own at all e.g. Malta, or are forced to make the difficult choice of between more staff/lower testing budget or less staff/higher testing budget e.g. UK.

At the same time, EU stakeholders, both from the supply side and the consuming side, have historically complained [7] of poor market surveillance and have repeatedly called for "more testing".

Much of the frustration voiced from the supplier side concerns the sporadic nature of EU market surveillance on energy consuming products. These are justifiable criticisms that are caused primarily through the legal requirement, contained in both the Ecodesign Directive (2009/125/EC) and the Energy Labelling Regulation (EU) 2017/1369), for each individual EU Member State to be responsible for enforcement in their country. This means more than 27 countries are each setting different levels of budget, different priorities, different interpretations, different approaches, different sanction regimes etc. For this alone, one should think of EU market surveillance as being more akin to herding chicken than the murmuration behaviour of starlings [8].

Challenges increase further for those market surveillance authorities dealing with ecodesign requirements rather than those for energy labelling as the Ecodesign Directive permits products to be placed on the EU market based on the suppliers' self-declaration of conformity and with no requirement for registration prior to placing on market. The authorities are therefore unaware of what products are in the market that they are responsible for. Their general approach, which will vary from country to country, is to conduct their own informal market research mainly through internet searches, reviews of trade literature, advertisements and visits to shops or distributors. From the results of this, most make risk assessments (of possible non-compliance) to assist targeting their scant resources on the product sectors, brands and models that they judge to be most likely to have some level of non-compliance.

Invariably, unless they are working in close collaboration with customs authorities, they are sampling from the vast numbers of products already dispersed in the EU market. And since the EU is a single 27+ nation market without borders, this is the largest single market in the world with 100s of millions [9] of energy consuming electrical products sold there every year.

The market surveillance authorities can be regarded as providing a quality control service on behalf of society. Yet no quality control professional would normally choose to apply such end-of-pipe quality techniques as practised by most EU market surveillance authorities. Instead, the quality control professional would be focussing quality control to take place as early in the production process as possible i.e. at front-of-pipe. To explore this analogy further, imagine trying to stop water spraying out from a leaking pipe. The best technique would not be to place buckets to try catch some of the water being scattered about. Instead, it would be far more effective to turn off the water supply to the pipe. All the time EU market surveillance authorities are largely using buckets to catch leaking water spraying out uncontrollably all over the place, so the supply side and the rest of society will continue to express disappointment with their performance.

Ultimately, success in market surveillance (aka market "policing") should not be all about being good at catching non-compliant products. Instead, it should be that of creating and maintaining a highly compliant market.

4 A Possible Future for Market Surveillance in the EU: The Paradigm shift [10]

Whilst the approaches of quality control professions and the approaches of disease control professionals have much in common, this paper focusses on the application of disease control. This is because production quality control techniques are more usually applied in a closed and supervised environment and not in the open uncontrolled environment of the marketplace.

What are the disease control techniques that have applicability to market surveillance aimed at reducing non-compliance of energy using products in the marketplace?

- Locating and eliminating the source;
- Stopping the spread by:

  Publicity campaigns;
  Teaching and supporting healthy habits, often by using cured patients to spread the message to their peers, leading to the adoption of new behavioural norms;
  Maintaining a constant watch for new outbreaks, moving fast if there are signs of any;
  Partnerships with other actors e.g. civil society, consumer and environmental NGOs;

- Treating those already suffering from the disease;
- Developing vaccines and similar immunization programmes.

4.1 Locating and Eliminating the Source

This is the most important change of all. It requires targeting to be refocussed on finding the source of the non-compliancc i.c. at front-of-pipe rather than at end-of-pipe. Sources are at manufacturers' premises, at importers warehouses and, ideally, at the point of import if effective cooperation with Customs can be developed. Identifying a complete batch of non-compliant products in a supplier's warehouse and thus preventing their distribution into the marketplace is exactly the equivalent of what health professions do when locating the sources of an outbreak of a disease.

4.2 Stop the Spread

Ensuring the entire supply-side community is aware of the issue of non-compliance requires communications to be a key tool in the armoury of the market surveillance authorities.

- *Publicity campaigns*—articles in trade association journals, presentations at trade association and chamber of commerce meetings, information stands at trade association exhibitions, guidance leaflets downloadable from market surveillance and customs' websites;
  - *Teaching and supporting healthy habits, often by using cured patients to spread the message to their peers, leading to the adoption on new behavioural norms*

- Teaching good (compliant) behaviour should be the first task in the market surveillance authorities' work list. Suppliers, like children, benefit from being taught how to behave properly. Regulations, for suppliers—especially the smaller less mature ones—can be a confusing mystery world. Helping them navigate and comprehend the regulatory jungle will always deliver improvements;

- Use offenders to evangelise compliant behaviour. Most suppliers have a good sense of who their competitors are and, indeed, may have a good sense of how compliant their competitor products are too. Instead of a financial sanction, give an offending non-compliant supplier the opportunity of meeting with other suppliers to spread targeted advice on achieving compliance as the new norm;
  - *Maintaining a constant watch for new outbreaks, moving fast if there are signs of any*

- This should be fairly easy for the market surveillance authorities to do as this is some of what they are doing currently. They still need to be seen to deal strongly with non-compliance seen in the marketplace. Repeat offenders and those that refuse to adopt a culture of complaint behaviour should be dealt with the types of sanctions normally applied to criminal activities;
  - *Partnerships with other actors e.g. trade associations, consumer and environmental NGOs*

- These, too, can be the eyes, ears and voices of market surveillance so working with them in an engaged and positive way can convert their negative opinions into constructive support. All these other actors can also be maintaining a constant watch for new outbreaks as well as providing an information conduit between the market surveillance authorities and the product supplier communities that need to be informed about how to achieve good compliant behaviour.

4.3 *Treating Those Already Suffering from the Disease*

If we are to borrow from the fight against Aids and the fight against violent crime in Chicago, then non-compliance should not be treated as "a problem with bad suppliers". Instead, it needs to be treated as a contagion that infects them. Once the infected parts have been cleaned i.e. the non-compliant products withdrawn from the market, the focus must become one of assisting the supplier to learn how to self-administer the necessary treatment i.e. developing and adopting procedures for ensuring all their future product supply is compliant.

4.4 *Develop the Equivalent of Vaccines and Similar Immunisation*

One form of vaccination, or at least a form of immunisation, has been introduced in the EU for the 16 product sectors that fall under the Energy Labelling Regulation [11]. As of 1 January 2019, suppliers (manufacturers, importers or authorised representatives) need to register their appliances that require an energy label in the European Product Database for Energy Labelling (EPREL) [12] before selling them on the EU market. Product information to be entered in the database relates to the energy label, technical documentation and that necessary for compliance monitoring. This data then being available to market surveillance authorities. Adoption of the equivalent of EPREL when the Ecodesign Directive is next revised can be expected to provide a substantial further benefit for EU market surveillance.

EPREL does not provide full immunisation. This would be better provided through a robust form of third-party certification, a requirement for which does not yet exist under the EU product energy efficiency regulations. Experience from elsewhere, such as the USEPA's Energy Star® program which requires such certification [13], suggests that third party certification delivers compliant products [14] thus ensuring consumer confidence and protecting the integrity of the applicable regulations.

Though third-party certification is not required in the EU, there are some schemes established by supplier groups e.g. Eurovent Certification [15] or by Standardization

Bodies e.g. European heat pump KEYMARK [16], that have the type of components [17] that are required to provide effective immunisation;

- Independent third-party testing
- Factory production control
- Quality control system

5 Concluding Remarks

Much could be learned by EU policy makers from schemes like the examples identified above. Developed society has learnt how to adopt and maintain good health practices so it should not be beyond the wit of suppliers to learn how to adopt and maintain good compliance practices. A culture of compliance will not be created overnight and will require leadership and strong continuing support from the market surveillance authorities. Since more can be achieved through cooperation than can be achieved through confrontation, it is time for the market surveillance authorities to change tack and move their focus from end-of-pipe activity to one of eliminating non-compliance at source.

References

1. http://www.eepliant.eu/images/Documents/WP1/EEPLIANT2014-Final-Report-12-12-2017.pdf
2. https://home.chicagopolice.org/wp-content/uploads/2014/12/2011-Murder-Report.pdf
3. http://assets.lapdonline.org/assets/crime_statistics/digest/1999/homicides_10yrs.htm
4. http://cureviolence.org/
5. http://cureviolence.org/results/scientific-evaluations/
6. Full details of market surveillance results are not normally available in the public domain. Instead, required practice is to record them in the European Commission's Information and Communication System on Market Surveillance (ICSMS), an IT platform facilitating communication between market surveillance bodies in the EU and in EFTA countries.
7. APPLiA, previously CECED, the EU trade association for home appliances, and ANEC, the EU consumer NGO, have been vociferous campaigners for improved market surveillance
8. https://www.youtube.com/watch?v=V4f_1_r80RY
9. https://applia-europe.eu/statistical-report-2017-2018/
10. A paradigm shift, a concept identified by the American physicist and philosopher Thomas Kuhn, is a fundamental change in the basic concepts and experimental practices of a scientific discipline.
11. https://ec.europa.eu/info/energy-climate-change-environment/standards-tools-and-labels/products-labelling-rules-and-requirements/energy-label-and-ecodesign/energy-efficient-products_en
12. https://ec.europa.eu/info/energy-climate-change-environment/standards-tools-and-labels/products-labelling-rules-and-requirements/energy-label-and-ecodesign/european-product-database-energy-labelling_en
13. https://www.energystar.gov/partner_resources/products_partner_resources/third_party_cert

14. https://www.gao.gov/new.items/d10470.pdf
15. https://www.eurovent-certification.com/en/certification-programmes
16. https://www.heatpumpkeymark.com/
17. ISO type 5 system according to ISO 17065
18. www.eepliant.eu, https://eepliant.eu/images/Documents/WP1/EEPLIANT2014-Final-Report-12-12-2017.pdf (2020).
19. EEPLIANT Fins Report, https://eepliant.eu/images/Documents/WP1/EEPLIANT2014-Final-Report-12-12-2017.pdf (2020).

A Residential End Use Energy Consumption and Appliance Ownership Patterns in India

Archana Walia, Tanmay Tathagat, Neha Dhingra, and Piyush Varma

Abstract Residential sector in India accounted for 24% of the total electricity consumption in 2016 and is projected to rise more than eight times by 2050. This increased energy demand would primarily be driven by appliances and equipment and attributable to several factors including better access to electricity and rising disposable income. However, at present, there is limited data and understanding of residential energy end-use and its variation with socio-economic strata and climatic zones. As the penetration of appliances and subsequently energy use expands in Indian households, it is becoming pertinent to establish a realistic end-use baseline for formulating informed energy policies and assessing the potential impact of these policies.

A 'first of its kind' comprehensive study was carried out integrating a pan-India survey of residential end-use across 5,000 urban households and a real time appliance energy use monitoring through Non-intrusive Load Monitoring (NILM) of 200 households in 13 cities, spanning different climate zones, socio-economic strata and demographics across the country. The objective of the study was to establish appliance ownership and usage patterns for urban households and develop a framework for collection and analysis of data on energy end-use in the India's residential sector for conducting such surveys in future.

The data indicates that appliance ownership and usage is on the rise significantly. With households shifting towards family nuclearisation, per capita energy use is also increasing. The results also provide good insights on variations in energy consumption across climatic zones, demographic parameters and socio economic strata for the major appliances. This could potentially influence the formulation of customised energy policy interventions to reduce energy consumption. This data can also promote better understanding of future electricity demand, thereby enabling better planning and demand side management programs. The data and the key

A. Walia · N. Dhingra (✉)
CLASP, New Delhi, India
e-mail: ndhingra@clasp.ngo

T. Tathagat · P. Varma
Environmental Design Solutions, New Delhi, India
e-mail: tanmay@edsglobal.com; piyush@edsglobal.com

P. Bertoldi (ed.), *Energy Efficiency in Domestic Appliances and Lighting*, Springer Proceedings in Energy, https://doi.org/10.1007/978-3-030-79124-7_6

findings are expected to be beneficial to academicians, think tanks, policymakers, utilities and consumers for research, modelling, planning and future projections.

Keywords Residential end use survey · Appliance ownership · Energy consumption · Non-intrusive load monitoring · Socio economic strata

1 Introduction

India is the fourth largest energy consumer in the World [1] and will constitute around one quarter of the total global energy demand by 2040 [2]. Appliances and equipment will contribute significantly to this increased energy demand [3]. The residential sector constituted 24% of total electricity consumption in India in 2016 (Fig. 1) and the energy demand from the sector is projected to rise significantly in the years to come.

India is expected to become one of the fastest growing consumer appliances markets driven primarily by rapid urbanization, household electrification, growing per capita income and female participation in the workforce [4].

Despite the substantial electricity consumption by the residential sector in India, there is no data on the household and appliance energy consumption and variations resulting from socio-economic factors and climatic zones. There is very limited information on appliance ownership and usage patterns. A few countries such as the United States [5] and Australia [6] carry out periodic surveys to understand the residential energy use for forecasting future energy demand, improving energy efficiency measures and developing successful demand side management programs. Thus far, no such surveys or studies have been carried out in India.

The National Sample Survey Office (NSSO) conducts nationwide household consumer expenditure surveys at quinquennial intervals [7]. These surveys provide

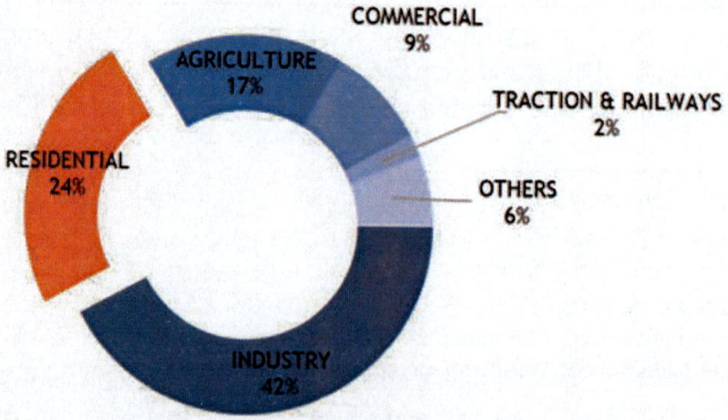

Fig. 1 India's sector-wise distribution of electricity consumption in 2016
Source: Ministry of Statistics and Programme Implementation (2017)

information on ownership of appliances, however, they do not capture variation across climatic zones or socio economic strata. Also, these surveys do not capture usage pattern or disposal behavior of appliances across households.

It has been established that insight into the human dimension of energy use is key to better understanding future energy trends and to effectively manage them [8]. Therefore, consumer usage data can also promote better understanding of future electricity demand, thereby enabling better planning and demand side management programs. With the growing electricity consumption and a steady progress on the policy development to improve the energy efficiency of the appliances and equipment, it is imperative to understand the trends related to the usage patterns and consumer behaviors. This will not only support formulation of effective energy efficiency policies but also help in establishing a realistic end-use baseline for assessing the potential impact of these policies, and enable better estimation of energy savings.

2 Objective

The primary objective of the study was to gather data on appliance **ownership** and **usage patterns** for urban households across climatic zones and socio economic strata, aimed at bridging the gap in existing understanding of residential energy end use in Indian households. The study also aimed to develop a **framework to gather data on energy end-use** in the India's residential sector for conducting such surveys in future.

3 Approach and Methodology

A 'first of its kind' pan-India study was conducted to include a residential electricity consumption survey (RECS) of 5,242 urban consumers with electricity end-use monitoring of 200 households. The study was divided into 3 phases as shown in the figure below (Fig. 2):

3.1 Phase 1: Household Survey

A secondary research was conducted on the methodology from national household census, market reports, journals, articles and government reports to design the survey. A survey questionnaire was designed which aimed to gather following information:

- Demographic characteristics such as household income, family type and size, occupation and educational qualification, socio economic status (SEC)

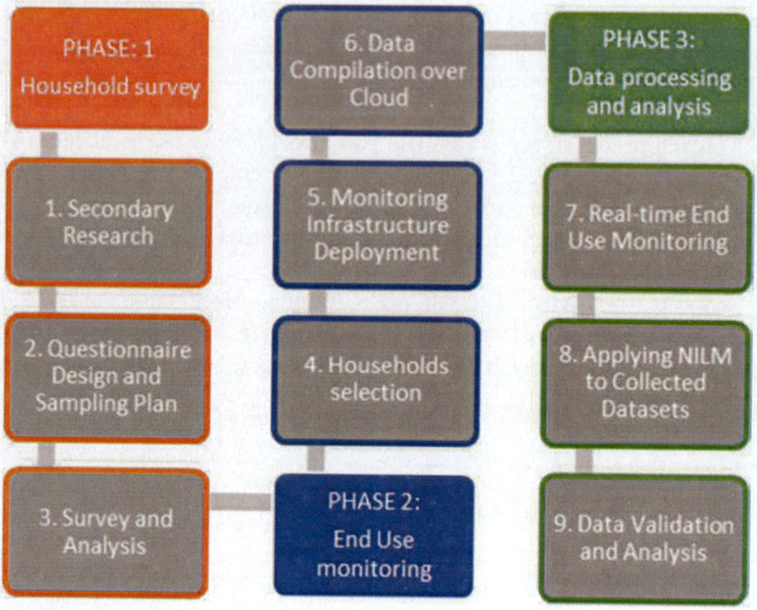

Fig. 2 Methodology of the study

- Appliance ownership, usage, and disposal behavior
- Electricity usage pattern

A representative sample of 8,448 households was derived from the census data, which included urban and rural households. As the urban sector accounts for almost two-thirds of the consumer durables market [9], the scope of the survey was limited to urban areas and the representative sample was adjusted to 5,242 households.

The distribution of sample size was based primarily on the different climatic zones, followed by urban agglomeration and by socio economic classification (see Fig. 3) leading to the selection of 21 representative cities.

The survey was conducted using Computer Assisted Personal Interview (CAPI) technique. The use of an electronic device facilitated data recording while also minimizing recording discrepancies. Willingness to participate in the detailed load monitoring exercise was also assessed during the survey.

3.2 *Phase 2: End Use Monitoring*

From the large survey, a sample of 200 willing households representing the climatic zones, urban agglomeration, dwelling type and socio economic strata across 13 cities covering 11 states and 2 union territories were selected (see Fig. 4). Non-Intrusive load monitoring (NILM) devices were installed for real-time electricity

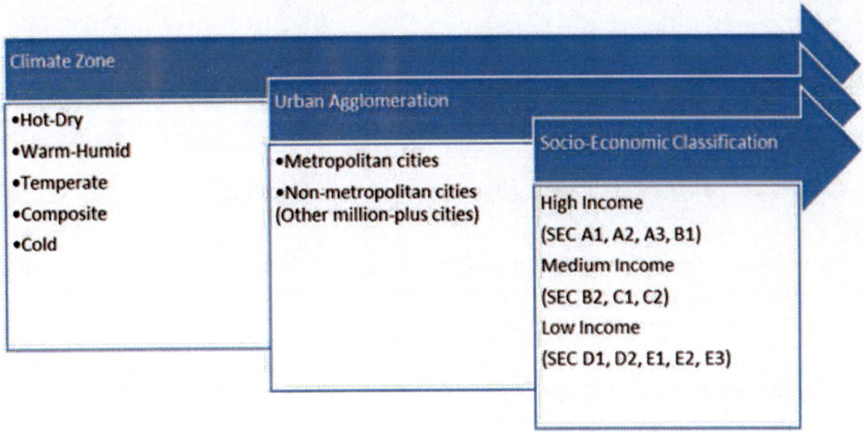

Fig. 3 Sample size distribution criteria

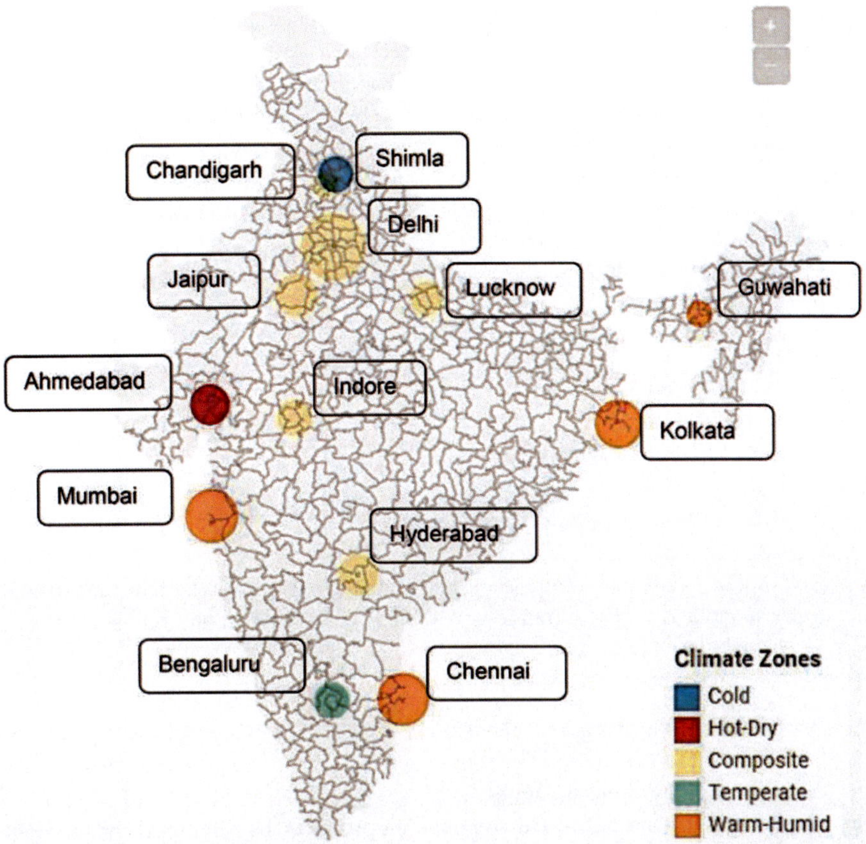

Fig. 4 Smart meter deployment across India

consumption across appliances in the household for a period of one year to capture any variations across seasons.

3.3 Phase 3: Data Processing and Analysis

Data collected from the survey and the energy metering was processed and analyzed using MySQL database and NILM algorithm, respectively.

The real-time energy use data for all monitored households was collated on a cloud server. The monitored data includes electricity consumption, voltage, current, power factor and frequency at 30 second intervals. This is the first time where data at such fine resolution has been recorded and analyzed for Indian residential sector. The monitored households represent different geographic, climatic, cultural and socio-economic aspects, and the data was aggregated and analyzed for median energy use and peak demand analysis. Considering all households and hourly data points, a representative household was constructed by taking median energy use and peak energy demand at an hourly interval.

This data was analyzed using NILM algorithms to provide detailed analysis of usage and operation of appliances. The NILM algorithms distinguish appliance use and operation by mapping the load signature from trained datasets onto monitored data. The NILM algorithm applied in the study was customized for Indian context and could successfully disaggregate data for air conditioners and water heaters, however for other major appliances such as refrigerator and washing machines, it needs to be further trained on India-specific datasets to identify the load usage signatures.

4 Key Results

4.1 Survey Outcomes

4.1.1 Ownership and Usage Pattern

The key outcomes from the survey classified into ownership and usage of the respondents across climatic zones, SEC classification and cities are shown in the table below (Table 1):

Across these categories:

- Cell-phone chargers, Television (TV) sets, set-top boxes, cooktops/stoves (fueled by liquid/gas fuel), refrigerators and ceiling fans have penetrated 80% or more of the surveyed urban homes.
- Washing machines, electric cooking products, lighting products, air coolers and air conditioners are expected to grow further and reach higher ownership levels.
- Ownership in percentage of total surveyed households and average annual usage hours for each appliance at the national level were derived from the survey data (see Fig. 5)

Table 1 Ownership and usage pattern of appliances across demography and climatic zones

Category	Family size	Income	Climate	City	Ownership
Home appliances[a]	No significant impact.	Ownership increases with income for products such as washing machine, vacuum cleaner. However, for products such as television, cell phone charger and set top boxes it does not vary with income	No significant impact	No significant impact	Cell-phone charger, television, set-top box and washing machine have maximum ownership in the category
Heating/ cooling[b]	Ownership increases with family size	Ownership increases with income except for fans, where income does not have any significant impact	Cold climate has highest ownership of space and water heating devices while ownership of appliances like Air-conditioner (AC) and room-heater is sensitive to climatic variation	Non-metros have marginally higher ownership while usage is similar except for ceiling fans	Highest penetration is for ceiling fans followed by desert cooler. Penetration of AC and storage water heater is increasing
Kitchen appliances[c]	Ownership increases with family size	Ownership increases with income for most products except essential ones such as cooktop, refrigerator and mixer grinder. A fifth of high-income respondents own electric cooking appliances, marking a shift to electricity as cooking fuel	No significant impact	No significant impact	Highest penetration is for cooktop, refrigerators followed by mixer grinder
Lighting products[d]	No significant impact	Light emitting diode (LED) penetration increase with income. In lower socio income strata, incandescent lamps are the most used products	Hot-Dry climate has lowest penetration in all lighting products except LEDs while Cold climate has highest ownership of Incandescent bulbs	No significant impact.	Highest penetration is for compact florescent lamps (CFL) followed by tubular fluorescent lamps (TFL) and LED. However, LED penetration is expected to take over CFLs

[a]Includes entertainment systems, IT equipment and some other productivity enhancing appliances
[b]Includes products that provide ventilation, space heating/cooling and hot water
[c]Includes cooking, food processing and storage appliances
[d]Includes products that aid illumination (general purpose lighting) in homes

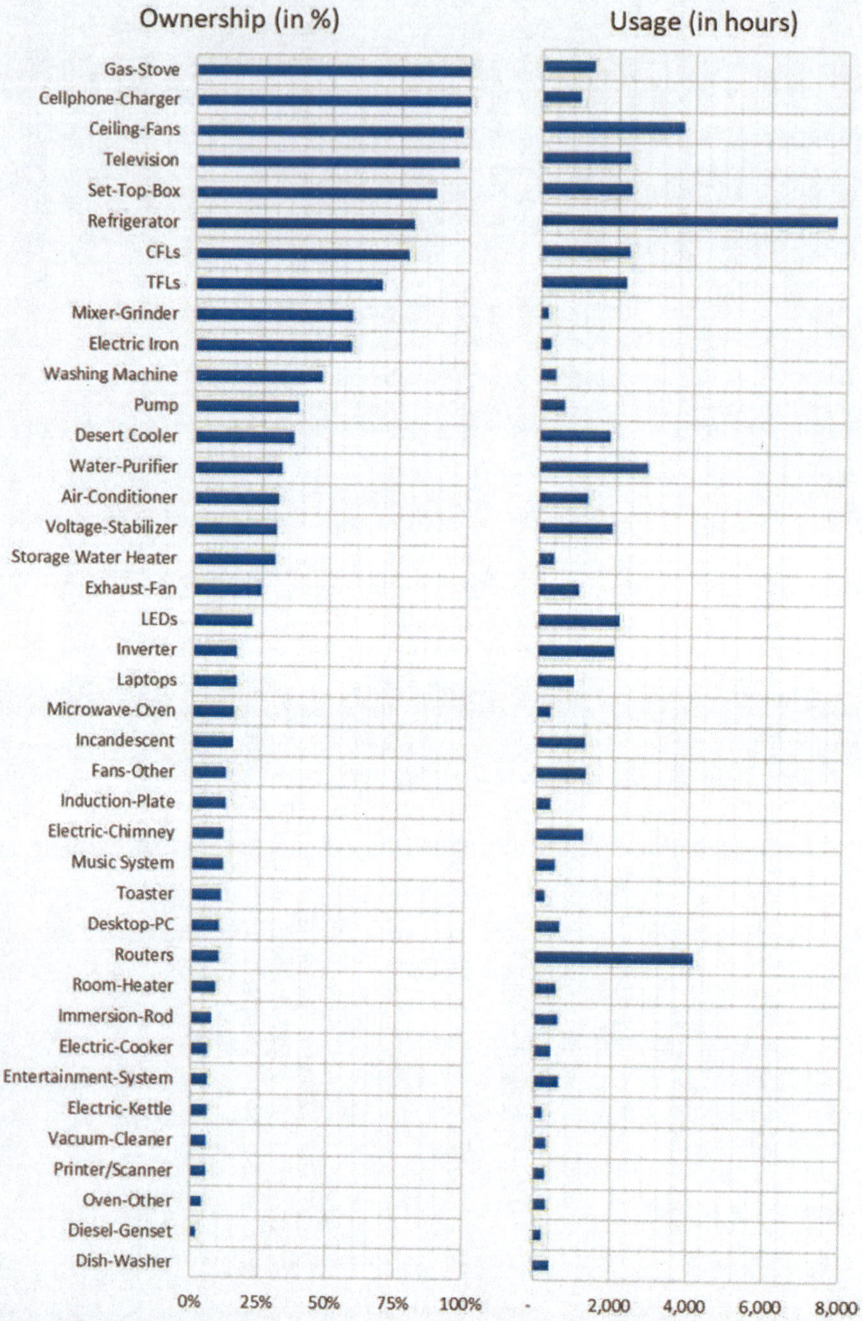

Fig. 5 Ownership (percentage out of total surveyed households) and average annual usage hours of appliances

4.1.2 Energy Use

India is a tropical country, where summers are intense and dominate 4 out of its five climate types. While climate is a distinguishing factor, family composition and the socio-economic status are significant factors as well. The survey captured energy use information from households. This information has been categorized by climate, family type and income for analysis (see Table 2). From the table below, it can be inferred that energy consumption was higher in summers as compared to winters across family types, socio economic strata, and climatic zones except cold climate. In addition, energy consumption increases with increase in family size and socio economic classification.

4.2 *End Use Monitoring and Data Analysis*

The shortlisting criteria for deployment of energy monitoring kits was based on ownership of major energy consuming appliances. The ownership pattern of key appliances in the households selected for monitoring is shown in Fig. 6 below:

Similar to the survey trends, the monitored households also show that across households,

- Energy consumption was highest in summer months (March and June) owing to air conditioners, and winter months (January–February) are characterized by low energy use across monitored households.
- Though energy use is lower in the winter months, high peak demand is observed on account of water and room heaters. Room heaters are used primarily in cold and composite climate, while water heaters are used across climatic zones (Fig. 7).

For detailed analysis, monthly energy use of an Air-Conditioned (AC) and Non-Air-Conditioned (Non-AC) households was disaggregated.

- Energy consumption of AC households was higher than non AC households throughout the year, not just in summers. Most households that own ACs belong to higher SEC groups and possess more appliances, which may explain higher energy consumption in AC households throughout the year.

Table 2 Variation in energy consumption with family size, socio economic strata and climatic zones

Category	Family size	Socio economic strata	Climatic zones
Energy consumption	Energy consumption increases with increase in family size and was highest in summers	Energy consumption increases with increase in socio economic classification i.e., higher the SEC, higher the consumption. Energy consumption was highest in summers	Composite and warm and humid climatic zone consume highest energy in summers, except cold climate, where highest energy is consumed in winners

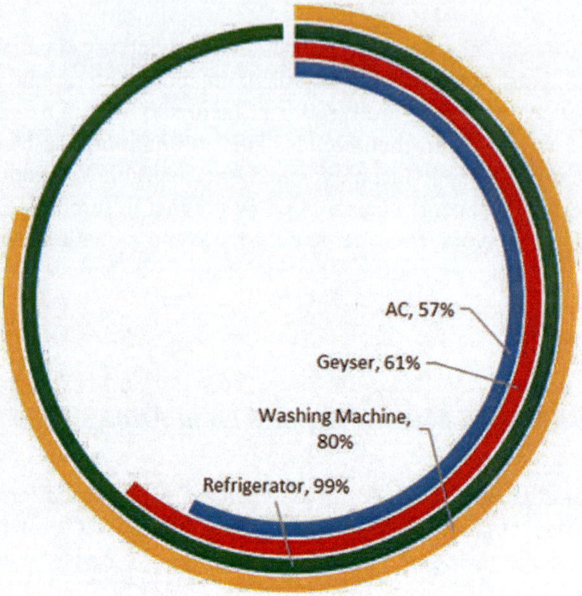

Fig. 6 Ownership of key appliances in metered households

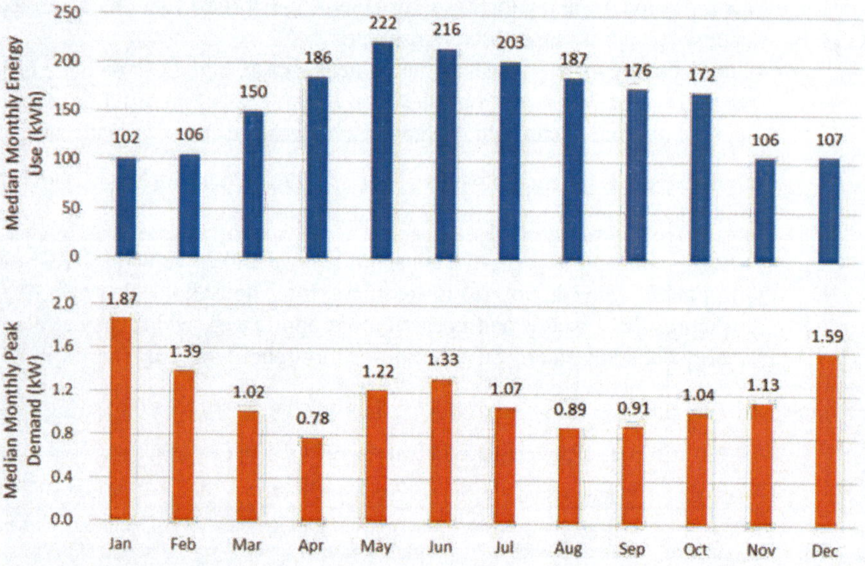

Fig. 7 Monthly energy use and peak demand across monitored households. Median monthly energy demand in metered households

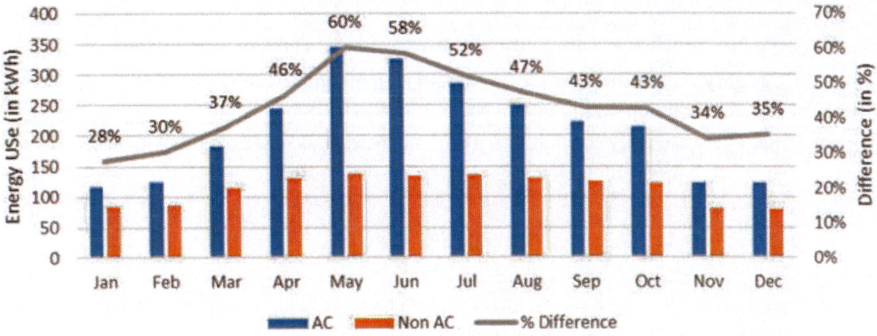

Fig. 8 Monthly energy use comparison for AC and non AC households

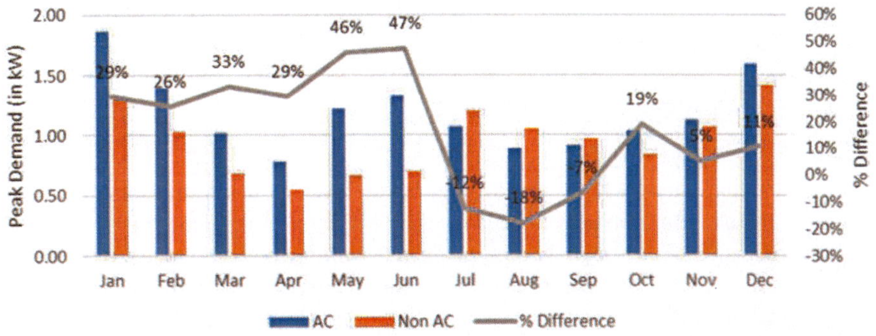

Fig. 9 Peak demand comparison for AC and Non-AC households

- AC households consume at least 50% higher energy as compared to non AC household during summers as shown in the figure below (Fig. 8):
- The magnitude of variation of peak demand (as shown in figure below) between AC vs non AC households is also highest in summers (Fig. 9).

It can therefore be concluded that AC is a significant contributor to overall energy use although the use varies with seasons.

The data was further analyzed to arrive at the energy use variation for AC and non AC households with family size, type of household, climatic zone as shown in the table below (Table 3):

Thus it can be concluded that the energy consumption increases with increase in size of the family and household. Composite climate zone had highest energy consumption.

Table 3 Variation in energy use with seasons for AC and non AC households

Category	Family type	Household size	Climatic zone
AC household	Joint family had highest energy consumption throughout the year, while nuclear family with elders had high energy consumption in monsoon and winters	Energy consumption increases significantly with increase in the size of household, the variation peaking in summers	Composite zone has highest energy consumption, while temperate zone has lowest
Non AC household	Energy consumption increased with increase in family size. Joint family had highest energy consumption throughout the year	Energy consumption increases marginally with increase in household size	Composite zone had highest energy consumption throughout the year particularly peaking in summers and monsoons

5 Conclusions and Way Forward

The nationwide study combining survey and end use monitoring provides deeper understanding on appliance ownership pattern and residential electricity consumption in India's urban homes to developing customized policy interventions. The results reveal appliance ownership and energy use patterns to provide behavioral insights appliances across socio-economic, cultural and climatic factors. The findings of the survey were also validated by the end use monitoring for energy consumption.

Appliances such as cell-phone chargers, TV sets, set-top boxes, cooktops/stoves (fueled by liquid/gas fuel), refrigerators and ceiling fans have penetrated 80% or more urban homes. Washing machines, electric cooking products, lighting products, air coolers and air conditioners are expected to continue growing and reach higher ownership levels.

Energy consumption is higher in summers as compared to winters across socio economic strata, dwelling type and climatic zones except cold climate. Space and water heating appliances are major contributors towards winter peak demand and summer peak is attributed to space cooling devices. Energy consumption increases with increase in family size and socio economic strata. It is notable that, the energy consumption in AC households is higher than Non-AC households not only in summers but throughout the year. This can be attributed to higher socio-economic status of AC households and consequently increased ownership and usage of appliances other than ACs. Peak demand as well as energy use varies by seasons, climatic zone, socio economic strata and dwelling type. These can help in developing targeted demand response strategies.

Since the study is nationally representative, the energy consumption by appliances and its growth patterns can be extrapolated to forecast future energy demand. If conducted periodically, the study would enable better understanding of usage patterns and trends over the years thereby helping analyze the impact of energy efficiency policies more realistically. Overall, the data generated can potentially advise demand response programs, energy efficiency policies for appliances and buildings, and strategies for consumer behavior.

Acknowledgements The authors would like to acknowledge John D. and Caterine T. MacArthur Foundation and Super-efficient Equipment and Appliance Deployment (SEAD) Initiative for funding the study. The authors would also like to acknowledge Market Xcel Data Matrix Pvt. Ltd. and Zenatix Solutions Pvt. Ltd. for conducting the survey and installation of meters, respectively.

References

1. US. Energy Information Administration. *International Energy Outlook 2019*. September 2019
2. International Energy Agency. *India Energy Outlook: World Energy Outlook Special Report*, 2015. OECD/IEA, Paris
3. https://www.eia.gov/todayinenergy/detail.php?id=33252
4. PricewaterhouseCoopers. *Future of consumer durables and electronics in India – the changing landscape***.** 2018
5. Government of Australia, *Residential baseline Study of Australia, 2000-2030*
6. Residential Energy Consumer Survey of USA
7. Schipper and Meyers, *Energy Efficiency and Human Activity*. Cambridge: Cambridge University Press, 1993
8. Indian Brand Equity Foundation. Consumer Durables, April 2018

Impact of Environmental Factors on Energy Efficiency of Room Air Conditioners in India

Neha Dhingra, Archana Walia, and P. K. Mukherjee

Abstract Room Air Conditioners (RAC) dominate the Air Conditioning (AC) space cooling market comprising almost 40% of total cooling energy consumption in 2017–2018. Though RAC penetration in households in India is only 8% currently, rising incomes, urbanization and increasing cooling degree-days are expected to raise RAC ownership to 40% in 2037–38. This would cause a significant increase in power and peak load demand, environmental impacts, and greenhouse gas (GHG) emissions. Energy efficiency policies for appliances are one of the most cost effective methods to reduce electricity consumption. India's Bureau of Energy Efficiency (BEE) initiated a labeling program for RACs in 2006, and subsequently revised the program periodically to increase efficiency and establish performance in terms of the Indian Seasonal Energy Efficiency Ratio (ISEER) based on climate conditions. The energy efficiency of an RAC is based on performance tests at standard rated conditions. However, under actual operating conditions, the RAC is exposed to several adverse climatic conditions such as polluted ambient and saline conditions around coastal areas which can impact energy performance. There is little information or data available in the public domain regarding the potential impacts due to such environmental conditions. A study of the impact of short-term environmental impacts such as salinity, dust, and humidity on energy efficiency of an RAC were measured under simulated environmental conditions. This paper presents the methodology along with key findings and forward looking research in assessing the impact of external environmental conditions.

Keywords Environmental factors · Indian Seasonal Energy Efficiency Ratio · Energy performance

---

N. Dhingra (✉) · A. Walia · P. K. Mukherjee
CLASP, New Delhi, India
e-mail: ndhingra@clasp.ngo; pkmukherjee@clasp.ngo

P. Bertoldi (ed.), *Energy Efficiency in Domestic Appliances and Lighting*, Springer Proceedings in Energy, https://doi.org/10.1007/978-3-030-79124-7_7

1 Introduction

The AC market in India is largely dominated by RACs, which comprise approximately 40% of total cooling energy consumption in 2017–2018, with a forecasted increase to 50% by 2037–38 [1].The annual sales of RACs in India have increased rapidly in the last 10 years, from 0.3 million in 2007–7.6 million in 2017 [2]. These trends have resulted in a significant increase in electricity demand and energy consumption in both the commercial and residential sectors. While Indian household RAC penetration is only 8% currently, rising incomes, urbanization and increasing cooling degree-days are expected to increase RAC ownership to 40% by 2037–38 [1]. This would require a significant increase in peak and base power generation capacity and increase GHG emissions.

Air conditioning can substantially increase the electricity consumption of a household – typically, a 1.5-ton capacity AC consumes about 1900 W of power equivalent to operating around 30 ceiling fans, which have been the traditional cooling appliance in India. However, the electricity demand from RACs can be reduced by increasing equipment efficiency. Energy efficiency policies for RACs are one of the most cost effective methods to reduce electricity consumption. In order to promote energy conservation and efficient use of energy, Government of India launched Energy Conservation Act in 2001 and established the BEE, a statutory body under Ministry of Power, for administration and implementation of the Act in 2002 [3]. BEE launched a Standards and Labeling (S&L) program to promote appliance and equipment efficiency in 2006, and brought RAC under the voluntary labeling program [4]. Currently, the labeling program for RACs is mandatory and includes both fixed and variable capacity (inverter) units.

Since 2018, the efficiency of a RAC in India is defined in terms of the Indian Seasonal Energy Efficiency Ratio (ISEER), which is the ratio of the cooling seasonal total load (CSTL) (in Watt hour) to cooling seasonal energy consumption (CSEC) (in Watt hour).This improved star rating methodology factors in the variance in temperature across the various climatic zones in India and annual operating hours.

For the labeling program, the energy efficiency of a RAC is tested under standard rated conditions in the test laboratory as per the Indian National Standard. However, ACs work under real life conditions, which are different from the standard rated conditions simulated in a test laboratory. In real operating environment, an AC is exposed to adverse climatic conditions such as polluted ambient (dust, vehicular and industrial pollution) and saline conditions around coastal areas. This may result in changes in efficiency due to one or more of the following:

(a) Reduced air circulation through heat exchanger,
(b) Increased power consumption,
(c) Longer operating times required to achieve desired cooling set points,
(d) Higher compressor failure rates due to condenser blockage causing higher condensing temperatures and refrigerant pressures,

(e) Increased refrigerant leakage to atmosphere leading to higher global warming potential, and
(f) Reduced effective useful life of equipment.

Considering the growing demand for RACs and their contribution to energy consumption, the in situ energy performance of the product must be tested to ensure performance is not impacted by adverse environmental conditions. There is very little information or data available in the public domain of any significant impact of such environmental factors. Therefore, this preliminary study was conducted to understand and measure the impact of short-term environmental factors on RAC energy efficiency performance.

2 Objective

The key objective of the study was to assess the impact of long-term environmental factors such as salinity, dust, temperature and humidity on energy efficiency of a RAC by exposing them to short-term environmental conditions simulated in the test laboratory. This report does not address differences between short-term tests and performance tests of units with long-term environmental exposures.

The findings of the study may inform consumers, policy makers and manufacturers of any likely impact of the short-term environmental exposure on energy efficiency of RACs. However, the study will need to extended for a longer duration to provide information regarding long term exposure to adverse environmental conditions. Accordingly, it would help to perform additional tests of units with long-term environmental exposures such as 4 or 5 years of adverse environmental conditions and identify potential solutions to improve quality, reduce operating and maintenance costs, and save energy.

3 Approach and Methodology

The study focused on RACs, accounting for about 60–70% of space cooling in India [1]. The scope of the study was limited to un-ducted single split and unitary/window units including fixed and variable speed compressors with up to 11 kW cooling capacity. The study included the following steps:

- Preparation of a representative sampling plan comprising of the following:

    (a) Major brands available in the Indian market. The manufacturers' names have been kept confidential, but ensuring representation of all major brands.

    (b) Categories (unitary/window and split) and type (variable and fixed type)

(c) Representative capacity of ACs sold in India (i.e., 5.27 kW or 1.5 TR (tons of refrigeration)
(d) Different type of heat exchangers

- Identification of relevant national/international test standard for the environmental and energy performance testing of RACs
- Identification of a nationally accredited test laboratory to conduct energy performance and environmental testing
- Following a stepwise approach for testing:

(a) Conduct testing for cooling capacity and power consumption to establish the baseline efficiency
(b) Subject samples to the different short-term environmental conditions (such as corrosive, dusty and saline) as per the relevant national/international standards
(c) Conduct performance tests again and assess change from baseline due to short-term tests
(d) Analyze the test results to compare cooling capacity, power consumption and ISEER based on short-term environmental test conditions
(e) Determine if short-term tests are consistent with performance tests of units with long-term environmental exposures

4 Test Results and Analysis

4.1 Sample Selection

A random sample consisting of major brands, type and technology of RACs, different star rating bands (represented by 3-,4- and 5-star), types of heat exchanger and most sold star rating was prepared as shown in Table 1 below:

4.2 National/International Test Standards

4.2.1 Identification of Environmental Test Standards

An analysis of national environmental test standards was conducted and the following Indian test standards were identified:

- IS 9000 part XII for dust test—This standard provides test procedure for the dust test on electronic and electrical items. The objective of this test is to determine

Table 1 RAC sampling plan

S. no.	Type of test	Type of AC	Heat exchanger	No of units
1	Dust test	Window	Copper	2
		Split (fixed speed)	Copper	2
		Split (variable speed)	Aluminum	2
2	Salt mist test	Window	Copper	2
		Split (fixed speed)	Copper	2
		Split (variable speed)	Aluminum	2
3	Composite temperature/humidity cycle test	Window	Copper	2
		Split (fixed speed)	Copper	2
			Aluminum	2
Total number of samples				**18**

the suitability of electronic and electrical items for use and/or storage under dust laden atmosphere.

- IS 9000 part XI as per procedure 3 for salt mist test—This standard deals with the determination of the corrosive effects of salt atmospheres on electronic and electrical items. The objective of this test is to determine the suitability of electronic and electrical items when used or stored under salt laden atmospheres. This test is intended mainly for the evaluation, the quality and uniformity of protective coatings.
- IS 9000 part VI (10 cycles) Composite temperature/humidity cyclic test—This standard deals with a composite temperature/humidity cyclic (moisture resistance) test procedure. The objective of this test is to determine in an accelerated manner the resistance of items to the deteriorative effects of high temperature/humidity and cold conditions.

4.2.2 Identification of Energy Performance Test Standards

The Indian test standards defined in BEE's labeling program for RACs were identified for efficiency testing:

- IS 1391 part 1 for unitary ACs
- IS 1391 part 2 for split ACs

Variable speed ACs were also tested using IS 1391 at full load only. The part load performance testing for variable speed RACs was not implemented.

The performance testing was performed using balanced ambient room-type calorimeter to measure cooling capacity and power consumption, in order to calculate the Indian seasonal energy efficiency ratio.

4.3 Selection of an Accredited Test Laboratory to Conduct Energy Efficiency and Environmental Testing

To select a nationally accredited independent test laboratory for efficiency and environmental testing, quotations were requested from all major test labs in India that have AC efficiency testing infrastructure in place. Sierra Aircon was selected to undertake the study based on a combined evaluation of the technical and financial proposals. Sierra Aircon partnered with SGS India Ltd. lab for environmental testing.

4.4 Test Results

The samples were procured as per the plan in Table 1 and shipped to Sierra Aircon Pvt. Ltd. The samples were provided a unique Unit ID to mask the brand name and for traceability. All samples were first tested to measure energy performance to establish a baseline. The samples were then subjected to environmental tests discussed above. After environmental testing, the samples were tested again for energy performance to assess any changes as a result of short-term exposure to environmental conditions.

Table 2 provides the performance and energy efficiency test results for both pre and post short-term environmental testing and the resultant variation in ISEER (as percentage) for all the samples.

The percentage variation in the cooling capacity, power consumption and ISEER resulting from post and pre-environmental testing for all types of ACs are shown graphically in Figs. 1, 2 and 3, respectively.

The key observations from the analysis of test results are:

- Reduced cooling capacity ranged from −3.79% to 0.57%, with an average of −1.28%.
- Reduced power ranged from −5.03% to 0.76%, with an average of 0.98%.
- Reduced ISEER ranged from −2.78% to 1.95%, with an average of −0.30%.

All these variation in the values are insignificant and within the tolerance limits prescribed in BEE's labeling schedule for all ACs (window, fixed and variable split).

4.4.1 Test Results as per the Type of Environmental Test

The results were further analysed to understand the impact of a specific environmental test t i.e., composite humidity, dust, salt mist test and combined environmental tests (sequential exposure) on energy performance. These trends are reflected graphically in Figs. 4–7 below:

Table 2 Summary of RAC pre and post performance test results

S. No.	Environmental test	Type of AC	Heat Exchanger type	Pre environmental test results			Post environmental test results			Percentage drop between post and pre environmental test results		
				Total cooling capacity (W)	Total Power Input (W)	ISEER	Total cooling capacity (W)	Total Power Input (W)	ISEER	Total Cooling capacity	Power input	ISEER
1	Composite humidity test	Window	Copper	4937	1602.9	3.08	4817	1586.2	3.04	−2.43	−1.04	−1.30
2				4944	1601.2	3.09	4888	1585.9	3.08	−1.13	−0.96	−0.32
3				5159	1465.7	3.52	5153	1453.5	3.55	−0.12	−0.83	0.85
4			Copper	4925	1430.4	3.44	4834	1412.8	3.42	−1.85	−1.23	−0.58
5		Fixed split		4840	1456.7	3.32	4804	1452.6	3.31	−0.74	−0.28	−0.30
6			Aluminum	5016	1510.5	3.32	5005	1505.9	3.32	−0.22	−0.30	0
Average percentage drop from composite humidity test										**−1.08**	**−0.77**	**−0.28**
7	Dust test		Copper	5143	1515.2	3.39	5014	1506	3.33	−2.51	−0.61	−1.77
8		Window		5127	1512.5	3.39	5073	1493.8	3.4	−1.05	−1.24	0.29
9			Copper	4935	1419.2	3.48	4963	1430	3.47	0.57	0.76	−0.29
10		Fixed split		5098	1417.1	3.6	4937	1409	3.5	−3.16	−0.57	−2.78
11			Aluminum	5786	2250.1	2.57	5594	2136.9	2.62	−3.32	−5.03	1.95
12		Variable split		5183	1797.3	2.88	5143	1772.9	2.9	−0.77	−1.36	0.69
Average percentage drop from dust test										**−1.70**	**−1.34**	**−0.32**

(continued)

Table 2 (continued)

S. No.	Environmental test	Type of AC	Heat Exchanger type	Pre environmental test results			Post environmental test results			Percentage drop between post and pre environmental test results		
				Total cooling capacity (W)	Total Power Input (W)	ISEER	Total cooling capacity (W)	Total Power Input (W)	ISEER	Total Cooling capacity	Power input	ISEER
13	Salt mist		Copper	5191	1625.3	3.19	5165	1614.6	3.19	−0.50	−0.66	0
14		Window		5096	1597.7	3.19	5047	1606.4	3.14	−0.96	0.54	−1.57
15			Copper	5248	1412.3	3.72	5220	1400.7	3.72	−0.53	−0.82	0
16		Fixed split		5295	1403.4	3.77	5232	1392.2	3.76	−1.19	−0.8	−0.27
17		Variable split	Aluminum	5579	1721.7	3.24	5562	1714.2	3.24	−0.30	−0.44	0
18				5473	1708.6	3.2	5468	1704.8	3.21	−0.09	−0.22	0.31
Average percentage drop from salt mist test										**−0.6**	**−0.4**	**−0.26**
19	Combined environmental tests	Fixed split	Copper	4925	1430.4	3.44	4844	1416.7	3.42	−1.64	−0.96	−0.58
20			Aluminum	5016	1510.5	3.32	4826	1457	3.31	−3.79	−3.54	−0.30
Average percentage drop from combined environmental tests										**−2.72**	**−2.25**	**−0.44**
Average percentage drop for all the samples										**0.98**	**−1.28**	**−0.30**

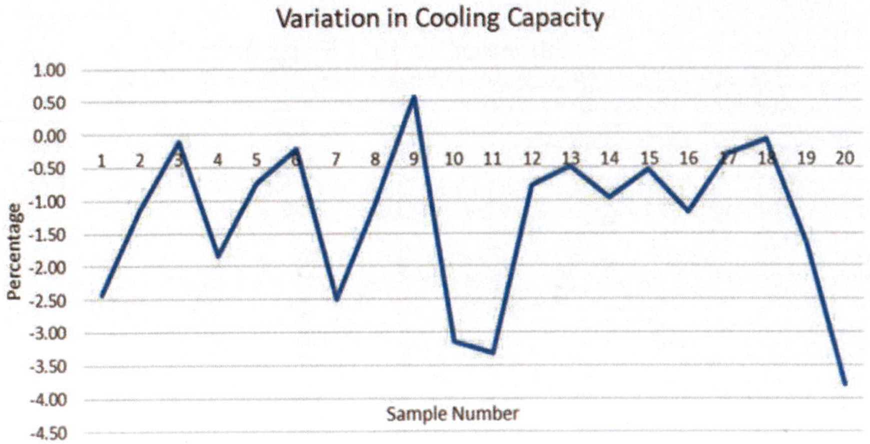

Fig. 1 Cooling capacity variation from pre and post environmental tests

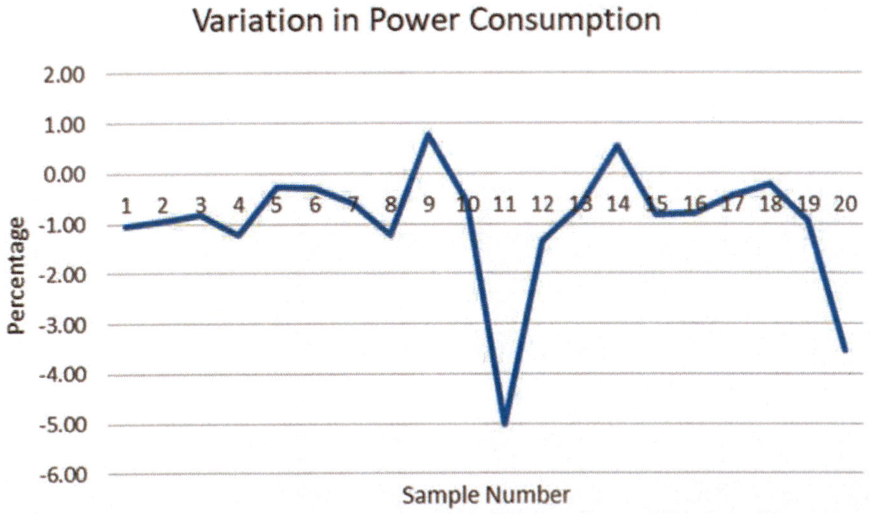

Fig. 2 Power consumption variation from pre and post environmental tests

The analysis of test results indicate that none of the environmental conditions individually or the combined effect when exposed sequentially had any significant effect on the energy performance of ACs. Minor variations observed with cooling capacity and power consumption were within prescribed tolerance limits due to short-term environmental exposure.

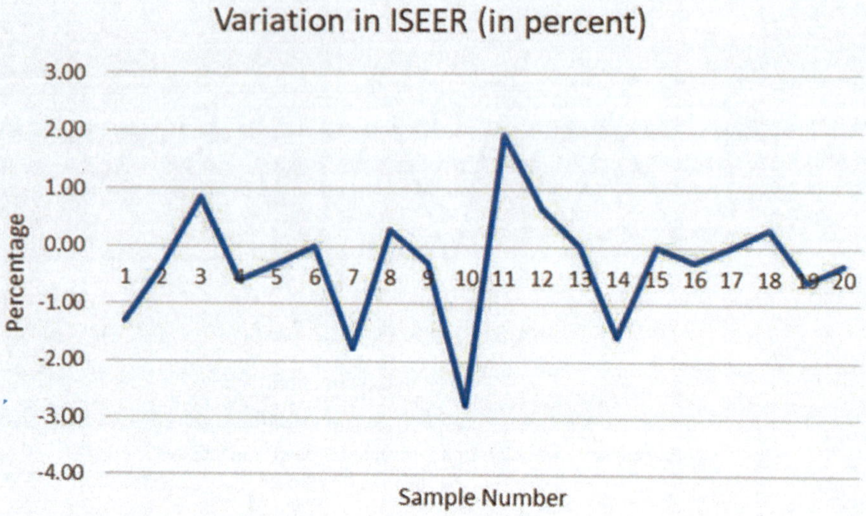

Fig. 3 ISEER variation from pre and post environmental tests

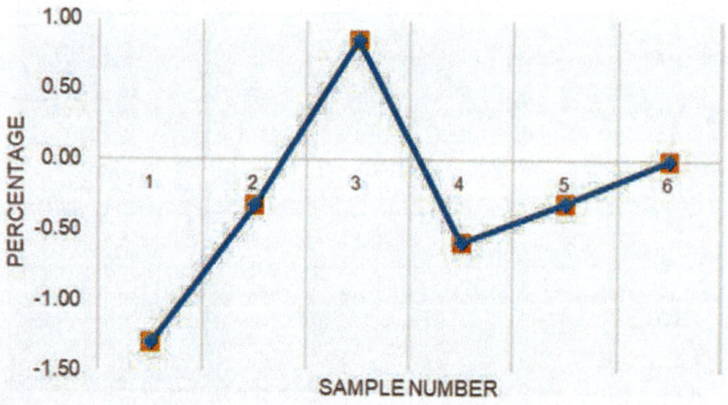

Fig. 4 ISEER variation from composite humidity environmental tests

4.4.2 Test Results as per the Type of Heat Exchanger Material

The results were further analysed to understand the impact of the short-term environmental tests on the basis of the type of heat exchanger material (Aluminum and Copper). The variation in ISEER post and pre environmental testing is shown in Figs. 8 and 9 below:

The analysis of test results indicate that none of the short-term environmental conditions individually or the combined effect when exposed sequentially had any

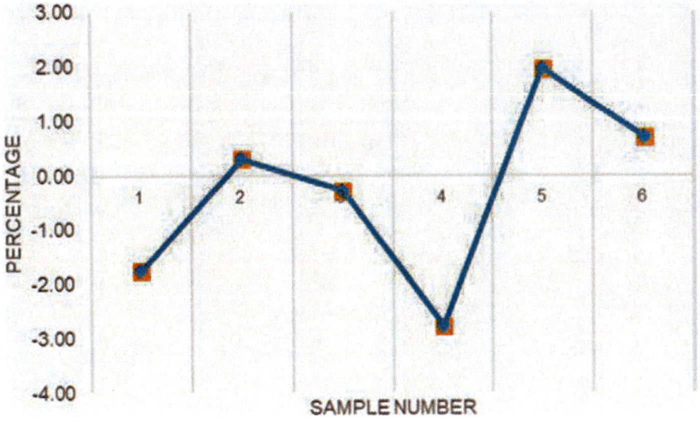

Fig. 5 ISEER Variation from dust environmental tests

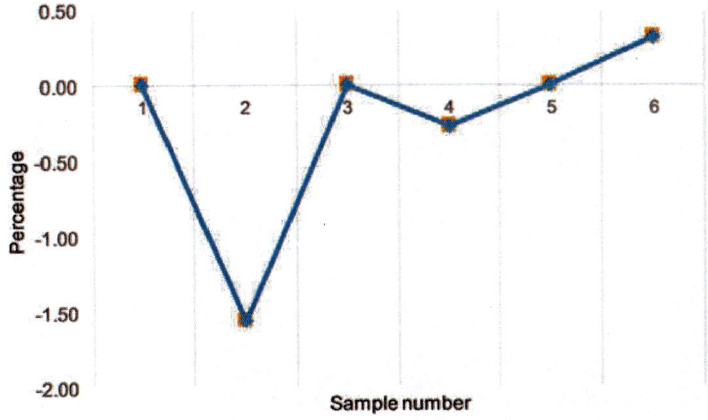

Fig. 6 ISEER variation from salt mist environmental tests

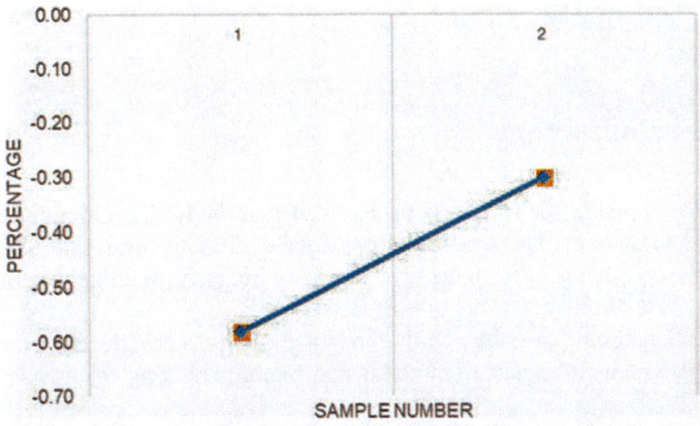

Fig. 7 ISEER variation from all environmental tests with sequential short-term exposures

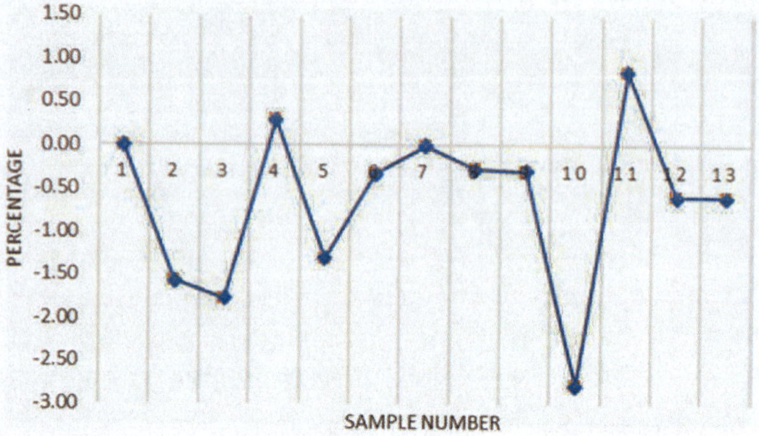

Fig. 8 ISEER variation for ACs with copper heat exchanger

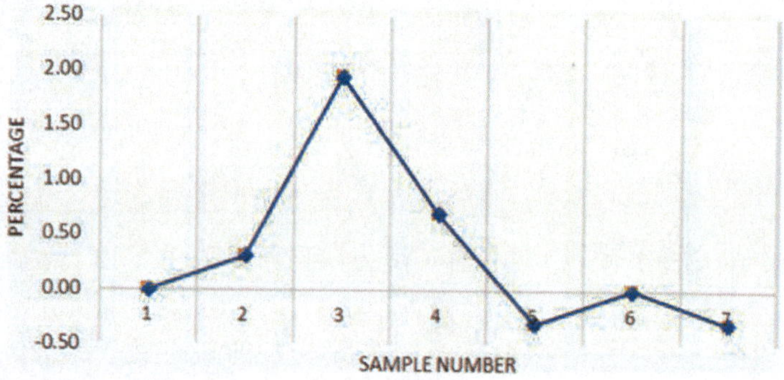

Fig. 9 ISEER variation for ACs with aluminum heat exchanger

significant effect on the energy performance of ACs with either Aluminum or Copper type heat exchanger.

5 Conclusion

Based on the test results, the performance of the RACs sold in India is not affected by short-term exposure to environmental conditions (dust, salt mist, and composite humidity) simulated in the test laboratory irrespective of the type of AC or the type of heat exchanger or the type of tests.

This could possibly, be due to some quality measures taken by manufacturers that safeguard against environmental factors resulting in little or no measurable impact on energy efficiency under standard testing conditions. These may include, but not limited to:

- Sound manufacturing technique that prevented galvanic corrosion of heat exchanger coils and thereby not affecting the life of aluminum fin and copper tube bonding and able to withstand the salt mist exposure.
- Thickness of the copper tube and aluminum fin is sufficient to take care of environmental exposure.
- The e-coating provided on heat exchanger surface gives adequate protection to withstand environmental exposure.

It should be noted that the conclusions are based on short term exposure by simulating environmental conditions in the test lab. However, in future, the study can be further expanded to expose the samples to salt mist test for a prolonged duration i.e. as defined in relevant Indian standard for heat exchanger to assess the impact of prolonged exposure to saline conditions on energy performance of RAC. In addition, samples from the field across various climatic conditions can be tested to assess the impact on energy performance in real life situation over a select period.

Acknowledgements The authors would like to acknowledge Institute for Governance and Sustainable Development (IGSD) and International Copper Alliance (ICA), India for funding the study and 50% cost of the samples respectively.

References

1. Ozone Cell, Ministry of Environment, Forests and Climate Change. *India Cooling Action Plan*, March 2019
2. Sachar, Sneha, Iain Campbell, and Ankit Kalanki, *Solving the Global Cooling Challenge: How to Counter the Climate Threat from Room Air Conditioners*. Rocky Mountain Institute, 2018
3. http://www.beestarlabel.com/
4. https://powermin.nic.in/en/content/energy-efficiency

Policy Measures and Impact on the Market for the Room Air Conditioners in India

P. V. N. Kishore Kumar, Sameer Pandita, Archana Walia, and T. P. Ashwin

Abstract Cooling is essential to human health and productivity and is becoming more important as India urbanizes, grows economically and increasing need for cooling due to extreme weather conditions. Room air conditioners (RAC) now account for up to 50% of peak load on the grid in major metropolitan areas of India. The RAC market in India has being growing enormously and these trends have resulted in a rapid increase in electricity demand in the commercial and residential sectors. Energy efficiency mitigates the individual risk of being saddled with inefficient, environmentally harmful and poor quality RACs.

India's Bureau of Energy Efficiency (BEE) launched the labelling program for fixed-speed Room Air Conditioners in 2006 with periodic revisions for increased stringency resulting in substantial efficiency improvement of 35% at the minimum energy performance standards. The most recent development covers both fixed and inverter units under a common rating plan based on an Indian seasonal energy efficiency metric. Over the years, the RAC policy and subsequent revisions has resulted in significant market transformations towards higher efficiencies comparable to the best in class globally.

Over the last decade, the annual sales of RACs in India have grown exponentially and with the improvements in efficiency have resulted in 38 million tons of carbon emission reduction as per the 2016–17 estimates. This paper analyses and discusses the trends in market growth, technology evolution and market transformation as a result of BEE's policy program and further explores possible efficiency improvements by 2030 and its climate impacts.

P. V. N. K. Kumar (✉) · A. Walia
CLASP, New Delhi, India
e-mail: kkumar@clasp.ngo

S. Pandita · T. P. Ashwin
Bureau of Energy Efficiency, New Delhi, India
e-mail: spandita@beenet.in

P. Bertoldi (ed.), *Energy Efficiency in Domestic Appliances and Lighting*, Springer Proceedings in Energy, https://doi.org/10.1007/978-3-030-79124-7_8

1 The Room Air Conditioner Market in India

1.1 Introduction

Room air conditioners (RACs) are a type of appliance used to dehumidify and lower indoor air temperatures, with the purpose of providing cooling comfort during hot weather. Air conditioners remove heat from enclosed spaces and discharge it outside. In India, air conditioning was once considered a luxury for affluent households, but the number of RACs is increasing rapidly due to extremely high ambient temperatures, urbanization and rising incomes and living standards.

The annual sales of RACs in India have grown exponentially in the last 10 years, from 0.3 million in 2007 to 7.6 million in 2017. These trends have resulted in a rapid increase in electricity demand and energy consumption in the commercial and residential sectors.

Air conditioning can substantially increase the electricity consumption of a household – typically, one AC of 1.5 ton capacity can consume an amount of electricity that is equivalent to operating 25 ceiling fans, which have been the traditional cooling appliance in India. In practice, electricity costs for running an AC depend upon the unit's energy efficiency, as designed by its manufacturer, the number of operating hours, and how efficiently it is operated and maintained.

Despite the fact that RACs consume significant amounts of electricity, demand continues to grow rapidly. Since RACs are used primarily during peak hours of the day, they now account for up to 50% of peak load on the grid in major metropolitan areas of India. There is thus an urgent need for energy efficiency policies for RACs to reduce electricity consumption and peak demand.

The Bureau of Energy Efficiency (BEE) launched the labeling program for fixed-speed RACs in 2006 as a voluntary initiative, and the program became mandatory in 2009 [1]. BEE revised the energy performance thresholds for RACs covered under the program on a biennial basis from 2009 to 2018. In 2015 BEE launched a voluntary labelling program for inverter RACs, and made the program mandatory in January, 2018. The labelling program for RACs now covers both fixed and inverter units under the same labelling scheme. These improvements in stringency have resulted in substantial efficiency improvement of 35% to the minimum energy performance standards for split units, the most popular RACs. Since the inception of the RAC labelling program, 46 TWh of electricity have been saved and avoided 38 Million tons of carbon emissions till 2016–17.

2 Indian Market Overview

2.1 Types of Room Air Conditioners in India

There are four types of RACs available in the Indian market:

Split-type air conditioners: This type of AC is comprised of two parts, an outdoor unit and an indoor unit. The outdoor unit, fitted outside the room, houses

Fig. 1 Types of room air conditioners
Source: Bureau of Energy Efficiency

components like the compressor, condenser and expansion valve. The indoor unit comprises the evaporator or cooling coil and the cooling fan.

Window air conditioners: In this type of AC, all the components, including the compressor, condenser, expansion valve or coil, evaporator and cooling coil are enclosed in a single box that is installed in a window in a household dwelling or commercial building.

Cassette air conditioners: Cassette units work the same way as wall-hung split system units, with the difference being that cassettes are installed into the ceiling instead of on the wall. The indoor unit sits flush to the ceiling and the outdoor unit is mounted outside as for a conventional wall mounted split system unit.

Floor mounted air conditioners: This type of AC is useful in indoor spaces that lack sufficient wall space to attach appliances, or within buildings constructed of fragile materials such as glass. Floor mounted units can look more discreet than their wall mounted alternatives, minimising the impact on a room's aesthetics (Fig. 1).

2.2 *Market Demand for RACs*

The air conditioning market in India is dominated by room ACs. The RAC market size in 2007–08 was 0.3 million, and has grown by 25 times in the last decade, to 7.6 million in 2017–18. The RAC market is expected to grow at a CAGR of 11% in the next 10 years [2] (Fig. 2).

The split segment comprising the majority share throughout the period and is at 87% in 2017–18. The window AC share has decreased gradually from 23% to 12% during the same period primarily because of consumer preferences for split ACs due to aesthetics, higher efficiencies, and their availability at comparable prices. The production of RAC in 2017 is represented graph in Fig. 3 below while the percent share of window and split RAC is represented by the column graph.

The RAC market is further categorized on the basis of cooling capacities (in tons, which is equal to approximately 3.5 kW), generally ranging from 0.5 to 2.8 tons.

As seen in Fig. 4 the market is well represented by RACs of all capacities. However, 1.5 ton capacity RACs have been consistently dominant for almost 6 years with share of 38% in 2017, followed by 23% of 1 ton capacity RACs.

3 Evolution of RAC Energy Efficiency Policy in India

In May 2006, Bureau of Energy Efficiency (BEE), a statutory body under the Government of India, Ministry of Power, launched the Standards and Labeling (S&L) Program for electrical home appliances with the objectives of regulating energy performance and assisting end users in making informed purchase decisions. BEE designed the program to rate appliances on a scale of 1–5, with the 5-Star rating going to the most efficient appliances on the market.

BEE launched the labeling program for RACs in 2006 as a voluntary initiative, and the program became mandatory on January 12, 2009. The program covers RACs of capacities up to 3TR with fixed speed technologies covered under the scope of Bureau of Indian Standards (BIS) – IS 1391(part I & II). The typical air conditioner energy label is shown in Fig. 5.

3.1 *RAC Labeling Program*

BEE developed distinct star rating plans for split and window/unitary type RACs. The split AC rating plan also includes the cassette and floor standing/ceiling mounted type RACs.

The efficiency of a RAC is defined in terms of the Energy Efficiency Ratio (EER), which is the ratio of the cooling output (in Watts) to the total power input (in

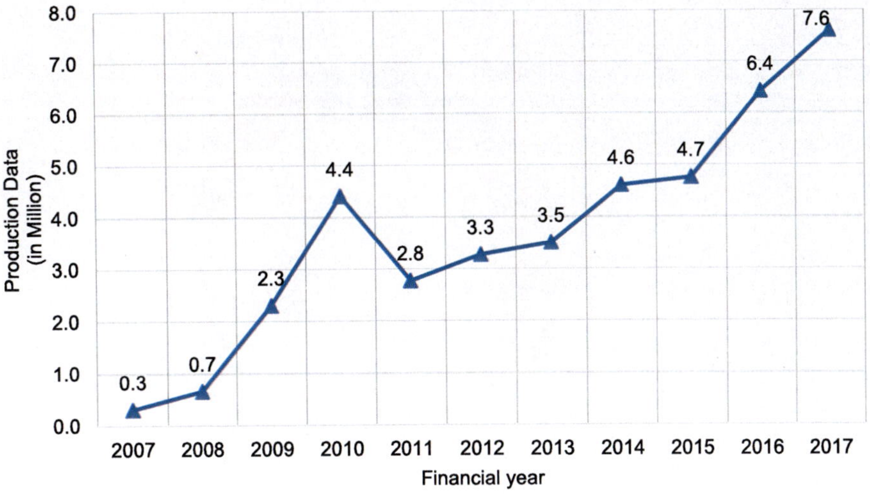

Fig. 2 RAC market growth from 2007 to 17
Source: Bureau of Energy Efficiency

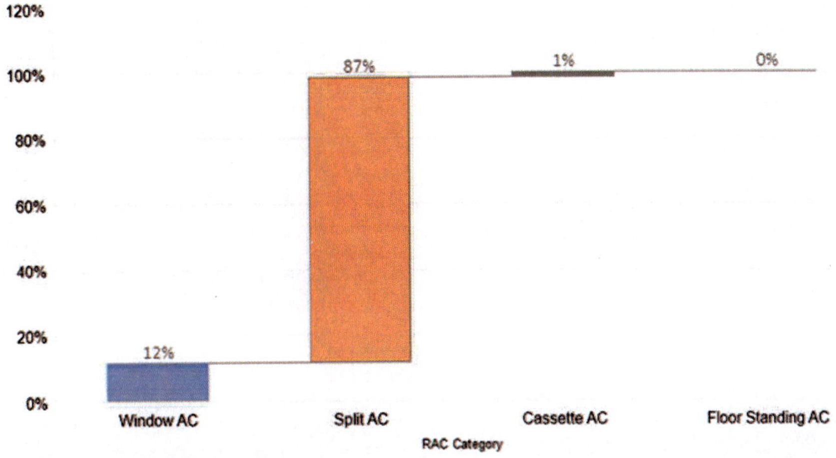

Fig. 3 RAC market segmentation in 2017
Source: Bureau of Energy Efficiency

Watts) at standard rating conditions. This means the higher the EER, the more efficient the air conditioner. In 2018, BEE adopted an improved rating methodology that factors in variance in temperature across the various climatic zones in India and operating hours. The new metric is called the Indian Seasonal Energy Efficiency Ratio (ISEER), which is the ratio of the cooling seasonal total load (in kWh) to cooling seasonal energy consumption (in kWh).

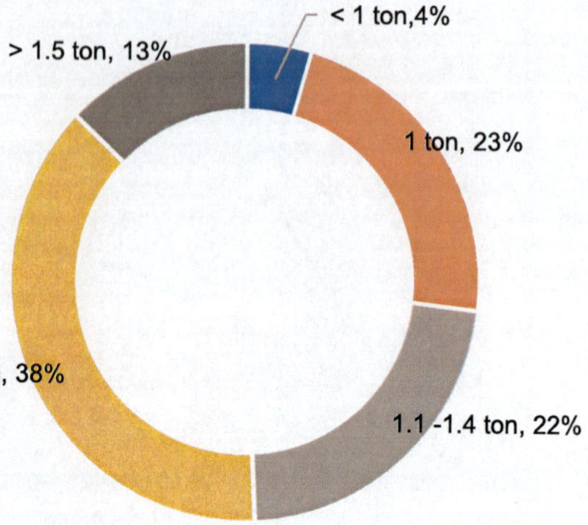

Fig. 4 RAC Segmentation by cooling capacity (in tons) in 2017
Source: Bureau of Energy Efficiency

Fig. 5 Star label of room air conditioner

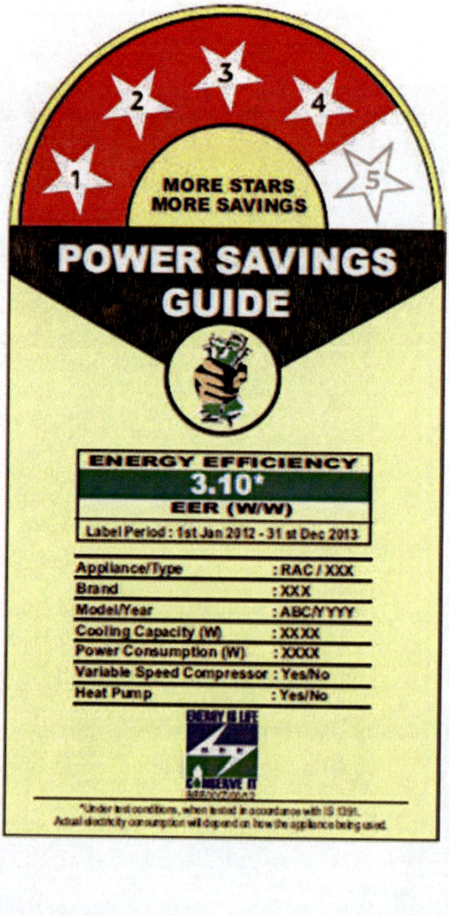

BEE has revised the star rating plans for window and split RACs to increase the stringency of the energy performance thresholds, as shown in Tables 1 and 2, respectively.

As per Table 1, the extent of efficiency improvements for window RACs has been limited due to technological and size constraints. As per the last revision in 2018, BEE does not allow registration of those models which would have been rated 1-Star.

In comparison, the split type RACs have seen more frequent and substantial revisions (see Table 2). For example, the existing 5-Star level in 2009 became 3-Star in 2015 and 1-Star in 2018 as per new star levels and ISEER methodology.

As shown in Fig. 6, the improvements in energy efficiency for window RACs resulted in marginal efficiency improvements of 9% to the minimum energy performance standard (1-Star) and 6% for 5-Star level. For split RACs, energy efficiency improvements resulted in overall improvement of 35% for minimum energy performance standards (1-Star) and 45% for the 5-Star level.

3.2 *Market Share by Star Levels and Overall Efficiency Improvement*

The majority of RAC sales over the last four years were of 3-Star and 5-Star models, with average market shares of 61% and 23% respectively. This trend points to a consumer preference for 3-Star RACs, possibly due to lower, more affordable upfront purchase costs.

Table 1 Revisions in star rating levels for window/unitary type RACs

Star level	1st January 2009–31st December 2011	1st January 2012–31st December 2013	1st January 2014–31st December 2015	1st January 2016–31st December 2017	1st January 2018–31st December 2019
	EER	EER	EER	EER	ISEER
1 Star	2.3	2.3	2.5	2.5	2.5
2 Star	2.5	2.5	2.7	2.7	2.7
3 Star	2.7	2.7	2.9	2.9	2.9
4 Star	2.9	2.9	3.1	3.1	3.1
5 Star	3.1	3.1	3.3	3.3	3.3

Source: Bureau of Energy Efficiency

Table 2 Revisions in star ratings plans for split type RACs

Star level	1st January 2009–31st December 2011	1st January 2012–31st December 2013	1st January 2014–31st December 2015	1st January 2016–31st December 2017	1st January 2018–31st December 2019
	EER	EER	EER	EER	ISEER
1 Star	2.3	2.5	2.7	2.7	3.1
2 Star	2.5	2.7	2.9	2.9	3.3
3 Star	2.7	2.9	3.1	3.1	3.5
4 Star	2.9	3.1	3.3	3.3	4.0
5 Star	3.1	3.3	3.5	3.5	4.5

Source: Bureau of Energy Efficiency

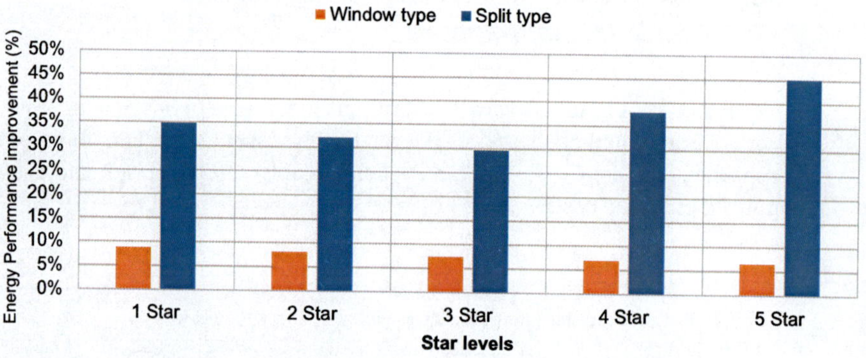

Fig. 6 Star level improvement of air conditioners from 2009 to 2018
Source: Bureau of Energy Efficiency

Considering production data available for RACs in 2017–18, and as shown in Fig. 7, 3-Star RACs dominated the market with 66% followed by 5-Star RACs with 18% market share.

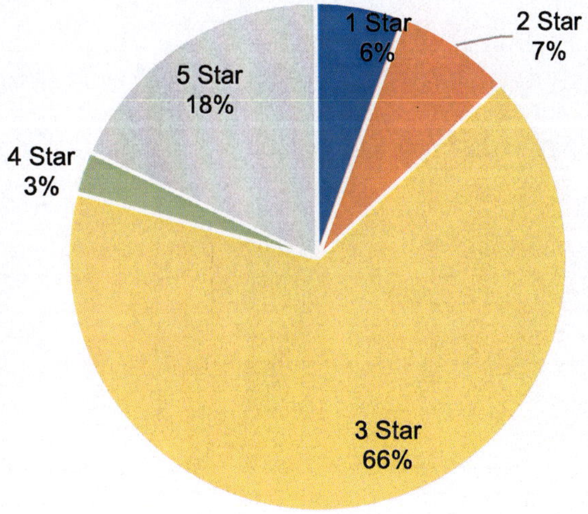

Fig. 7 RAC market by Star Level in 2017 **Source: Bureau of Energy Efficiency**

4 Market Transformation of Fixed and Variable Speed Labeled RAC

4.1 Overview of Labeling Program for Variable Speed RACs

In 2015, the market share of variable speed (commonly known as inverter) RACs was less than 1%. In June of that year, BEE introduced a voluntary labeling program for variable speed RACs with a new star rating methodology called Indian Seasonal Energy Efficiency Ratio (ISEER)[1] [3].

In January 2018, BEE mandated the labeling program for variable speed RAC and introduced a single star rating plan for variable and fixed speed RACs. The overall market for RACs reached 7.6 million units by 2017–18, the highest volume of sales recorded under the RAC labeling program, while the share of variable speed RACs increased to 30% (see Fig. 8).

[1] The ISEER method of evaluation is based on bin hours of the Indian climatic zone, bin temperature range of 24–43 °C and 1600 operating hours for cooling per year. BEE derived this methodology from the ISO 16358-1 standard for calculating the seasonal performance metric for both fixed-speed and inverter ACs and modified the temperature bin distribution to account for hotter weather in India.

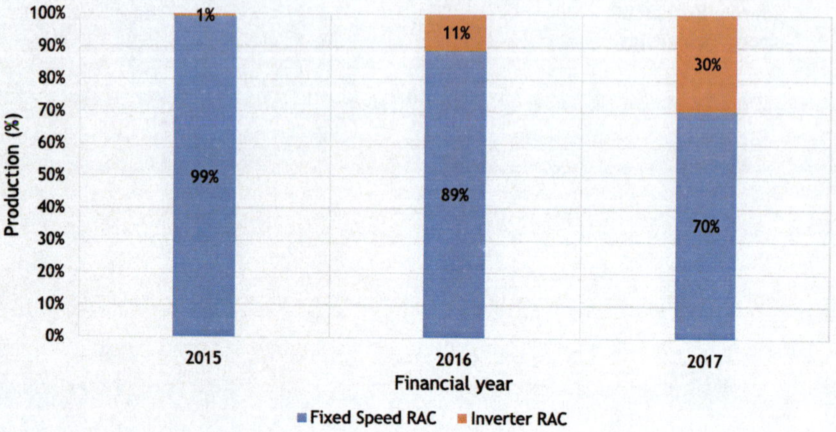

Fig. 8 Market segmentation of fixed speed vs. inverter ACs, 2015–2017
Source: Bureau of Energy Efficiency

5 Impact at the National Level

5.1 Energy Saving and Emission Reduction

BEE estimates the actual energy saving attributable to S&L program by factoring in verified annual/quarterly productions for all equipment/appliances models produced by the manufacturers registered with it and actual energy consumption, annual hours of usage and accounting the average life of each appliance. This energy saving approach is adopted for mandatory & voluntary households' appliances/equipment.

The baseline energy performance level for each appliance is established with following criteria:

- Minimum efficiency levels at the time of launch of appliance are used in case of appliance where energy performance parameters have been defined for each star level without any variable function. (example: Air Conditioner etc., where minimum level is fixed, which is 1-star)

5.1.1 Energy Savings Calculation Procedure

$$Annual\,Energy\,Savings_{(Appliance\ Name)} =$$
$$\left(Baseline\ Value - Actual\ Value\right) \times Production \times Operation\ Usage \times \left(1 - T\ \&\ D\,Losses\right.$$

$$For\ Operation\ Usage\ of\ each\ appliance$$

5.1.2 In Billion Units

Since the inception of the RAC labelling program, 46 TWh of electricity have been saved and 38 Million tons (MT) of carbon emissions have been avoided cumulatively by 2016–17. The highest carbon emission reduction was recorded in 2016–17 at 9.7 MT CO_2, as the overall market of efficient RACs increased significantly (see Fig. 9).

6 Policy Options and Results

6.1 Policy Options

For split type RACs, the minimum energy performance standards (MEPS) (1-Star) has increased by 3% on yearly basis (since 2009 to 2020), which resulted in annual average energy saving of 9%. The energy performance levels were revised or upgraded biennially.

6.2 Increasing MEPS by 4% and Star Ratings Levels

Given that India's MEPS efficiency levels increased gradually with biennial revisions and has resulted in 3% improvement each year, it can be considered as baseline scenario. Alternatively, for accelerating the policy revision, MEPS could be increase by 4% on yearly basis for the period of 2021–2030, with periodic revision of every three years instead of two years, as India's RAC market has technological transformed with efficient products over the last decade. Thus, under this

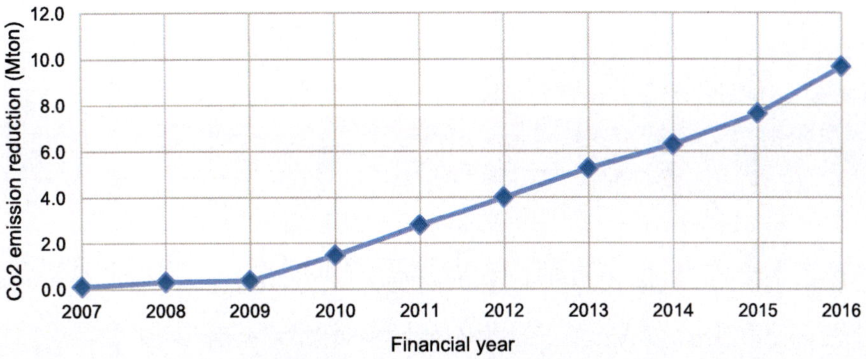

Fig. 9 Carbon emission reduction since inception of labeling program
Source: Bureau of Energy Efficiency

accelerated policy scenario, the MEPS will become ISEER 3.3 in 2021, with gradually increase of ISEER 4.5 in 2030 (see Fig. 10).

6.3 Energy Saving and Emissions reductions

Under the accelerated policy scenario, 4% increase in MEPS, will result in energy savings of over 6.9 TWh of electricity in 2030, equivalent to GHG emission reduction of 5.6 MT CO_2. Over ten-year period, the cumulative energy savings could reach nearly 56 TWh of electricity from 2021 to 2030, which is equivalent to GHG emission reductions of 46 MT CO_2. Whereas in the business as usual scenario (increasing MEPS by 3%), the cumulative GHG emissions reduction of 40 MT CO_2 can be achieved, as shown in Fig. 11. Therefore, accelerated scenario would result in additional GHG emission reduction of 6MT CO_2 by 2030.

6.4 Policy Implementation Assumptions

The standard year, or year when the policy is implemented, is set at 2021. The analysis focuses on the impacts of a policy implemented from 2021 to 2030.

- The RAC market estimated to transform towards variable speed technology over this period at an annual growth rate of 6%.

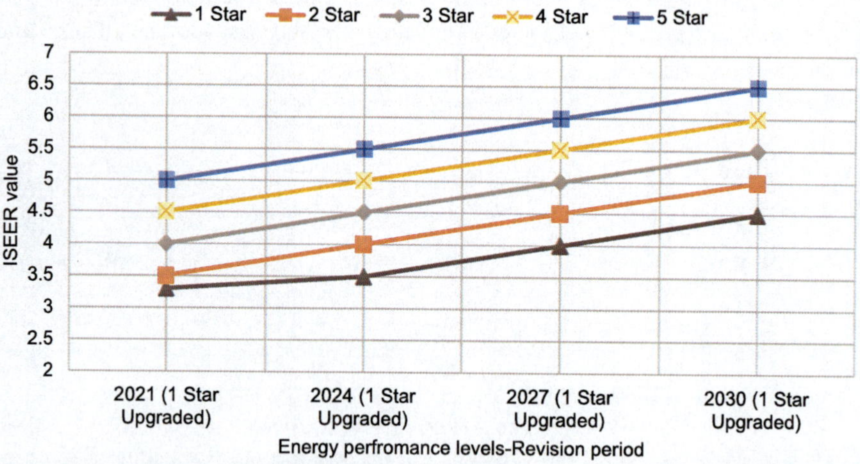

Fig. 10 Energy performance revision level in 2021–2030

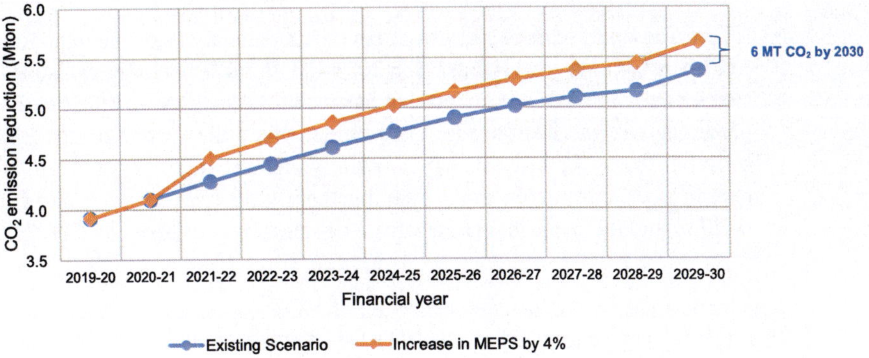

Fig. 11 Avoided CO_2 emissions over time and in 2030

7 Conclusions

The paper analyses RAC market growth, technology evolution and market transformation as a result of BEE's labeling program and further estimates the electricity saving potential with accelerated revision by 2030. The Indian RAC market is ready for an increase in MEPS by 4% and further revision of the labeling ratings in 2021. This paper can be used to revise efficiency level for RACs, quantify potential energy and GHG emissions savings in support of national energy efficiency targets or national determined contribution (NDC) commitments, and estimate other potential benefits from revising the S&L program.

The RAC market in India has grown 25 times in the last 10 years and is expected to grow further at a CAGR of 11% in the next 10 years. The overall production for RACs reached 7.6 million units in 2017–18, the highest volume recorded thus far under the RAC labeling program.

BEE's labelling program has effectively moved the Indian RAC market to higher efficiencies through star rating plans that revise periodically and increase the stringency of energy performance thresholds. The introduction of labeling for variable speed RACs also facilitated the transition to more efficient RACs. The market share of inverter RACs went from less than 1% in 2015 to 30% in 2017–18.

Following the analysis of RAC market and the findings presented above, it is recommended to

- **Ratchet the energy performance standards for room air conditioner by January 2021**:
- As the existing star rating plan for split type RACs is valid until December 31, 2020, there is strong evidence to increase the stringency of the RACs labeling program before or at the time of the next revision scheduled for January 2021. The current market for variable speed RACs has already transformed towards higher efficiencies, with 30% penetration in 2017–18.

- It is proposed to revise star rating plan based on analysis of the data, to **increase MEPS by 4% on yearly basis for 2021–2030, with periodic revision of every three years** as India's RAC market has technological transformed with efficient products over the last decade. Thus, under this policy scenario, the MEPS will become ISEER 3.3 in 2021, with gradually increase of ISEER to 4.5 in 2030, this results in the cumulative energy savings nearly 56 TWh of electricity from 2021 to 2030 and the projected cumulative emissions reduction of 46 MT CO_2.

Acknowledgements The authors would like to acknowledge Bureau of Energy Efficiency for support in the developing this paper, providing the market data and for review of data analysis.

References

1. Gazette notification of Room air conditioner labeling program published by Government of India, Bureau of Energy Efficiency, India http://beestarlabel.com/Content/Files/AC_Notification.pdf
2. Ministry of Environment Forest and Climate change (MoEFCC), India cooling action plan published in 2019, Roadmap for cooling activities in India, http://www.indiaenvironmentportal.org.in/files/file/DRAFT-India%20Cooling%20Action%20Plan.pdf
3. Technical schedule of Variable speed room air conditioner labeling program, introduced in Standard and Labeling program in 2015, by Bureau of Energy Efficiency, India, http://beestarlabel.com/Content/Files/Inverter%20AC%20schedule%20final.pdf

When the Party's Over, Don't Turn Off the Lights! Making Donor Funded S&L Programs Sustainable

Theo Covary and Stephane de la Rue du Can

Abstract Residential Standards and Labelling (S&L) Programs have come a long way since first being introduced in the US in the early 1970's; and are now a proven approach to reduce electricity consumption. Indeed, a 2014 study identified over 80 countries with S&L programs, many of which in certain mid-income countries and emerging economies, are donor funded Examples include South Africa, Russia, Turkey, Cape Verde, and Liberia. And here the benefit of repetition and experience has meant that project outcomes for most such country programs are well known and include label design, selection of appliances, testing facilities, awareness and compliance. Typically, high penetration and large electricity consuming appliances are selected; namely, refrigerators, laundry, lighting, cooling (fans and AC) and more recently televisions – with some slight variations due to cultural preferences (rice cookers in Asia) and income. Certainly, this was the case for South Africa, which in 2011 received a $4.3 million GEF grant to implement a residential S&L program. However, implementation comes with challenges; and key amongst these is sustainability – an issue likely to be faced by most countries whose programs' continued viability, and in certain instances, very existence, is heavily reliant on donor funding. What happens to the program when the donor funding and time limits approach expiry? This paper thus communicates the South African experience, which may provide insights to assist countries that are developing and implementing a donor funded S&L program.

Initially in South Africa, the absolute newness of such a program and certain unforseen implementation delays that this resulted in, led the GEF to grant the S&L

Theo Covary writes in his personal capacity and the view expressed are his own and do not necessarily reflect those of the South African Government or the UNDP.

T. Covary (✉)
Johannesburg, South Africa
e-mail: theo@unlimitedenergy.co.za

S. de la Rue du Can
Lawrence Berkeley National Laboratory, Berkeley, CA, USA

P. Bertoldi (ed.), *Energy Efficiency in Domestic Appliances and Lighting*, Springer Proceedings in Energy, https://doi.org/10.1007/978-3-030-79124-7_9

program two extensions. This ultimately proved to be the correct decision, as the project's implementation revived due to renewed commitment and stakeholder prioritization. The continuation of these successes and other critical work-in-progress outcomes, such as compliance, however came under threat as the project deadline loomed. Indeed, current Minimum Energy Performance Standards (MEPS) are already becoming obsolete as appliance efficiencies improve; and of greater concern has been the real risk that the program would not expand beyond the 11 original appliances selected in 2011.

To mitigate these risks, the "*Next Set of Appliances*" study was commissioned, using existing project resources. This was to: (1) Evaluate the existing MEPS, and identify which could be strengthened; and, (2) Identify new electrical equipment for the potential introduction of MEPS (expanded to light commercial equipment). In this, a broad approach was adopted, where the consultants identified as many as 96 potential pieces of electrical equipment and then narrowed these down to 8, based on mutually agreed criteria (penetration rates, opportunity for electricity savings, global MEPS implementation, ease of adoption and technological/other barriers). The maximum electricity savings and benefits were demonstrated to the Department of Mineral Resources and Energy's (DMRE) Clean Energy Branch Director; placing the unit in a position to make informed decisions, in support of an application for an allocation in the upcoming internal budgeting process (outcome pending at time of writing). Moreover, the study provides reliable, robust, up-to-date information and data, needed for new donor funding applications.

The paper explains the approach followed, as well as industry responses, while reporting on the program's outcomes and highlighting the next steps to integrate the program into national planning. In so doing, it provides an approach for other countries to consider, when faced with the inevitable issue of sustainability of their S&L program.

1 Introduction to South Africa's Residential Appliance S&L Program

The issue of residential appliance minimum energy performance standards (MEPS) and compulsory energy efficiency (EE) labelling, features prominently in the Energy White Paper (1998) [1] and was specifically identified as a key intervention of the Department of Minerals and Energy Resources' (DMRE) 2005 Energy Efficiency Strategy [2], which set an overall energy intensity reduction target of 12% by 2015 and a 10% reduction in the residential sector. This initiative also features in the Industrial Policy Action Plan (IPAP) [3] updated in 2010 by what is now the Department of Trade, Industry and Competition (dtitc), an implementing partner of the program.

Here, a formal project was deemed necessary, to address the policies, information and technology involved, as well as the financial barriers preventing widespread

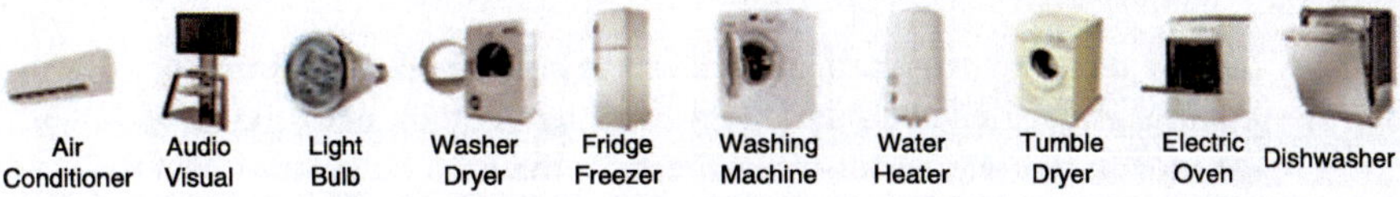

Fig. 1 Residential Appliances selected for South Africa's S&L program

introduction and uptake of efficient appliances. A successful government application to GEF, through the UNDP, was thus endorsed, and a budget of US$ 4.375 million allocated. This was geared to provide assistance to Government, as well as national agencies and the private sector, to introduce and implement a mandatory S&L program. The official start date of the five-year program was September 2011; and in facilitating comprehensive market transformation towards the use of EE (electrical) residential appliances, this GEF-funded project ultimately aims to accelerate national decarbonization and reduce household energy demand. Currently, the project targets large electrical appliances and lighting, as detailed in the South African National Standard 941 [4] and shown in Fig. 1.

The S&L project funding application, commonly referred to as the ProDoc, submitted to GEF, estimated that as much as 388 GWh of electricity would be saved per year, leading to 4.6 Mt of direct CO_2 emissions reduction (over the lifetime of the appliances covered) and indirect CO_2 emissions reduction of 11.5 Mt CO_2 [5]. However, achieving these savings would require the removal of the most significant and persistent barriers impeding the widespread uptake of energy efficient residential appliances. Consequently, the GEF-funded S&L project was designed accordingly and consisted of six outcomes:

Outcome 1:	Policy and regulatory framework for the S&L program
Outcome 2:	Define labelling specifications and MEPS thresholds for the 12 products considered by the DoE & DTI for S&L regulation
Outcome 3:	Strengthen the capacity of institutions and individuals involved in the S&L program
Outcome 4:	Awareness-raising campaign for standards and labels—targeting manufacturers, distributors, retailers and end-users
Outcome 5:	Implementation of S&L Market Surveillance & Compliance (MSC) regime to ensure energy performance standards is met
Outcome 6:	Development of Monitoring and Evaluation (M&E) capacity

2 S&L Program: 2011 Until 2015

As is customary for all GEF-funded programs, a Mid-Term Review (MTR) is undertaken to evaluate progress and effectiveness. This effort was delayed due to the program's relatively slow beginning, and the report was only issued in May 2016, less than 18 months before program closure. Here, a high-level summary of the MTR [6] findings is detailed in Table 1. In presenting this summary, the objective is to outline the most important issues initially facing the project, most of which are not unique or particular to South Africa. By no means does this table serve as indictment of the performance of the UNDP or the

Government of South Africa; nor does it seek to dwell on the underlying causes, but to rather demonstrate typical program implementation challenges, through the use of a credible case study.

Undoubtedly, the most pertinent consequence of challenges faced, is that program implementation had fallen behind schedule; leading to a mutually agreed revision of the end date between the UNDP and the Government of South Africa, to

Table 1 Summary of mid-term review findings of South Africa's S&L program

Strengths and major achievements	
Highly relevant with respect to national priorities	Program was initiated during national power outages, benefiting from broader EE initiatives
Country ownership shown through co-financing contributions	DMRE and SABS (national test laboratory) funded studies and upgraded test laboratories in excess of US$1.2 million. National standards developed and regulation passed to make specifications mandatory
Strong willingness among industrial sector to comply with new EE regulations	Excluding electric water heaters, most appliances met the new MEPS in advance of the regulations – acting in advance to ensure compliance
Highly qualified and skilled enabling stakeholders	Government and industry representatives are knowledgeable and participatory. A highly noteworthy outcome which significantly increases the likelihood of the program being effectively implemented
Strengthened institutional and individual capacities, and facilitated strategic partnerships	Implementing partners had undertaken multiple international visits to build capacity (Korea, UK, Australia and Brazil) and created a strategic partnership between SABS and a UK test laboratory
Major shortcomings and opportunities for improvement	
Lower-than-expected GHG, based on midterm estimations	Primarily due to the delayed start of the project, the GHG would be lower than originally forecast. It was recommended that the estimations be -re-calculated
Delays associated with cash disbursement have significantly impacted project effectiveness	A failure between the UNDP and DMRE to agree on a payment modality, left the implementing partner without funds. The project could not procure and became reliant on DMRE funding
Inconsistent project governance	Although Project Steering Committee met regularly, decisions and recommendations were not enacted
Unclear management arrangements	Stronger working relationship needed between the UNDP and the national implementing partners
Unsatisfactory delivery	The delay in appointing a PM, the funding modality issues, low level of readiness of the national test laboratory and Regulator – resulted in only 8% of the GEF funds spent at the project's mid-point

(continued)

Table 1 (continued)

Likelihood of achieving project objective diminished by lower-than-expected co-financing	An international donor withdrew US$four million of funding, citing unsatisfactory progress, and so did the national utility for a similar amount, due to funding constraints. This shortfall was somewhat offset by contributions made by the SA Government as detailed above
Stakeholder engagement has not been sufficiently integrated into existing structures	Deemed that industry consultation was lacking
Industry frustrated by lack of communication from regulator	Ever increasing backlog in the approval process of compliance approvals by the Regulator, compounded by insufficient communication and preparedness
Questionable financial sustainability of market surveillance, control and enforcement	Concerns raised regarding the Regulator's ability to fund its activities, demonstrated by the low level of compliance activities and slow approval process
Awareness is low, largely due to delay in implementing communication strategy	Activities not initiated
Unsatisfactory progress with respect to developing and implementing incentive programs	Weak knowledge management and limited online information regarding the project. Preliminary study to identify possible incentive programs undertaken, but an inadequate report delivered by the service provider raised more questions that it answered

March 2019. The next section now details the actions taken to address implementation since then, as well as the progress achieved during this period.

3 Project's Second Half: 2015 up to March 2019

The South African S&L program is undoubtedly a story of two halves. Following the release of the MTR, a concerted effort to achieve the program's objectives was further underpinned by the national urgency for electricity savings, as the country continued to experience power outages due to the vertically integrated utility's inability to meet national demand. Herein, actions taken can be categorized into two – direct and indirect – with the former being discrete actions to address specific program issues, while the latter were more strategic in nature, as they were intended to effect change within a broader sphere. Some of the key actions taken, and the outcomes thereof, are provided below.

3.1 UNDP Organization Changes (Indirect)

A new country director (CD) and program manager for the energy portfolio were appointed; and an outcome of the findings of the MTR (Table 1) that were released soon after the CD took on his new role, was that it precipitated new discussions between the UNDP and the DMRE. This led to renewed commitment to achieve the objectives of the program, while a more flexible approach was adopted, to allow the project team to respond more effectively and efficiently. The program manager was replaced in the second half of 2016, and the program's new project manager, a direct intervention, was then appointed in early 2017. Here thus, new people fostered renewed relationships and adopted different approaches, which in this case proved to be a key success factor.

3.2 Procurement (Direct)

Recognizing the program's high procurement requirements, the DMRE contracted a state agency to manage the process. The UNDP also agreed to procure on behalf of the project, thus providing two procurement channels. These actions also resolved the cash disbursement and funding modality issue which had hampered delivery, as detailed in the MTR (Table 1).

3.3 Project Governance and Private Sector Involvement (Direct)

Although Project Steering Committee (PSC) meetings were being held, they were largely ineffective. This was corrected by: (1) The frequency of meetings being increased from 3 or 4, to a minimum of six per year; and, (2) The UNDP CD and senior officials from the DMRE attending the meetings. The combination of these two actions led to greater accountability and significantly increased performance. Also, industry association representatives at the PSC were asked to play a more active role, and to assist the project by fulfilling their mandate with greater enthusiasm and commitment.

From 2017 to 2019, under the new project manager appointed, over 30 service contracts were awarded. By March 2019, 98% of the GEF allocated budget, compared to 8% reported in the 2015 MTR, had been spent or encumbered. This performance value, brought by the outputs of the contracts, was recognized by all stakeholders as a major achievement [7]. And from the implementing agency's (UNDP) perspective, it pointed to a resuscitated project, which could now move towards closure, having met most of its objectives. From the perspective of the S&L program, however – what happens next?

4 Projects Have a Start and End Date: Programs Do Not

In theory, the benefits of an S&L program to the nation – including substantial bills' savings for consumers – should justify government funding for the program internally, without donor funding. However, without donor funding, important climate and sustainability programs in many emerging economies may have a low priority, with the implication that they are often not considered, delayed indefinitely, or enjoy limited political will and resources if they are actioned; with their post-funding being equally crucial, if they are to remain effective and sustainable. Also, institutions (private, public and international agencies) have performance metrics against which they are measured. To satisfy the requirements of, and maintain its status as a GEF implementing agency, the UNDP in our case for example, provided operational support to the S&L project. This included oversight, a dedicated project manager and assistant, procurement, accounting, payment access to GEF funding and international technical expertise for the duration of the project; with contributions from the Clean Energy Ministerial SEAD[2] initiative. And more recently, USAID.

Indeed, regarding such operational support, the expectation usually is that the required skill transfer occurs during the prescribed project duration, (here a five-year period, extended to eight), thus allowing for a smooth transition on termination date. However, such an approach must be carefully managed from both sides if it is to succeed, as stakeholder priorities are not necessarily aligned. On the one hand, the implementing agency is under pressure to 'deliver' or spend the allocated available budget within the prescribed date, in donor parlance. Whilst on the other hand, the government's involvement may vary from fully engaged to arm's length. Regardless of the level of participation, government's primary objective is to ensure that project activities are appropriate and meet their requirements. This may be complicated further if the ProDoc, which is the de facto guideline on what activities should be undertaken for each outcome, becomes outdated, because of technological advances or a change in government perspective, due to political or market forces, or both. If not carefully managed, such competing objectives have a potential negative impact on the program, such as implementation delays or stagnation. A suitably experienced and competent project manager however, should be able to navigate such pressures, if all parties remain rational and flexible. But a program-creating project that experiences such issues as it approaches its closure date however, will exert additional pressure on all parties, which are likely to be prioritized over the important and critical actions needed for a smooth transition - delaying those, until it is too late. A counterfactual analysis of the South African S&L project is thus undertaken below, to demonstrate the steps taken by the project management team, to ensure to the greatest extent possible, a smooth transition and the long-term sustainability of the S&L program.

4.1 Project Performance Considerations Leading up to Project Closure

Having received a sub-par MTR rating, the UNDP and national government were under pressure to ramp up delivery; with the result that multiple assignments were undertaken concurrently. Indeed, during certain periods, the project manager was overseeing more than ten service contracts. This placed significant demands on the project team (drafting ToR's, evaluation of proposals, project management); on the UNDP administration who oversee procurement, payments and accounting; as well as state entities whose cooperation and participation is essential. Ultimately, progress is reliant on unfettered access to government counterparts and their tacit support of proposed assignments, together with their participation in proposal evaluation, in deliverable and invoice sign-off, as well as the shared understanding of the project's strategic direction. In this, an outcome of the constant interaction with government colleagues who are not dedicated resources and have other duties, is that it leads to periods of deprioritization of the project implementation and/or fatigue. This manifests in long turnaround and response times, unavailability for meetings on short notice, and in the extreme, a blunt request to slow down. Matters are further complicated by the fact that the project must engage with implementing partners in other government ministries or state-owned entities, which do not fall under the control of the DMRE; namely the national test laboratories (SABS) and the Regulator (NRCS). Here, inter-ministerial collaboration becomes key; but as priorities tend to differ, accountability can be blurred. This can result in funding of government apportioned activities that has not materialized or been delayed, becoming a stumbling block, due to the knock-on effects on dependent activities funded by the project. By way of example, the training of national laboratory testing officials (donor funded activity) can only occur if the laboratory is functional (government obligation). If the government's effort to bring the test laboratory to a functional state occurs after the project termination date, the training cannot occur, with the likely implication that the donor funding is withdrawn, and government either having to forego or pay for the training. Both outcomes inherently imply delays and are therefore undesirable.

4.2 Funding Sources Post Project Closure

For transitions to be sustainable, they should at the very least include: (1) Updated policy objectives; (2) A business plan; (3) Formal agreement amongst implementing partners; (4) Handover processes; and (5) Revision and expansion of the Program; all of which are listed below. As can be seen, a definition of the action is followed by a short description of the steps taken by the S&L project to meet the requirement. This is done for the first four; followed by a more detailed explanation for item 5 – Revision and Expansion. (The Super-efficient Equipment and Appliance Deployment

(SEAD) initiative provided technical support to the S&L program implementation in this, through its main contractor Lawrence Berkeley National Laboratory (LBNL).)

1. *Update Policy Objectives:* With the termination of the GEF-funded S&L project and its objectives, the responsible ministry (DMRE) would need to identify and formalize its future S&L program policy objectives and targets.

   This is being dealt with, in two steps. The first is to quantify the electricity savings achieved by the project up to its termination date. Here, data, collected from the Regulator's database, as well as studies and industry reports, is to be modelled by the energy planning department within the Ministry; and the results will be compared to the 2011 projections. This effort is being supported by the LBNL and USAID. The second step is to then use the outputs to inform new and updated targets that align with the National Energy Efficiency Strategy, a Cabinet approved strategy, which mandates the DMRE's Clean Energy Branch to meet its objectives and targets.

   Ultimately, although this has not been possible for the South African program yet, an EE law should institutionalize the S&L program - including a mandatory requirement to review standards within a specified time period, (ideally every 3–5 years, but not more than six).

   – This requires constantly assessing the need to revise and upgrade existing tests and energy performance standards. By way of example, in the United States "*Beginning with the Energy Policy and Conservation Act of 1975 (EPCA), a series of congressional acts have directed the U.S. Department of Energy (DOE) to establish minimum energy conservation standards for a variety of consumer products and commercial and industrial equipment. The EPCA, as amended, requires the DOE to update or establish standards at levels that "achieve the maximum improvement in energy [or water] efficiency ... which the Secretary determines is technologically feasible and economically justified." EPCA defines "economically justified" standards as those for which benefits exceed the costs, given a number of factors, including impacts on consumers and manufacturers and the nation's need to save energy or water.*" [8]

2. *Business Plan:* An approved business plan for a dedicated S&L unit to implement the policy objectives and targets, is needed. The plan would seek to institutionalize and operationalize S&L, by providing it with the resources (human and financial) to build on what has been achieved during the program's first project phase.

   Recognizing that the shift from program-implementing project to ongoing program, signifies a move to the next lifecycle stage (from introduction to growth), a five-year business plan has been developed by the project office and was submitted to the DMRE's Clean Energy Branch for inclusion in its 2019/20 internal funding allocation.

Generally, the business plan should Identify internal (government) capacity, to ensure general operation and expansion of the program, as well as external support (aid agency), to complement and integrate international best practice to the SA program.

3. *Formal Agreement:* A formalized working relationship, outlining each implementing party's obligations, is necessary – if not already in place. This measure ensures that momentum is not lost, as success relies heavily on one ministry's policy (DMRE) being implemented by another ministry or its agencies; in the case of South Africa, the dtic.

   A meeting was convened by senior representatives of all government ministries and agencies involved, and it was unanimously agreed for a Service Level Agreement (SLA) to be developed by an external lawyer, in collaboration with legal representatives from each institution, to govern the project going forward.

4. *Handover:* A transition period of three to six months is crucial; so as to allow for an orderly transition between the UNDP project manager and the newly formed S&L unit. Knowledge and project documentation transfer will increase their effectiveness.
5. *Revision and Expansion of the Program*: The project appointed a consultancy to undertake two reports. The first was to assess, through industry consultation, whether the appliances regulated under VC 9006 and 9008 (Fig. 1) could have their MEPS improved. The second, aimed to identify the next set of electrical equipment to be regulated.

4.3 Report 1: Review of Existing Appliance Energy Classes and Recommended Changes to Existing MEPS

The assessment of existing energy classes was undertaken to ascertain whether there is sufficient scope to improve (strengthen) the current standards; and to simultaneously identify possible effects on existing testing capacities, national standards and the regulator. The research was informed by industry reports (specifically Euromonitor), local market research data, interviews with industry representatives and in-house desktop research, which consisted of web-crawling, as well as product brochures and reports/data made available to the consultants by the S&L research team. The research findings were then presented to industry representatives at a public consultation meeting, to note their response and gather any additional information or feedback on the new standards and MEPS being proposed. Moreover, this data resource, complemented with additional data, was compiled in a bottom-up model developed by the DMRE's energy planning unit, with assistance from LBNL, to assess the potential impacts of revising and expanding the program. The South African Energy Demand Resource (EDR) model is a bottom-up end-use model, specifically developed to assess the impact of the S&L program. It projects energy demand in order to calculate the impact of proposed and/or possible additional

Table 2 Residential end-use and proposed energy efficiency standards

Appliance	Current MEPS	Proposed MEPS	Date	GWh/a Savings in 2030
Audio visual	Standby <1 W	<0.5 W	By 2020	40.1
Electric ovens				
1. Small & medium	B	A	By 2020	165.4
2. Large	None	B		
Washer dryers	A	No change	–	–
Washing machines	A	A+	By 2022	131.1
Tumble dryers	D	C	By 2020	9.4
Dishwashers	A	No change	–	–
Refrigerators	B	A A+	By 2020 By 2026	667.0
Freezers	C	B A	By 2020 By 2026	266.6
Air conditioners	B	A	By 2021	202.7
Electric water heaters	B	No change	–	–
Additional Electricity Savings (GWh) in 2030				**1493.9**

policies; with energy consumption projected by end use from 2015 (base year) to 2040. The strategy of the model is to first project end-use activity, which is represented by the sales of equipment, driven by increased ownership of household appliances. In this, the total stock of appliances is modeled according to penetration of ownership in the base year and then per unit sales projections. Electricity consumption, or intensity of the appliance stock, is then calculated according to estimates of the baseline intensity of the prevailing technology in the local market. Finally, the total ultimate energy consumption of the stock is calculated. This is done by modeling the flow of products into the stock and the marginal intensity of purchased units, either as new additions or as replacements of old units, based on equipment retirement rates.

More details about the model and detailed assumptions can be found in the South Africa EDR Report [9]. Results for the residential sectors are summarized in Table 2.

As demonstrated by Table 2, large electricity savings remain in the residential sector, if standards reach efficiency levels already implemented in economies that are major trade partners with South Africa. For example, it is estimated that South Africa could save as much as 667 GWh/year in 2030, if by 2026 it adopts the refrigerator standard that came into effect in the EU in July 2014 (A+ level).

Table 3 provides more detail on the assumptions for the proposed standards as available in [7].

Additionally, the DMRE has also developed mandatory technology neutral technical specifications to remove inefficient light bulbs. Initially, lighting had not been part of the residential S&L program, because of Eskom CFL distribution programs. However, the utility then suspended its EE program, leaving confusion in the market [10], while CFL bulbs were also been overtaken in efficiency by LCD technology.

Table 3 Baseline and MEPS scenarios used in the residential energy demand resource model

	Product sub-type	UEC Stock (KWh)	UEC Baseline (kWh)	Rating	UEC Proposed (year: kWh)	Rating
Appliances	Refrigerator-Freezer	344	308	B	2021: 281	2021: A
					2027: 243	2027: A+
	Refrigerator	280	250	B	2021: 228	2021: A
					2027: 197	2027: A+
	Freezer	423	406	C	2021: 366 2027: 330	2021: B 2027: A
	Clothes Washers	190	185	A	2023: 162	2023: A+
	Dryers	294	275	D	2021: 271	2021: C
	Dishwashers	291	285	A	No change	No change
	Ovens	119	112	B	2021: 101	2021: A
Water heating	Electric Water Heaters	1351	1042	B	No change	No change
Entertainment	TV	213	213	–	No change	No change
	Standby TV	5.08	5.08	0.58 W	2021: 4.38	0.5 W
	Other plug load	5.08	5.08	0.58 W	2021: 4.38	0.5 W
Space conditioning	Split cooling only	993	960	B	2021: 900	A
	Split Reversible	2056	1988	B	2021: 1864	A
	Evaporative air coolers	804	804	–	2021: 804	–

Source: South Africa EDR [9]

Thus, by targeting performance rather than a specific technology, all lamp types need to comply with the minimum lumens per watt energy efficiency requirement. Here, the proposed MEPS regulations specified that minimum energy-efficiency requirements be introduced in two phases. The first phase became effective in 2020, with a limit of 80 lm/W, and the second phase will be 95 lm/W [11]. It bears reemphasis that this approach makes the regulation non-discriminatory toward specific technologies and avoids the need to develop additional regulations should a new lamp technology enter the market. A report [11] was been commissioned to assess the cost benefits of implementing the standard for lighting; and the data collected was used in the EDR to calculate the energy savings

Results from the bottom-up analysis show that the proposed set of new energy standards could limit growth of electricity demand to 2.3% instead of 2.7% [9]. Indeed, if international best practices standards are adopted sooner and across multiple end-uses in the residential sector, the bottom-up modeling shows that residential electricity growth could be limited to 1.9%. It is important to note that S&L impact will allow for reduction of energy consumption, but not the level of energy services made available, which stay the same in all scenarios. EE means using less energy to provide the same service. For example, as already pointed out, an LED bulb is more efficient than a traditional incandescent bulb and CFL, as it uses much less electrical energy to produce the same amount of light.

More than anything, it is abundantly clear that in a South African context, where power reliability is uncertain and electivity prices are increasing sharply for consumers, (electricity tariffs increased by 300% from 2007 to 2015, whilst inflation over this period was 45% [12]), EE represents a cheap and sustainable resource of energy, which must be prioritized by government.

4.4 Report 2: Identification of the Next Set of Electrical Appliances for Inclusion in the National S&L Program

The objective of this study was to identify up to ten new items of electrical equipment that would deliver meaningful savings if MEPS were introduced; and was not limited to the residential sector. A market and engineering analysis, underpinned by international practices and experience, was used to identify suitable equipment and to then eliminate the non-suitable, through the following screening process:

1. Identifying all possible options (long list)
2. Removing equipment which is out of scope e.g. non-electrical, covered by MEPS
3. Proceeding with equipment that is regulated in at least two countries, thus ensuring that international standards are available and new standards do not need to be developed
4. Ranking equipment according to estimated future electricity savings
5. Scanning for any potential adoption, implementation and operational issues
6. Ensuring that choices are nationally appropriate – possible impact on local manufacturing, increased purchase costs, rate of market change etc.

Following this approach, the initial long list comprised of 96 possible candidates. The second, third and fourth screening criteria, then reduced the number to 72, 24 and finally to eight, as shown in Table 4.

5 Conclusion

Standards and labeling (EESL/S&L) programs are highly effective policy instruments with which to save energy and support growing markets for EE products. They are cornerstone of reducing energy demand worldwide and have been implemented in more than 80 [13] countries, covering over 50 different types of products in the commercial, industrial and residential sectors. These programs encourage removal of inefficient technologies from the market; avoid the dumping of older,

Table 4 Equipment considered for program expansion

Equipment	First screen	Appliance Profiling	Decision
Computers	Medium	High potential rates	Include
TV's	High	High potential—globally implemented	Include
Electronic power supplies	Medium	High potential and standard product	Include
Electric motors	High	High potential and standard product	Include
Pool pumps	Low	Limited international interest but straightforward	Include
Commercial refrigerators	High	Complicated due to customized sizes & complex testing	Include, but specific types
Chillers	High	Large savings & low numbers but complex testing	Include
Transformers	High	National priority	Include
Cooktops/hobs	High	Straightforward but limited savings	Exclude
Microwave ovens	Medium	Limited savings	Exclude
Electric irons	Low	Limited savings	Exclude
Electric heaters	Low	Low savings	Exclude

less-efficient technologies from more advanced economies; and empower consumers to make informed purchasing choices. And they are essential in transforming markets toward more advanced technologies, while also fostering innovation, and thus contributing to the improvement of technology in a country. Indeed, EE standards should be regularly revised to more stringent levels, to reflect rapid technological evolution and relevant market changes.

Ultimately, if energy savings are to be maximized, significant and dedicated resources are required to ensure that development, enforcement and revision of standards is maintained. But this investment is truly dwarfed by its multiple national benefits, which *inter alia* include avoided electricity generation, reduced electricity bills, reduction of GHG emissions and a more globally competitive national economy.

Internationally, S&L programs that yield high energy savings, consist of a dedicated team of technical experts, who foster global collaboration to leverage the program's performance. This, for example, includes support and technical advice from specialists, who conduct analyses to upgrade or expand the program to new products, or who help develop specific tools and in-house capacity to evaluate and improve the program. And it extends to providing consistent guidance on compliance approaches, together with the ongoing sharing of information.

Within this global context, South Africa's S&L program is nascent. As an intervention it is burgeoning, and its potential is proving to be vast. The program is now running and operating with well-established

tools and resources; and most importantly, has successfully communicated the value of EE standards and labels to consumers and industry stakeholders. Herein, the UNDP project has provided a solid foundation for the program to grow and improve. Now, S&L needs to continue the transition from a GEF funded and UNDP implemented project, to a formalized Government program. And its sustainability and effectiveness can only be maximized if the responsible ministry builds on the initial project's robust foundation, by placing a sizable value on the benefits that the program delivers to consumers and the economy. To do so, it must provide dedicated resources, both human and financial, to institutionalize the program and allow for its expansion in the long term, while substantially increasing its visibility, through political will and committed prioritization.

Finally, as an insight from a donor perspective, be it GEF or any other, it is equally crucial that the task of transitioning from donor project to government program is recognized and allocated sufficient time and financial resources from the outset, when designing the scope of work. This allows for the inevitable transition from project to program to become a natural outcome of the process, which assures the seamless sustainability, growth and evolution of the program, once the initial project's "booster rocket" has fallen away.

References

1. Department of Minerals and Energy (1998): *White Paper on the Energy Policy of the Republic of South Africa.* Available at: http://www.energy.gov.za/files/policies/whitepaper_energypolicy_1998.pdf
2. Department of Minerals and Energy (2005): *Energy Efficiency Strategy of the Republic of South Africa.* Available at: www.gov.za/sites/default/files/gcis_document/201409/energy- efficiencystrategy051.pdf
3. Department of Trade and Industry (2010): *Industrial Policy Action Plan 2010/11 – 2012/13*
4. South Africa Bureau of Standards (2012): *Energy Efficiency of Electrical and Electronic Apparatus SANS 941 Edition 1*
5. UNDP. (2010): *Market Transformation Through the Introduction of Energy Efficiency Standards and the Labelling of Appliances in South Africa.* Project document submitted to GEF to approve funding for South Africa's S&L Project. Available at: https://www.thegef.org/project/market- transformation-through-energy-efficiency-standards-and-labeling-appliances-south
6. Lenoci, J. (2015): Mid-Term Review Report of the Market Transformation of Energy Efficiency Standards and Labelling of Appliances in South Africa
7. de la Rue du Can, S. and M. McNeil (2018). *South Africa: Energy Savings Estimates from New Standard and Labeling Program Policy Brief.* http://eta- publications.lbl.gov/sites/default/files/south_africa_energy_savings_estimates_from_new_standard_a nd_labeling_program_0.pdf
8. Meyers, S., Williams, A., and Chan, P. (2014): *Energy and Economic Impacts of U.S. Federal Energy and Water Conservation Standards Adopted From* 1987 *Through 2013.* Lawrence Berkeley National Laboratory. Available at: https://www.osti.gov/servlets/purl/1163738
9. Department of Energy, LBNL, USAID (2019): *South Africa Residential Energy Demand Resource Modeling*
10. Theodore Covary, Stéphane de la Rue du Can, and Mercy Shuma-Iwisi (2019): A review of the residential efficient lighting programme rollout in South Africa, ECEEE Summer Study 2019
11. Nova Economics, (2019): *Cost Benefit Analysis of technology neutral regulations to introduce minimum energy performance standards for general lighting"*, Report prepared for the United Nations Development Programme
12. Moolman, S. (2015) Eskom tariff increases vs inflation since 1988. Available at: www.power-optimal.com/infographic-eskom-tariff-increases-vs-inflation-since-1988-projections-2017/
13. Harrington, L.; Brown, J. and Caithness, M. (2014): *Energy Standards and Labelling Programs Through the World in* 2013. Study funded by SEAD, UNEPDTU and IEA 4E

Final Impact Assessment of a Small-Scale Biomass Gasifier Fuel-Cell CHP System for Clean On-site Power Generation

Thomas Goetz, Lena Tholen, Jan Kaselofsky, and Thomas Adisorn

Abstract The EU Horizon 2020 funded project 'FlexiFuel-SOFC' (Grant Agreement n° 641,229, 2015–2019) developed a new and highly efficient small-scale biomass Combined Heat and Power (CHP) system for clean on-site cogeneration. It shall replace traditional systems based on fossil fuels, being at the same time fuel flexible for utilizing solid biomass residues (e.g. wood chips or olive stones), robust, cost efficient, and distinguish itself by high electric and overall efficiencies as well as almost zero emissions. In particular small-scale CHP technologies suitable for micro-generation are challenging, but biomass gasification and solid oxide fuel cells (SOFCs) offer significant potentials and important co-benefits, such as security of energy supply as well as emission reductions in terms of greenhouse gases or air-quality related pollutants.

This paper presents final impact assessment results from the development of the novel CHP system, consisting of a fuel flexible small-scale fixed-bed updraft gasifier, a compact gas cleaning unit and an SOFC for electricity generation. System efficiencies and emissions of solid fuel combustion and grid electricity effects were evaluated. Gasifier-fuel cell CHP technologies produce significantly less fuel-related emissions compared to traditional heating systems and also produce electricity with fewer emissions than traditional grid electricity generation systems.

Such new developments are also influenced by several national and international policies and measures, which can prevent or incentivize the potential market of the new technology. Therefore, complementary to the results of the final impact assessment, this paper also addresses selected policies on EU and Member States level with relevance for small-scale CHP technologies. In doing so, this paper asks from an innovation point of view how the current policy mix hinders or supports the market uptake of such small-scale CHP technologies. The paper factors in relevant elements of the policy package such as the CHP Directive, the Renewable Energy Directive and the Energy Performance of Buildings Directive.

T. Goetz (✉) · L. Tholen · J. Kaselofsky · T. Adisorn
Wuppertal Institute for Climate, Environment and Energy, Wuppertal, Germany
e-mail: thomas.goetz@wupperinst.org; lena.tholen@wupperinst.org; jan.kaselofsky@wupperinst.org; thomas.adisorn@wupperinst.org

P. Bertoldi (ed.), *Energy Efficiency in Domestic Appliances and Lighting*, Springer Proceedings in Energy, https://doi.org/10.1007/978-3-030-79124-7_10

1 Introduction

Traditionally, CHP systems are often based on fossil fuels, such as natural gas or heating oil, but the transition to efficient energy systems using renewable energy sources is urgently needed in particular in the heating sector to achieve world-wide sustainability targets. Based on EU classifications combustion plants are rated >50 MW_{th} for large systems and < 1 MW_{th} for small appliances (residential heaters and boilers), with medium combustion plants (MCP) in between. Today, CHP is mainly realised in the medium and large-scale sector, especially for renewable biomass fuels. However, the applied traditional technologies have restrictions regarding fuel flexibility and electric efficiencies.

In contrast, dedicated biomass integrated gasification fuel cell systems (B-IGFCs), are deemed to achieve much higher efficiency levels [1]. Adaptation to small scale generation applications based on renewable energy sources (such as solid biomass) is specifically challenging, because of feedstock compositions and heat integration. Small-scale cogeneration systems are typically intended to replace or complement traditional heating equipment in residential buildings. In addition to space heating or domestic hot water supply, a part of the fuel energy is used to generate electricity for consumption at the building or to be sold to the electric power grid. In addition to efficiency potentials, B-IGFCs also offer important co-benefits, such as security of energy supply as well as emission reductions in terms of greenhouse gases (GHG) or air pollutants. In particular, solid residual biomass as renewable local energy source is best suited for decentralised operations such as micro-grids to avoid inefficient long-haul fuel transports to centralized power plants.

Against this background, the EU Horizon 2020 project 'FlexiFuel-SOFC' (Grant Agreement n° 641,229, see also http://flexifuelsofc.eu) developed a new, innovative, highly efficient and fuel-flexible biomass CHP technology. The new technology integrates a small-scale fixed-bed updraft gasifier, a novel and compact gas cleaning concept (covering particle precipitation, removal of HCl, H_2S and other sulphur compounds as well as tar cracking), and a high temperature solid oxide fuel cell (SOFC) for electricity generation. The technology was developed for the residential sector with a capacity range of 25–150 kW (fuel power) and the utilisation of cost efficient residual biomass feedstocks to enlarge the applicable fuel spectrum. Overall the new system shall achieve an operation with significantly reduced emissions (regarding CO, OGC, NO_x, HCl, SO_x, PAH and PM and due to the utilisation of biomass also regarding CO_2), in combination with high electric and overall efficiencies. A two-phase approach for the construction of testing plants, the performance of test runs, and accompanying assessments provides a significant advance in the technology performance of small-scale biomass based CHP systems.

This paper provides an overview of environmental aspects relevant for biomass gasifier fuel cell systems in general and also presents a generic approach for assessing related impacts. All results are based on the final work performed in the 'FlexiFuel-SOFC' ('FF-SOFC') project. Following the opportunities and challenges to be addressed, the development of respective technologies for clean on-site heat and power generation is also strongly linked to sufficient incentives. Besides technology aspects, a well-aligned and comprehensive energy efficiency and environmental policy framework is crucial. Accordingly, the obtained results are set into relation with the EU policy framework and recommendations are provided for aligning technical and policy evolution to harness maximum environmental and overall benefits.

2 Environmental Considerations of Small Scale Biomass CHP

In advance of any large-scale future deployment of new technologies, such as efficient CHP systems fostered e.g. by polices and measures, the potential environmental impacts have to be adequately assessed. Accordingly, a systematic approach is needed to evaluate the specific environmental and policy implications. The following section describes a generic assessment method, typically used e.g. before the implementation of polices and measures for the EU or other markets, as well as the most relevant parameters and effects to be considered. The described approach is applied by Wuppertal Institute for small-scale biomass gasifier fuel-cell CHP systems for the residential sector, small enterprises, hospitals and hotels. Respective findings of the final environmental impact analysis from the EU Horizon 2020 project 'FlexiFuel-SOFC' are presented based on the data inputs from all project partners.

2.1 Assessing Environmental Impacts

Based typically on consecutive results from comprehensive market studies (provided for the FlexiFuel -SOFC project by partner Utrecht University) and data from techno-economic analyses (provided for the FlexiFuel-SOFC project by partner BIOS in cooperation with all other partners), the subsequent environmental Impact Assessments (IA) follows a well-defined structure. For the presented final impact assessment, the main aspects, as defined by the Impact Assessment Guidelines of the European Commission [2], are used as basis and further modified for the purpose of the analysis. A holistic method is described based on the following steps:

Step 1: Problem Definition ➔ Step 2: Define Objectives ➔Step 3: Develop Options ➔
Step 4: Impact Analysis ➔ Step 5: Comparison of Options

2.1.1 Step 1: Problem Definition: Energy and Resource Efficiency

Besides general objectives of world-wide sustainability targets, such as mitigating global warming by reducing greenhouse gas and air pollutant emissions as well as the dependence on fossil fuels, other technology specific aspects have to be tackled. For CHP using renewable energy sources, this applies especially to constraints regarding the availability of biomass feedstocks.

Global trends concerning population, crop yields, diet, climate change, etc. usually suggest an expansion of cropland – if at all – only for the purpose to feed the world population. Further land requirements for dedicated energy crops would come on top, whereby the sustainable availability of arable land is the essential limiting factor. If land is converted from natural habitats to agricultural areas, there is significant risk for severe biodiversity loss as well as other negative environmental impacts. For example, if major carbon sinks, such as forests, grass- and peatlands are destroyed to provide space for cultivation, further negative consequences on greenhouse gas balances are the inevitable effect. As long as the overall demand for cropland grows for the needs of food production, any land use for crop production for material or energy purposes will lead to additional direct and indirect land use change [3]. If not strictly controlled, this might also lead to unintended and inefficient long-haul fuel transports, e.g. from tropical countries, where conditions for cheap feedstock production are most favourable. Availability of water is another limiting factor for growing biomass feedstocks, both in terms of quality and quantity, as agriculture already uses about 70% of fresh water globally [3]. Any expansion of intensive energy crop cultivation would be adding to this. In particular in water scarce regions, this may lead to another form of competition with food production. Thereby, extreme weather events due to climate change might further increase uncertainties in terms of available water resources.

The above-mentioned exemplary environmental impacts related to the 'water, energy and food nexus' apply to many 'first generation biofuels'. However, there are also new pathways for more sustainable production and alternative use of biomass for energy purposes that can help to reduce potential pressures on the environment.

2.1.2 Step 2: Define Objectives: Efficiency First

Overall, demand side energy efficiency should provide the 'first fuel' for any future economic development [4]. On the supply side, any use of fuel also for renewable biomass, should be as efficient as possible. In this context, in particular energy recovery from waste and residual biomass can save significant GHG emissions without requiring additional land use change. Specifically, the inevitable part of municipal organic waste and residues from agriculture as well as forestry provide significant energy potentials, which are still largely untapped worldwide. In the same vein, the cascading use of biomass to produce (construction) material first, then recovering the energy content of the resulting waste, can further maximize the carbon dioxide (CO_2) mitigation potential of biomass.

Thereby, comprehensive further research is still required, especially concerning the proper balance of residues remaining on-site for soil fertility and removal for energy provision, as well as with regard to nutrient recycling e.g. by ash utilization. Nevertheless, promising approaches exist or are under development to maximize benefits and to minimise negative environmental effects. In this context, the presented final results from the 'FlexiFuel-SOFC' project concentrate on the principles for efficient use of solid biomass fuels from agricultural or forestry residues in small-scale CHP systems based on B-IGFCs during the operation phase, when energy efficiency and pollution control during energy recovery has to be addressed in particular.

2.1.3 Step 3: Develop Options: System Application Cases

Before starting an impact assessment, framework conditions have to be established, in particular the geographical scope (e.g. EU-28) and time horizon (e.g. 2050) of the analysis. Based on market studies and techno-economic analyses, the most promising fields of application for the new technology need to be defined. For the analysed systems, decentralised operation close to fuel feedstocks is envisaged to avoid increasing levels of transportation of biomass with market penetration, which could otherwise offset emissions reduction benefits to a certain extent. Accordingly, based on the results of the 'FlexiFuel-SOFC' project, the following specific application cases have been identified for the European market (with focus on Central Europe):

- Application A is a system with about 70 kW_{th} nominal heat output and 20 kW_{el} electric power at nominal load to be used typically for base load heat and electricity production for small district heating networks (micro grids). It can also be applied to hotels, hospitals, or enterprises with permanent electricity and heat demand over the whole year. It uses olive stones (Application A1) or wood chips (Application A2) as biomass solid fuel and is characterized by 8000 effective full load hours annually for electricity generation.
- Application B is a system with about 21 kW_{th} nominal heat output and about 5 kW_{el} electric power at nominal load, to be used typically for space and process heating as well as domestic hot water supply for large apartment buildings or public buildings with a buffer storage system. Olive stones (Application B1) or wood chips (Application B2) are used as the fuel. The system is optimized for heat-controlled operation (electricity and heat production in winter and transitional period; heat supply without electricity production in summer). It is characterized by 4000 effective full load hours annually for the electricity generation part.

For each of the application cases, the new FF-SOFC technology is compared to state-of-the-art technologies that have the same nominal heating capacity. Four different technologies or technology combinations are modelled and compared to the new FF-SOFC technology. Due to the envisaged decentralised operation and consumption strategy no general limitations in terms of electricity grid feed-in capacities are assumed.

- Biomass boiler with grid electricity (BBwGRID): In this scenario a biomass boiler is employed for heat production, while the electricity demand is supplied from the grid. The biomass boiler has a nominal heating capacity of about 70 kW_{th} (Application A) and 21 kW_{th} (Application B), respectively. In the scenarios, the biomass boiler is either fuelled with olive stones (Applications A1 and B1) or wood chips (Applications A2 and B2).
- Biomass fired small scale CHP (BCHP): In this scenario a small biomass CHP is used to produce electricity and heat. The CHP has a nominal heating capacity of about 70 kW_{th} (Application A) and 21 kW_{th} (Application B), respectively. The gross electric capacity is 31 kW_{el} (Application A) or about 9 kW_{el} (Application B). The biomass CHP is fuelled with wood chips across all application cases (Applications A1, A2, B1 and B2).
- Natural gas fired CHP (NGCHP): Likewise, a small CHP is used to produce electricity and heat. The CHP has a nominal heating capacity of about 70 kW_{th} (Application A) and 21 kW_{th} (Application B), respectively. The gross electric capacity is about 42 kW_{el} (Application A) or about 9 kW_{el} (Application B). The natural gas fired CHP is fuelled with natural gas across all application cases (Applications A1, A2, B1 and B2).
- PuroWIN with photovoltaics (PWINwPV): In this scenario, the Windhager PuroWIN ultra low emission boiler is used to supply heat. The PuroWIN boiler is combined with a photovoltaic system that supplies electricity to the boiler. Surplus electricity is fed into the grid. The boiler has a nominal heating capacity of about 70 kW_{th} (Application A) and 21 kW_{th} (Application B), respectively. The photovoltaic system is seized to have a gross electric capacity of 144 kW_{el} (Application A) and about 22 kW_{el} (Application B). The PuroWIN boiler is either fuelled with olive stones (Applications A1 and B1) or wood chips (Applications A2 and B2).

The environmental performance parameters from the FlexiFuel-SOFC technology have been compared within an environmental performance analysis with respective data from other state-of-the-art systems in order to evaluate and quantify the relative performance and improvement potentials of the new technology on a single product level. Results for Total Suspended Particles (TSP), also referred to as 'total dust', and energy efficiency (%, based on fuel input in terms of net calorific value 'NCV'/lower heating value (LHV) / lower calorific value (LCV), as well as combined useful heat and electricity output) are presented in Figs. 1 and 2.

Taking the available results from the FF-SOFC project into account, the environmental performance analysis shows significant technical emission saving potentials of the FF-SOFC, PuroWIN and CHP scenarios compared to the standard biomass boiler. Though the absolute differences are small and, due to rounding, cannot be seen in Figs. 1 and 2, the TSP emission intensity of the new FF-SOFC technology is lower than those found for the biomass fired CHP and the PuroWIN boiler. The large difference between the emission intensity of the biomass boiler between

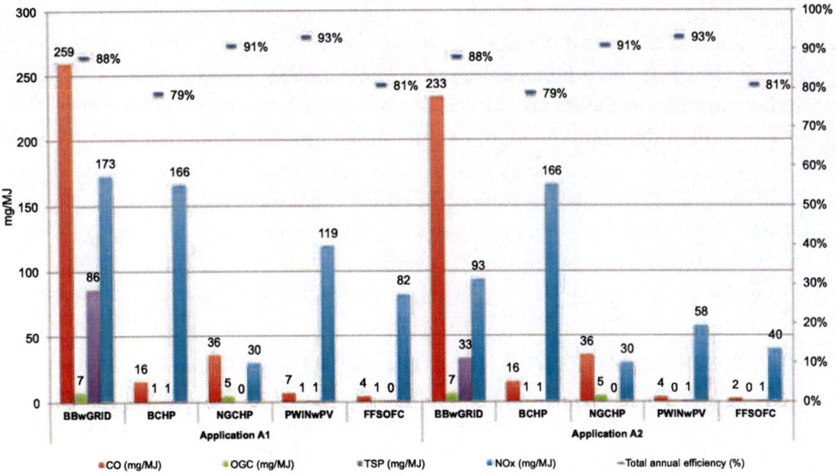

Fig. 1 Application A emission factors and energy efficiency compared
Source: Own illustration

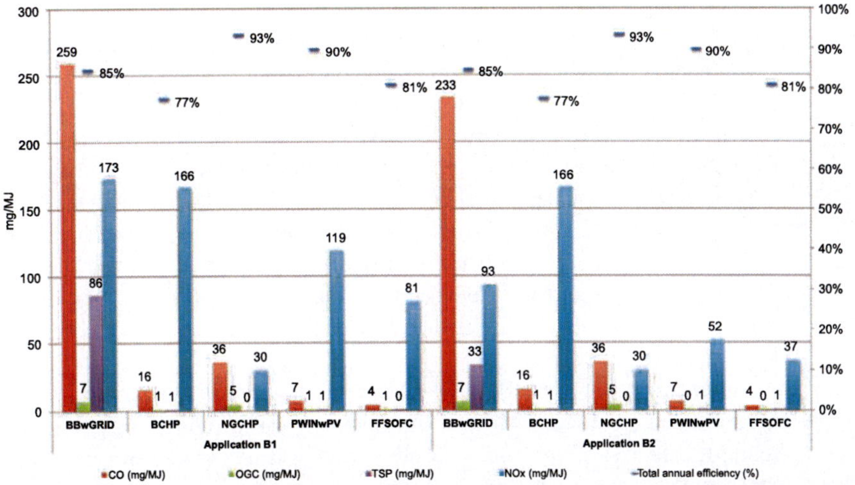

Fig. 2 Application B emission factors and energy efficiency compared
Source: Own illustration

Applications A1 and A2 and Applications B1 and B2, respectively, can be explained by the fuel type (olive stones in the case of Applications A1 and B1, wood chips in the case of Applications A2 and B2). For other air pollutants, such as organic gaseous compounds (OGC) and carbon monoxide (CO), the FF-SOFC technology is associated with considerably lower emission intensities than the biomass boiler as well as the biomass and natural gas fired CHP. The FF-SOFC also constitutes an improvement compared to the PuroWIN boiler, though the improvement is not as large as those of other technologies.

Accordingly, the FF-SOFC technology has the potential to reach considerable on-site emission reductions in a short time period if broad market diffusion rates can be achieved in the future. Based on the current results and input data it can also be concluded that even stringent future Emission Limit Values (ELVs) in the EU for new installations (e.g. as part of a future revision of the EU Ecodesign Lot15 regulation on solid fuel boilers), should be no constraint for the FF-SOFC system. The environmental performance analysis also revealed that in contrast to the very low emission levels, there remains a further technical optimisation potential especially for the total annual efficiency levels of FF-SOFC technology. Nevertheless, even now, the FF-SOFC technology constitutes a small, but considerable, improvement compared to the biomass fired CHP. The biomass fired CHP is an adequate case for comparison as it also co-generates heat and power. For the other biomass technologies, the total annual efficiency of the FF-SOFC is lower than those of the biomass boiler and, even more so, the PuroWIN boiler. This can be explained by the more complex CHP system operation of the FF-SOFC compared to the heat only operation of the biomass boiler and PuroWIN boiler, respectively. Nevertheless, further increasing the total annual efficiency of the FF-SOFC will remain a major goal of future research and development.

Based on the previous definitions and findings, for the macro-scale EU wide impact assessment, five technology scenarios have been modelled. In every technology scenario the demand for heating systems estimated in the market study is supplied exclusively by the technology giving the scenario its name, i.e. in the FF-SOFC scenario every heating system sold for the application case under consideration is a FF-SOFC appliance. As such, the scenarios provide insights into 'extreme' pathways with 100% of sales switch to a given technology. This approach has the advantage of only requiring total sales and stock data for each application case, and no market share split is required at this stage. This provides upper and lower limits for the available corridor for the emission saving potential, which is especially relevant for new technologies that are still under development, and for which only very preliminary data is available.

As an essential part of this step, the application of a dynamic stock model is needed to calculate scenarios for the development of future stock sizes for the different technologies. Generally, based on market study data, stock data can be computed with a sales-driven model combining the last known or reconstructed stock volume data for applications, historical and expected total sales, and average life spans for the different applications and system components. Stock data are calculated successively for each year of the simulation period, using classical stock dynamics model equations. Accordingly, as relying on preliminary values for several key parameters (such as emission intensities) of a technology still under development, this assessment does not seek to dwell into every last detail regarding absolute amounts, e.g. of emitted pollutants. The emphasis is put on comparing the general dynamics of different options and explaining the results for model calibration with the aim to give recommendations regarding the general future technical or policy evolution. The stock is assumed to have a size of zero in the first year of the period under consideration (2023). The model assumes that appliances that are decommissioned after their lifetime will be replaced by a new appliance of the same type.

2.1.4 Step 4: Impact Analysis

The analysis in step 4 quantitatively evaluates the operation-related impacts of the options identified in the previous step. Each technology option is modelled on its own and then compared to the other technologies. Following this approach, the most relevant impact categories and associated indicators for the analysis are presented.

Air emissions are the most pertinent use-phase environmental indicators for B-IGFCs, including CO_2 as the relevant greenhouse gas. Regarding harmful emissions, particulate matter (PM), given in this paper as 'Total Suspended Particles' (TSP), organic gaseous compounds (OGC) and carbon monoxide (CO) are parameters typically addressed for biomass combustion systems as well as by related standards and regulations. The derived absolute emission levels depend on assumed stock volumes and product lifetimes, which dictate the pace of (re-)investment cycles. Total annual efficiencies determine fuel requirements for a given energy output. The fuel type is an important influencing factor in terms of combustion processes and technology requirements, as the two fuel types (i.e. olive stones and wood chips) considered in the model do amount to different air pollutant emission intensities.

As peculiarity for CHP systems, the emissions need to be treated as a combined result of direct on-site fuel combustion and grid electricity effects. Avoided grid electricity consumption is taken into account by subtracting the product of the grid's emission intensity and the gross electricity generation of the CHP or PV from the emissions caused by burning the fuel. The following results are calculated with basic emission values per fuel type for solid fuel combustion and average emission intensities per type of electricity generation of conventional power generation in Europe [5, 6] (GHG emission intensities do include Life Cycle Analysis (LCA) aspects for fuel processing and transportation). Due to this method, all technology scenarios except for the biomass boiler with grid electricity scenario may lead to negative emissions if the emissions avoided by not consuming grid electricity overcompensate the emissions caused by fuel combustion. In the case of non-GHG emissions, every technology scenario that implies an on-site electricity generation results in negative net emissions (see Figs. 3–6, exemplarily for Application A2). The biomass boiler with grid electricity and the natural gas CHP scenarios produce positive net GHG emissions, while the other scenarios lead to negative net GHG emissions (see Fig. 7). This is due to the GHG emission intensity of grid electricity being significantly higher than the GHG emission intensity of the solid biomass fuels even in the year 2050 and under consideration of the life cycle. A side effect of this is that scenarios which model technologies with a higher gross electrical capacity, and consequently generating more electricity, lead to lower net GHG emissions. The biomass fired CHP has a gross electric capacity that is more than 10 kW_{el} (Application A) and about 4 kW_{el} (Application B) higher than the

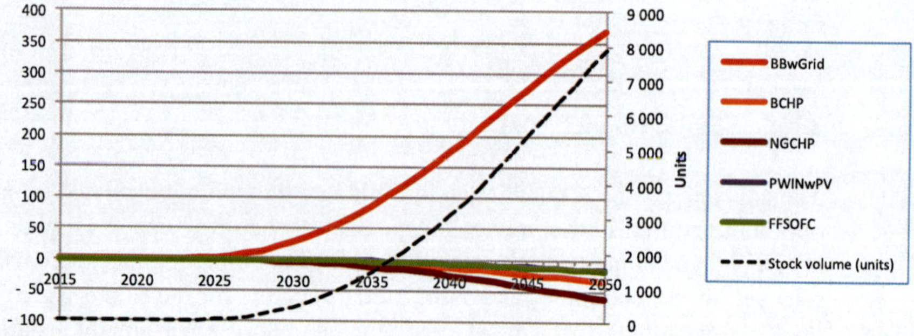

Fig. 3 Total net TSP emissions (t) and stock volume (units), EU-28, Application A2
Source: Own illustration

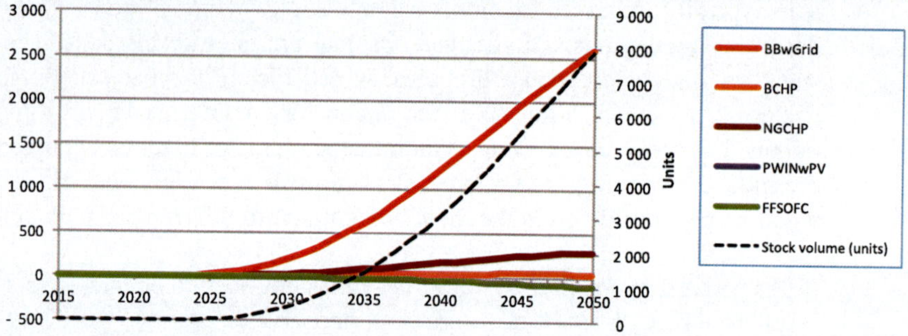

Fig. 4 Total net CO emissions (t) and stock volume (units), EU-28, Application A2
Source: Own illustration

FF-SOFC. Accordingly, the net GHG emissions seen in the BCHP scenarios are lower than those seen in the FF-SOFC scenarios.

An important driver of future fuel consumption and, consequently, emissions is the future development of energy demand per building. Presuming all other aspects being equal, the total stock emissions may decrease in the long run even with an increasing stock. The model assumed that the typically required nominal output of heating appliances will decrease as expected effect of improved insulation and energy performance of buildings (e.g. in Europe, based on the European Performance of Buildings Directive 'EPBD' [7]). Consequently, this would mean that less fuel input per unit is required, resulting directly in less fuel related emissions.

2.1.5 Step 5: Comparison of Options

Based on the market study, Application B has a market potential about twice as high as Application A. Yet, due to the larger nominal system capacity and a higher number of annual full load operating hours, absolute values (irrespective of the sign) for fuel and grid electricity consumption, as well as air pollutant and GHG emissions

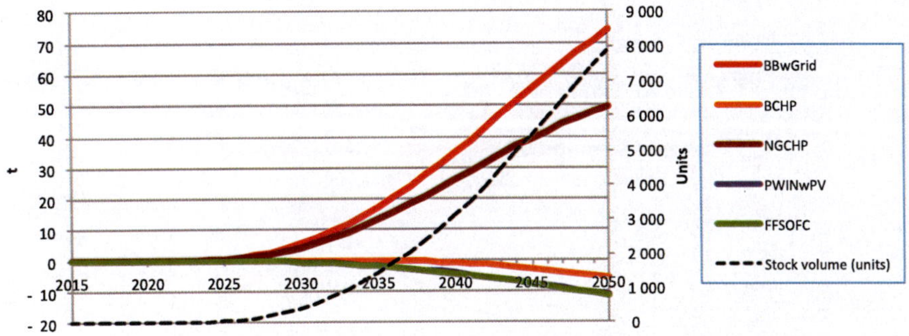

Fig. 5 Total net OGC emissions (t) and stock volume (units), EU-28, Application A2
Source: Own illustration

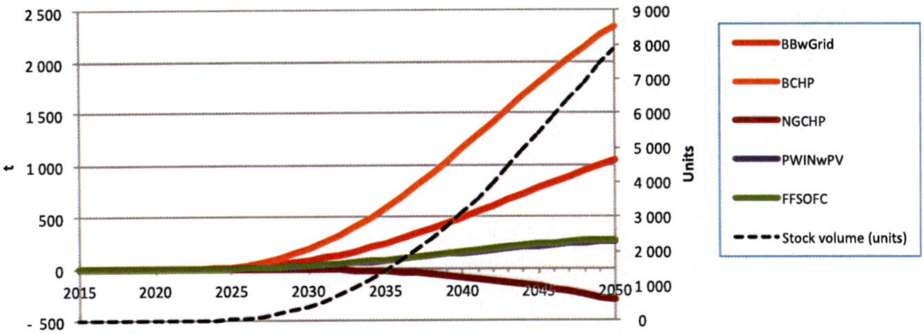

Fig. 6 Total net NO_X emissions (t) and stock volume (units), EU-28, Application A2
Source: Own illustration

tend to be considerably higher in the Application A scenarios. Comparing the technology options with each other, one finds that, when avoided emissions are considered in the calculation of net emissions, the FF-SOFC technology allows for net negative GHG emissions. This distinguishes the FF-SOFC scenario from the biomass boiler with grid electricity and natural gas fired CHP scenarios, which show positive net GHG emissions. At the same time, the PuroWIN boiler with photovoltaics and the biomass fired CHP scenario lead to even lower (i.e. more negative) emissions. This is mainly due to a higher amount of electricity being generated by these technologies for a given heat demand. All results are dependent on the development of the emission intensity of grid electricity assumed for the future.

As with GHG emissions, air pollutant emissions are calculated by adding the emissions caused by on-site solid fuel combustion to the emissions caused off-site by generating grid electricity. If electricity is generated on-site by a CHP or PV, the emissions avoided by replaced grid electricity generation are added to the emissions caused by fuel combustion. Consequently, air pollutant emissions become negative when avoided off-site emissions overcompensate on-site emissions. The scenarios with the new FF-SOFC technology lead to air pollutant emissions that are lower or similar to those seen in the scenarios with the state-of-the-art biomass fired

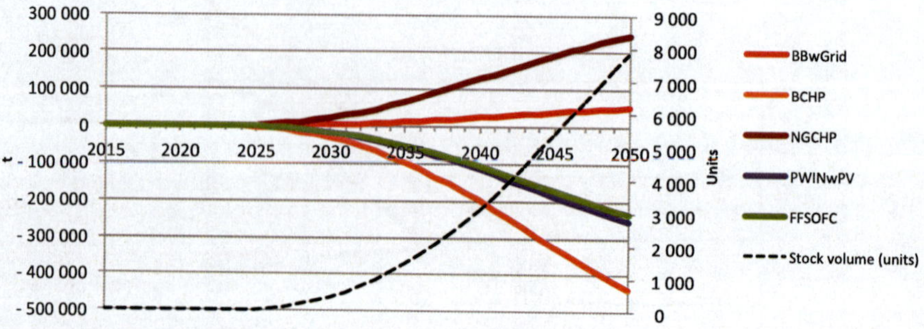

Fig. 7 Total net GHG emissions (t) and stock volume (units), EU-28, Application A
Source: Own illustration

technologies. Compared to a state-of-the-art natural gas fired CHP, the FF-SOFC scenario produces higher NO_X emissions (due to the fuel's higher nitrogen content), slightly higher TSP emissions, and lower CO and OGC emissions.

Besides the environmental aspect, (fuel) efficiency is a very important parameter, both regarding energy costs and energy security. Compared to the most similar technology options, the biomass fired CHP, the FF-SOFC technology has a slightly higher total annual efficiency. As there is no general financial reward for the superior emission reduction performance of the ultra low emission FF-SOFC technologies at the moment, higher efficiencies and thus lower energy costs may become an important selling point. Furthermore, efficiency is typically also the most relevant ranking criteria when regulators implement Minimum Energy Performance Standards (MEPS). The same applies to product energy labels, which allow a better visibility and the active promotion of innovative technologies. Related to this, also the selection for product-specific incentive programmes to foster a voluntary early retrofit or replacement of old installations with much better new products depends usually on (very) high efficiency levels. The high relevance of such policies and measures for CHP is therefore further addressed in the following section.

3 A Policy Package for Small-Scale CHP Technologies

The new and highly efficient FlexiFuel-SOFC technology is distinguished by high electric and overall efficiencies as well as almost zero emissions. However, many barriers (financial barriers, information barriers, regulatory barriers etc.) still exist and market forces alone are often unlikely to initiate broad market diffusion. Even if well thought-out and long-term business plans are available, it is often challenging for companies to sell a new technology to a large number of customers. Therefore, policy is needed to overcome these barriers and to exploit the existing high potential. That is why policy makers have to develop adequate strategies to influence the market development. Experience shows that several instruments need to interact and reinforce each other in a comprehensive policy package. Every policy is tailored to overcome one or a few barriers, but there is no one-fits-all policy [8]. Due to scope limitations in this paper, it is not possible to describe and analyse all relevant policies and measures in detail. Therefore, the next section focuses on a few key instruments at EU level.

3.1 A Policy Package on EU Level

Currently, there are several efforts in the European Union to implement policies and measures to address cogeneration. The aim is to enhance the technology, increase energy efficiency, and improve air quality. Two Directives with a high influence on the development of cogeneration systems are the Renewable Energy Directive (RED) and the Ecodesign Directive. Other Directives primarily address air quality or influence cogeneration indirectly, e.g. by setting higher energy prices. The Energy Efficiency Directive (EED) strongly influences cogeneration as the CHP-Directive 2004/8/EC was repealed in 2014 and replaced by the EED. Figure 8 illustrates the policy package on EU level for cogeneration with a focus on three directives. It is important to notice that this is not to say that other policies such as the Energy Performance of Buildings Directive (EPBD) and the Energy Labelling Directive do not have a relevant direct or indirect influence on the technology. In the scope of this paper, however, the focus is on the illustrated instruments.

The Ecodesign Directive 2009/125/EC [10] (also referred to as 'Energy related Products' or ErP Directive) forms a framework to set minimum performance requirements for specific product groups. Several Ecodesign implementing measures address cogeneration, e.g. regulation 2015/1189 applies to solid fuel systems with a nominal heat output of 500 kW_{th} or less as well as to solid fuel cogeneration boilers with an electrical capacity of less than 50 kW_{el}. In addition to energy performance criteria, the regulation entails emission limits for PM, OGC, CO and NO_X. Both energy and emission standards depend on the boiler's rated heat output, on the boiler type (automatically or manually stoked) or on the fuel type (biomass or fossil fuel). Manufacturers have to meet the standards, which will become valid as of January 2020. Some types of boilers are excluded from current regulation including non-woody biomass boilers, which includes fuels such as straw, grains, olive stones, or nut shells. Other exceptions include boilers generating heat exclusively for providing hot drinking or sanitary water; boilers for heating and distributing gaseous heat transfer media such as vapour or air (Commission Regulation 2015/1189, Art. 1, 2). A review of Commission Regulation 2015/1189 is scheduled prior to 1 January 2022. For the long-term perspective, the review will include an assessment to set potentially stricter requirements beyond 2020 for energy efficiency and emissions regarding PM, OGC and CO. In addition, the review will factor in whether it is appropriate to also regulate "non-woody biomass boilers with Ecodesign requirements for their specific types of pollutant emissions." Ecodesign can be considered as a driver for innovative technologies such as

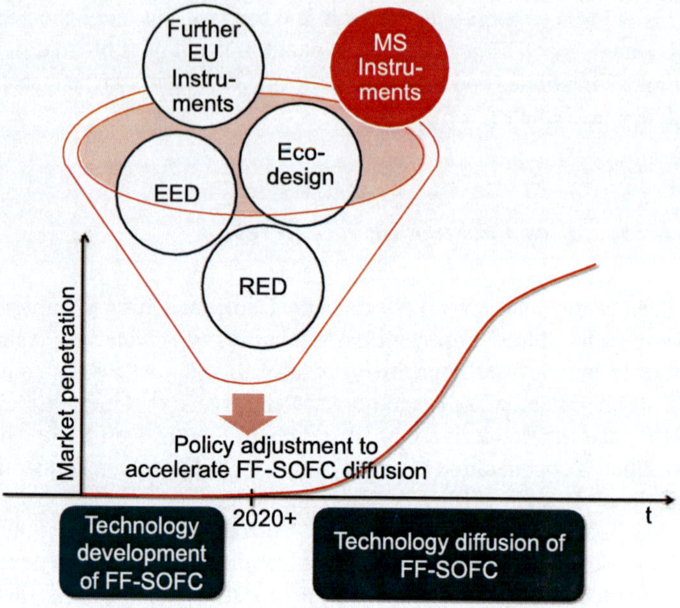

Fig. 8 Towards a policy package for fuel flexible CHP technology
Source: [9]

FlexiFuel-SOFC. Only highly efficient solid-fuel boilers meeting the standards are allowed to be put onto the EU single market. However, as FlexiFuel-SOFC boilers can also make use of non-woody biomass, which remains unregulated under the current Ecodesign implementing measure, there remains some market uncertainty.

Besides Ecodesign, the Renewable Energy Directive 2018/2001/EU (RED) builds a general policy framework [11]. The RED contributes to facilitating the use of renewable energy for heating by establishing objectives to expand the share of renewable energy (including biomass) in the energy mix of Member States. Generally, the RED may draw increasing attention of policy makers to the FlexiFuel-SOFC technology as a means to promote the use of renewable energy in the heating sector. In January 2019, the RED was revised (RED II). It provides, among other things, a new target of 32% renewable energy in final consumption for the EU by 2030 and a sub-target of an indicative 1.3% yearly increase of renewable energy in heating and cooling installations, calculated on a period over 5 years starting in 2021. RED II does also set sustainability criteria for solid biomass.

Another relevant Directive is the Energy Efficiency Directive (EED), which seeks to ensure the EU headline target on energy efficiency, paving the way for more efficient heating technologies in residential buildings [12]. The EED requires Member States to develop strategies for the implementation of high-efficient CHP and district heating. The Directive was revised in December 2018. The recast includes a new energy efficiency target for the EU for 2030 of 32.5%, with an upwards revision clause by 2023.

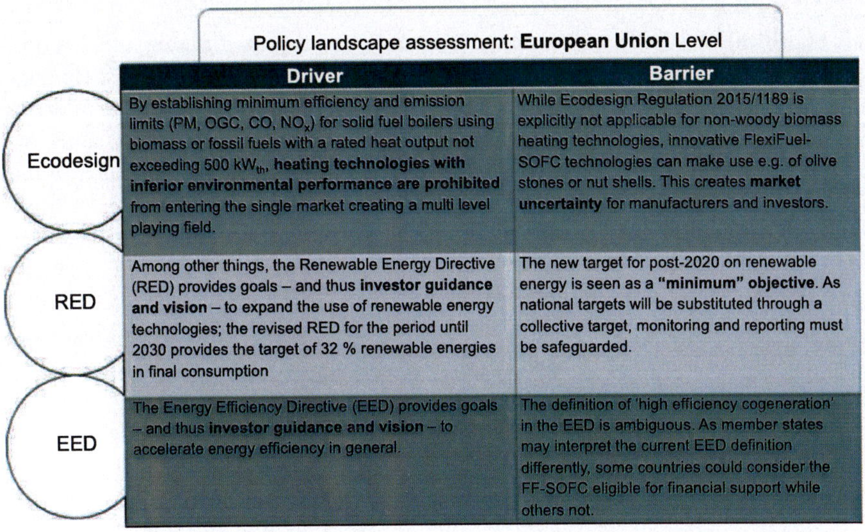

Fig. 9 Drivers and barriers of selected policies
Source: [9]

3.2 *Drivers and Barriers of Selected EU Policies*

All the mentioned regulations constitute both positive drivers as well as negative barriers for uptake of the FlexiFuel-SOFC technology. Figure 9 gives a brief overview of these drivers and barriers.

4 Conclusions and Outlook

This paper provides insights into the most essential environmental aspects of small-scale gasifier fuel cell CHP systems, based on final environmental impact assessment results for the technology developed in the EU Horizon 2020 project 'FlexiFuel-SOFC'.

Two specific application cases were investigated, which represent the most promising fields of application for the new technology in the European market: Application A for hotels, small enterprises, hospitals and small district heating networks; Application B for public buildings and multi-family homes. Furthermore, for each of the application cases, the new FF-SOFC systems are compared to four state-of-the-art technology options: A biomass boiler (with electricity demand supplied from the grid), a biomass fired CHP, a natural gas fired CHP and the Windhager PuroWIN boiler that is supplemented by a photovoltaic system for electricity generation. Thereby, on-site air emissions as well as grid electricity consumption effects have been jointly taken into account. The impact assessment for the

FlexiFuel-SOFC demonstrates general dynamics of different application cases and their sensitivities to technical parameters and other modelling assumptions. This modelling shows the impacts a wide-scale introduction of the FlexiFuel-SOFC technology might mean for air pollutant emissions, GHG emissions and thereby supports decision-making processes regarding the general direction of the technology development and for policy making.

The presented results clearly identify the main emission drivers for the different technologies considered. In all scenarios, greenhouse gas (GHG) emissions are driven by grid electricity consumption effects. Since the CHP technologies and the PuroWIN boiler with PV technology package generate their own electricity, avoided off-site grid electricity emissions quickly overcompensate direct on-site emissions from fuel usage in these scenarios, meaning that net GHG emissions are negative. For air pollutant emissions, a similar overcompensation of on-site emissions by avoided off-site emissions is seen for certain technologies and air pollutants as well. Compared to the other scenarios with biomass-firing technologies, the FF-SOFC technology scenarios lead to lower or at least similar air pollutant emissions. It has to be mentioned that such results are sensitive to several crucial technical parameters of the FF-SOFC technology, such as e.g. emission intensities of the different solid fuels used by the application cases. Furthermore, assumptions regarding the future development of EU grid electricity emission intensities and heat energy demand (driving thermal output, hence fuel requirements) are also very relevant for the overall behaviour of the model.

Additionally, the gained results have to be set in relation to policy developments addressing the CHP sector. Especially for smaller CHP systems targeting the residential sector, policies that address the typically required size of heating appliances may considerably affect the development and usage profiles of such systems. Regulation, such as the Ecodesign Directive, the Energy Efficiency Directive and the Renewable Energy Directive further incentivise the market of residential scale B-IGFC systems. Overall, in this context FF-SOFC systems may provide one of the essential key technologies for efficient decentralised power and heat generation based on renewable energy sources to pave the way towards a decarbonized energy system.

Acknowledgements The research leading to the presented results has received funding from the European Union Horizon2020 Programme (H2020/2014-2020) under Grant Agreement n° 641229. The authors also express their gratitude to the other project partners and all the members of the consortium for their good cooperation.

References

1. Biomass-Based Fuel Cell Systems. (2016). Efficiency of Biomass Energy, 633–672. doi:https://doi.org/10.1002/9781119118169.ch16
2. EC (2015). *Guidelines on Impact Assessment, gathering and analysing evidence to support policy making*, http://ec.europa.eu/smart-regulation/guidelines/ug_chap3_en.htm
3. UNEP (2009): Bringezu, S.; Schütz, H.; O'Brien, M.; Kauppi, L.; Howarth, R. W.: *Assessing biofuels: towards sustainable production and use of resources*; UNEP Nairobi, United Nations Environment Programme, 2009. http://www.unep.fr/scp/rpanel/pdf/Assessing_Biofuels_Full_Report.pdf
4. IEA (2016): Energy Efficient Prosperity: *The 'first fuel' of economic development*, International Energy Agency, Paris, 10 October 2016, https://www.iea.org/newsroom/news/2016/october/energy-efficient-prosperity-the-first-fuel-of-economic-development.html
5. Kemna, R. et al. (2019): *"Ecodesign Impact Accounting - Status Report (2018)"*, prepared by VHK for the European Commission, p. 171, published January 2019, Delft, The Netherlands
6. Own calculation based on European Environment Agency data on air pollutant emissions (LRTAP convention) and energy balances compiled by Eurostat
7. *Directive 2010/31/EU on the energy performance of buildings*. http://eur-lex.europa.eu/LexUriServ/LexUriServ.do?uri=OJ:L:2010:153:0013:0035:EN:PDF
8. bigEE (NA): How policies need to interact in packages. bigEE recommendations on which measures to combine for effective policy
9. Goetz, Thomas; Adisorn, Lena; Tholen, Lena; Moser, Raphael (2018): Evaluating EU's legislative framework for fuel-flexible and highly efficient biomass heating. Poster presentation at the International Energy Policy & Programme Evaluation Conference
10. *Directive 2009/125/EC establishing a framework for the setting of ecodesign requirements for energy-related products*. http://eur-lex.europa.eu/LexUriServ/LexUriServ.do?uri=OJ:L:2009:285:0010:0035:en:PDF
11. *Directive 2018/2001/EC on the promotion of the use of energy from renewable sources*. https://eur-lex.europa.eu/legal-content/EN/TXT/PDF/?uri=CELEX:32018L2001&from=DE
12. *Directive 2018/2002 on energy efficiency*. https://eur-lex.europa.eu/legal-content/EN/TXT/PDF/?uri=CELEX:32018L2002&from=EN

Mainstreaming Energy Productivity Lending Practices among Indian Financial Institutions (FIs)

Janardhana Anjanappa and Juthathip Jongwanich

Abstract A number of studies have addressed issues on Energy Efficiency (EE) pertaining to improving productivity, employment generation, and increased energy security. Investing EE reduces demand for electricity generation, reduces fiscal deficit and energy deficits. In this context, very few studies focused on scaling up EE efforts, the present approach is fragmented and lack holistic approach to make finance available from financial institutions. Therefore, this study makes an effort to promote mainstreaming energy productivity lending practices among financial institutions through five voluntary principles framework. The proposed framework is in line with India's National Mission on Enhanced EE (NMEEE) as a part of the National Action Plan on Climate Change (NAPCC). The outcome of this study will provide a boost to the country's efforts on enhancing EE efforts by strengthening national action plans, strategies and policies. Thus, a greater collaboration, mitigating institutional barriers and develops a favorable environment for the fiscal and regulatory landscape to attract investments for EE efforts.

Keywords Energy efficiency · Five voluntary principle's

1 Introduction

Climate Change (CC) has been shown as having a potential to affect all sections in the economy. At present, it is a quite challenge for countries to allocate resources effectively, so that economies and societies can be placed on climate resilient pathways. In developing countries there are a number of national policies and institutions which address Climate Change (CC). According to the Intergovernmental Panel on Climate Change (IPCC) estimates, 67% of global GHG emissions are now covered by some form of legislation or national strategy, but these policies are under early stages of implementation [1]. To keep global temperature below 2 °C, there is

J. Anjanappa (✉) · J. Jongwanich
Faculty of Economics, Thammasat University, Bangkok, Thailand
e-mail: juthathip@econ.tu.ac.th

P. Bertoldi (ed.), *Energy Efficiency in Domestic Appliances and Lighting*, Springer Proceedings in Energy, https://doi.org/10.1007/978-3-030-79124-7_11

a need of additional investment USD 190–900 billion per year through to 2050 in the energy sector alone [2]. The fifth review of the financial mechanism of UNFCCC, suggests that in order to limit temperature increase from pre-industrial level to no more than 2 °C, it is essential to consider pattern of investment [3]. Thus, there is need for different economic actors and climate friendly models. Therefore, decision made by International Financial Intuitions (FIs) will play a major role in addressing issues related to CC and overall development and growth in developing countries.

Considering the above, public finance institutions like Global Environment Fund (GEF), Climate Investment Fund (CIF) and Asian Development Bank (ADB) are investing in climate change projects and programs. The GEF is an operating entity of the financial mechanism of the UNFCCC and the Cooperation of Parties (COP) advocates GEF for appropriate future work. The investment criteria of GEF emphasize on global environment problems, global environmental benefits and its potential to scaling up the project.[1] Similarly, CIF main investment criteria are based on long-term GHG mitigation potential from the project [4, 5]. The policies and procedures pertaining to the design and implementation of projects and programs rely on Multilateral Development Banks (MDBs). On the other hand, regional MDBs such as Asian Development Bank (ADB) provides investment based on country partnership strategy and country operation business plan and in-house feasibility study on sector, technology and business model(ADB).[2] Evidence suggests that there is biasness towards supply-side investment compared to demand-side efficiency projects.[3]

Public-sector FIs are making efforts to mainstream EE financing initiatives across the institutions but the scale of efforts varies across the institutions [6]. In contrast, private-sector FIs found little evidence of dedicated activities towards mainstream Initiatives and commercial finance is in the early stage of developing innovative financing methods. This is due to lack of time and resources to assess opportunities and develop appropriate financing products and EE financing operations are relatively costly and time-consuming to develop and implementation made less attractive [7, 8].

Hence, there is a missing component of institutional strategy and policies across the FIs board and assessment of impact of CC on its lending portfolio (risk from climate) is not systematically mainstreamed within institutional operations [9]. Though, there is a huge potential for scaling up these efforts with programs and projects, but need to address barriers such as lack of external marketing to clients, internal marketing across other parts of financial institutions, lack of track record of such as financing projects among public institutions, and lack of symmetry in information sharing. All in all, a mixed picture indicates that EE improvements remain relatively low priority for FIs [10]. The lack of cohesive and comprehensive energy productivity policies within in FIs will impact economy by increase in energy used in long-term and thereby increase in GHG emissions [11, 12]. Therefore,

[1] GEF, Annual Report of various years.

[2] ADB, Annual Report of various years.

[3] Project database of GEF, CIF and ADB.

mainstreaming strategies and policies within FIs will play effectively role in identify risk and opportunities to mitigate risks in the economy.

This is true in the Indian context as well where the investments in EE fall behind its true potential. Investing in EE is vital to increase the economic productivity and competitiveness, increase energy security, mitigate GHG emissions and reduce costs in short-term to medium term [13]. As per Asian Development Bank (ADB) investment in EE as an overall energy sector investment, can meet 25% of the projected increase in primary energy consumption in developing Asian countries by 2030. India clearly has a huge potential for cost-effective investments in EE and as per ADB estimates about 4.5 billion USD required per year through 2020 in order to meet India? Established national energy-saving targets [14]. Though, there is huge potential for investment at national and state level but rate of investment in EE technologies and projects is lagging behind its potential. One of the key reasons is financial barriers. The experience from using financial instruments and mechanisms for EE, at the national and state levels, and with international donor support suggests that there is a need for significantly scale up financing for EE technologies and projects. But, new or innovative mechanisms are needed to help realize the country's full potential for EE [15].

2 Literature Review

Investing in EE is vital to increase the economic productivity and competitiveness, increase energy security, mitigate GHG emissions and reduce costs in short-term to medium term [13]. As per Asian Development Bank (ADB) investment in EE as an overall energy sector investment, can meet 25% of the projected increase in primary energy consumption in developing Asian countries by 2030.India clearly has a huge potential for cost-effective investments in EE and as per ADB estimates about 4.5 billion USD required per year through 2020 in order to meet its established national energy-saving targets [14]. Though, there is huge potential for investment at national and state level but rate of investment in EE technologies and projects is lagging behind its potential. This is due to lack of external marketing to clients, internal marketing across other parts of financial institutions, lack of track record of such as financing projects among public institutions, lack of uniformity of EE finance across FIs and lack of symmetry in information sharing. Further, EE financing operations are relatively costly and time-consuming to develop and implementation made less attractive. The experience from using financial instruments and mechanisms for EE, at the national and state levels, and with international donor support suggests that there is a need for significantly scale up financing for EE technologies and projects. But, new or innovative mechanisms are needed to help realize the country's full potential for EE [15].Fostering EE efforts are vital to meet target of India NDC under UNFCCC to keep global temperature well below 2 °C by end of the century. Enhancing EE remains one of the cheapest options to produce energy, as the efficiency of many energy systems has a large scope for improvement. As these options plays an important part in enhancing the country energy security. In addition provides other co-benefits towards economy by reducing air pollution, cost and energy savings and provides employment opportunities. This is true in the context of India as well, where rapid urbanization and increase in per capita income has increased the demand for various energy products and services among Households (HHs). As Indian stands out as one of the fastest growing economies in the Asian region due to its rapid economic development, high population, rapid urbanization and large industrial sectors.

As per IEA database,[4] over a time, the India's net energy imports has increased from 149.0 Mtoe to 330.0 Mtoe between 2007 to 2017, the energy imports constitutes composition of oil, coal and natural gas. During the same period total CO_2 emissions has increased from 1261.0 Mt. to 2162.0 Mt. There by, investing in EE initiatives is a good option to reduce burden of energy imports for India and meet NDC target. Hence, EE has a potential to anchor efficient and cleaner technologies in the supply-chain covering production, distribution and consumption.

EE is one of the approaches to drive low carbon energy transition that leads to low carbon societies at low cost as these measures and technologies delivers range of benefits. As this is the important objective for the Government of India for meeting target of NDC within the framework of Energy Conservation Act 2001. In line with this Act it has prioritized several national programs. As these demand side measures complement with supply side measures of Renewable Energy Technologies (RETs). As a result, EE is one of the important strategies to achieve low emissions targets at the same time decoupling the economic growth from high emissions, pollution and resources [16].

Improving EE is one of the options for India to meet commitment of Nationally Determined Contributions (NDC), which in line with Paris agreement. In medium-term (2030) and long-term (2050), EE measures will reduce CO_2 emissions, promote cleaner production by reducing fossil-based energy and deliver benefits of energy security and improved air quality. Therefore, there is need of policy roadmap to drive transitions towards clean and energy efficient technology in the medium and long-term.

In summary, it is possible to meet NDC targets and improve in energy intensity under EE strategy. In long term EE coupled with RETs will shift towards cleaner and sustainable electricity. As a result, it has greater implications on sectors like residential, industry and transport sectors. Particularly, under EE strategy, industry sector has large scope of improvement. However, in the transport sector EE measures complement with other strategies like, SE4ALL and 2 °C scenario and it can greatly reduce energy intensity in transport sector. In general coupling EE initiatives with SE4ALL strategy would deliver multiple co-benefits [17].

Considering the above, various approaches are being implemented to achieve the improvement of EE. The approaches vary, as in some cases the government

---

[4] https://www.iea.org/data-and-statistics?country=INDIA&fuel=Imports%2Fexports&indicator=Coal%20imports%20vs.%20exports.

identifies and prioritize the technologies and provide subsidies and capacity building to use these technologies. There is Perform, Achieve & Trade (PAT) framework, which is market, based regulatory mechanism, in which more efficient technologies have been incentivize within industries and in order to raise awareness among consumers, Bureau of Energy Efficiency has introduced standards / codes and labeling that provides the information about the energy performances of the technology. All in all, these approaches have yielded positive results and succeed in market penetration but lacks in achieving its full potential. The extent of EE efforts that is achieved in different key sectors, it is found that the majority of the cases, policies and regulations have been used to promote a technology but there is need for scaling up these efforts. Therefore, it is recognized that there is a need to overcome barriers like market imperfections and institutional barriers [18].

Therefore investing in EE play vital role in alleviating climate change in short term by investing on EE technologies and clean technology investments. Such opportunities lie in all sectors for India, especially in industry, which has high carbon intensive infrastructure (electricity supply, industry and transport). Therefore, with deep understanding of challenges and opportunities there is need for mainstreaming EE efforts across all sectors of the economy, particularly a combination of EE investments with technical support and coupling with donor finance with suitable financial instruments.

Many of the financial barriers to Industrial Energy Efficiency (IEE) arise from the unique characteristics of such projects relative to traditional investment projects. While EE reduces energy costs and improves the "bottom line" of enterprises, it does not increase the "top line," which may make it somewhat difficult for corporate or government executives and managers as well as bankers and other members of the financial community to clearly perceive the benefits of IEE. Further, IEE projects are typically much smaller than conventional projects for new plant construction, plant expansion, new product development, research and development (R&D), or facility modernization. Other salient characteristics of IEE projects include their high project development and transaction costs; utilization of new or innovative technologies; the relatively small value of their project assets; and, in some cases, utilization of new business models involving performance contracting and third-party implementation (by ESCOs or other types of energy service providers).

A study by Institute for Industrial productivity (IIP) suggests to include EE loans as priority sector lending by banks and financial institutions, which is in line with efforts of National Mission on Enhanced EE (NMEEE) of the National Action Plan on Climate Change (NAPCC).Further, this study also highlights that in order to scale-up efforts of EE it is essential to eliminate barriers like limited availability of commercial financing from banks and financial institutions. Therefore, this study recommends classifying EE Loans as Priority Sector Lending by Banks and Financial Institutions [19].

3 Approach and Methodology

The five voluntary principle mechanism is developed by the World Bank Group and Agence Francais de Development (AFD) on behalf of, and in collaboration with other banks like the Africa Development Bank (AfDB), Asian Development Bank (ADB), CAF Latin American Development Bank (CAF), Credite Agricole S.A., Cassie des Depots Group (CDC), Development Bank of South Africa (DBSA), European Investment Bank (EIB), European Bank for Reconstruction and Development (EBRD), Inter-American Development Bank (IADB), International Bank for Reconstruction and Development (IBRD), International Finance Corporation (IFC), Japan International Cooperation Agency (JICA), KfW Bankengruppe (KfW), and the Multilateral Investment Guarantee Agency (MIGA). The objective of the mechanism is to direct investments and place the economy on low-carbon, climate resilient development pathways. In this context, financial institutions play a key role in mitigating climate change risks; therefore there is a need for FIs to adapt this mechanism to financing shifts in the market. As a result, this mechanism will help to deliver financing at scale and integrate climate change considerations systematically and explicitly across all levels of its strategies, programs and operations. All in all, mainstreaming this mechanism will help FIs to deliver more sustainable, short-term and long-term results.

In line with the above discussion, this study adopts five voluntary principles called commit, manage, promote, improve, and account for mainstreaming EE solutions within FIs. Firstly, COMMIT to climate strategies; senior management will demonstrate institutional commitments to address EE finance through explicit strategic priorities, policy commitments and objectives that allows the integration of climate change considerations in lending and advisory activities of the financial institution over time. Secondly, MANAGE climate risks; assess your portfolio, pipeline, and new investments. Work with clients to determine appropriate measures for building resilience to climate impacts and improving the long-term sustainability of investments.

Thirdly, PROMOTE climate-smart objectives; promote approaches to develop instruments, tools, and knowledge on how best to overcome risks and barriers to investment in low carbon and resilient investments. This could include mobilizing and catalyzing additional financing and developing specialized financing vehicles/products, such as green bonds, risk-sharing mechanisms, or blended finance. Engage clients and other stakeholders such as rating agencies, accounting firms on climate change risks and resilience, and share lessons of experience to help further mainstream climate considerations into activities and investments. Fourthly, IMPROVE climate performance; establish operational instruments to improve the activities' climate performance. Financial institutions monitor and monitor climate change priorities-related indicators including reporting of GHGs, lending and consultancy volumes in support of green investment, allocations of climate-related assets, and climate footprint of the institution.

Finally, ACCOUNT for your climate action; be transparent and report as far as possible on your institution's climate performance, including increased energy efficiency financing or other climate-related activities and investment. Be transparent and report, wherever possible, the climate footprint of the institutions' own investment portfolio, and how the institution is addressing climate risk. Bonnel and Swann [20] (Table 1).

4 Discussions

The five voluntary principle mechanism is an important one compared to other mechanisms, business models and other financial instruments in various ways, but some of the unique characteristics s are discussed here. This mechanism helps to streamline CC considerations as a core part of financial institutions operations and make projects climate-resilient. The approach is multidimensional and in line with the national action plan and strategy to streamline, thereby, ensuring EE investments are sustainable in the long-term.

This mechanism plays an important role in raising awareness among FIs, particularly among Local Financial Institutions (LFIs). For example, in developing countries capital is scarce, LFIs may not have adequate expertise in financing EE projects and priority would be given to power plants and its related investments. Consequently, financing EE projects are considered to be risky. Therefore, if this mechanism is supported or complemented by the regulations, policy support and behavior change initiatives by the government, then it will be easy to implement at the national level.

The principles are designed in such a way that the mechanism suits local market conditions since market structure would be different and it helps to tailor the financial mechanism to local market barriers and cultural barriers. Further, it is effective enough to build suitable products and services for consumers and investors. In the end, financial mechanism also helps to develop financial risk mitigation instruments to eliminate risk perception from various stakeholders about the project.

5 Conclusions

In short and medium term EE is a cost-effective option to reduce energy demand, increase productivity, competitiveness and mitigate greenhouse gases (GHG) emissions. It is evident that from the above discussion that India has a huge potential to improve EE efforts. However, there is a need to eliminate barriers like institutional, policy and financial, which limits the adoption of EE measures and technologies. There are mechanisms developed and deployed by the national, state governments and donor agencies but they at the moment are not enough to achieve the full potential of scaling-up EE. In this context, this study makes an effort to scale-up EE

Table 1 Five voluntary principles for mainstreaming EE within FIs operations

Principles	Objective	Activity	Methodology	Outcomes
Commit to climate strategies	Assess institutional commitments	Assess senior management leadership, explicit strategies, policies, commitments and targets	Asses' plan that integrate a systematic mainstreaming of EE into business and strategic plans. Positions that are built in institutional structure that promotes and are accountable for meeting EE objectives. Incentives structure, performance management systems or other systems that track the delivery of systems at all levels	A EE strategy to promote -relevant strategic priorities, policy commitments, plans and/or targets at institutional level
Manage climate risks	Education on non-financial risk mitigation	Asses EE risk of new investment, pipeline or an existing portfolio	Assessing EE risk of existing portfolio of projects. Developing approaches to assess and manage project risks	Identify the linkages with approaches for screening the new investments
Promote climate smart objectives	Identify EE considerations in routine business development	Share learning lessons and experience to help further mainstream EE considerations into activities and investments	Identify dedicated mechanisms that provide financing for projects, specific strategies to increase lending related EE	Promote approaches to generate instruments, tools and knowledge. Encourage knowledge sharing and dissemination for creative and smart partnership between departments, thereby, mainstreaming and implementation of investment at scale

(continued)

Table 1 (continued)

Principles	Objective	Activity	Methodology	Outcomes
Improve climate performance	Identify operational tools that improve the climate performance of activities	Identify operational tools and result framework	Assess the GHG impact of investments, EE finance flows. Screen risk and opportunity for the same	Develop appropriate operational tools for tracking, monitoring and incorporate EE considerations into day-to-day operations
Account for your climate action	Transparency and reporting	Link EE considerations in financial reporting	Identification of institutional reporting of EE of investments and related policies and potential reporting on additional funds leveraged or mobilized	Improve overall institutional governance

Source: Bonnel and Swann [20]

efforts through mainstreaming energy productivity lending practices among financial institutions by five voluntary principles framework. As a result, a greater collaboration, mitigating institutional barrier and developing a favorable environment for the financial, fiscal and regulatory landscape to attract investments for EE measures.

References

1. Navroz K. Dubash, Markus Hagemann, Niklas Höhne & Prabhat Upadhyaya (2013), Developments in national climate change mitigation legislation and strategy, Clim. Pol., 13:6, 649-664, DOI: https://doi.org/10.1080/14693062.2013.845409
2. IEA (2012), Energy Technology Perspectives 2012, Pathways to Clean Energy Systems, International Enrgy Agency, Paris, France
3. UNFCCC (2014), Fifth Review of the Financial Mechanism (Technical Paper): Standing Committee on Finance
4. Climate Investment Funds (CIF) (2014), Delivering at Scale Empowering Transformation (Annual Report), Washington DC: Climate Investment Funds (CIF)
5. Climate Investment Funds (CIF) (2015), Climate Investment Funds: Accomplishments, Transformational Impact and Additionally in the Climate Finance Architecture, Washington D.C: Climate Investment Funds (CIF)
6. Kirsty Hamilton (2009), EE and the Finance Sector, UNEP Finance Initiative
7. Robert P. Taylor, Chandrasekar Govindrajalu, Jeremy Levin, Anke S.Meyer, WIlliam A. Ward (2008) Financing EE Lessons from Brazil, China, India and Beyond: Washington DC USA, The World Bank
8. Rezessy, S., & Bertoldi, P. (2010). Financing energy efficiency: forging the link between financing and project implementation. *Report prepared by the Joint Research Centre of the European Commission*

9. Dr Rory Sullivan (2011), Investment-grade climate change policy: financing the transition to the low carbon economy, Institutional Investors Group on Climate Change (IIGCC), the Investor Network on Climate Risk (INCR), the Investor Group on Climate Change Australia/New Zealand (IGCC) and the United Nations Environment Programme Finance Initiative (UNEP FI)
10. Bertoldi, Paolo, et al (2020). "How to Finance Energy Renovation of Residential Buildings: Review of Current and Emerging Financing Instruments in the EU." *WIREs Energy and Environment*, Wiley, June 2020. *Crossref*, doi:https://doi.org/10.1002/wene.384
11. Rory Sullivon (2014), Climate Change implications for investors and Financial Institutions: University of Cambridge, Institutional Investors Group on Climate Change (IIGCC), UNEP Finance Initiative
12. Ajai Chopra (2015), Financing Productivity and Innovation Led Growth in developing Asia; International Lessons and policy Issues, Washington DC: Peterson Institute for International Economics
13. USAID (2011), Energy Trends in Developing Asia: Priorities for a Low-Carbon Future. USAID Regional Development Mission for Asia
14. ADB (2013), Asian Development Bank, Same Energy, More Power, Manila
15. USAID (2013),Financing EE in India: A review of current status and recommendations for innovative mechanisms. USAID Regional Development Missions for Asia
16. Lee C.T., Hashim H., Ho C S., Klemes J.J., (2016), Low-carbon Asia: technical contributions to an ambitious goal for sustainability, Clean Techn. Environ. Policy, 18, 2335-2336
17. Subash Dhar, Minal Pathak, P R Shukla, (2018), Role of EE in a low carbon transformation of India, Chem. Eng. Trans., 63, 307-312 DOI:https://doi.org/10.3303/CET1863052
18. Garg, A., Dhar, S., Kankal, B., & Mohan, P. (Eds.) (2017), Good Practice and Success Stories on EE in India
19. IIP (2012), Financing Industrial EE In India: Lessons Learned And Directions For The Future
20. Alexis Bonnel and Stacy Swann (n.d.) Mainstreaming Climate Action within Financial Institutions Emerging Practices, Washington DC: The World Bank Group and Agence Francais de Development (AFD) on behalf of, and in collaboration with, the Africa Development Bank (AfDB), Asian Development Bank (ADB), CAF Latin American Development Bank (CAF), Credite Agricole S.A., Cassie des Depots Group (CDC), Development Bank of South Africa (DBSA), European Investment Bank (EIB), European Bank for Reconstruction and Development (EBRD), InterAmerican Development Bank (IADB), International Bank for Reconstruction and Development (IBRD), International Finance Corporation (IFC), Japan International Cooperation Agency (JICA), KfW Bankengruppe (KfW), and the Multilateral Investment Guarantee Agency (MIGA)

The Effectiveness of Gas-Boosted Solar Water Heaters in an Australian Monitored Low-Energy Housing Development

D. M. Whaley, S. R. Berry, R. Liddle, and E. Halawa

Abstract This paper reports on the effectiveness of gas-boosted solar water heaters in dwellings within an eco-friendly and comprehensively monitored Australian housing development. Unlike some electrical appliances, including the oven, dishwasher, kettle etc., that show unique energy signature patterns, the solar water heaters examined in this paper do not show obvious energy usage patterns. The auxiliary energy consumed by these water heaters depends on the type of booster, i.e. either an instantaneous or the more traditional storage type, as well as the great variety in individual household demand. Given the recent decline in popularity of solar water heating in Australia, and the inability of householders to determine if their gas-boosted solar water heater is operating correctly, this paper determines the effectiveness of 4 gas-boosted solar water heaters (SWHs) within the Lochiel Park Green Village estate. This is achieved by calculating the auxiliary energy used and hence the solar fraction of 2 instantaneous, and 2 storage, type solar water heaters, based on analyses of monitored hot water and gas usage data, collected every minute over an 8-year period. Furthermore, the paper compares a small sample of early monitored data, with modelling used to determine energy savings and compliance in Australia, to highlight the spread of performance outcomes. Finally, the paper describes various installation and design issues found when some householders have suspected their solar-water heater was not operating effectively.

1 Introduction

The uptake of renewable energy technologies has exploded around the Globe, yet while at a domestic scale in many countries rooftop solar photovoltaics (PV) has experienced a phenomenally high growth rate, solar water heaters have struggled to grow their share of the water heating market.

D. M. Whaley (✉) · S. R. Berry · R. Liddle · E. Halawa
University of South Australia, Adelaide, SA, Australia
e-mail: david.whaley@unisa.edu.au; stephen.berry@unisa.edu.au; raymond.liddle@unisa.edu.au; edward.halawa@unisa.edu.au

P. Bertoldi (ed.), *Energy Efficiency in Domestic Appliances and Lighting*, Springer Proceedings in Energy, https://doi.org/10.1007/978-3-030-79124-7_12

For many years various governments, often driven by the desire to decarbonise their economy, have incentivised the uptake of low carbon hot water systems such as solar thermal and heat pump technologies. In Australia, the peak annual number of installation of solar water heating systems (SWHs) occurred in 2009 [1] when the Australian government introduced a Solar Hot Water Rebate for systems "installed on or after 3 February 2009 until 30 June 2012" [2]. Since 2013, this downtrend has halted and market share has stabilised thanks to the rebate made available under the "One million solar roofs" program" [3]. Despite the pause of this downtrend, SWHs are facing new competition from heat pumps which are easier to install and potentially more energy/carbon efficient.

This paper draws on the evidence from four monitored hot water systems in Lochiel Park, South Australia to expand our knowledge of the energy and carbon performance of various types of solar water heaters, and to answer the following research question: given the variability in household hot water demand, is an instantaneous boost or storage type solar water heating system more energy and carbon efficient in a warm temperate climate.

2 Literature Review

The use of direct solar radiation for water heating has occurred commercially for more than a century, with devices commercially popular in California and Florida by the end of the nineteenth century [4]. Solar water heating has been successfully utilised for supplying general hot water needs and hydronic space heating, in locations with substantial space heating requirements [5–7].

In Australia, a country rich in solar irradiation, a range of technical, social, economic and policy barriers have meant that solar technologies have not yet come close to becoming a mainstream method of heating water in the residential sector [8]. Policy initiatives have impacted the uptake of solar thermal and other low carbon water heaters from time to time (see Fig. 1), and although there are regional variations, the residential market share for solar water heaters remains below 10% [8, 9].

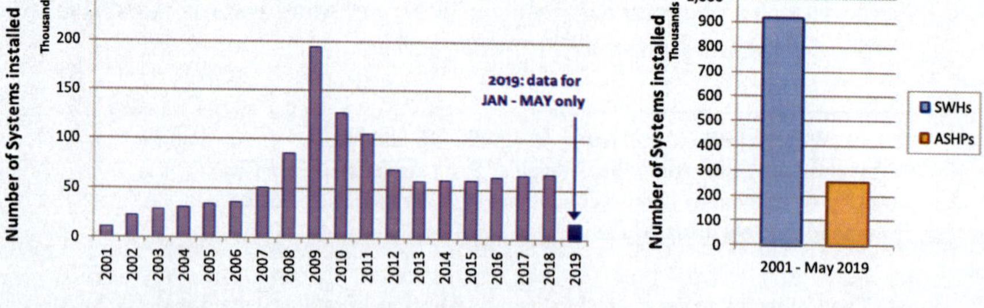

Fig. 1 Number of solar water heaters (SWHs) and air source heat pump (ASHPs) systems installed in Australia, 2001–31-May-2019, by (**a**) year, and (**b**) type. Data obtained from [1]

The widespread application of solar thermal water heaters can provide various benefits including economic and energy security impacts, environmental impact reduction, and the potential for peak load reductions and other energy management benefits to local energy networks [10]. For example, studies have indicated the potential for solar thermal water heaters to reduce peak electrical network demand [11].

Policies have provided both incentives for the uptake of solar thermal technologies and disincentives. For example, European product energy labelling standards for water heating systems have assumed high primary energy and carbon costs for electrical back-up when compared to gas, but as we decarbonise our generation capacity with renewables, solar-electric water heaters may provide hot water (a) at a relatively lower carbon cost; and (b) which provides a schedulable load to help manage otherwise excess renewable energy [12].

The energy efficiency of water heating systems is relative to many factors including the water inlet temperature, schedule control, outlet temperature setting, the use of thermal insulation, the type of boost, and the volume demanded by consumers [13]. For solar thermal water heaters performance affecting factors include collector design, collector tilt angle, coating of pipes, fluid flow rate, thermal insulation, integrated collector storage, thermal energy storage, the use of phase change materials, and the type of boost [14].

In this study we will focus on the energy and carbon performance impact related to the type of water temperature boosting system for solar thermal water heaters, that is, the energy using system that ensures the end-use temperature is adequate and that there are no health related issues due to the collection and heating of water. Some studies have found that instantaneous electric or gas back-up are more efficient than storage systems due to the elimination or reduction of storage tank energy losses, but this may not be valid for all system arrangements and climates [15, 16].

In Australia, the energy efficiency (performance) of the SWH systems is currently not subject to any minimum performance standards [17, 18]. Instead, the emphasis has been on maximizing the contribution (production) of renewable energy in domestic water heating sector through the allocation of small-scale technology certificates (STC) which can be traded to provide a financial incentive [18]. As such, the approach favours the increase of the renewable energy production but may not necessarily lead to an increase in system efficiency or reliability [19]. The system efficiency will in fact have direct impact on the household in the form of the energy cost over the entire life of their systems.

The findings reported in this paper will explore whether predicted energy savings match those monitored in the field [19]. While the residents of Lochiel Park have access to in-home displays and monitoring systems that show them the total household gas usage as well as the volume of hot water drawn, this information may not always lead to the proper detection of the system operating issues.

3 Case-Study: Lochiel Park

3.1 The Lochiel Park Green Village

Developed between 2008 and 2017, Lochiel Park was conceived as a leading example of nearly zero carbon housing for a warm, mostly temperate climate, with seasonal heatwaves. Within a predominantly owner-occupied housing market, the State Government of South Australia designed, developed and marketed Lochiel Park as a mixed tenure 'green village' in partnership with the local private sector housing industry [20]. Within the estate there are 103 dwellings, mostly two storey, three or four bedroom family houses, other than 23 apartments and 2 terrace apartments that are either owned by a public housing agency or sold to designated low income households.

Each dwelling includes grid-connected solar photovoltaic and solar thermal water heating technologies, i.e. for general electricity generation and water heating purposes. The home include elements of passive solar design, intended to promote passive heating in the winter, to protect against the summer sun and to encourage cross ventilation. High efficiency appliances and equipment were also used. Under Australian 'Nationwide House Energy Rating Scheme (NatHERS)' energy performance criteria, all dwellings achieve a minimum 7.5 Stars out of 10 and are considered relatively low energy homes. Although most homes are not fully net zero energy or net zero carbon in operation, they are a substantial improvement on previous practice.

3.2 In-Home Monitoring System

Each home is fitted with an 'EcoVision' brand monitoring system, which incorporates an in-home feedback display, a programmable logic controller, and an array of intelligent meters and sensors [21]. This equipment measures and displays the residents' electricity, water and gas use, plus solar PV electricity generation, in real-time. All energy figures are based on monitored electricity delivered to the house and generated at the site as well as delivered natural gas, rather than the calculation of 'primary energy' which includes losses that occur in energy generation, transmission, and distribution.

3.3 Water Heater Systems Used in this Study

Table 1 lists some properties of the gas burners associated with the four solar water heater systems examined in this study. Note that the maintenance rate and start up energy applies to STOR and INST systems, respectively. These are further discussed in methodology section.

4 Methodology

The performance evaluation presented in this paper is based on the data recorded from gas boosted SHWs - comprising 2 gas-storage (STOR) and 3 instantaneous (INST) systems, installed in the Lochiel Park green village estate. The data recorded consists of minute by minute hot water and gas consumption, for each household, and in-situ measured maximum indoor water temperatures. This system data monitoring was part of the early initiative of low or zero-net energy homes [22]. In addition, some information regarding faults came from informal discussions with residents.

Hot water, gas and electricity consumptions were recorded each minute. This resolution of data was required to calculate the number of daily start up events (SUE) for the INST systems, however the daily load calculations were performed on a daily basis; Microsoft Excel was used to convert minuted to daily data, and to perform the load calculations (MJ), in accordance with AS/NZS 4234:2008 [23].

4.1 Heat Energy Required

The total water heat load (energy) is based on the quantity of heat energy required, for a given hot water demand (volume of hot water), plus additional loads that are associated with the maintenance rate and start up events of STOR and INST systems, respectively. These additional calculations are separately addressed in the following sections, however, the common heat energy required, Q_{Water}, is shown below.

Table 1 Properties of the water heater systems examined in this study

Lot	Type	Tank Vol (L)	Hot water capacity (L @ 25 °C)	Gas consumption			Burner efficiency (%)
				Heating (MJ/h)	Maintenance rate (MJ/h)	Start up energy (MJ)	
L04FO	STOR	260	–	26	1.0	–	80.3
L20FF	STOR	260	–	26	1.0	–	80.3
L26ST	INST	215	16	125	–	0.189	81.0
L23SS	INST	215	16	125	–	0.189	81.0

Instantaneous (INST) systems highlighted in grey

This was calculated for each households based on the individually measured hot water usage volume, outlet water temperatures, and a series of assumptions.

$$Q_{Water} = mc_p\left(T_{out} - T_{in}\right)$$

where:

m = the mass, m, determined by the volume of hot water consumed V, and the water density, ρ, i.e. m = ρV.
c_p = the specific heat capacity, Cp, used in the calculation is that at the mean temperature of the water heater, i.e. 0.5*(T_{out} − T_{in}).
T_{in} = monthly cold water inlet temperature provided in AS/NZS234:2008 [23]. Although results from a previous study [24] that showed that the measured water inlet temperatures did vary slightly from those assumed in this standard, these were used for our calculations.
T_{out} = measured maximum indoor hot water temperatures.

4.1.1 Additional Loads for Instantaneous Systems

Each time that hot water is demanded from an instantaneous system, the gas burner must first ignite and start heating, causing a slight delay between the time the water heater turns on and delivers hot water. This start up event (SUE) consumes gas and the system experiences a startup gas loss (SGL); values have been determined through testing. The number of SUE has been determined from the analysis of hot water demand from minute data, and the resulting total daily heat load, is given below.

$$\mathrm{Q}_{daily(INST)} = \mathrm{Q}_{water} + SGL \times SUE$$

4.1.2 Additional Loads for Gas Storage Systems

To comply with the requirements of AS/NZS 3666: 2011 [25], storage systems use gas burners to maintain the tank temperature at 60 °C, and hence attract an additional load, i.e. maintenance rate.

$$MR = 24 \times M\left(\frac{\Delta T_{tank}}{45}\right)$$

where:

MR = Maintenance rate of monitored system (MJ)
M = Maintenance rate of system at AS4552:2005 [26] test conditions.
ΔT_{tank} = Temperature difference between the ambient air temperature (T_{amb}) and the average water temperature within the tank (T_{tank}). T_{amb} was the average daily tem-

perature recorded at the closest Bureau of Meteorology (BOM) station. T_{tank} is assumed to always be 60 °C.

The resulting total daily heat load for a gas storage water heater system, is calculated to include the water heating energy required and the maintenance energy.

$$Q_{daily(STOR)} = Q_{water} + MR$$

4.2 Solar Fraction

The solar fraction is the percentage of the water heating load provided by the solar collectors. It is calculated below, from the monitored daily gas consumption (G), which takes into account the thermal efficiency of the gas burner; these details are listed in Table 1.

$$Solar\ Fraction = 1 - \frac{G}{Q_{daily}}$$

5 Results

In this section a number of observations from the monitored data are presented, coupled with the ways the performance issues arising were detected and dealt with. The data recorded provides important information on how system operational performance deviates from the ideal conditions.

Energy calculations made in this paper are sensitive to many parameters, including water inlet temperature, hot water outlet temperature, gas burner efficiency, start up energy (INST systems) and maintenance rates (STOR) systems. Given that some of these parameters have been assumed, e.g. based on relevant Australian Standards, it is important to show how some of these vary.

5.1 Observation of Water Heater Settings and Patterns of Usage

Although solar water heater systems are modeled to determine their energy displacement and hence the number of STCs, there is very little evidence of how these systems actually perform in-situ, and how these systems are setup by the trades people who install them. This paper sheds some light on this, by highlighting some in-situ monitored parameters that are otherwise assumed by the Standards used in the modeling; this is important given how sensitive the energy load is to some of these.

Figure 2 shows a summary of the monitored maximum indoor water temperatures for a number of instantaneous (INST) and storage (STOR) solar water heaters, along with average of each type. Note that each house is fitted with a thermostatic mixing valve, which is a regulated requirement for solar water heaters in Australia, as these mix heated and cold water such that the internal temperature is limited to a temperature range that cannot cause human scalding. These valves can be adjusted manually and normally limit the indoor water temperature to the range 35–50 °C, although Fig. 2 does show a number of instances where the monitored internal temperature exceeds 50 °C. This is possible if the tank feeding the home has water stored in it at very high temperatures and the cold water inlet is sufficiently warm to be unable to adequately reduce the water temperature by mixing. Similarly, if an INST system is used where stored water is below 55 °C then the instantaneous burner will heat the water up to 70 °C, to reduce health risks by killing Legionella and other bacteria.

The recorded water temperature boosting times for the STOR systems are shown in Fig. 3. Each of the system has its own boosting window period(s), some of which follow recommended manufacturers suggested boosting time. Interestingly, two of the systems have 24-hourly boosting time spread. Boosting time can affect the energy use but also depends on the users' daily need of hot water.

When queried directly, some residents reported that they believe their boosting window had been setup in accordance with the manufacturer's recommendation, which is somewhat seen for some systems, whilst others claimed that their window was always ON. The latter effectively causes the solar water heater to operate as a standard gas storage system whenever the sun is not present, consuming gas unnecessarily which frustrated some residents. In some of these cases, monitored gas data contradicted this claim, where the boosting windows of between 1 and 2 h per day appeared. This can be easily identified by interacting with the in-home monitoring system.

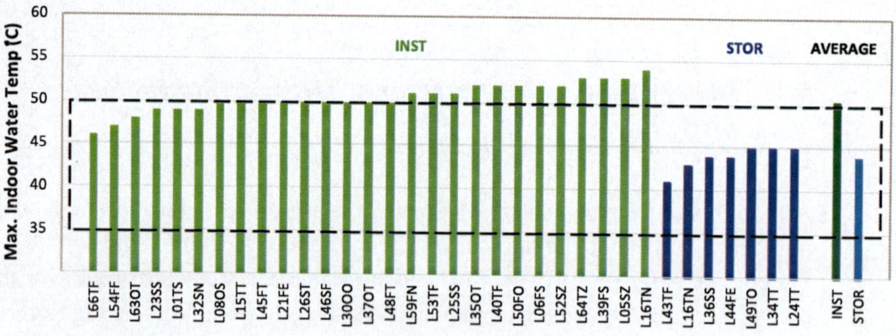

Fig. 2 Measured maximum household (internal) water temperatures

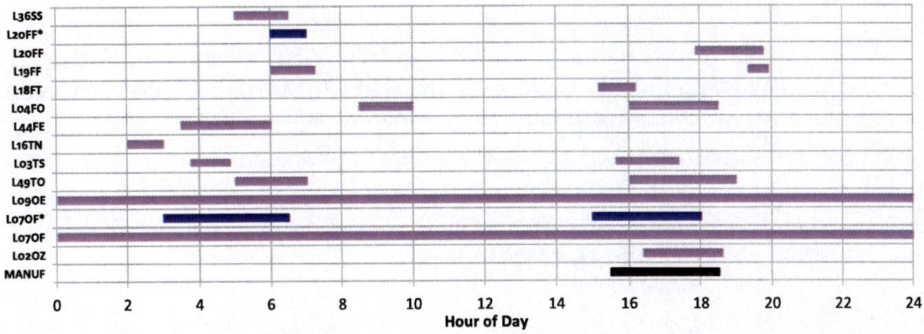

Fig. 3 Daily STOR system boosting window (0–24 H). Dark purple* indicates boosting window changed throughout monitoring period. Black window is the manufacturer's recommendation

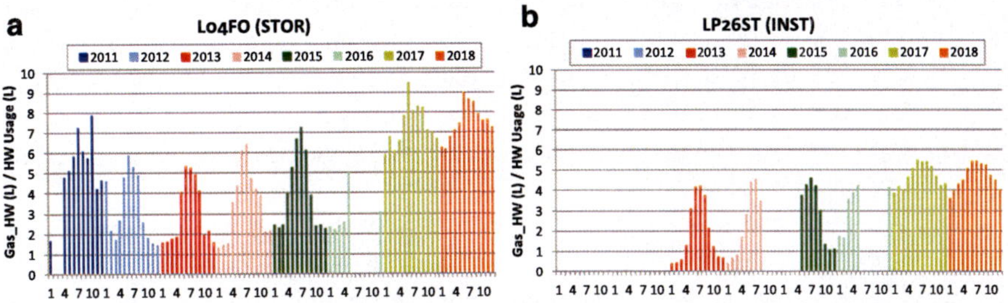

Fig. 4 Monthly water heater gas consumption, normalised by monthly hot water consumed, for **(a)** STOR and **(b)** INST water heater systems

5.2 *Assessment of Water Heater from Monitoring System Data*

5.2.1 Monthly Gas to Hot Water Patterns

Figure 4 shows the ratio of monthly gas consumed by the water heater and the monthly hot water volume over the entire monitoring period of 2011–2018, for one storage (STOR) and instantaneous (INST) water heater. This is useful metric to identify changes in patterns, or to suggest faults, i.e. where gas may increase if the contribution of thermal energy provided by the solar collectors decreases. This is shown in Fig. 4 (a) as of January 2017, and in Fig. 4(b) as of March 2016.

Consider house *L04FO*, with the STOR system. Figure 4 shows 2 distinct changes in yearly patterns. Firstly, the monthly usage profiles for 2011 are higher than those for 2012–2016. In contrast, the second change in gas consumption pattern was a sudden increase in the monthly gas usage starting in early 2017. This gas usage in 2017 and 2018 exceeds that of any other year including 2011, where initially some issue causing high usage was addressed.

Similarly, consider house *L26ST*, with an INST water heater. Although much of the data is missing across some of the years 2014–16, it is clear that some issue or fault occurred just prior to 2017 that has resulted in significantly higher gas to hot water ratios, compared with earlier years of operation.

5.2.2 Daily Gas to Hot Water Ratios

A more insightful metric is to plot the daily gas consumed by the water heater as a function of the hot water used. This is shown in Fig. 5 for 2013 and 2018, for the same STOR and INST water heaters. These periods were selected due to complete data sets for both houses and both years. Solar water heaters are designed to provide a high contribution of the heating, which is shown along the x-axis where zero or little gas is consumed whilst meeting the households' daily hot water demand. This is offset for the STOR system, due to the maintenance rate of about 200 L/d, shown by the black box. The data also shows a straight line starting at the origin, which represents the days of no solar gain, i.e. all of the hot water demand is met by the gas burners (boosting).

Both water heaters show that their respective gas usage is higher in 2018 compared with that in 2013, verifying the increased gas consumption trends seen in Fig. 4. However, the figure additionally shows that little solar gain is now experienced by both water heaters, as no data points exist along the x-axis for 2018. In fact, there is a significantly higher concentration of data points along the line emanating from the origin, indicating the systems are starting to operate as more traditional gas water heaters without solar thermal contributions. To verify this claim, the solar fraction is determined from the measured and predicted energies that are calculated from the monitored gas and hot water data, and hot water system parameters (Table 1).

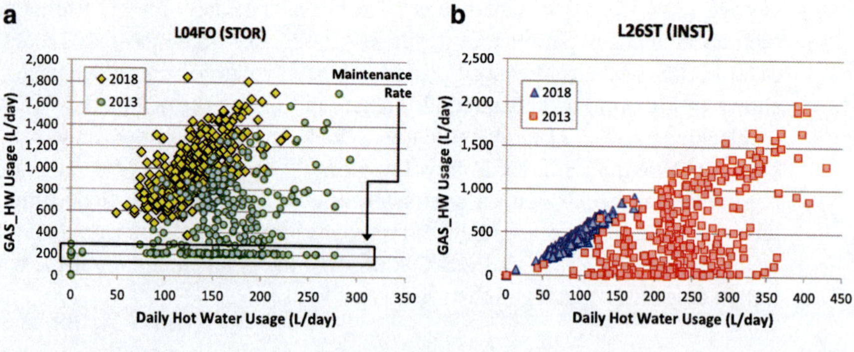

Fig. 5 Daily gas used by the water heater vs. daily hot water demand, for 2013 and 2018, for houses (**a**) L04FO [STOR] and (**b**) L26ST [INST]. The black horizontal box in (a) represents days with good solar gain; the gas used is for the maintenance rate (200–220 L/d)

5.3 Calculated Energy Savings and Solar Fractions

Figure 6 shows the daily calculated energy consumed by the gas burners as a function of the predicted daily load, for 2013 and 2018. The black lines represent the case where no solar gain is available, which is seen for both water heaters in 2018, unlike 2013. In addition, there are a small number of data points above the line in 2018 for the STOR system, which is likely caused by some of the assumptions used to calculate the energies, which are sensitive to several parameters. However, given that the bulk of the data points lie on or below the line, we are confident in the methodologies applied here to calculate the energy from the measured data.

The above figure shows there several days in 2013 where both water heaters use zero gas to boost the water temperature to the desired target, yet meet a hot water demand (data points along the x-axis), which indicates that the sun has provided all of the heating energy required. Although this occurs more frequently for the INST water heater, it was observed in 2013 for the STOR system for about one month. This observation does not appear in 2018 for either system and suggests the solar components of the systems are no longer operating as intended. This is more clearly seen in Fig. 7, which shows the solar fraction for each water heater in 2013 and 2018.

Figure 7 shows that both types of water heaters were working well in 2013, with high solar fractions of 80–100% for the STOR system, and 90–100% for the INST system. However, these solar fractions are significantly reduced in 2018, where this varies between 0–40% for the STOR system, and 0–20% for the INST system. Note that in some cases the solar fraction is shown to be less than 0%, which is possible based on the definition and methodology used to calculate this. This occurs when the energy calculated from the measured gas meter exceeds that predicted for a day's hot water demand, that is, there is a net loss of energy above that needed to provide the hot water, and was seen in Fig. 6 where data points were above the black line that represents zero solar gain.

Figure 8 shows a *boxplot* (statistical summary) of the daily solar fraction of these two solar water heating systems, along with two others, in 2013 and 2018. It should be noted that the two additional systems include one STOR and one INST type that

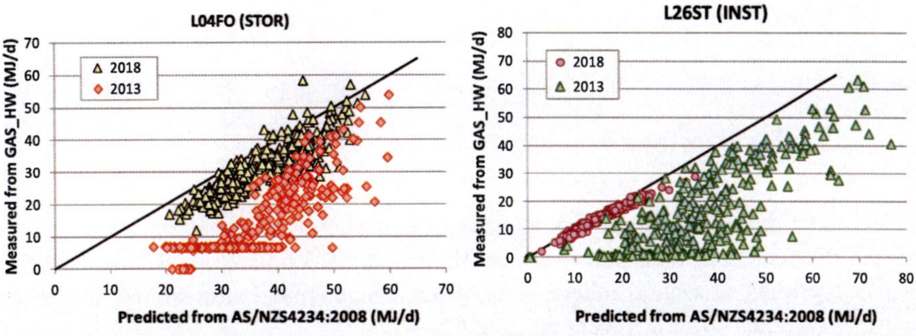

Fig. 6 Daily consumed energy vs. that predicted from AS/NZS 4234:2008 [23]. The black line represents the case where no solar gain is available

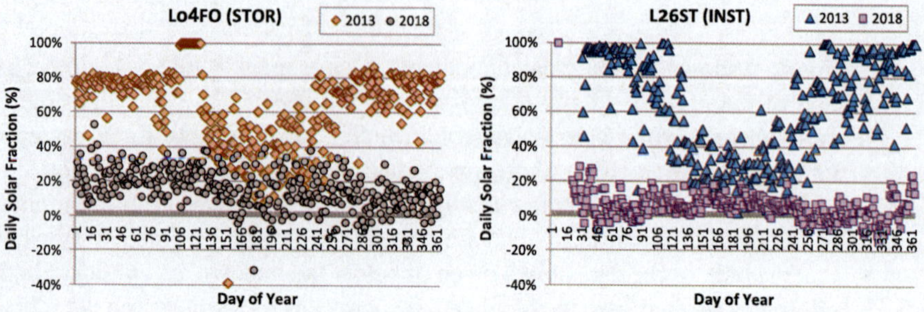

Fig. 7 Daily solar fraction of water heater in houses L04FO and L25ST, for 2013 and 2018

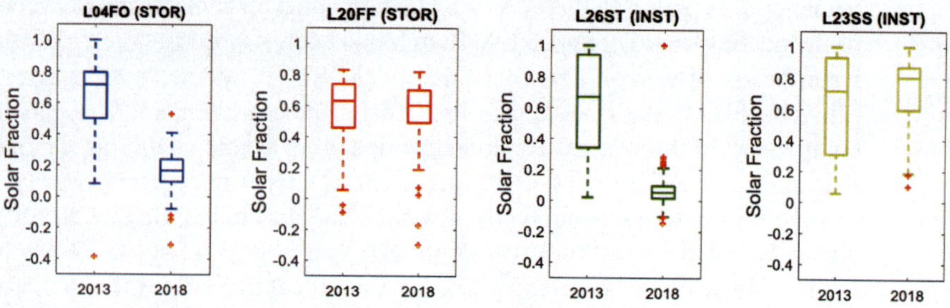

Fig. 8 Solar fractions of 4 Lochiel Park solar water heaters

are identical to those in houses L04FO and L26ST, respectively (see Table 1). These other systems (L20FF and L23SS) appear to be working better than L04FO and L26ST, given their significantly higher solar factions in 2018.

Each system is shown to be operating correctly in 2013, however by 2018 two of the four systems have shown a significant reduction in solar fractions, suggesting a major fault with them. These warrant further investigation, given that some residents are consuming more gas than a standard gas water heater, but also as the amount of STCs awarded for a system (which is linked to a previous rebate scheme) is based on a system's ability to displace energy over a 10-year period.

6 Discussions

The data alone could not suggest any reasons why the performance of the water heaters examined here degraded between 2013 and 2018. The residents of these systems, as well as others, were hence contacted and after some initial discussions, a number of issues were discovered.

6.1 L04FO (STOR)

The resident of house L04FO had a background in IT and was one of the residents within Lochiel Park who actively and regularly engages with their monitoring system. This resident suspected, after reviewing the information available from the in-home display, that the gas usage was higher than expected and requested a service technician to inspect the solar water heater. The technician found a corroded impeller, and replaced this in early February 2012, which resulted in a significant drop in monthly gas to hot water ratios. This reduction occurred as the impeller was able to pump more water to the roof-mounted thermal collectors resulting in a significant reduction of gas. This is seen by the monthly gas to hot water ratios of Fig. 4.

In contrast, the resident was unaware of the sudden increases in monthly gas usage from 2017, when compared to that of previous years. This is difficult to determine from the monitoring system as it is limited in how much information it can display, i.e. it can only show monthly totals, and it is not capable of comparing this with previous monthly totals. The change in usage patterns was only seen in 2019 when the analysis presented here was performed, some 30 months after the issue first appeared. The resident was presented with a copy of the analysis and has since organized to have a service technician investigate. It was also revealed that a custom-built enclosure was screening the water heater warning systems, and when presented with this data, the resident removed the screen and discovered that a fault alert LED was flashing on their water heater control system.

When queried about the any manual interventions, it was revealed that the gas supply was physically switched off for about one month in 2013, which explains why the hot water demand was met whilst consuming zero gas. By definition, the solar gain is 100% during this time, where it otherwise appears to have a steady upper limit of between 80–85%, due to the maintenance rate. Finally, the resident revealed that at times when the water heater temperature was insufficient, they activated the manual override which turns the gas burner (boost) for one hour. It is possible that during these events the gas burner was activated for more time than it would otherwise have run automatically, explaining how more gas was consumed then predicted from the hot water demand. This follows on to explain how the solar fraction is less than 0% on some days using the applied formula.

6.2 L26ST (INST)

The residents of house L26ST are fairly new to Lochiel Park and moved into their house in early 2016, where they were presented with a house full of monitoring equipment, smart sensors and solar technologies. Despite having a PV system in a previous property, neither the monitoring system nor solar water heater was explained to the new residents. The resident did inspect the solar thermal collectors after one of these leaked, creating a waterfall from their roof. They noticed the pipe

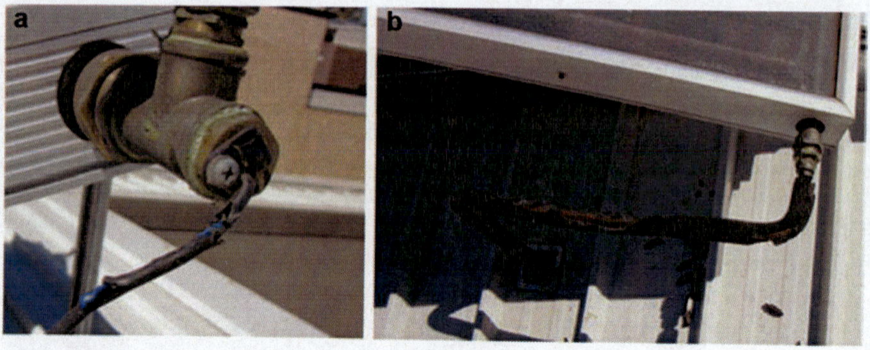

Fig. 9 Example of (**a**) control cables and (**b**) lagging that has been chewed by native birds

work was cold to touch when it should have been warm. The leak was fixed by the resident, and within one week of this, they were contacted by the authors who showed them the results presented in this paper. The residents were grateful to receive this information, however they were not surprised. They were unaware that their solar water heater had not been operating when they moved in to the house.

6.3 Other Issues

Finally, discussions with other Lochiel Park residents revealed that a large population of native Australian birds frequent household rooftops. These birds have been known to chew through power and control cables, pipe thermal lagging and even conduits that house power and control cables. An example of this is shown in Fig. 9, where a control cable has been damaged to the point where it is unable to tell the ground mounted pump to start pumping up to the relatively warm thermal collector.

7 Summary and Conclusions

Policy makers seeking to address the energy and carbon emission impact of domestic hot water use have typically encouraged the application of solar water heaters. The evidence available from this small study ($n = 4$) raises serious concerns about the reliability and overall energy performance of various types of solar water heaters.

Our research shows that the solar contribution is greatly affected by many factors such as volume and timing of hot water draw, the configuration of the system, and attacks from local wildlife. All systems are subject to issues of reliability that can result in lower solar contributions than expected.

This paper reported on current market trends, and mechanism in place for solar water heaters, in the Australian context. It focused on the real-world performance (solar fraction) of four solar gas-boosted water heaters in a low-energy housing estate. The latter was conducted for the years 2013 and 2018, by analysing the gas and hot water usage data collected every minute over these periods. The conclusions from this work include:

- Solar thermal water heaters are not gaining in market share relative to other products, and that market share is threatened by products such as air-source heat pumps, given the simpler and less expensive installation.
- Instantaneous type solar water heaters offer higher solar fraction than storage types, as these avoid the need for a pilot flame and hence the associated daily maintenance gas use.
- Despite the small sample size (n = 4), each solar water heater was operating correctly in 2013, the performance of half of these systems had deteriorated drastically by 2018.
- In addition to component corrosion, local wildlife has interfered with the equipment by chewing thermal pipe lagging, control cables and conduits that house cables.
- Although each solar water heater is fitted with a thermostatic mixing valve, it was seen in a number of houses that indoor water temperatures exceeded the valve's temperature limit.
- The majority of residents that have access to household monitoring systems but are unable to easily assess the performance of, or identify any faults with, their solar water heater.
- Although one resident did suspect and correctly identify a fault with their solar water heater based on their monitored gas usage, they did not detect a second and more significant fault.

Given the evidence from this small study, product developers should consider opportunities in this digital age to facilitate performance awareness for end-users, and to provide appropriate alerts when systems are grossly underperforming. This should not lessen the importance of providing a more robust product that achieves a more reliable and relatively high solar contribution.

References

1. Clean Energy Regulator. Small-scale installations by postcode. Clean Energy Regulator website - http://www.cleanenergyregulator.gov.au/RET/Forms-and-resources/Postcode-data-for-small-scale-installations#Smallscale-installations-by-installation-year – accessed 2 July 2019
2. Department of the Environment, Water, Heritage and the Arts - Energy Efficient Homes Package - Solar Hot Water Rebate – Guidelines and Application Forms - http://www.sjwilson-plumbing.com.au/shw-application-booklet.pdf - accessed 12 July 2019
3. Intelligent Energy Systems (IES). Small-scale technology certificates forecasts (2014 to 2016) e prepared for clean energy regulator by intelligent energy systems (IES). 2014. http://www.cleanenergyregulator.gov.au/DocumentAssets/Documents/Intelligent%20Energy%20Systems%20%E2%80%93%20February%202014.pdf

4. Butti, K., & Perlin, J. (1980). A golden thread: 2500 years of solar architecture and technology. New York: Cheshire Books
5. Baechler, M., Gilbride, T., Ruiz, K., Steward, H., & Love, P. (2007). High-Performance Home Technologies: Solar Thermal & Photovoltaic Systems. Golden, Colorado: National Renewable Energy Laboratory
6. Kroll, J., & Ziegler, F. (2011). The use of ground heat storages and evacuated tube solar collectors for meeting the annual heating demand of family-sized houses. Sol. Energy, 85(11), 2611–2621. doi: https://doi.org/10.1016/j.solener.2011.08.001
7. Thür, A., Furbo, S., & Shah, L. (2006). Energy savings for solar heating systems. Sol. Energy, 80(11), 1463–1474. doi: https://doi.org/10.1016/j.solener.2006.03.007
8. Djordjevic, S., Baverstock, G., Walker, E., & Urmee, T. (2014). Identification of the barriers for the slow uptake of solar water heaters in Australia. Paper presented at the Solar 14, the 52nd Annual Conference of the Australian Solar Council, Melbourne
9. Green Energy Markets. (2016). Small-scale technology certificates data modelling for 2016 to 2018: Report to the Clean Energy Regulator. Melbourne: Green Energy Markets
10. Ibrahim, O., Fardoun, F., Younes, R., & Louahlia-Gualous, H. (2014). Review of water-heating systems: General selection approach based on energy and environmental aspects. Building and Environment, 72, 259-286. doi: https://doi.org/10.1016/j.buildenv.2013.09.006
11. Hohne, P., Kusakana, K., & Numbi, B. (2019). A review of water heating technologies: An application to the South African context. Energy Reports, 5, 1-19. doi: https://doi.org/10.1016/j.egyr.2018.10.013
12. Armstrong, P., Bhagavathy, S., Kang, R., & McCulloch, M. (2019). Pitfalls in decarbonising heat: A misalignment of climate policy and product energy labelling standards. Energy Policy, 131, 390-398. doi: https://doi.org/10.1016/j.enpol.2019.04.012
13. Nel, P., Booysen, M., & Van Der Merwe B. (2018). Saving on household electric water heating: What works best and by how much? Paper presented at the 2017 IEEE Innovative Smart Grid Technologies - Asia: Smart Grid for Smart Community, ISGT-Asia 2017
14. Sadhishkumar, S., & Balusamy, T. (2014). Performance improvement in solar water heating systems — A review. Renewable and Sustainable Energy Reviews, 37, 191-198. doi: https://doi.org/10.1016/j.rser.2014.04.072
15. Saman, W., Babovic, V., Whaley, D., Liu, M., & Mudge, L. (2011). Assessment of electricity displacement due to installation parameters of solar water heaters. Adelaide: University of South Australia
16. Sangoi, J., & Ghisi, E. (2019). Energy efficiency of water heating systems in single-family dwellings in Brazil. Water (Switzerland), 11(5). doi: https://doi.org/10.3390/w11051068
17. Australian Government. Solar water heaters e MEPS and labelling requirements. Energy Rating website - http://www.energyrating.gov.au/products/water-heaters - accessed 2 July 2019
18. Halawa, E., Chang, K.C.., Yoshinaga, M..(2015). Thermal performance evaluation of solar water heating systems in Australia, Taiwan and Japan - A comparative review Renewable Energy 83, 1279–1286
19. Commonwealth of Australia (Department of Industry). Product profile: solar hot water heaters. Commonwealth of Australia (Department of Industry) - Equipment Energy Efficiency (E3) Committee; 2014 - http://energyrating.gov.au/sites/new.energyrating/files/documents/Product-Profile_Solar-Water-Heaters_August-2014.pdf - accessed 2 July 2019
20. Berry, S., Whaley, D., Davidson, K., & Saman, W. (2014). Do the numbers stack up? Lessons from a zero carbon housing estate. Renew. Energy, 67, 80–89. doi: https://doi.org/10.1016/j.enpol.2014.05.011
21. Whaley, D., Saman, W., Halawa, E., & Mudge, L. (2010). Lessons learnt from implementing intelligent metering and energy monitoring devices in a new housing development. Australian Solar Energy Society Conference, Canberra, Dec. 2010
22. Saman, W.Y., Whaley, D., Mudge, L., Halawa, E., and Edwards, J. (2011)"The Intelligent Grid in a New Housing Development" Final Report, Project P6,CSIRO Intelligent Grid Research Cluster, University of South Australia

23. AS/NZS 4234:2008 (Australian/New Zealand Standard), Heated water systems - calculation of energy consumption, Standards Australia
24. D. Whaley, R. Liddle, L. Mudge, E. Harmer, W. Saman, 2014: In-situ evaluation of water consumption and energy use in Australian domestic water heaters, World Renewable Energy Congress, London, Aug 2014
25. AS/NZS 3666 Set:2011, Air-handling and water systems of buildings Set, Standards Australia
26. AS 4552–2005, Gas fired water heaters for hot water supply and/or central heating, Standards Australia

Modelling Residential End-Use Electricity Consumption Using Statistical and Artificial Intelligence Approaches

Ebru Ada and Merih Aydınalp Köksal

Abstract Both appliance ownership and characteristics determine residential electricity consumption, the exact electricity consumption of each appliance depends on the usage patterns of the occupants. Various approaches have been used to determine the appliance specific electricity consumption at household level. One of the statistical approaches used since 1980s is the conditional demand analysis (CDA) which takes into account appliance ownership, appliance characteristics (such as volume, size, power, etc.), weather, and market data to disintegrate the billing data into appliance specific form. Since 1990s, neural network and fuzzy logic concepts which are artificial intelligence-based modelling approaches have been used to determine the appliance specific electricity consumption using various types of data available on appliance and occupant characteristics. An artificial intelligence-based approach which combines the prediction performance capabilities of neural networks and fuzzy logic is adaptive network based fuzzy inference system (ANFIS) models have been used such as for the water quality, hourly electricity load demand, and air pollutant emission estimation studies. According to the investigations and researches, it has been observed that ANFIS approach has not been used to model end-use electricity consumption of the residential sector, yet. The aim of this study is to compare the prediction performances of CDA based and ANFIS based approaches to determine the electricity consumption based on the appliances at household level. An extended survey data which covers detailed information about 92 different types of appliances including all domestic and minor appliance properties, occupant characteristics, and billing information of 260 homes is used for developing the CDA and ANFIS models. It has been found that while CDA model is developed with the mean absolute percent error (MAPE) of 12%, ANFIS model has been finalized with the mean absolute percent error (MAPE) of the testing data as 17%. Besides the MAPE values, based on the error distribution graphs of each approach, it has been observed that CDA model has significantly high prediction performance than ANFIS approach for the appliance-based electricity consumption estimation. As a last step of the study, different scenarios have been applied to esti-

E. Ada · M. A. Köksal (✉)
Department of Environmental Engineering, Hacettepe University, Çankaya, Ankara, Turkey
e-mail: aydinalp@hacettepe.edu.tr

P. Bertoldi (ed.), *Energy Efficiency in Domestic Appliances and Lighting*, Springer Proceedings in Energy, https://doi.org/10.1007/978-3-030-79124-7_13

mate the appliance impacts on household electricity consumption. Scenarios are applied by considering the usage patterns of lighting, dishwasher and washing machine and percent decrease values in overall electricity consumption are observed as 0.7%, 4.3%, 1.4%, respectively.

1 Introduction

In recent years, global electricity usage increased substantially due to the increase in population and life standards. The demand for electricity also rapidly increases in parallel with the global economic growth. Electricity is one of the most important parameters of socio-economic development. Thus, providing a reliable, sufficient, good quality and environmental-friendly electricity is essential. Apart from electricity supply, it is also important to determine a strategy for efficient electricity usage methods to sustain the supply of electricity requirement. In every aspects of life, efficiency precautions can be implemented such as industrial demands, transportation and residential use.

The increase in electricity consumption for residential sector is another important concern and promotes to increase efficiency studies by applying End-Use Electricity Consumption Modelling approaches in residential areas, as well. Many factors can be listed which effect the increase in electricity use at homes. Number, education level, income and age of the occupants, characteristics and usage patterns of the appliances can be listed as the main parameters which effects the consumption level of the residential electricity use.

In many countries, residential sector significantly contributes to the increase in the total national electricity consumption due to the increase in population and also in innovative technological appliances, which aim to increase the comfort level of the occupants. Focusing on residential electricity consumption to obtain the reduction measures is thus crucial to reach determined sustainability targets to decrease the total electricity usage. Besides, apart from the major appliances (such as refrigerator, freezer, clothes washer and dryer, dish washer and oven), electrical devices regarding information communication technology (ICT), such as TVs, computers and related auxiliaries, are responsible for about 15% of global residential electricity consumption including stand-by mode consumption [1]. Currently, it is estimated that approximately one quarter of total electricity consumption is from the residential sector in Turkey [2]. Almost two thirds of the generated electricity in Turkey is from coal and natural gas fuelled power plants.

Since the residential end-use electricity consumption has a high importance among the overall electricity usage in the world, there are many approaches that are studied in the literature to identify the level of consumption of each end-use and impact of factors affecting the residential electricity consumption. There are many end-use electricity efficiency and electricity demand improvement studies applied to reduce the electricity consumption and determine the efficient usage methods in residential sector. Engineering Methods, Artificial Neural Networks (ANN) and

Conditional Demand Analysis (CDA) models are the major approaches which have been performed for the electricity modelling studies for the residential electricity consumption [3].

Engineering methods are generally applied by using a simulation program to estimate the end-use electricity consumption by utilizing a collected data from an extensive survey including detailed information on both occupant behaviours and appliance properties. Engineering method basically depends on mathematical equations and correlations as well as the engineering sense to solve the problems by minimizing the risks in an engineering design. Thus, as a disadvantage of this method, preparation of the input data takes long time and needs very sensitive study period. On the other hand, even though end-user behaviours and socio-economic structure parameters have significant impacts on the residential electricity usage, these items cannot be evaluated in terms of numerical manners to implement into an engineering method. Therefore, uncountable variables which affect the electricity consumption should also be taken into consideration to prevent the difficulties which can be faced during the model development and result evaluation period [3].

Another applicable important model development study for electricity consumption in residential sector is ANN approach. ANN is inspired by the biological neural system of human brain. The human brain can be thought as a perfect computer that can work fast and able to adopt any changes in the environment. The most amazing capability of a human brain is to figure out unclear and incomplete data, analyse fuzzy information, and provide a logical meaning by making own decisions. To implement this ability on estimation studies, a kind of artificial intelligence method is used by researchers that is called as artificial network based adaptive neuro fuzzy inference systems, ANFIS, which is able to integrate both Neural Networks (NN) and fuzzy logic principles. ANFIS is one of the well-known approaches for estimation studies that can adapt the fuzzy systems to the existing data. Apart from the electricity consumption estimate studies, ANN based ANFIS approach is also being applied for modelling of any type of non-linear processes such as load forecasting, predicting weather for wind farms and solar radiation, *etc*. In contrast to engineering models, it is possible to receive more versatile results with ANN studies by evaluating all environment conditions during data evaluation of the model [3, 4].

The development of the concept of fuzzy logic is based on the way the human brain works. It is provided to develop approximate and predictive logic separating human from machines with fuzzy logic. Fuzzy logic operations are the whole process of defining the problem and interpreting the model with fuzzy sets. It can be applied to solve the complexity in detailed analysis of real-time operations where there are no mathematical model exists. Fuzzy logic concept basically enables the formation of a system with very complex and nonlinear, traditional methods [5]. With the help of fuzzy logic, problems which contain uncertainties, nonlinear and missing data can be modelled easily. Because of this structure, fuzzy logic is widely used in many areas, primarily control, decision making and prediction problems. With ANN approach, fuzzy logic can be learned and thus the result achieved by neural networks can be understood easily. In fuzzy neural networks, fuzzy logic enhances the information presentation capabilities of classical neural networks. In this way, the ability to learn, to express, adapt, and to process information inaccurate data and results are provided.

ANFIS is one of the best-known fuzzy systems. It was developed by Jang in 1993 based on the Sugeno model [6]. It is a hybrid artificial intelligence method that uses parallel computing and learning ability of artificial neural networks and the inference feature of fuzzy logic. It consists of learning algorithm of fuzzy inference system and back-propagation algorithm with least squares method. It is used to identify the system using environmental information and system data. It also includes continuous self-updating data analysis techniques such as numerical grouping and rule setting. ANFIS program allows you to assign the rules by experts as well as your own. In this way, it allows ANN to benefit from expert opinions in many estimation problems and gives better results. ANFIS approach has been applied by the researchers to estimate day-ahead electricity prices, electricity and energy load demands in selected regions. ANFIS also has been used in various areas for modeling studies [7–12].

CDA, which is a regression-based method, can be applied especially for the estimation studies of residential Unit Electricity Consumption (UEC) by using a survey data [13]. Thus, total electricity consumption of a household can be divided into its end-uses. Different variables which are responsible for electricity consumption can be easily adapted to the model to increase the prediction performance efficiency. CDA model can be performed even with a fewer detailed data of end-user and less detailed data of appliance characteristics and usage. Nevertheless, as a disadvantage of CDA, even if it is performed by using less detailed data, due to the multicollinearity problem, in some cases the results can be suspicious. Additionally, socio-demographic impacts on appliance usage can be estimated through CDA when high range of dataset is available, as well [14].

CDA was first performed by Parti and Parti (1980) [15]. They were focused on the determination of electricity demand functions with respect to the particular appetences by using a covariance framework analysis. Electricity billing data regarding more than 5286 households in San Diego Country, weather data, appliance ownership and demographic variables are combined for empirical analyses for individual households. It was observed that the CDA results were more accurate to reach appliance-based electricity demand and straight forward than the engineering methods. Additionally, another advantage to apply CDA model rather than other engineering methods is that it has been proved that the CDA model helps to estimate accurate electricity consumption even in variable price, income, and size of households. CDA has then been applied by various researchers to estimate the electricity and gas end-use consumption of homes in various regions and countries [13, 14, 16–19].

The objective of this study is to determine the approach with high prediction performance in determining the contribution of major residential end-uses and the effects of major factors affecting the residential end-use electricity consumption using a detailed survey database. Within this study, CDA and ANFIS based models are developed to identify the electricity consumption of major end-uses of homes located in Ankara, Turkey, and the prediction performances of CDA and ANFIS models are compared.

This paper consists of four major sections. After presenting introductory information about the study, data sources used in the study is briefly explained in the second section. Afterwards, the methodology for the model development for both CDA and ANFIS approaches have been presented in detailed within the third section. The paper is concluded by presenting the major results regarding the applied model performances.

2 Sources of Data Used

Data from an extended survey is used for the development of models. Survey includes sections with information on the dwelling (ownership, footage, type, rooms, etc.), socio-demographic properties, education level, gender, number, income of the all occupants, main and secondary heating systems, cooling systems, and appliance properties (type, size, model, age, capacity, power, usage, etc.) and ownership of all appliances (refrigerator, deep freezer, oven, cooker, microwave, fan, television, receiver, washing machine, dryer, dishwasher, private computer, vacuum cleaner, iron, and all other small house appliances) and lighting. At the last section of the survey, the average annual billing data and consumption information of the households are reported. The questionnaire form included forty-five questions and conducted to 260 households in Ankara province of Turkey in 2012.

According to the survey results, information on 92 different types of appliances are reported. There are 20 detached type houses among 260 dwellings. The largest household is 500 m^2 net area and the smallest one is 40 m^2. Every house has at least one refrigerator and only 17 of them have second refrigerators. Except a few numbers of them, almost all of the households have vacuum cleaner, washing machine and iron. Regarding the lighting data, there are high number of incandescent lamps preferred by the households and some of them have almost 50 lamps.

3 Methodology of the Study

After all survey data are collected for each household and analyzed in detailed, the main appliances are determined by calculating the stock data which are used during the model development stage. As a first step, CDA model is developed and approach performed. Then, reliable data is obtained for further model development steps. After input and output data are obtained from the CDA model at the first stage, ANFIS model development has been started and approach is performed accordingly. The prediction performances of CDA and ANFIS models are compared to determine the most efficient method. Different scenarios are applied to determine the saving measures based on the model results.

3.1 CDA Model Development

As stated previously, detailed survey data for 260 houses and 92 appliances are used for the CDA model development. Total annual electricity consumption (kWh/y) of all appliances is calculated using the survey data and usage information from manufacturers. Then, stock data of each appliance is determined using the ownership information. Afterwards, appliance-based percent electricity consumption is calculated to determine the percent electricity consumption of each appliance.

The analyses show that the appliances with high contribution (>1.5%) to the total annual electricity are main refrigerator, lighting, dishwasher, iron, main television, vacuum cleaner, washing machine, oven, boiler, desktop computer, secondary television, receiver, notebook, deep freezer, and secondary refrigerator. These appliances account for about 87% of the total average household electricity consumption that are taken into consideration as a first step of the model development and rest of the appliances which represent 13% of the total average household electricity consumption are assumed to be represented as the base load of the consumption.

In the next step of the study, UEC equations are developed for these appliances by selecting parameters, which affect the consumption by these appliances. The CDA model is then developed by combining all UEC equations as shown below:

$$HEC_{it} = \sum_{j=1}^{J} \left(UEC_{ijt} * S_{ij} \right) \tag{1}$$

where,

HEC_{it}: electricity consumption by household i in period t, kWh
UEC_{ijt}: end-use j unit electricity consumption of household i in period t, kWh
S_{ij}: household i's binary indicator ownership as per ownership of appliance j

Before the first regression analyse trial based on the UEC equations, multicollinearity of the selected parameters has been checked based on the determined appliances and related variables. If one independent variable is highly correlated with another independent variable the marginal contribution of that independent variable is influenced by other independent variables. These relationships between the variables cause problem for the model efficiency and called as multicollinearity. Multicollinearity problem of the determined variables can be checked by taking into consideration the variance inflation factor (VIF) values which received from the coefficients table by performing co-linearity diagnostics analysis. VIF detects the multicollinearity in a regression analysis. If the VIF is greater than 5, it means that the variables are highly correlated. In accordance with the co-linearity diagnostics analysis with respect to the determined appliances and related variables. It has been found that there was no multicollinearity problem. SPSS Statistics – Version25 program has been used during the multicollinearity check and regression analyses. If the VIF is greater than 5 the variables are highly correlated [20].

Regarding the evaluation of the model output, Adjusted R^2 and significance (Sig.) values are also taken into consideration. The smaller Sig. value means greater the evidence of statistically significant difference. To interpreting the Sig. values, ranges are given below shall be evaluated as a rule of thumb [20];

- If Sig. is between 0.01 and 0.05 the output has statistically significant difference
- If Sig. is between 0.001 and 0.01 the output has moderately statistically significant difference
- If Sig. is less than 0.001 the output has high statistically significant difference

Besides, in statistical analysis, Durbin-Watson result shall also be considered to check the autocorrelation that refers lag of correlation or serial correlation in a regression analysis. Durbin-Watson value is in the range between 0 and 4. The optimum Durbin-Watson value is accepted as 2 that indicates there is no autocorrelation within the selected variables in the model [20].

In regression analyses, it is also necessary to examine all regression coefficients and overall constant to assess the relative influence of independent variables. By evaluating the magnitude of coefficients, the statistical significance of coefficients is tested. If the coefficient for a particular variable is significantly greater than zero, it means that the variable contributes to the predictive ability of the regression equation. In this way, it is possible to distinguish the variables that are more useful for prediction from those that are less useful. Therefore beta (B) values are also analysed and less useful items are removed from the model during the development period.

According to the first trial, it was observed that the B value of the constant is less than zero which directly refers to the multicollinearity problem. Therefore, the independent variables that are thought to be highly correlated to each other based on the Sig., VIF and B values are removed from the first regression analyse trial. Sig. values of main refrigerator variables are received as 0.301 and 0.092 for volume of main refrigerator and age of main refrigerator, respectively. Mentioned values has not enough statistically significant difference as explained above. As same as main refrigerator, Sig. values of the main television screen size and hourly usage per day have been received as 0.401 and 0.847, respectively. Additionally, since almost all of the homes have main refrigerator and main television, related variables are removed from the model to prevent the multicollinearity.

After various attempts to develop a model with significant parameters and no multicollinearity, and removing 10 outliers, a CDA model with an intercept and 12 variables is developed. The coefficients of the finalized CDA model is presented in Table 1.

The abbreviations used in the above table are.

DW: Dishwasher ownership (Dummy Variable: 1 = Yes, 0 = No).

CYC_{DW}: Sum of dishwasher cycles in short and long programs per week.

IR: Iron ownership (Dummy Variable: 1 = Yes, 0 = No).

$HUSE_{IR}$: Hourly usage of iron per week.

VC: Main vacuum cleaner ownership (Dummy Variable: 1 = Yes, 0 = No).

$HUSE_{VC}$: Hourly usage of main vacuum cleaner per week.

WM: Washing Machine ownership (Dummy Variable: 1 = Yes, 0 = No).
CYC_{WM}: Cycle of washing machine use per week.
OV: Main oven ownership (Dummy Variable: 1 = Yes, 0 = No).
$HUSE_{OV}$: Hourly usage of main oven per day.
BOIL: Boiler ownership (Dummy Variable: 1 = Yes, 0 = No).
AREA: Footage of the dwelling, m^2.
PC: Main desktop computer ownership (Dummy Variable: 1 = Yes, 0 = No).
$HUSE_{PC}$: Hourly usage of desktop computer (active use + st.by use) per day.
TV2: Secondary television ownership (Dummy Variable: 1 = Yes, 0 = No).
SCR_{TV2}: Screen size of secondary television, inch.
REC: Main receiver ownership (Dummy Variable: 1 = Yes, 0 = No)
$HUSE_{REC}$: Hourly usage of main receiver (active use and standby use) per day.
REF2: Secondary refrigerator ownership (Dummy Variable: 1 = Yes, 0 = No).
VOL_{REF2}: Volume of secondary refrigerator, L.
LGHT: Lighting ownership (Dummy Variable: 1 = Yes, 0 = No).
$WINCA_{LGHT}$: Power consumption of incandescent lamps, watt.
$WCFL_{LGHT}$: Power consumption of compact fluorescent lamps, watt.

As can be seen from the above Table 1, the variables are all significant and have positive coefficient values. Also, VIF and the Sig. values of all the variables are between in the reliable ranges. Thus, it can be assumed that these variables are acceptable in terms of multicollinearity. After complete the significant statistical inferences, regression analyses are performed to check the CDA model prediction performance efficiency.

Table 1 Coefficients of the CDA model

Variables	Unstandardized coefficients		Standardized coefficients	t	Sig.	Collinearity statistics	
	B	Std. error	Beta			Tolerance	VIF
(Constant)	662.500	90.842		7.293	0.000		
$DW*CYC_{DW}$	91.334	13.342	0.248	6.845	0.000	0.734	1.362
$IR\text{-}HUSE_{IR}$	208.803	29.080	0.264	7.180	0.000	0.717	1.394
$VC*HUSE_{VC}$	91.100	24.125	0.131	3.776	0.000	0.798	1.253
$WM*CYC_{WM}$	108.797	31.327	0.127	3.473	0.001	0.725	1.380
$OV*HUSE_{OV}$	167.038	31.699	0.172	5.269	0.000	0.913	1.096
BOIL*AREA	2.056	0.379	0.174	5.423	0.000	0.940	1.064
$PC*HUSE_{PC}$	18.729	2.865	0.209	6.537	0.000	0.942	1.062
$TV2*SCR_{TV2}$	2.917	0.942	0.102	3.097	0.002	0.885	1.130
$REC*HUSE_{REC}$	31.729	6.159	0.166	5.152	0.000	0.934	1.070
$REF2*VOL_{REF2}$	2.236	0.283	0.260	7.901	0.000	0.895	1.117
$LGHT*WINCA_{LGHT}$	0.283	0.122	0.082	2.315	0.021	0.762	1.312
$LGHT*WCFL_{LGHT}$	0.608	0.273	0.082	2.224	0.027	0.708	1.412

3.2 ANFIS Model Development

After CDA model is finalized, ANFIS approach is performed with the 12 variables that are obtained from the finalized CDA model and 250 households. MATLAB software is used for the ANFIS model performance.

ANFIS is a hybrid artificial intelligence method that uses parallel computing and learning ability of artificial neural networks and the inference feature of fuzzy logic. Thanks to its hybrid learning ability, ANFIS helps to create optimum membership function distributions between the inlet and outlet data. It consists of learning algorithm of fuzzy inference system and backpropagation algorithm with least squares method. It is used to identify the system which is using the environmental information and system data. It also includes continuous self-updating data analysis techniques such as numerical grouping and rule setting. In a fuzzy inference system, basically there are two types of partitioning which are grid and sub-clustering partitioning. Grid partition method is called as GENFIS1 and it generates rules by enumerating all possible combinations of membership functions of all inputs. SUBCLUST technique which is called as GENFIS2 is used to produce scattering partition. Both GENFIS1 and GENFIS2 use a given training data set to generate an initial fuzzy inference system that can be fine-tuned via the ANFIS command. GENFIS1 always needs subsequent optimization by ANFIS command, while the one generated by GENFIS2 can sometimes have a good input-output mapping precision already. In general, if there are inputs less than 6 and a large size of training data grid partition method is followed. Otherwise, subtractive clustering method is performed to decrease the number of rules.

According to this study, as a first step of the model development phases, since 12 inputs with a large size of training dataset are considered, GENFIS2 is used for the model estimation performance. For ANFIS model trial, dataset is divided into training and testing groups. In the light of previous studies approximately 75% of the data set has been used as training data which includes the appliance variables for 200 households and 25% which includes the appliance variables for 50 households has been selected as testing data. Training data includes the maximum and minimum values for each determined input variable. Testing data is selected randomly after training data determined. Afterwards, mentioned training and testing data are imported to the system and model is performed.

ANFIS model development finalized after too many trials are performed by increasing the input numbers to reach the optimum MAPE values that are received from both the overall model output and calculated in accordance with the testing data. As a consequence, the most accurate ANFIS model is generated with 24 inputs.

Configuration of the finalized ANFIS model can be described as following;

```
Input_Daa = ANFISModel1(1:200,1:24);
Output_Data = ANFISModel1(1:200,25);
Train_Data = ANFISModel1(1:200,:);
Test_Data = ANFISModel1(201:250,:);

RangesOfInfluence = 0.5;
fismat = genfis2(Input_Data,Output_Data,RangesOfInfluence);
epoch_n = 5;
error_tolerance_rate = 0.005;
trnOpt = [epoch_n, error_tolerance_rate];
dispOpt = ones (1,4);

[out_fis, trnErr] = anfis(Train_Data, fismat, trnOpt, dispOpt);
MAPE = trnErr(length(trnErr));
anfis_result = evalfis(Test_Data(:,1:24), out_fis);
c                    o                    m                    p
= [Test_Data(:,25), anfis_result, Test_Data (:,25) anfis_result];
```

4 Results and Discussion

After developing end-use electricity models based on CDA and ANFIS approaches using the detailed data of 250 homes, the prediction performances of these models are compared based on various parameters. Afterwards, scenario applications are performed to estimate the appliance-based electricity consumption saving methods.

The statistical results of the developed CDA model are presented in model summary table in Table 2.

According to the above table Durbin-Watson and Adjusted R^2 values are sufficiently reliable presenting that the model can successfully estimate the annual electricity consumption of the households. The MAPE of the estimates is calculated as 12% for the CDA model.

The distribution of the model error which is the difference between the estimated and actual annual electricity consumption of 250 homes is presented in Fig. 1.

Table 2 Model summary of the CDA model

Description	Value
R	0.878
R Square	0.771
Adjusted R Square	0.759
Std. Error of the Estimate	419.35702
MAPE	12%
Durbin-Watson	1.778

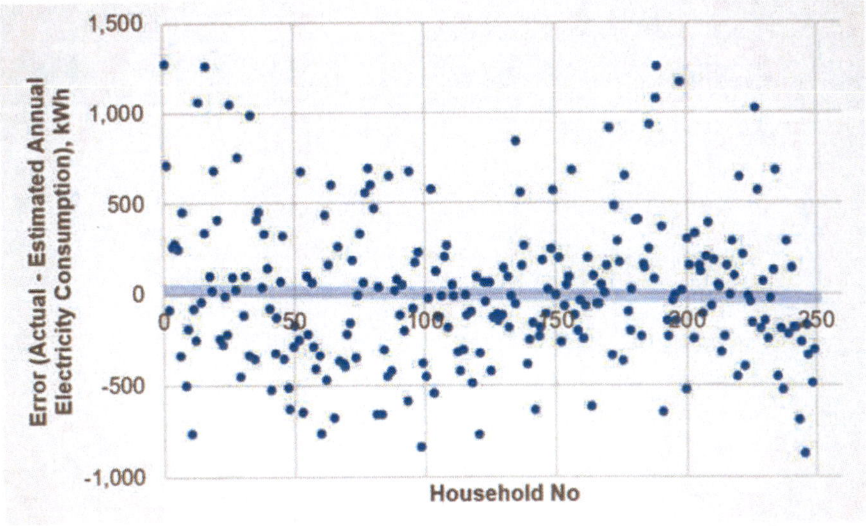

Fig. 1 The distribution of the annual electricity consumption errors of the households estimated by the CDA Model

```
ANFIS info:
        Number of nodes: 10027
        Number of linear parameters: 5000
        Number of nonlinear parameters: 9600
        Total number of parameters: 14600
        Number of training data pairs: 200
        Number of checking data pairs: 0
        Number of fuzzy rules: 200
```

Fig. 2 ANFIS model information output

As can be seen from the figure the errors are randomly distributed and are within −1000 and +1500 kWh range.

After ANFIS model is performed with the model configuration as given in previous section, ANFIS performance information is received as given in Fig. 2 below.

MAPE values of each epoch for the finalized ANFIS modes is given in Table 3 below. Since ANFIS Model has tried to create a new matrix system with the testing dataset which it has never been seen before, the percent error value became significantly high with respect to the training dataset.

In Fig. 3, the actual and estimated electricity consumption of 250 homes obtained from the CDA Model is presented. The graph also presents that the estimates of the CDA model are in good agreement with the actual electricity consumption for 250 homes (Fig. 3).

Table 3 MAPE of ANFIS model output

Epoch No.	MAPE
1	0.227731
2	0.258013
3	0.272949
4	0.24602
5	**0.252061**

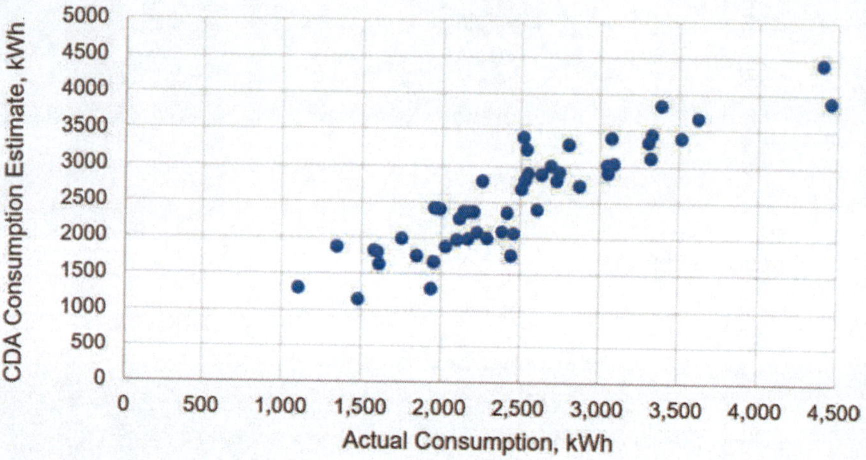

Fig. 3 The distribution of the actual electricity consumption and estimated consumption results by the CDA Model

Actual electricity consumption and estimated results by ANFIS model distribution is presented in Fig. 4. When the training data is adopted to the ANFIS Model, the system is able to learn the configuration and formulization easily. Therefore, the error percentage is almost 0% for the training dataset. The distribution chart given below presents only for the testing dataset, which resulted in a MAPE of 17%.

CDA and ANFIS approaches are provided reliable estimation results for appliance-based electricity consumption. However, as it can be seen visually from the above graphs, CDA approach results are better than ANFIS approach. In that case, CDA can be considered to be a guide for the next studies to provide retrospective information.

After regression analyses and model development stages are completed by using CDA and ANFIS approaches, respectively, different scenarios are applied to determine and discuss the appliance efficiency methods. Since the prediction performance efficiency of CDA model is quite better than ANFIS model, scenarios are applied by using the generated model with CDA.

Applied scenarios are described step by step as following;

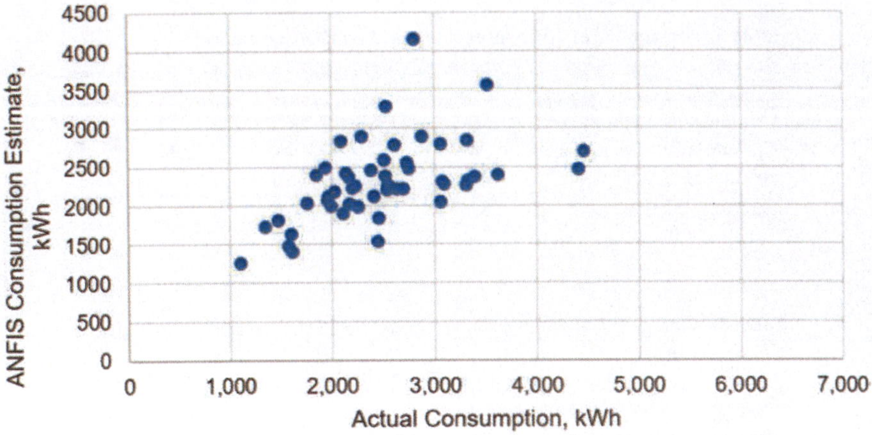

Fig. 4 The distribution of the actual electricity consumption and estimated consumption results by the ANFIS Model

Table 4 Average electricity consumption (EC) of 195 households based on CDA Scenario-1

	Lighting EC	Total EC
Original consumption, kWh	176	2549
Consumption after scenario application, kWh	159	2532
Percent reduction, %	9.5	0.7

1. It is assumed that all households use compact fluorescent lamps instead of incandescent lamps. Wattage values are taken into consideration during the scenario application. The wattage value of the compact fluorescent lamps is taken as 20 W with respect to the average wattage value of the compact fluorescent lamps that are used by all households. Number of incandescent lamps are added to the number of compact fluorescent lamps by multiplying with 20 W. First scenario is applied for 195 households. For household selection, it is checked that the selected households have at least 1 incandescent lamp. Both appliance-based and overall electricity consumption values are compared to explain the efficiency patterns. Related results are given in Table 4.
   According to the results which are shown in above table, using compact fluorescent lamps instead of incandescent is helped to decrease electricity consumption approximately 1% in overall usage per household.
2. It is assumed that all households are use their dishwashers maximum three times in a week. 197 households are taken into consideration for the second scenario application. The ownership of the dishwashers for the selected households have been checked, as well. Cycles of the dishwashers are changed to three times in a week for the households which are using their dishwashers more than three

Table 5 Average EC of 197 households based on CDA Scenario-2

	Dishwasher EC	Total EC
Original consumption, kWh	360	2752
Consumption after scenario application, kWh	242	2634
Percent reduction, %	32.7	4.3

Table 6 Average EC of 230 households based on CDA Scenario-3

	Washing machine EC	Total EC
Original consumption, kWh	224	2633
Consumption after scenario application, kWh	186	2595
Percent reduction, %	16.9	1.4

times. Cycles of the dishwashers that are less than three times in a week are stayed same. Both appliance-based and overall electricity consumption values are compared to explain the efficiency patterns. Related results are given in Table 5.

According to the results which are shown in above table, reducing the dishwasher use to maximum three times in a week is helped to decrease the electricity consumption approximately 4% in overall usage per household.

3. It is assumed that all households are use their washing machines maximum two times in a week. 230 households are taken into consideration for the third scenario application. Cycles of the washing machines are changed to two times in a week for the households which are using their washing machines more than two times. Cycles of the washing machines that are less than two times in a week are stayed same. Both appliance-based and overall electricity consumption values are compared to explain the efficiency patterns. Related results are given in Table 6

   According to the results which are shown in above table, reducing the washing machine use to maximum two times in a week is helped to decrease electricity consumption approximately 1% in overall usage per household.

As a consequence, decreasing the number of cycles of dishwasher is provided good results to decrease the overall electricity consumption. It can be considered as a governmental policy that should be followed and implemented to the people.

5 Conclusion

In this paper, total residential electricity consumption and the electricity consumption of each appliance had been investigated with respect to the usage patterns of the occupants based on the appliance ownership and characteristics. Among the various types of approaches, regression based CDA approach and one of the major fuzzy-inference systems which is called as ANFIS approach had been considered during the model development phase. An extended survey data which includes appliance ownership information, physical properties and detailed characteristics of 260 houses and 92 appliances had been used for the development of models. For the CDA model development, appliance-based percent electricity consumption is calculated to determine the percent electricity consumption of each appliance. Appliances with low contribution, less than 1.5%, are not included in the model development. Main refrigerator, lighting, dishwasher, iron, main television, vacuum cleaner, washing machine, oven, boiler, desktop computer, secondary television, receiver, notebook, deep freezer, and secondary refrigerator are responsible 87% of total average household electricity consumption and rest of them which represent 13% of the total average household electricity consumption were ignored.

In the next step UEC equations created for the obtained appliances by selecting variables which cause electricity consumption to perform regression analyses. After CDA model finalized, ANFIS approach is performed with the same variables. Estimation results and related MAPE values of each model had been compared to decide more efficient approach. The MAPE of the developed CDA model is determined as 12%. Finalized CDA model variables are used for the inputs of ANFIS approach. After several trials are performed, the most accurate prediction results are obtained with the MAPE of 17% for testing data of ANFIS model. Afterwards, error distribution graphs have been plotted by considering the selected households for the testing data of ANFIS approach for both CDA and ANFIS model estimates. The results are also quite better visually for CDA approach and CDA has significantly high efficiency performance than ANFIS approach for the appliance-based electricity consumption estimation.

As a last step, different scenarios have been applied to observe the appliance-based efficiency methods and saving measures in annual electricity consumption. Type of lamps which are used for lighting, weekly dishwasher and washing machine usages are taken into consideration for the scenario application. Percent values of decrease per household in overall electricity consumption are observed as 0.7%, 4.3%, 1.4% for lighting, dishwasher and washing machine, respectively as per CDA model scenario applications.

According to the results of this study it has been observed that both two models provided accurate estimation performances to interpret the appliance-based electricity consumption. Even though the estimation methodology of the ANFIS approach is in much more sensitive manner than the regression based CDA approach, more accurate appliance-based electricity consumption estimation results are obtained with the CDA model.

Therefore, to increase the ANFIS model performance efficiency, some improvement methodologies can be implemented to the ANFIS approach in the future studies. For instance, by implementing the genetic algorithm to ANFIS and performing the estimation study with GA-ANFIS method as applied in the literature previously would be very useful to optimize the model. Another recommendation, Principal Component Analysis (PCA) methodology can be implemented to the database to analyze the raw data in detail. It will help to find the significant variables for each approach.

References

1. Mary P., Richard F.H., Konstantinos J. C. ICT Entertainment Appliances' Impact On Domestic Electricity Consumption. Renewable and Sustainable Energy Reviews. 2017, pp. 843–853
2. Meryem N. M. and Hatice T. M. Understanding Residential Electricity Consumption Considering Energy Efficiency Policies and The Impact on the Electricity System. M.Sc. Thesis, Industrial and Systems Engineering, Istanbul Şehir University, İstanbul, Turkey, 2014
3. Merih A. A New Approach for Modeling of Residential Energy Consumption. Ph.D. Thesis, Mechanical Engineering, Dalhousie University, Halifax, Nova Scotia, Canada, 2002
4. Harsh K., Bharath N., Siddesh C.S., and Kuldeep S. An Introduction to Artificial Neural Networks. International Journal of Advance Research and Innovative Ideas in Education. 2016, pp. 3–7
5. Hugh C. Fuzzy Logic – The Genius of Lotfi Zadeh "Father of Fuzzy Logic". IEEE Industrial Electronics Magazine. September 2017
6. Jyh-Shing, Roger J. ANFIS: Adaptive Network Based Fuzzy Inference System. IEEE Transactions on Systems Man Cybernetics. June 1993, pp. 665–685
7. Jin-Liang Z., Yue-Jun Z., De-Zhi L., Zhong-Fu T. and Jian-Fei Ji. Forecasting Day-Ahead Electricity Prices Using A New Integrated Model. Electrical Power and Energy Systems. September 2018, pp. 541–548
8. Li-Chih Y. and Mei-Chiu P. Using Adaptive Network Based Fuzzy Inference System to Forecast Regional Electricity Loads. Energy Conversation and Management. August 2007, pp. 205–211
9. Özkan D., Adnan K. and Mehmet T. Electric Energy Load Forecasting Using ANSI and ARMA Methods. J. Fac. Eng. Arch. Gazi Univ. 2010, pp. 601–610
10. Yeşim O. and Mehmet A. Medium Term Electricity Energy Demand Prediction in Turkey with Adaptive Neuro Fuzzy Inference System. M.Sc. Thesis, Industrial Engineering, Institute of Science and Technology, Gazi University, Ankara, Turkey, 2010
11. Muhammad F. and DoHyeun K. A Prediction Methodology of Energy Consumption Based on Deep Extreme Learning Machine and Comparative Analysis in Residential Buildings. Electronics September 2018, 7, 222
12. B. KanagaS, D. D., R. Narmatha B. and V. Agnes Idaya S. A Hybrid Wind-Solar Energy System with ANFIS based MPPT Controller. Journal of Intelligent & Fuzzy Systems. 2018, pp. 1579–1595
13. K. H. T. Using Conditional Demand Analysis to Estimate Residential Energy Use And Energy Savings. Energy. 2007, pp. 1279–1283
14. Shigeru M. How Do Household Characteristics Affect Appliance Usage? Application of Conditional Demand Analysis to Japanese Household Data. Energy Policy. April 2016, pp. 214–223
15. Micheal P. and Cynthia P. The Total and Appliance-Specific Conditional Demand for Electricity in the Household Sector. The Bell Journal of Economics. Spring 1980, pp. 309–321

16. Dennis J. A., Cyrus S. and Pamela K. Conditional Demand Analysis for Estimating Residential End-Use Load Profiles. The Energy Journal. July 1984, pp. 8–97
17. Merih A. K. and V. Ismet U. Comparison of Neural Network, Conditional Demand Analysis, And Engineering Approaches for Modeling End-Use Energy Consumption in The Residential Sector. Applied Energy. 2008, pp. 271–296
18. Guy R. N. and Cara L. D. A Model of Residential Energy End-Use in Canada: Using Conditional Demand Analysis to Suggest Policy Options for Community Energy Planners. Energy Policy. April 2013, pp. 133–142
19. Bodil M. L. and Runa N. Household Electricity End-Use Consumption: Results from Econometric and Engineering Models. Energy Economics. 2004, pp. 179–200
20. United Nations Conference on Trade and Development, 2018 Handbook of Statistics. 2016

Going Digital! South Africa's Energy Labeling Program Adopting Digital Tools of Database, QR Code and Smart Phone APP

Theo Covary and Jiayang Li

Abstract Legislation to make energy labelling mandatory for 11 appliances, selected by South Africa under its Standards and Labelling (S&L) program, came into effect in 2015. As would be expected, the number of models requiring regulatory approval grew annually, and it became apparent to national Government and the Regulator that effective oversight would require a dedicated fit-for-purpose product registration and database system. Thus, technical specifications were developed, followed by the appointment of an ICT company to build the system towards the end of 2018. Simultaneously, the S&L project team evaluated the viability of introducing a QR code and smartphone APP to the program. They were quickly convinced (based on the experience of China) that incorporating them would realize sizable benefits to the consumer, the regulator, relevant ministries and the appliance industry. Primarily, a database and online registration system would significantly reduce the processing time needed by the Regulator to issue manufacturers with an approval certificate (Letter of Authority – LoA). It would also enable the efficient tracing of data for both public and private sectors, while assisting the Department of Mineral Resources and Energy (DMRE) to track penetration rates of energy classes, so as to reliably inform future policy decisions. Ultimately, this would all enhance the Regulator's oversight capabilities. Moreover, by utilizing the energy label in the digital domain, the QR code and associated mobile smart device APP are capable of also combining to raise consumer awareness on energy conservation. They would thus play an important role in consumer education and the promotion of energy efficiency (EE) related programs.

This paper presents the process that could be considered in establishing a database, setting up a registration system, and developing QR code features and smartphone APPS. It also discusses the benefits that such digitalization brings to South

T. Covary (✉)
Unlimited Energy, Johannesburg, South Africa
e-mail: theo@unlimitedenergy.co.za

J. Li
Independent Consultant, Toronto, ON, Canada

P. Bertoldi (ed.), *Energy Efficiency in Domestic Appliances and Lighting*, Springer Proceedings in Energy, https://doi.org/10.1007/978-3-030-79124-7_14

Africa—thus encouraging more energy efficiency labeling programs around the globe to adopt these digital methodologies, in leveraging the potential that energy labels provide.

1 Introduction of Regulatory Framework of South Africa's Standards and Labelling (S&L) Program

Aligned to global best practice in reducing energy consumption and carbon intensity, while enhancing consumer protection, South Africa has committed to pursuing improved efficiency in household appliances—establishing this commitment in relevant legislation and frameworks.

Here, the National Energy Act, Act 34 of 2008 [1] empowers the Minister of what is now the Department of Mineral Resources & Energy (DMRE) to regulate the sale of appliances that consume wasteful amounts of electricity.[1] And the National Regulator for Compulsory Specifications Act, Act 5 of 2008 [2] then empowers the National Regulator for Compulsory Specifications (NRCS)—an agency of the current Department of Trade Industry and Competition (DTIC)—to administer and maintain mandatory specifications in the interest of public safety and health, or for environmental protection. The NRCS thus issues a Letter of Authority (LoA) for compliant products that is valid for a 3-year period; without which, products may not be sold on the South African market.

The DTIC is also the parent ministry of the South African Bureau of Standards (SABS) which develops national standards and operates the country's testing laboratories.

The DMRE , as policy owner of residential energy efficiency (EE), thus collaborated with the DTIC and the above-mentioned agencies—NRCS & SABS—in implementing the program "*Market Transformation through the Introduction of Energy Efficiency Standards and the Labelling of Household Appliances in South Africa*". The initial project to establish the Standards & Labeling (S&L) program as it is now known, was supported by the Global Environment Fund (GEF) and the United Nations Development Program (UNDP). And the overall ongoing objective

[1] ...by notice in the Gazette make regulations regarding—

(a) the publication of energy statistics or information;
(b) the type, manner and form of energy data and information that must be provided by any person;
(c) the form and manner of the link between the energy database and information system to any other system within the public administration;
(d) minimum levels of energy efficiency in each sector of the economy;
(e) steps and procedures necessary for the application of energy efficiency technologies and procedures;
(f) labelling for energy efficiency purposes of household appliances, devices and motor vehicles;
(g) prohibition of the manufacture, or importation or sale of electrical and electronic products and fuel burning appliances for reasons of poor energy efficiency;

of the S&L program is to remove inefficient residential electronic appliances from the South African market, while encouraging the adoption of efficient technologies. This includes implementing measures/interventions such as Minimum Energy Performance Standards (MEPS), as well as labeling and incentive programs; and it is regulated by compulsory specifications, VC9006 and 9008 for residential appliances. These initially included refrigeration, ovens, laundry and tumble dryers, dishwashers, electric hot water systems, air conditioners, and standby power (AV only), while being expanded to other items, such as lighting and electric motors.

2 Compliance Procedures: Obtaining a Letter of Authority

Established in 2008, the NRCS' mandate is to promote public health and safety (H&S), environmental protection and fair trade. Thus, in seeking a mechanism for mandatory oversight of the S&L program, the DMRE (then known as the Department of Energy - DOE) determined that the NRCS was the most appropriate entity to regulate the program. This is because it had the enabling legislation (NRCS Act 2008), and was familiar with all the appliances selected for the S&L program, as they were already required to apply for an H&S LoA. It was rationalized that henceforth a 'one-stop' service would be offered to applicants for both the H&S and EE LoA's.

In regulating the market, the NRCS applies two separate mechanisms: (1) Pre-certification—certifying a product prior to market entry, through the issuance of a LoA, and (2) Monitoring, Verification and Enforcement (MVE)—Monitoring the market to identify and penalize non-compliant products. This compliance approach aligns with that of many countries; but it is the robustness of its implementation that ultimately determines the effectiveness of compliance. Indeed, international studies [3, 4] show that effective market regulation is only possible if both these components are present, and that without MVE, compliance levels average at around 40%. These increase to 80% when both instruments are in place and operating effectively.

Currently, NRCS emphasis is on the pre-certification process, with limited focus on the MVE component, which is in the process of being bolstered. In terms of certification, a compulsory standard, or VC, requires NRCS approval for 'every type and model' before 'offered for sale' in the country [5], as demonstrated by the NRCS' issuance of an LoA. Previously, the NRCS issued LoAs within 30 days. This time period had steadily increased over time though, with the procedural cap making allowance for a maximum of 120 days [5], but which could take even longer. International experience has however proved that it is possible to streamline this pre-certification component. By way of example, Australia issues registration certificates within a fortnight, while in China, a successful application results in a system-generated certificate being issued within a 2-week period. NRCS backlogs on the other hand [6–8], (as reported in Annual Reports and identified for intervention by the Parliamentary Portfolio Committee on Trade and Industry) had led to extended turnaround times, with only 37% of applications processed within the prescribed 120 days in 2016/2017. Turnaround times did however improve in 2018, due to intense pressure from industry. Indeed, at the South African Investment Conference in September 2018, an EU Delegation specifically raised the inefficiencies of the LoA process as one of three "*main constraints to potential FDI (and trade) that would benefit from the government's urgent attention*". The delegation emphasized the importance of certification procedures being "proportional to the risk at stake and efficient enough so as not to inhibit existing business operations, as well as new investment" [9]. This intense pressure led to some improvement; and in 2018, turnaround times had dropped to around the 100-day mark, which were still higher than international norms.

In addressing LoA issuance delays, and to get a better understanding of the LoA application approval process, the S&L project undertook an Appliance Registration Database Study [10]. It sought to ascertain what constraints existed, and to determine what actions could be taken to improve the situation. As stated earlier, in terms of EE, an LoA verifies that the particular appliance conforms to the MEPS specified for that particular category of appliance. The applicant (entity) must also be registered with the NRCS' Electro-technical Division[2] Customer Relationship Management (CRM system), which is a hybrid of on-line and downloadable paper-based forms, before any LoA can be issued.

In terms of this process, the applicant, once registered, could then proceed to complete an application form for each product. Here, part of the application process required the uploading of electronic copies of certain documentation, such as test reports and declarations. An NRCS evaluating officer then reviewed the application and either rejected, approved or returned it for amendment. Below, the overall process is detailed in Fig. 1.

As the NRCS CRM system was originally designed to manage registration approvals for H&S applications, which simply require a record of whether a product is compliant or not, it was not geared, and therefore unsuitable, for the purposes of S&L regulation. Consequently, this '*not fit for purpose*' system had many shortcomings; most notably, that it was not capable of modification to facilitate the requirements of the S&L program. Indeed, the NRCS was forced to add a paper-based form to the application process, in capturing additional S&L specific data, which could not be captured electronically. These paper-based forms were, however, deficient in terms of the scope of data collected, while including no internal checks as a means for reducing processing time. They were also administratively cumbersome and, most significantly, did not allow for the electronic capture of data into a central database.[3] Furthermore, the paper-based EE form also duplicated much of the information already captured in the CRM system, such as the applicant's details and some product particulars. But it did not capture relevant technical detail of the

[2] Foreign (international) entities must apply for registration through a South African agent

[3] In theory, the data captured in the paper-based forms could be manually keyed into an electronic database, however, this does not appear to occur. Even if this was to occur, the dataset available from the paper-based forms is inadequate and would be subject to re-keying errors.

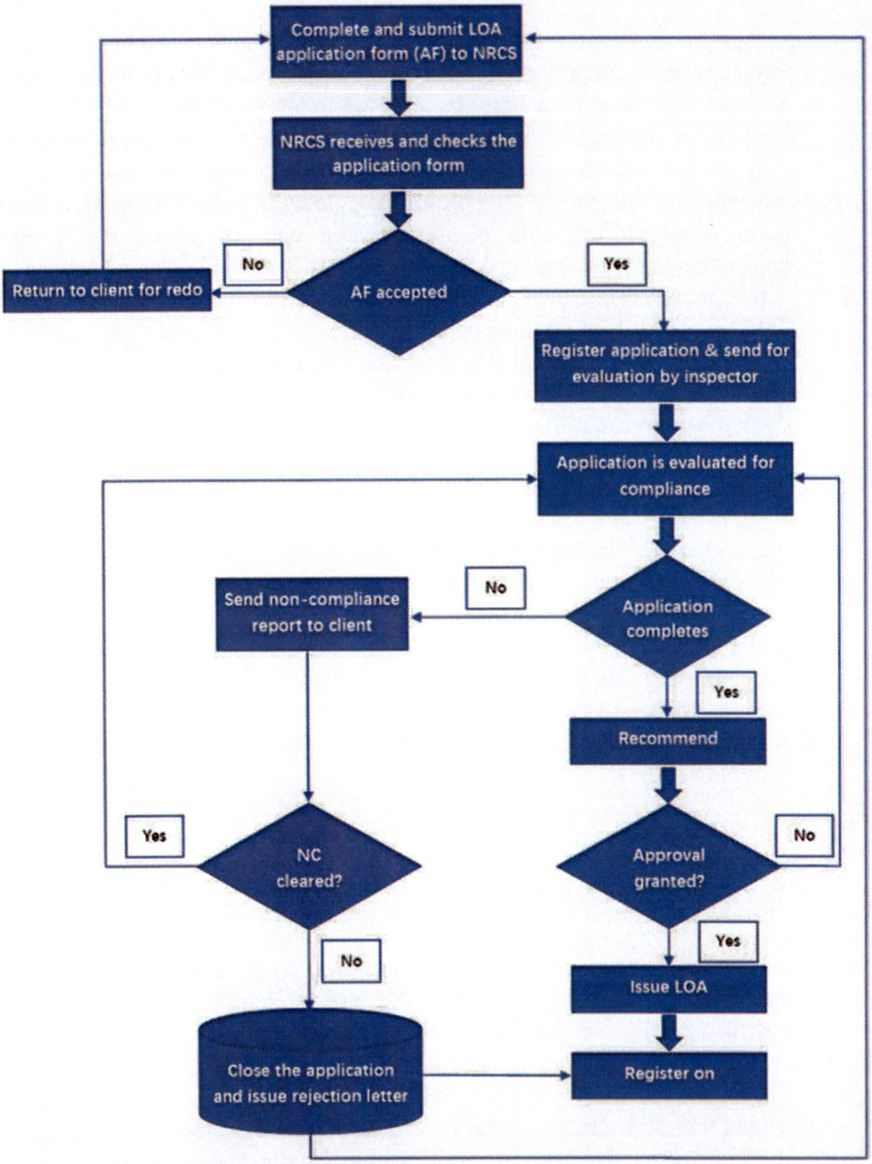

Fig. 1 NRCS LoA Application Process (Source: NRCS)

product being registered, such as its energy consumption, energy rating or capacity. These key data sets all needed to be captured, to facilitate crucial components of the S&L program [10].

The inefficiencies of the initial pre-certification system were soon acknowledged; and the NRCS identified the need for a streamlined system. Thus, to counter the uncertain delivery time period and expedite the S&L program objectives, the project team driving the program's implementation, initiated the development of a product registration database. Its aim was to automate the registration process for household appliances, with respect to the MEPS component of compliance. Registration database development rapidly reached an advanced stage, with user acceptance testing and a pilot project completed in 2019, and the system going live soon thereafter. Capable of modification and expansion, the database was then extended to include registration and certification of lighting products. The system's success is also evidenced by its further current adaptation to also deal with the H&S aspect of LoA application processing.

As stated earlier though, another challenge is that the Regulator's MVE capacity can be described as less than optimal since MEPS were made mandatory by VC9008 and 9006 in 2016 and 2017 respectively. A "blitz" was undertaken in 2017 when approximately 20 retailers were visited nationwide, but the inspection was limited to the label being displayed prominently on the front of appliances. Since then, MVE efforts have been plagued by inconsistent site inspections, insufficient accredited test facilities in critical locations and the inability to issue fines for non compliance, amongst others. These issues too, have been acknowledged and are being addressed as part of the ongoing efforts of the S&L program—involving the collaboration of the Regulator, other government entities and industry. Indeed, a primary reason for the lack of MVE surveillance taken by the Regulator, (amongst other factors), was the lack of accessible and accurate data, due to the deficient system used to process LoA applications. And this too, the registration database has sought to address.

3 Consumer Awareness

Consumer interest and public awareness are crucial to driving the process of EE appliance market penetration. Thus, the S&L program developed a marketing strategy to tactically stimulate and proliferate these.

Activities range from traditional media advertising, such as newspaper advertisements and radio competitions, to national road-shows and store activation campaigns with key retail partners, then as well as a dedicated website, targeted social media presence through various platforms, and the development of a smart mobile device APP. Key to the success of consumer awareness campaigns from inception, was the clear and unambiguous public support for the S&L program by Government, which the Minister of what was then the Department of Energy, very clearly gave. This was witnessed through frequent inclusion of EE in his media statements and his highly publicized presence at the public launch of the new compulsory South African EE label.

Most importantly, the rapid development and launch of the Appliance Energy Calculator mobile device APP, then tangibly enabled consumers to make much more informed purchasing decisions, in terms of appliance EE. Freely available on both Android and iOS platforms, the APP was ultimately developed to support retail staff and consumers. And it requires nothing more than the input of information provided on the compulsory energy label affixed on the appliance, as well as the price. The APP then provides data on the appliance's 10-year running cost (including both upfront cost and long-term electricity spend), as well as greenhouse gas emissions—allowing the consumer to compare appliances and make an informed buying decision, while empowering retail staff to provide sound advice on efficient product selection (Fig. 2).

Having touched on the challenges of effectively applying EE regulation and assuring compliance through the South African S&L project experience—and given the global prevalence of digital technology—the next section details how digital

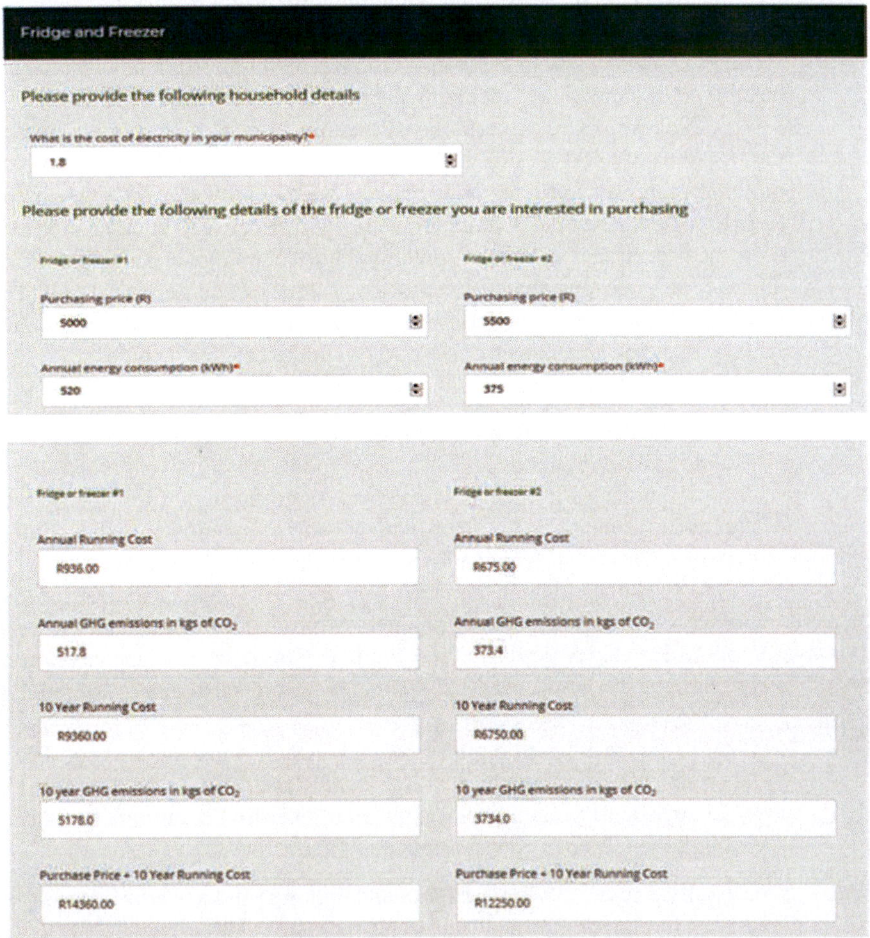

Fig. 2 The South African Appliance Energy Calculator (APP)

measures could assist MVE efforts, while raising consumer awareness and improving information access among appliance manufacturers and consumers.

4 Digital Measures to Improve Efficiency and Effectiveness of Energy Labeling Programs

4.1 Overview of Digital Measures for Energy Labeling Programs

Information and communication technologies have dramatically improved over the past decade. They have become ubiquitous due to their universal popularity—be they devices such as smart phones, or 4G (and soon 5G) mobile networks, and WIFI (Wireless Fidelity) amongst others—with rapid digitalization now allowing these to be applied to many fields, including energy labeling programs (ELPs).

At their core, ELPs are established mainly to inform consumers of the energy performance of energy-using and-related products (EuP and ErP). Simultaneously, these programs also provide policymakers with improved market insight; allowing them to regulate the market more efficiently and effectively. In order to do so though, it can be argued that in light of contemporary technological developments, specific digital tools not only enhance the process, but have become crucial to it. These primarily take the form of a database, QR (quick response) code and smart phone APP. In this, key advantages of adopting digital measures include:

- A database enhances the efficiency and effectiveness of product registration;
- QR code labels can present additional helpful information to consumers, retailers and market regulators;
- A smartphone APP enables consumers to go through all available models of appliances; and provides them with a "scan" function to extract information stored in QR codes. It also allows the dissemination of the latest news and policies to consumers, which may include market surveillance results, incentive programs for efficient appliances, product maintenance reminders, responsible disposal advice, and the like. The APP can thus assist consumers throughout the lifecycle of the appliance—from selection, to use, maintenance and recycling;
- Finally, database-based digital labels are easier to manage and regulate. Unlike information printed on physical labels, databases can be corrected and updated as frequently as needed—with updated information being reflected in real time through the QR code and smartphone APP.

Globally, a few ELPs for appliance and lighting products have adopted, or are on their way to adopting, digital tools and measures. These include China's Energy Label, EU Eco-design, Australian Energy Rating program and South Africa's Energy Labeling Program.

The particular digital measures involved, will now be discussed in greater individual detail.

4.2 Database

A database hosts data and information electronically; thus significantly reducing the use of hard copy (paper-based) materials for registration and tracing of data—therefore greatly enhancing efficiency and effectiveness of processing applications. As such, it is a preferable solution for handling large and continuously increasing volumes of product registrations. It is also the fundamental prerequisite for further digitalization, as all digital measures are based on extracting data from the database.

A key advantage is that a database system can incorporate an automated validation function, which, for example, can automatically compare product data against standard requirements set up in the system, and thus verify compliance. The automated validation function therefore greatly reduces the need for physical compliance checks—further increasing the efficiency and speed of processing applications. Furthermore, a database also hosts all communication correspondence, market surveillance results and product advertisement material; again reducing ELP management time and effort when compared to a manual and hard-copy-based system.

Finally, stored information and data can be made easily accessible to all stakeholders, such as consumers, market regulators and ELP managers, through other integrated digitalization tools. These tools include, but are not limited to, QR codes and smart mobile device APPS, which interact directly with stakeholders and present them with information they request and/or may be interested in viewing; such as in the South African S&L experience outlined in previous sections.

4.3 QR Code

A Quick Response (QR) code is a technology that provides scanners with information stored in a particular space, such as a cloud server. Users request information by scanning the code via their smart device, which then transports them to the specific webpage, based on the scanned code. Here, the primary benefit of QR code technology is that it is a gateway to vast amounts of information—far more than can be communicated in conventional physical forms of hard copy print and media.

In this, each QR code is uniquely created, and directs scanners to its own URL (Uniform Resource Locator) link, which allows all scanners of a specific QR code to see the same information. Of course, because a QR code directs scanners to an online link, internet access is needed to acquire information. This truly makes QR

Fig. 3 Example of a QR Code

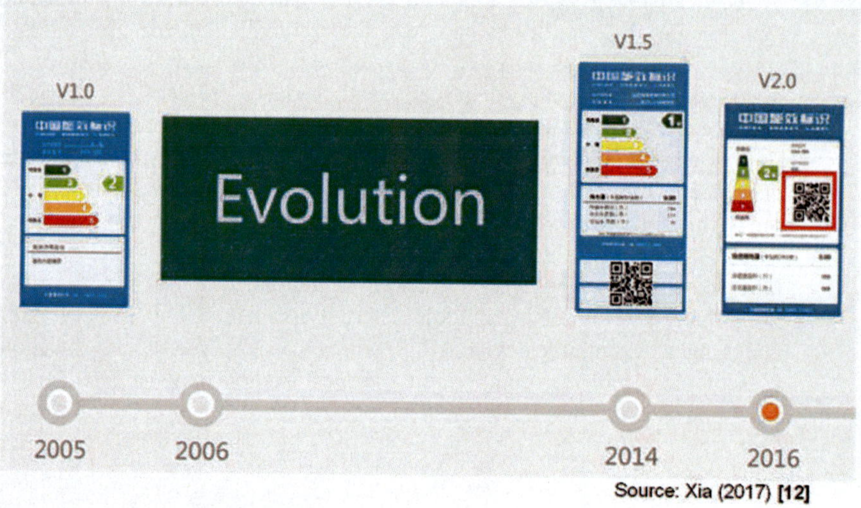

Fig. 4 Evolution of China Energy Label with QR code (Source: Xia (2017) [11])

coding a perfect tool for its time, given the increasing global prevalence of internet access, which allows it to present much more information to consumers. And it creates huge potential to extend the capacity of physical energy labels (as can be seen in Figs. 3 and 4 respectively, which show a generic example of a QR code and the evolution of its incorporation into the energy label used in China).

When consumers in China scan the QR code above, they are directed to a webpage, which contains various types of content; and Table 1 below, showcases some of the sections of information available.

It must be reiterated that each of the pictures above is a link to further data and discourse. Therefore, when consumers select one of these images, they will be directed to the webpages available on that specific issue or focus area—containing more detailed information and particulars.

Table 1 Contents in QR code landing page of China Energy Label [12]

<table>
<tr><th>1</th><th>2</th></tr>
<tr><td>
</td><td>
</td></tr>
<tr><td>Basic product information, such as the particular registration number and manufacturer, and energy efficiency level (this is a level 1 product[a])</td><td>Information about a project that aims to expose products that over-declare their energy performance</td></tr>
<tr><th>3</th><th>4</th></tr>
<tr><td>
</td><td>
</td></tr>
<tr><td>Details about a basic consumer survey, designed to draw dynamic consumer attention to energy saving</td><td>The latest news on appliance recycling—highlighting the serious environmental problems illegal recycling would cause</td></tr>
</table>

[a]China's energy efficiency standard and labeling program categorizes appliances' energy efficiency using numbers rather than letters or star symbols. Some types of products have 5 levels of energy efficiency and some have 3. Level 1 means the most efficient and Level 3 or 5 signifies entry level efficiency i.e. the least efficient products permitted on the market

To illustrate this further, Fig. 5 below, provides an example of the "next level" of information that can be accessed by clicking on the Basic Information image in Picture 1 of Table 1—showing the escalation that is possible, once a particular selection is made.

<table>
<tr><td>Information box shown on the landing page after scanning the QR code, (as seen in Table 1 above)</td><td>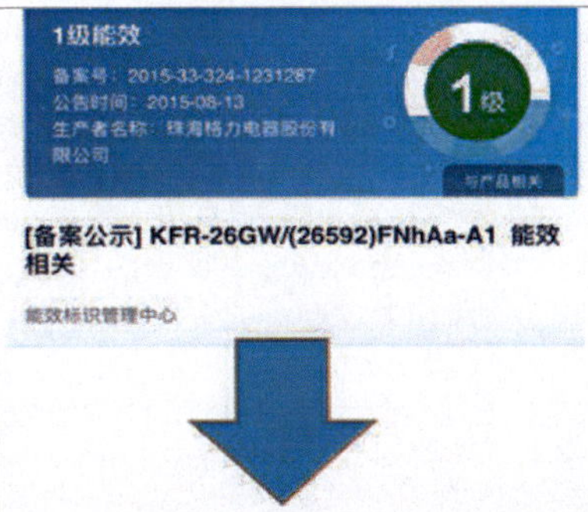
</td></tr>
</table>

<table>
<tr>
<td>When going to the next level of information, an electronic certificate issued by the China National Institute of Standardization (CNIS) is shown
CNIS manages the China Energy Label, with the authority to approve manufacturers' registration; and this certificate indeed certifies that the product has been duly registered</td>
<td>
</td>
</tr>
<tr>
<td>This section then shows the registered product's performance parameters
When clicking on the arrows to the right, one can access a brief explanation of what specific parameters mean, explained in simple and understandable terms, so as to paint a clear and detailed picture of expected performance</td>
<td>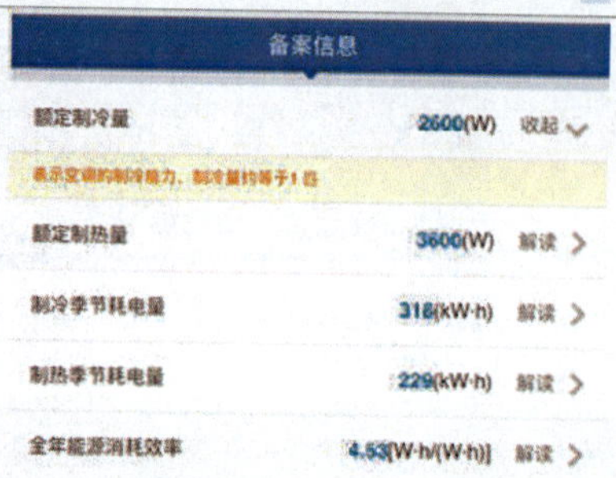
</td>
</tr>
</table>

As can be seen, the access to vital information is significant; and the energy label QR code system in China continues to be updated [12].

5 Smart Phone APP

The current South African Appliance Energy Calculator APP enables consumers to compare usage costs of appliances with their counterparts—providing a powerful tool in purchasing decision-making, by showing the long-term financial advantages of products with higher EE levels. But it lacks an intuitive user interface (UI), as well as better and more useful features that may attract more users. Nonetheless, it has already revealed its potential for further market impact, especially in raising consumer awareness of the South Africa Energy Labeling Program and the need for energy conservation. Indeed, the APP has been downloaded to, and installed in, thousands of devices since its release, and received an average rating of 4.43 (out

Information box shown on the landing page after scanning the QR code, (as seen in Table 1 above).

When going to the next level of information, an electronic certificate issued by the China National Institute of Standardization (CNIS) is shown.

CNIS manages the China Energy Label, with the authority to approve manufacturers' registration; and this certificate indeed certifies that the product has been duly registered.

This section then shows the registered product's performance parameters.

When clicking on the arrows to the right, one can access a brief explanation of what specific parameters mean, explained in simple and understandable terms, so as to paint a clear and detailed picture of expected performance.

Fig. 5 Example of Deeper Level of Information provided by the QR Code

of 5) from users.[4] This success can then be further expanded on, by developing a more functional and powerful APP, which works closely with QR codes, when those are in place. Indeed, an upgraded smartphone APP may be needed, for consumers and other potential scanners, such as market regulators, to scan and receive feedback information from QR codes.

Taking one step back, however, a powerful smartphone APP is also extremely helpful, even without the QR code. Indeed, it shares common functionalities with the QR code, such as providing more information about the appliance than physical

[4] Information received from APP developer and hosting company Urban Earth—10 July 2019

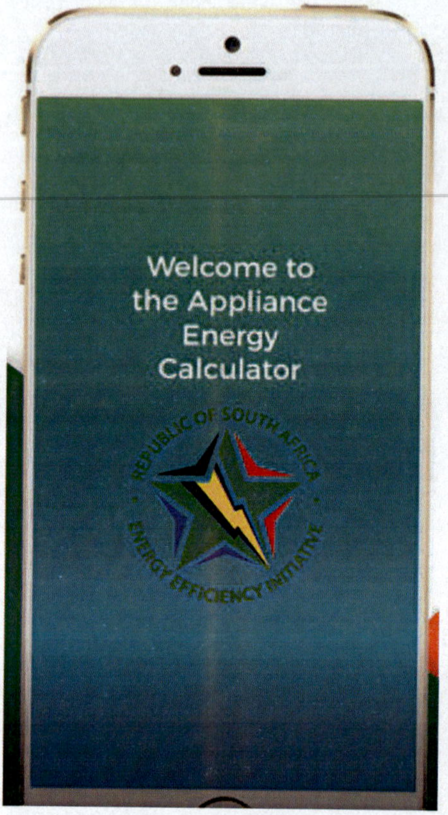

Fig. 6 South Africa Smartphone APP (Appliance Energy Calculator)

labels do. In this, the APP could also assist with tackling some of the disadvantages of the QR code, namely that it is a reactive too, i.e. data is only available on scanning. A smartphone APP on the other hand, as already pointed out in previous sections, can be proactive. It allows policymakers to push notifications to millions of consumers, such as updates on market surveillance activities and promotional programs of efficient appliances. Smart mobile device APPs also enable consumers to have an overview of all products in the database, and to conveniently select products for comparison; while QR codes are more focused on showing detailed information of specific products that are scanned (Fig. 6).

6 Conclusion

Ultimately, the tactical triangulation of database, QR Code and smartphone APP, appears to offer the optimal combination of digital measures; with significant potential to provide ELPs with multifaceted tools that impact the entire value chain. This ranges from seamless product registration and manufacturer liaison, to improved

Table 2 Summary of benefits when going digital for energy labeling programs

Aspects that can be improved	Owner	Process with traditional hard copy materials	Process with digital measures (database, QR code and smart phone APP)	Improvements
Application for LoA	Manufacturer	Mailing of hard copies and filling out of forms online	Only need to fill out the forms online	**Saves time** **Provides certainty**
Check application material against standard requirements	Authorities	Review hard copies and online material against standard requirements	The database can automatically complete most of the review and verification	**Saves time and improves accuracy of review and verification**
Communication	Both manufacturers and authorities	Mostly through mails	Mostly online	**Saves time** **Ensures far-reaching dissemination**
Amount of useful information that should/can be presented to consumers	Authorities	Limited due to restriction of the size of physical labels	• As much information as consumers need; • Detailed (targeted) information that can guide consumers towards energy saving and creates a circular economy; • Promotional information to millions of consumers at very low cost, which "spreads the word" for policymakers about their energy efficiency policies and activities	**Raises crucial consumer awareness and encourages engagement in ELPs**
Support to market regulators	Market regulators	Very hard for market regulators to verify data on the label against registration data	Market regulators only need to scan a QR code or input product model numbers into a smart device APP and they will be able to verify product registration, and if labeled information is aligned to that of registration	**Enables rapid and convenient market surveillance and MVE, regardless of location**

MVE capacity and enhanced consumer decision-making, which all translate to a lot more energy savings.

Indeed, it is this rationale that drives the ongoing further integration of the online database, with a QR code system and the APP in South Africa, based on global best practice and experiences.

By way of explanation, Table 2 below, lists the benefits of a digital ELP, which is particularly relevant to the South African context, given the combined circumstances of the appliance market expected to grow as income levels increase [13], a regulator seeking to significantly improve MVE and management of rising volumes of LoA applications and Government's recognition of the need to modernize its operations [14].

More than anything, given the clear advantages of digital tools when utilized for ELPs, the authors of this paper would encourage more ELPs to explore digital solutions such as those outlined here. They have been conclusively proved to significantly leverage and enhance the potential of energy labels and thus achieve greater energy savings.

References

1. Department of Minerals and Energy (2008): *National Energy Act No 34*. Available: www.energy.gov.za/files/policies/NationalEnergyAct_34of2008.pdf
2. Department of Trade and Industry (2008): *National Regulator for Compulsory Specifications Act No 5*. Available: www.thedti.gov.za/business_regulation/acts/national_regulator_act.pdf
3. Zhou, N., Romankiewicz, J., Fridley, D., Zheng, N. (2012). *International Comparison of Product Certification and Verification Methods for Appliances*. Lawrence Berkeley National Laboratory.
4. Electromagnetic Compatibility (2011): *Final report on the 4th joint cross border EMC market surveillance*.
5. National Regulator for Compulsory Specifications (2019): *ElectroTechnical—Letter of Authority (LoA)*: Available: www.nrcs.org.za/siteimgs/Electrotec/LOA%20ADMIN%20Revised%2012_06%20May%202019.pdf
6. National Regulator for Compulsory Specifications: *NRCS Annual Report 2015/16*. Available: https://www.nrcs.org.za/content.asp?subID=7156
7. National Regulator for Compulsory Specifications: *NRCS Annual Report 2016/17*. Available: https://www.nrcs.org.za/content.asp?subID=7156
8. NRCS (2019): *NRCS Energy Efficiency Enforcement*. Presentation made to the S&L Project Steering Committee on June, 20 2019
9. European Union Discussion Paper (2018): *South Africa's Investment Conference*. Available at: https://www.tralac.org/documents/news/2197-eu-discussion-paper-on-south-africas-investment-conference-september-2018/file.html
10. Energy Efficient Strategies (2018): *South African Product Registration Database/Database Study*. South African Standards and Labelling Project—available on request.
11. Xia. Y, and Aang.G. (2017): *Best Practices of QR Code Application in China Energy Label*. Resource and Environment Branch, China National Institute of Standardization. Available at: http://mddb.apec.org/Documents/2017/SCSC/WKSP2/17_scsc_wksp2_008.pdf
12. Integrated content of the China Energy Label through the scanning of a QR code. (Note: As versions of QR code content and features update irregularly without version control, the

contents listed here may have been changed by the time this paper is published. However, the content examples showcase the potential of how a QR code could be used for detailed communication and building consumer awareness.)

13. Li. J, Yu. Y, Zeng. L: 2014 *Market Analysis of China Energy Efficient Products (MACEEP 2014)*. Available at https://clasp.ngo/publications/new-clasp-study-analyses-changes-in-the-market-for-energy-efficient-products-in-china
14. National Regulator for Compulsory Specifications: *NRCS Annual Report 2017/18*. Available: https://www.nrcs.org.za/siteimgs/publications/2018/K-14717%20NRCS%20AR%202017-18.pdf

New Trends of Energy Efficiency Requirements for LED Lighting Products in China, EU and US

Xiuying Liang, Jiayang Li, and Michael Scholand

Abstract Around the world, lighting markets are becoming more efficient, abandoning old traditional light sources in favour of new light-emitting diode (LED) lamps and luminaires. China is the world leader in the manufacturing of LED lighting products, building on a long legacy of national expertise in electronics, optics and high-technology industries. Chinese policy makers are currently working to revise and develop the national minimum energy performance standards (MEPS) for LED products, seeking to expand the regulatory scope of product coverage and ambition of China's lighting policy.

This paper is offered as an input to that work, providing an overview of current and future MEPS for LED lighting in China, the requirements of United States's DesignLights Consortium (DLC) and the new European Commission's one-lighting regulation. The paper provides a summary of each of these three quality and performance standards and provides a comparison between them along with some current market data for reference. The paper concludes by making some suggestions for consideration by policy makers on coverage, efficacy, temporal light artefacts and lifetime.

1 Introduction: LEDs Getting Mature as Leading Lighting Technology Globally

Over the last decade, LED lighting products have transitioned from niche applications such as exit signs and traffic signals to become the mainstream general light sources, fulfilling popular and demanding lighting applications such as household

X. Liang
China National Institute of Standardization (CNIS), Beijing, China
e-mail: liangxy@cnis.ac.cn

J. Li (✉)
Toronto, ON, Canada

M. Scholand
CLASP Europe, London, UK

P. Bertoldi (ed.), *Energy Efficiency in Domestic Appliances and Lighting*, Springer Proceedings in Energy, https://doi.org/10.1007/978-3-030-79124-7_15

general lighting, office lighting, outdoor/street lighting and stadium flood lighting. LED light sources have successfully made this transition to mainstream light sources because, *inter alia*, they offer high efficacy, long lifetime, compact form-factor, instant brightness, are easily dimmable and vibration resistant. Over the last 20 years, LED technology innovations and breakthroughs have been delivered to markets around the world. As LED technology continues to improve, it will capture more of the general illumination market, and in the process, reduce operational costs and CO_2 emissions.

LEDs are global-traded lighting products, and many governments around the world either have established quality and performance standards on LED lighting products or are planning to do so. In October 2019 for example, the European Commission adopted a new lighting regulation that sets ambitious new requirements for LED lamps and luminaires, and schedules the phase-out of the most common T8 linear fluorescent tubes in 2023. In China, policy makers are currently studying the LED market to determine appropriate quality and performance requirements for China's domestic market. Given that China is the world's largest manufacturer of LED lamps and luminaires, it's expected that any domestic lighting policy measure adopted will benefit both China and abroad.

The global phase-out of incandescent, halogen, fluorescent and other inefficient lighting technologies has created a golden opportunity for the LED lighting industry. Many major national and regional economies are adopting policy measures, incentive schemes and development plans to accelerate the take-up of quality, energy-efficient LED lighting products into their markets. China was one of the first to take such action, initiating the National LED Lighting Project in 2003. This pioneering initiative helped to guide the development of China's LED lighting industry from a variety of angles, such as R&D roadmaps, financial support, demonstration projects, and quality and performance standards. Now, 16 years later, the foresight of the National LED Lighting Project has been demonstrated, as China enjoys a vertically integrated supply chain, and is the largest global supplier of LED lighting products.

In addition to supporting energy conservation and environmental protection, LED lighting will also continue to grow its market share because it can also be 'smart'. Adding digital and intelligent features helps to enhance the lighting experience, with colour adjustable, occupancy sensing, daylight dimming and other critical features incorporated directly into the lights. Researchers are now even able to conduct digital communication over the lighting system—LiFi, which is indicative of this trend of cross-sector integration and collaboration, especially for smart lighting.

2 Energy Efficiency Standards (MEPS) for LED Lighting Products in China

China's first national (GB) energy efficiency standard for LED was issued in 2013 and took effect in 2014. It regulates LED light bulbs for general purpose. As LED lighting products are developing fast to apply in more situations, Chinese policy makers started to develop new regulations to cover them as well.

2.1 *Framework of MEPS for LED Lighting Products*

In China, MEPS for LED lighting products is in addition to other mandatory standards already in place for LED lighting that cover safety and electromagnetic compatibility. The Chinese regulations establish energy efficiency grades and criteria for each grade, and also set requirements for certain quality parameters such as colour rendering index (CRI), correlated colour temperature (CCT) and lifetime (i.e., lumen maintenance over the product's service life). The Chinese standards (GB 30255-2013, GB 30255-2019, GB 37478-2019) each define three grades of energy efficiency, with Grade 1 being the most efficient and Grade 3 the minimum required performance. If China were to update the 2013 standard, increasing the stringency of the Grade 3 requirement, they would remove approximately 10% of the least efficient LED lighting products on the market. The Grade 2 requirement is defined as the "energy efficient" level, and only allows approximately 15–20% products to pass this level. Grade 1 is the highest requirement, with just 5–10% of products passing. When setting up criteria for different efficacy grades, China decided to group products by CCT, and sub-categorized by power, lighting distribution type, and size.

China's MEPS for LED also standardises test methods, either referring to existing ones, or setting new ones where there are no references to make. The test methods form the basis for comparison of test results and criteria (Fig. 1).

2.2 *Development of More MEPS for LED Lighting Products*

China's first energy performance standard of LED lighting products *Minimum allowable values of energy efficiency and energy efficiency grades of non-directional self-ballasted LED lamps for general lighting services* was issued in December 2013 and became effective in September 2014. Later, the China National Institute of Standardization (CNIS) carried out research on additional LED lighting products, including LED roadway lighting, tunnel lighting, downlights, directional integrated lamps and flat panel luminaires.

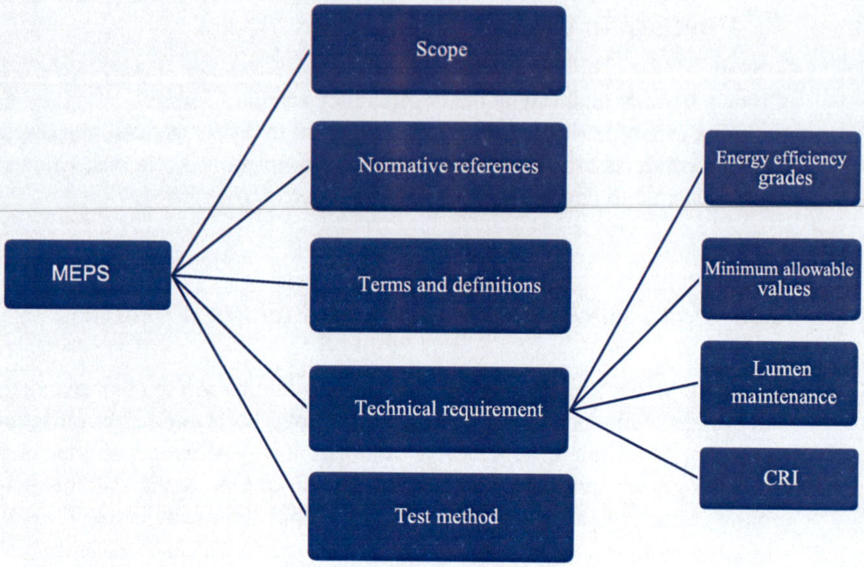

Fig. 1 Framework of MEPS for LED lighting products in China

On April 4, 2019, the two new energy efficiency standards for LED lighting products were issued. One is *Minimum allowable values of energy efficiency and energy efficiency grades of LED products for indoor lighting* (GB 30255-2019 [1]), the other is *Minimum allowable values of energy efficiency and energy efficiency grades of LED luminaires for road and tunnel lighting* (GB 37478-2019 [2]). These two MEPS will take effect on May 1, 2020, when GB 30255-2013 [3] will be outdated. Moreover, the draft standard of *Minimum allowable values of energy efficiency and energy efficiency grades of LED flat panel luminaires* have been developed and submitted for approval.

Table 1 is a summary of existing, upcoming and foreseeable MEPS for LED lighting.

2.3 Changes in the New MEPS for LED Lighting Products

While the two new standards inherit the original tradition of the existing MEPS for lighting products in terms of structure of framework, energy efficiency rating system and principle of product categorization, they present some key changes as well.

- While MEPS has been considered as mandatory standards, they actually mandate only minimum values of energy efficiency, and requirements in all other sections, e.g. evaluating values of energy conservation and lumen maintenance, are voluntary. However, the requirements are mandatory throughout the two

Table 1 China's existing and upcoming and foreseeable energy efficiency standards (MEPS) for LED lighting

Standard name	Standard No.	Date of issue	Date of implementation
Minimum allowable values of energy efficiency and energy efficiency grades of non-directional self-ballasted LED-lamps for general lighting services[a]	GB 30255-2013[a] (will be outdated in May 2020, and the new version of -2019 will take effect)	December 2013	September 2014
Minimum allowable values of energy efficiency and energy efficiency grades of LED products for indoor lighting[a]	GB 30255-2019[a]	April 2019	May 2020
Minimum allowable values of energy efficiency and energy efficiency grades of LED luminaires for road and tunnel lighting	GB 37478-2019	April 2019	May 2020
Minimum allowable values of energy efficiency and energy efficiency grades of LED flat panel luminaires	Draft reported for approval, standard number and issue date is pending		

[a]The new version of this standard is re-named as it is incorporating more types of LEDs for indoor use

newly issued MEPS for LED. In terms of standard implementation, the role of mandatory standards as the basis of supervision, inspection and administrative enforcement has been further strengthened.

- Previously before 2016, each MEPS for LED lighting applies to only one type of product. However, each of the two new MEPS, applies to several types of LED lighting products. This shows a new trend that China is combining multiple MEPS into one.

These changes present the new trend and requirements in deepening the reform of China's standardization work in overall.

2.4 Key Parameters of LED Lighting MEPS

As indicated above, there are four existing and upcoming MEPS for LED lighting. Table 2 is a summary of the requirements of key parameters covered by these standards.

Comparing the requirements of GB 30255-2013 and -2019 versions, China will capture additional energy savings through increasing the efficacy requirements of non-directional LED lamps and expanding the product scope to include more products. Also, the upcoming standards also included color quality metrics to ensure user acceptance of LED lighting products.

Table 2 Summary of China's energy efficiency requirements

Standard No.	Product type	Category		Parameters and requirements[a]				
				EEG[a] and thresholds (lm/W)				
				1	2	3	CRI[b]	Lumen maintenance[c]
GB 30255-2013	Non-directional self-ballasted LED lamp	Omnidirectional light distribution	CCT: 3500/3000/2700 K	100	80	59	–	Depending on rated lifetime, LM requirements vary. Requirements cover @3000 h and @6000 h.
			CCT: 6500/5000/4000 K	110	90	63		
		Quasi-omnidirectional/semispatial light distribution	CCT: 3500/3000/2700 K	105	85	65		
			CCT: 6500/5000/4000 K	115	95	70		

Standard No.	Product type	Category		Parameters and requirements[a]				
				EEG[a] and thresholds (lm/W)				
				1	2	3	CRI[b]	Lumen maintenance[c]
GB 30255-2019	Non-directional self-ballasted LED lamp	Omnidirectional light distribution	CCT < 3500 K	105	85	60	$R_a \geq 80$ $R_9 > 0$ Tolerance: −3	Must be greater than $70\%^{(3000/t0)}$. t_0 is rated lifetime. LM requirement @3000 h.
			CCT ≥ 3500 K	115	95	65		
		Quasi-omnidirectional/ semispatial light distribution	CCT < 3500 K	110	90	70		
			CCT ≥ 3500 K	120	100	75		
	Directional integrated LED lamp	PAR16/20	CCT < 3500 K	95	80	65		
			CCT ≥ 3500 K	100	85	70		
		PAR30/38	CCT < 3500 K	100	85	70		
			CCT ≥ 3500 K	105	90	75		
	LED downlight	P ≤ 5 W	CCT < 3500 K	95	80	60		
			CCT ≥ 3500 K	100	85	65		
		P > 5 W	CCT < 3500 K	105	90	70		
			CCT ≥ 3500 K	110	95	75		
GB 37478-2019	LED luminaire for road and tunnel lighting	P ≤ 60 W	CCT < 3500 K	125	115	95	CRI ≥ 70 Tolerance: −3	
			3500 K ≤ CCT ≤ 5000 K	130	120	100		
			CCT > 5000 K	–	–	125		
		P > 60 W	CCT < 3500 K	130	120	100		
			3500 K ≤ CCT ≤ 5000 K	135	125	105		
			CCT > 5000 K	–	–	130		

(continued)

Table 2 (continued)

Standard No.	Product type	Category	Parameters and requirements[a]				
			EEG[a] and thresholds (lm/W)			CRI[b]	Lumen maintenance[c]
			1	2	3		
TBD	LED flat panel luminaire	CCT < 3500 K	110	95	60	$R_a \geq 80$ $R_9 > 0$ Tolerance: −3	
		CCT ≥ 3500 K	120	105	70		

[a] *EEG* energy efficiency grade
[b] EEG thresholds for LED indoor lighting products that have CRI ≥ 90 are 10 lm/W lower than the values specified in the table above
[c] For LED luminaires, test data of lumen maintenance collected according to LM-80 test data of LED package is acceptable

3 European Lighting Regulations Under Ecodesign

3.1 Overview of Recent Efforts That Have Been Taken in European Union

In Europe, the Commission worked to set minimum quality standards for directional and non-directional LED lamps in regulation EU No 1194/2012. This regulation established minimum quality requirements for both directional and non-directional LED lamps on the following: lifetime, rapid switching cycles, start time, run-up time, premature failure rate, CRI, colour consistency and power factor. This regulation set a limit for energy use per unit light for directional LED lamps which took effect over three stages (in September 2013, 2014 and 2016) and pushed the market towards higher ambition at each stage. Importantly, the third stage took effect in September 2016 and phased-out mains-voltage (230 V) halogen directional lamps. Thus, for the last 3 years, mains-voltage spot lamps including GU-10 and E27 base types spot lamps have to be LED, and halogen is no longer allowed in that form factor. Europe was the first economy to adopt such a measure, moving faster than California which took a similar step in 2018.

Starting in 2016, the European Commission initiated a comprehensive review process of all of its lighting regulations—for domestic and professional lighting—and made a decision to establish one regulation that covered all lighting technologies. This new regulation was finalised in 2018 and voted on by Member States in December 2018. It was adopted by the European Commission on 1 October 2019 and is expected to be published in the Official Journal of the European Union in November 2019, and enter into force 20 days later. The new lighting regulation has some measures that come into effect in September 2021 and others that are introduced later in September 2023.

Table 3 presents the current and future European ecodesign regulation for LED lighting products, including the day it was issued and the effective dates when the different stages of requirements took effect. Please note that the future European lighting regulation does not yet have a standard number because it has not yet been published in the Official Journal of the European Union, although this is expected to happen in December 2019.

Table 3 Europe's current and upcoming LED lighting energy efficiency standards

Standard name	Standard No.	Issue date	Effective dates
Ecodesign requirements for directional lamps, light emitting diode lamps and related equipment	EU No 1194/2012 [4]	12 December 2012	September 2013; September 2014; September 2016
Ecodesign requirements for light sources and separate control gears	Commission Regulation (EU) .../... [5]	Expected December 2019	September 2021; September 2023

3.2 Newly Adopted Lighting Regulation in EU

The European single lighting regulation contains requirements for efficacy and light quality (colour and consistency), power quality and limits on flicker. It introduces a new lifetime test which combines a measure of lumen maintenance and the switching cycle endurance test. It also sets out requirements for information on the packaging and the light source.

As a measure of efficacy, the regulation sets a maximum power consumption for a given level of luminous flux. The equation is given in the regulation as follows:

$$P_{onmax} = C \cdot \left(L + \Phi_{use} / \left(F \cdot \eta\right)\right) \cdot R$$

where:

- The values for threshold efficacy (η in *lm/W*) and end loss factor (L in *W*) are specified in Table 1, depending on the light source type. They are constants used for computations and do not reflect true parameters of light sources. The threshold efficacy is not the minimum required efficacy; the latter can be computed by dividing the useful luminous flux by the computed maximum allowed power.
- Basic values for correction factor (C) depending on light source type, and additions to C for special light source features are specified in Table 2.
- Efficacy factor (F) is:

  1.00 for non-directional light sources (using total flux)
  0.85 for directional light sources (using flux in a cone)

- CRI factor (R) is:

  0.65 for CRI ≤ 25;
  (CRI + 80)/160 for CRI > 25, rounded to two decimals.

Taking the above variables into account, the equation is then populated with values from a table of threshold efficacy, end-loss factors and correction factors, based on the light source technology and product. When all the calculations are made, an LED lamp that emits 800 lm (the most popular lamp type in Europe) with 80 CRI must be 91 lm/W starting in September 2021.

The regulation sets requirements for separate control gear that operate LED Lamps, defining minimum energy efficiency (power out/power in) of separate control gear at full load.

The regulation also sets a series of quality requirements that ensure good quality light will be delivered to the consumers. The functional requirements included in the regulation are given in Table 4.

There are correction factors that are given for different special features, such as high CCT, high CRI, HID with a second envelope and other aspects. These correction factors are cumulative, meaning that lamps or luminaires that incorporate more than one of these features for which power allowances are made are additive.

Table 4 European Lighting Regulation—functional requirements for light sources

Parameter	Requirement in the regulation
Color rendering	CRI ≥ 80 (except for HID with $\Phi_{use} > 4000$ lm and for light sources intended for use in outdoor applications, industrial applications or other applications where lighting standards allow a CRI < 80, when a clear indication to this effect is shown on the light source packaging and in all relevant printed and electronic documentation).
Displacement factor[a] (DF, cos f_1) at power input P_{on} for LED and OLED MLS	No limit at $P_{on} \le 5$ W DF ≥ 0.5 at 5 W < $P_{on} \le 10$ W DF ≥ 0.7 at 10 W < $P_{on} \le 25$ W DF ≥ 0.9 at 25 W < P_{on}
Lumen maintenance factor (for LED and OLED)	The lumen maintenance factor X_{LMF}% after endurance testing shall be at least $X_{LMF,MIN}$% calculated as follows: $X_{LMF,MIN}\% = 100 \times e^{\frac{(3000 \times \ln(0.7))}{L_{70}}}$ where L_{70} is the declared $L_{70}B_{50}$ lifetime (in hours). If the calculated value for $X_{LMF,MIN}$ exceeds 96.0%, an $X_{LMF,MIN}$ value of 96.0% shall be used.
Survival factor (for LED and OLED)	Light sources should be operational as specified in row “Survival factor (for LED and OLED)” of Annex IV, Table 6, following the endurance testing.
Color consistency for LED and OLED light sources	Variation of chromaticity coordinates within a six-step MacAdam ellipse or less.
Flicker for LED and OLED MLS	Pst LM ≤ 1.0 at full-load
Stroboscopic effect for LED and OLED MLS	SVM ≤ 0.4 at full-load (except for HID with $\Phi_{use} > 4000$ lm and for light sources intended for use in outdoor applications, industrial applications or other applications where lighting standards allow a CRI < 80).

[a]Displacement factor (also called Fundamental Power Factor, or cos Φ_1)

4 LED Lighting Certification and Qualification Programs in the USA

4.1 Energy Star and Quality Product List (QPL)

In United States, ENERGY STAR is well known amongst consumers because of its long-established and high-profile promotions. Its eligibility criteria were once the leading benchmark for global lighting S&L programs. However, its most recent update was made in 2016 and 2017, for fixtures and lamps respectively, thus failing to capture the latest progress of the rapid technological changes in LED technology. For example, current version of ENERGY STAR eligibility criteria (version 2.1) still accepts applications of compact fluorescent lamps (CFLs). The criteria establish a minimum efficacy requirement for non-directional lamps at 80 lm/W, and fails to seize on substantial energy saving potentials from high efficacy LED lamps that are 100 lm/W and greater.

Table 6 Key metrics of recently updated requirements for LED lighting in China, the US and the US

Metrics	Units	China MEPS (2019 Versions)	United States—Design Lights Consortium (2019 Draft v5.0)	Europe's Ecodesign One-Lighting Regulation (2019)
Effective date	n/a	1 May 2020	Expected in April 2020	1 September 2021
Product Scope (as specified in standards)	n/a	LED lighting: • Downlight • Directional integrated light • Non-directional self-ballasted lamp • Street light • Tunnel light • Flat panel light **NOT** including dimmable or colour tuneable LED lamps for indoor use **NOT** including separately-installed controls for internet connection or other gears that not related to lighting functions	LED lighting: • Luminaire (indoor, outdoor) • Retrofit kit (indoor, outdoor) • Lamps (linear, screw-based, pin-based) Note: with further sub-categorization by lumen output, tube diameter, shape, etc. Also categorized products by application in detail. Most indoor products are dimmable	All light sources, including LED lighting; sets requirements for lamps and modules used in luminaires (but not LED dies or LED chips or packages); as well as LED drivers; light sources have a defined x,y chromaticity coordinates, a surface area emission of <500 lm/mm^2, flux between 60 and 82,000 lm and CRI > 0

(continued)

Table 6 (continued)

Metrics	Units	China MEPS (2019 Versions)	United States—Design Lights Consortium (2019 Draft v5.0)	Europe's Ecodesign One-Lighting Regulation (2019)
Efficacy	lm/W	Indoor: Non-directional self-ballasted LED lamp, omni-directional: 60–65 lm/W or semi-directional: 70–75 lm/W; Directional lamps, PAR 16/20: 65–70 lm/W; PAR 30/38: 70–75 lm/W LED downlight: $P \leq 5$ W: 60–65 lm/W; $P > 5$ W: 70–75 lm/W; LED flat panel luminaire: 60–70 lm/W	Indoor Luminaires: 80–120 lm/W Indoor Retrofit Kit 110–120 lm/W Linear Replacement Lamps: 120 lm/W Four Pin-base Replacement Lamp: 85–120 lm/W	Varies with equation given in Sect. 3.2 of this paper. Efficacy requirements on all light sources increase with higher levels of light output
		Outdoor: LED luminaire for road and tunnel lighting, $P \leq 60$ W: 95–125 lm/W $P > 60$ W: 100–130 lm/W	Outdoor Luminaires and Retrofit Kits: 105 lm/W Mogul Screw-Base Replacement for HID lamp: 105 lm/W (120 lm/W for high-bay)	
	Allowance	Please note that 10 lm/W allowance is given for indoor products with CRI ≥ 90	Allowance to efficacy may accrue up to 15% based on colour and light quality	Directional light sources, certain lamps with very high CCT, high CRI values, directional lamps, colour tuneable sources, and others. Please view Annex II of the regulation when it is published

(continued)

Table 6 (continued)

Metrics	Units	China MEPS (2019 Versions)	United States—Design Lights Consortium (2019 Draft v5.0)	Europe's Ecodesign One-Lighting Regulation (2019)
Colour Rendering Index	Ra	Indoor: Ra ≥ 80	Indoor, except high-bay: Option 1: ANSI/IES TM-30-18: • IES $R_f \geq 70$ • IES $R_g \geq 89$ • −12% ≤ IES $R_{cs,h1} \leq +23\%$ Option 2: CIE 13.3-1995: • $R_a \geq 80$ • $R_9 \geq 0$	Indoor: CRI ≥ 80 (except for HID with Φuse > 4 klm, outdoor, industrial or where lighting standards allow CRI < 80)
		Outdoor: Ra ≥ 70	Outdoor and high-bay: Option 1: ANSI/IES TM-30-18: • IES $R_f \geq 70$ • IES $R_g \geq 89$ • −18% ≤ IES $R_{cs,h1} \leq +23\%$ Option 2: CIE 13.3–1995: • $R_a \geq 70$ • $R_9 \geq -40$	Outdoor: No requirement
Displacement Factor	DF	No requirement	No requirement	No limit at $P_{on} \leq 5$ W, DF ≥ 0,5 at 5 W < $P_{on} \leq 10$ W, DF ≥ 0,7 at 10 W < $P_{on} \leq 25$ W DF ≥ 0,9 at 25 W < P_{on}
Lifetime	Hour	No requirement		

(continued)

Table 6 (continued)

Metrics	Units	China MEPS (2019 Versions)	United States—Design Lights Consortium (2019 Draft v5.0)	Europe's Ecodesign One-Lighting Regulation (2019)
Lumen maintenance	% of initial	Must be greater than $70\%^{(3000/t0)}$ t_0 is rated lifetime LM requirement @3000 h	$L_{70} > 50{,}000$ h ($L_{90} > 50{,}000$ h, additional requirement for DLC Premium)	See equation defining Lumen Maintenance Factor in Table 4. Example: 93.11% at 3000 h for 15,000 h declared lifetime
Other parameters included		No additional parameters required	• Spectral quality • Spectral power distribution • Backlight, Uplight and Glare (BUG) • Flicker • Controllability (e.g. dimming, colour tuning) • Driver in-situ temperature measurement test • Colour consistency • Colour maintenance • Zonal lumen distribution	• Colour consistency • Flicker (PstLM) • Stroboscopic effect (SVM) • Survival factor

There is another qualification program for lighting products in North America which is maintained by a non-profit organisation called the Design Lights Consortium (DLC). The DLC has managed a Qualified Products List (QPL) for over 10 years, providing a one-stop resource which contains high-efficiency, high-quality LED lamps and luminaires. Across several categories of LED lighting products, this program updates their technical requirements regularly, based on the market data and models in their database. The QPL works to promote high-efficiency, high-quality LED-based lighting products primarily for commercial and outdoor applications, whereas ENERGY STAR focuses mainly on residential lighting applications.

DLC uses technical requirements (TR) to define criteria for the QPL. DLC-qualified lighting products are mainly used for commercial and outdoor purposes. These products are grouped by product types, such as various types of luminaires, retrofit kits and lamps. The luminaires are subdivided by application, such as tunnel lighting, roadway lighting, parking garage lighting, replacement lamps for high bay luminaires, etc. The requirements for different types of products may vary significantly. This paper reviews the latest version of DLC's Technical Requirements for Solid-State Lighting, the upcoming update of Version 5.0, second draft, requirements which are scheduled to be adopted in January 2020. It could represent the trend of energy performance levels and focus of parameters, such as light quality and controllability.

4.2 DLC Technical Requirements for QPL

DLC's current version of the Technical Requirement for Solid-State Lighting is Version 4.4. The next version is Version 5.0, which is proposed to be finalized and released in January 2020 and take effect in April 2020. Table 5 presents a summary of the key requirements from the second draft of Version 5.0, which was issued for public comment on 30 September 2019. Please note that some of these values may change based on the public consultative process. It was indeed noted that DLC increased the ambition of the efficacy requirements when updating Version 4.4. The efficacy values increased on average around 10.8% from Version 4.4, with some categories as high as 23%.

5 Summary of Key Metrics in the Requirements of the Three Economies

The purpose of this paper is to present the recent improvements in three important economies, being China, EU and the US, which represent the biggest markets and manufacturing base of lighting products in the world. Their experiences could help and encourage policymakers in other economies to establish strong regulations.

Table 6 is a summary of structure and key metrics of China's new energy efficiency standards, the newly adopted EU directive and DLC QPL's qualification requirements.

Figure 2 presents a comparison of the new European standard (assuming 80 CRI) with the new GB standard (assuming omni-directional lamp, CCT $\geq$ 3500 K). LED lamps taken from product registry databases in Australia, Singapore, China and Korea have been added to the diagram to show the current market relative to the ambition of the MEPS. DLC does not cover non-directional, self-ballasted LED lamps therefore it has been omitted from the figure. At 800 lm of light, the approximate equivalent of a 60 W incandescent lamp, Europe has a requirement of 91 lm/W and China's requirement is 65 lm/W. However, it is worth noting that China's paying more and more attention to the implementation assessment of MEPS, and the process of updating and revising LED MEPS will be accelerated.

Figure 3 presents a comparison for LED streetlighting, and includes the new standards from China, the US and Europe. The datapoints presented represent all the models of streetlight registered between October 2017 and June 2019 in the DLC database (n = 43,800 models). The models are plotted by CCT vs. efficacy, as this is how the Chinese GB standard sets requirements on streetlighting. China has two requirements, one for luminaires that are less than or equal to 60 W and one that is slightly higher for models that are greater than 60 W. Within those two categories, GB 37478-2019 provides further breakdown by CCT, into three categories. The DLC and European requirements do not vary with CCT—they are constant across

Table 5 Overview of the improvements in DLC's technical requirements of Version 5.0

TR Versions	Version 5.0 (second draft)
Release date	January 2019, first draft September 2019, second draft January 2020, final version (proposed)
Summary of main changes	• Increased efficacy • Additional requirements on quality of light, e.g. color quality, glare, flicker, etc. • Additional requirements on controllability, e.g. dimming, integral controls and compatibility; • Allowance in efficacy as trade-off of higher light quality (low glare, low CCT, high CRI) • Removal of requirements of Total Harmonic Distortion (THD) and Power Factor (PF) • Widened range of Chromaticity to max 6500 K for both indoor and outdoor products
Efficacy	On average, efficacy requirements in Version 5.0 increases by 10.8% from Version 4.4, and is as much as 23% for some products.
CRI	Indoor products (except high-bay) must be capable of meeting one of the following criteria: • *ANSI/IES TM-30-18*: – IES $R_f \geq 70$ – IES $R_g \geq 89$ – $-12\% \leq$ IES $R_{cs,h1} \leq +23\%$ • *CIE 13.3-1995*: – $Ra \geq 80$ – $R9 \geq 0$
	Outdoor and high-bay products must be capable of meeting one of the following, criteria: • *ANSI/IES TM-30-18*: – IES $R_f \geq 70$ – IES $R_g \geq 89$ – $-18\% \leq$ IES $R_{cs,\,h1} \leq +23\%$ • *CIE 13.3-1995:* – $Ra \geq 70$ – $R9 \geq -40$
New testing and reporting requirements in Version 5.0	• Spectral quality • Spectral power distribution • Backlight, Uplight and Glare (BUG) • Flicker • Controllability (e.g. dimming, colour tuning) • Colour consistency • Colour maintenance

Note: DLC sets a higher level of criteria named DLC Premium Requirement. It requires an additional 15 lm/W for the same categories of products that also fall into the Premium scope

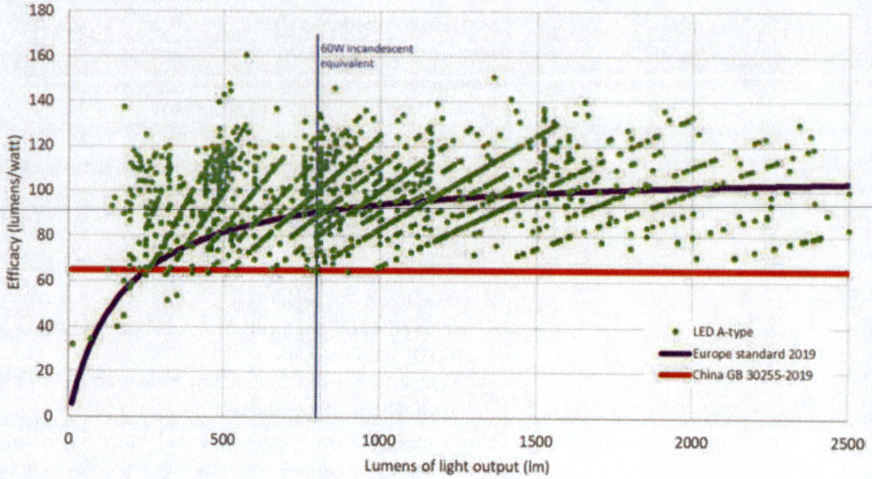

Fig. 2 Comparison of MEPS for non-directional LED lamps

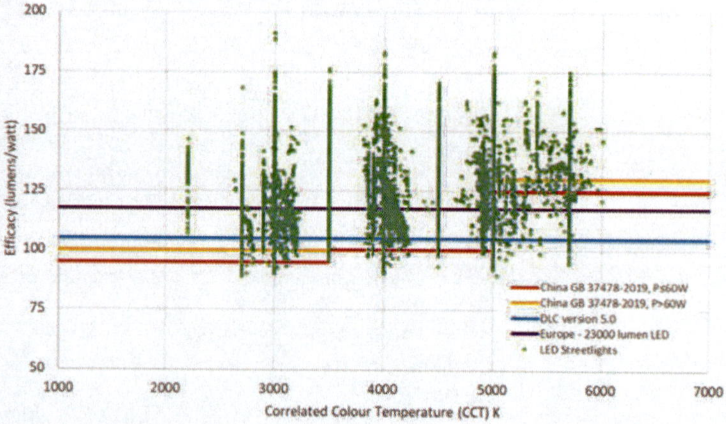

Note: The Chinese standard for >60W between 3500 and 5000K (105 lm/W) is the same as the DLC value (105 lm/W), thus the yellow line is not visible at those CCT values in Figure 3.

Fig. 3 Comparison of street lighting requirements

the CCT values. The European line has been plotted for a 23,000 lm streetlight (the average of the models in the DLC database) with a CRI of 70.

Note: The Chinese standard for >60 W between 3500 and 5000 K (105 lm/W) is the same as the DLC value (105 lm/W), thus the yellow line is not visible at those CCT values in Fig. 3.

6 Implications on Regulations Trend Toward LED Lighting Based on the Latest Experiences from China, EU and the US

The comparison of the requirements of the three quality and performance requirements in China, the US and Europe illustrates that there is a trend toward increasing energy performance and light quality of LED lighting products. Furthermore, there are other important quality and performance requirements which are included in the standards which help to ensure good quality lighting that protects human health.

The following are our key recommendations for consideration:

- **Expand coverage**. All LED products need to be covered and regulated, to ensure that the quality and performance are maintained and that human health and well-being is protected.
- **Increase efficacy**. The performance of LED products is constantly increasing and it is very difficult for policy requirements to keep pace with the rate of industry innovation. It is important therefore that policy-makers regularly review the requirements of LED products, and when setting the requirements take into consideration the effective date of the requirements, and the projected improvement in the market.
- **Consider Requirements for Temporal Light Artefacts**. The introduction of light emitting diode (LED) lighting products has brought renewed concern about possible adverse human health effects including migraine, eyestrain, seizures, vertigo, anxiety and fatigue. These health effects can be triggered by exposure to flickering, cyclic or transient variations in light which are collectively known as temporal light artefacts (TLA). The main TLAs of concern to US and European policy makers are flicker and the stroboscopic effect. Both the DLC and EU policy makers have set requirements that significantly reduce any health risk from TLAs. Both specifications set requirements for PstLM (IEC TR 61547-1) and SVM (IEC TR 63158) that will help to protect people from poor-quality lighting that may affect their health and productivity.[1]
- **Minimum Lifetime Requirements**. Ensuring that consumers can enjoy the full lifetime that they have been promised on the package is important. Policy-makers may consider adopting minimum lifetime requirements that are in line with consumer expectations and assess the use of a new lifetime test adopted by Europe as an alternative test to determine both switching cycles and lumen maintenance at the same time. Currently, none of the three standards examined in this paper set out minimum lifetime requirements.

[1] Short term flicker visibility measure PstLM—measured using: IEC TR 61547-1:2015 "Technical Report: Equipment for general lighting purposes—EMC immunity requirements—Part 1: An objective voltage fluctuation immunity test method."

Stroboscopic visibility measure, SVM—measured using: IEC TR 63158-1:2018 "Technical Report: Equipment for general lighting purposes—Objective test method for stroboscopic effects of lighting equipment."

LED lighting products are being applied to more and more applications thanks to their flexibility and "smartness". Due to the wide application and expanding market share, policy makers, certification program managers and many other stakeholders are putting more emphasis on LED lighting performance parameters not only for higher efficacy, but high quality of light and colour, comfortability and controllability. The more that global energy-efficiency standards are aligned, the better it will benefit manufacturers and consumers alike, realising the promotion of high-performance, high-quality light for the world.

References

1. GB 30255-2019 *Minimum Allowable Values Of Energy Efficiency And Energy Efficiency Grades Of LED Products For Indoor Lighting.*
2. GB 37478-2019 *Minimum allowable values of energy efficiency and energy efficiency grades of LED luminaires for road and tunnel lighting.*
3. GB 30255-2013 *Minimum Allowable Values Of Energy Efficiency And Energy Efficiency Grades Of Non-Directional Self-Ballasted LED-Lamps For General Lighting Services.*
4. Commission Regulation EU No. 1194/2012 of 12 December 2012 implementing Directive 2009/125/EC of the European Parliament and of the Council with regard to ecodesign requirements for directional lamps, light emitting diode lamps and related equipment (Text with EEA relevance).
5. Commission Regulation (EU) … /… of 1.10.2019 laying down ecodesign requirements for light sources and separate control gears pursuant to Directive 2009/125/EC of the European Parliament and of the Council and repealing Commission Regulations (EC) No 244/2009, (EC) No 245/2009 and (EU) No 1194/2012.

Smart Efficient Fan Controller® with Fault Detection Diagnostics

Robert Mowris

Abstract Heating, Ventilating, Air Conditioning (HVAC) consumption in the US accounts for 33% of average summer peak-day electricity loads, 13% of total electricity use, and 44% of total natural gas use. Research studies indicate that most existing residential HVAC systems are oversized and improperly installed causing short cycling and reduced efficiency. Furthermore, most systems have no fan-off delay or short delays leaving significant unrecovered energy in the system at the end of each cycle. This paper describes a patented Smart Efficient Fan Controller® (EFC®) with Fault Detection Diagnostics (FDD) improves cooling and heating efficiency by providing variable fan-off delays to correct thermostat short cycling faults, adjusting delays based on low capacity faults, and reducing continuous or intermittent fan-on operation by 50–75% or turning off the fan during unoccupied periods. Patented Smart EFC® algorithms can be licensed and deployed on smart communicating thermostats. Tests of the Smart EFC® were performed in the field and at a third-party ISO-certified laboratory used by manufacturers and USDOE to test HVAC equipment for compliance with Federal efficiency standards. Based on field and laboratory tests and building simulations, the Smart EFC® with fan-on correction can provide 25% ± 3% cooling plus fan savings and 17.4% ± 2.6% heating savings. About 80% of electricity savings are from reducing continuous or intermittent fan-on operation. The Smart EFC® can potentially save about 1% ± 0.1% of total California energy use, and 1.3% ± 0.2% of total US annual energy use.

1 Introduction

Residential and commercial heating, ventilating, and air conditioning (HVAC) consumption in the United States accounts for 30% of average summer peak-day electricity loads, 13% of total electricity use, and 44% of total natural gas use [1]. A 2002 study published by the Hewlett Foundation indicates that improving HVAC

R. Mowris (✉)
Verified® Inc., Olympic Valley, CA, USA
e-mail: robert@verified.co

P. Bertoldi (ed.), *Energy Efficiency in Domestic Appliances and Lighting*, Springer Proceedings in Energy, https://doi.org/10.1007/978-3-030-79124-7_16

cooling and heating efficiency represents one of the largest economically achievable opportunities for energy efficiency and peak demand savings [2]. Market research data indicates that about 77% of existing air conditioners in the United States have no fan-off delay and 23% have fixed fan-off delays ranging from 30 to 90 s [3]. Direct Expansion (DX) vapor-compression refrigerant-based Air Conditioning (AC) systems with no fan-off delay or fixed fan-off delays leave water and unrecovered evaporative cooling energy on the evaporator coil at the end of each cycle. Field studies indicate that it takes about 15–30 min for water left on the coil to flow down the condensate drain or evaporate [3]. Gas furnace, heat pump, and hydronic heating systems in the US operate with fixed fan-off delays of zero to 120 s, which leaves unrecovered heat in the Heat Exchanger (HX). For DX cooling systems, the Smart Efficient Fan Controller® (EFC®) recovers latent energy from the evaporator coil by providing a variable fan-off delay after the AC compressor turns off to evaporatively cool the conditioned space, satisfy the cooling thermostat setpoint longer and lengthen the off cycle.[1] For heating systems, the Smart EFC® provides a variable fan-off delay after the heating system turns off to overshoot the thermostat setpoint and lengthen the off cycle. For gas furnace heating systems, the EFC® can operate the fan at a higher speed to satisfy the thermostat sooner. The Smart EFC® adjusts variable fan-off delays to correct thermostat short cycling and low capacity faults and reduces continuous or intermittent fan-on operation by 50–75% or turns off the fan during unoccupied periods.

This paper provides field and laboratory test results of a Smart EFC® installed on residential split and packaged HVAC systems with DX cooling and gas furnace, heat pump, or forced-air hydronic hot water heating [4].[2] The Smart EFC algorithms can be deployed on smart communicating thermostats and smart fans to improve energy efficiency.

Field tests were performed on two DX AC gas furnace split-systems with (hereafter units #1 and #2) at a single-family residential building located in Reno, Nevada. Table 1 provides a description of units #1 and #2. Laboratory tests were also performed on one DX AC gas furnace split-system unit #3, one DX AC gas furnace packaged unit #4, one DX Heat Pump (HP) split-system unit #5, and one DX AC Hydronic (HYD) heating split-system unit #6. Table 2 provides a description of laboratory test units #3, #4, #5, and #6. The laboratory equipment was set up in two chambers to simulate typical indoor and outdoor conditions per ANSI/AHRI 210/240 [5]. Test conditions differ from those used to rate cooling and heating systems to match typical installations in California.[3]

[1] Latent energy is the quantity of heat absorbed or released by air undergoing a change of state, such as water vapor condensing out of the air as water onto a cold evaporator coil or cold water evaporating to water vapor which will cool the air.

[2] US Patent 8763920C1, US Patent 9328933, US Patent 9500386. US Patent 9671125, US Patent 10533768. US Trademark Efficient Fan Controller® Reg. No. 5,163,211 (First Use 03-01-2012), EFC® Reg. No. 5,198,335 (First Use 03-01-2012).

[3] Cooling tests were performed at Outdoor Air Temperature (OAT) Dry-Bulb (DB) 95 °F (35 °C) and Indoor Air Temperature (IAT) DB 75 °F (23.9 °C) and Wet Bulb (WB) 62 °F (16.7 °C). Gas

Table 1 Description of field test units #1 and #2

Description	Unit #1: 3.5-ton[a] split AC gas furnace	Unit #2: 5-ton split AC gas furnace
Indoor AC model	C23-41(FC)	RCF6024STAMCA
Rated SEER/EER	10/8	14/11.7
Rated heating efficiency	80% AFUE	80% AFUE
HX/coil	AC/gas HX	AC/gas HX
Rated total and sensible cooling capacity, airflow, and static pressure	40,600 Btu/h (11.9 kW) total and 27,608 Btu/h (8.09 kW) sensible, 1200 cfm (566.34 lps) at 0.5 IWC (124.54 Pa)	58,000 Btu/h (17.0 kW) total and 41,500 Btu/h (12.16 kW) sensible, 1188 scfm (560.67 lps) at 0.5 IWC (124.54 Pa)
Outdoor AC model	HS23-461-2P	RA1460AJ1NA
Fan speed	Low, Med, High	Low, Med, High
Refrigerant charge	R22 117 ounces (3.32 kg)	R410A 162 ounces (4.593 kg)
Duct leakage @ 25 Pa	21%	21% and 6%
Heating model	GUA120A020AIN	R801SA125524MSA
Rated heating capacity, airflow, and static pressure	100,000 Btu/h (29.3 kW) and 1012 scfm (477.61 lps) at 0.5 IWC (127.3 Pa)	100,000 Btu/h (29.3 kW) and 1080 scfm (509.7 lps) at 0.8 IWC (201.8 Pa)
Fan-off delay cooling	0 s cooling	0 s cooling
Fan-off delay heating	120 s heating	30 and 45 s heating

[a]One ton of cooling is defined as heat energy removed from one short ton of water (2000 lb or 907.1847 kg) to produce one ton of ice at 32 °F (0 °C) in 24 h. Energy required for phase change of liquid water at 32 °F (0 °C) into solid ice at 32 °F is referred to as heat of fusion equal to 144 Btu/lb times 2000 lb of water or 288,000 Btu of energy over 24 h period or 12,000 Btu/h to make one ton of ice in 1 day. British thermal unit (Btu) is heat required to raise temperature of 1 lb (0.454 kg) of water 1 °F (0.556 °C). Btu is equivalent to 1055.06 J or 251.997 cal

2 Test Equipment Laboratory Setup

Tests were performed at Intertek®, an AHRI-certified laboratory, located in the United States. The laboratory is used by manufacturers to certify air conditioners and heat pumps for AHRI equipment efficiency testing for the U.S. Department of

heating tests were performed at OAT DB 47 °F (8.3 °C) and IAT DB 72 °F (22.2 °C) and WB 53 °F (11.7 °C). Heat pump tests were performed at following OAT DB 17 °F (−8.3 °C), 35 °F (1.7 °C), 47 °F (8.3 °C), and 62 °F (16.7 °C) and IAT DB 70 °F (21.1 °C) and WB 55 °F (12.8 °C). Hydronic heating tests were performed at OAT DB 47 °F (8.3 °C) with Hot Water Temperature (HWT) 130 °F (54.4 °C) and 140 °F (60 °C) and IAT DB 70 °F (21.1 °C) DB and WB 55 °F (12.8 °C). ANSI/AHRI 210/240 Standard EER_A test conditions are OAT DB 95 °F (35 °C) and IAT DB 80 °F (26.67 °C), WB 67 °F (19.44 °C), EER_B test conditions are OAT DB 82 °F (45.6 °C) and IAT DB 80 °F (26.67 °C), WB 67 °F (19.44 °C). SEER test conditions are: OAT DB 82 °F (45.6 °C), IAT DB 80 °F (44.2 °C), and WB 57 °F (31.7 °C).

Table 2 Description of laboratory test units #3, #4, #5, and #6

Description	Unit #3: 3-ton split AC gas furnace	Unit #4: 3-ton Pkg AC gas furnace	Unit #5: 1.5-ton split heat pump	Unit #6: 1.5-ton split AC HYD heat
Indoor AC model	CNRHP3617ATA	GPG1336070M41BA	ARUF25B14AA	19CDX-HW
Rated SEER/EER	13/11.2	13/11	14	13/11.7
Rated heating efficiency	80% AFUE	80% AFUE	3.76 COP	78% efficiency
AC/HX coil	DX AC/gas HX	DX AC/gas HX	DX AC/HP	DX AC/HYD coil
Rated total and sensible cooling capacity, airflow, and static pressure	33,800 Btu/h (9.9 kW) total, 25,660 Btu/h (7.52 kW) sensible, 1200 cfm (566.34 lps) at 0.5 IWC (124.54 Pa)	35,800 Btu/h (9.9 kW) total, 28,547 Btu/h (8.37 kW) sensible, 1188 cfm (560.67 lps) at 0.5 IWC (124.54 Pa)	17,300 Btu/h (5.07 kW) total and 12,283 Btu/h (3.6 kW) sensible, 525 cfm (247.77 lps) at 0.4 IWC (101.8 Pa)	17,500 Btu/h (5.13 kW) total and 12,425 Btu/h (3.64 kW) sensible, 550 scfm (259.57 lps) at 0.3 IWC (74.72 Pa)
Outdoor AC model	24ABS336A300	GPG1336070M41BA	GSZ140181KD	MHH-19-410
Fan speed and RPM	Low 1050, Med 1080, High 1100	Low 850, Medium 980, High 1040	1043 RPM	1550 RPM
Refrigerant charge	R22 86.4 oz (2.5 kg)	R410A 70 oz (2 kg)	R22 92 oz (2.64 kg)	R410A 102 oz (3 kg)
Duct leakage @ 25 Pa	6%	6%	6%	6%
Heating model	58STA070-12	GPG1336070M41BA	ARUF25B14AA	R801SA125524MSA
Rated heating capacity, airflow, and static pressure	54,000 Btu/h (15.83 kW) and 1140 scfm (538.02 lps) at 0.5 IWC (127.3 Pa)	55,200 Btu/h (16.18 kW) and 1173 scfm (553.59 lps) at 0.5 IWC (127.3 Pa)	18,000 Btu/h (5.28 kW) and 555 scfm (261.93 lps) at 0.47 IWC (119.7 Pa)	18,000 Btu/h (5.28 kW) and 550 scfm (259.57 lps) at 0.4 IWC (101.8 Pa)
Fan-off delay cooling	0 s cooling	0 s cooling	0 s cooling	0 s cooling
Fan-off delay heating	90 s heating	0 s heating	0 s heating	0 s heating

Energy (DOE) compliance and enforcement program to meet energy conservation standards required by the Energy Policy and Conservation Act of 1975 (as amended) [6]. The test facility consists of climate-controlled indoor and outdoor chambers where ducts, evaporator, condenser, furnace or hydronic heating equipment and forced air units are located. The HVAC systems and standard test equipment were assembled and installed in the test chambers by laboratory technicians. The AHRI 210/240 cooling verification tests were performed according to ANSI/AHRI 2008 Standard for Performance Rating of Unitary Air-Conditioning and Air-Source Heat Pump Equipment Standard 210/240 and ANSI/ASHRAE Standard 37-2009 [2, 7]. Thermal Efficiency verification tests were performed according to ANSI Z21.47-5th Edition 2006/CSA 2.3-5th Edition 2006 [8]. The psychrometric room meets ASHRAE 41.2-1987 standard specifications [9]. Calibration for all equipment at the laboratory test this facility is conducted in accordance with ISO 17025 requirements by an ILAC accredited calibration provider. Gas furnace heating equipment performance and AFUE tests were performed per ANSI Z21.47 specifications.

The DX cooling laboratory tests were performed under non-steady state field conditions to measure base cooling energy use, sensible capacity, and efficiency with no fan-off delay or fixed 60-s delay for the 3-ton packaged unit #4 or no delay and 90-s delay for the 3-ton split-system unit #3. Non-steady state cooling tests were also performed with the Smart EFC® providing a variable fan-off delay based on the duration of the cooling on/off cycle. Gas furnace heating lab tests were performed to measure base heating energy use, capacity, and efficiency with fixed 90-s and 120-s fan-off delays for units #3 and #4. For unit #3, non-steady state heating tests were performed with the Smart EFC® providing increased fan speed from low-to-high or medium-to-high after 4 min of furnace operation and variable fan-off delay based on the duration of the heating on/off cycle. Lab tests of the 1.5-ton split-system heat pump unit #5 and 1.5-ton split-system hydronic unit #6 were performed under non-steady state conditions to measure base energy use, sensible cooling or heating capacity and efficiency with no delay or 65-s fixed delay for unit #5 or 60 s for unit #6 after the cooling or heating system turned off. For unit #5 and #6, non-steady state cooling and heating tests were also performed with the Smart EFC® providing a variable fan-off delay based on the cooling or heating on/off cycle.

3 Cooling Test Data and Energy Savings Analysis

The Intertek® laboratory performed 27 split-system and packaged unit cooling tests and 24 heat pump cooling tests with and without the Smart EFC®. Tests were performed at 75 °F (29.3 °C) return air DB and 62 °F (16.7 °C) return air WB temperatures and 95 °F (35 °C) DB outdoor air temperature. Tests measured the additional sensible cooling capacity provided by the Smart EFC® extended variable fan-off delay compared to the baseline system with no delay or a fixed fan-off delay. The laboratory tests measured energy input and sensible cooling capacity output (Btu or J) with and without the EFC® for compressor operating times from 2 to 50 min. The

Table 3 Intertek® cooling tests unit #3—Smart EFC® v. no-delay and 90-s delay

Compressor on time (min)	5	5	10	15	30	Ave.
Base no delay wet coil tests	**201**	**202**	**203**	**204**	**205**	
No delay sensible cooling (Btu) [a]	1006	1396	3264	5381	10,995	4409
No delay AC energy use (kWh) [b]	0.265	0.271	0.544	0.828	1.673	0.717
No delay sensible efficiency (EER*) [c = a/b/1000]	3.79	5.14	6.00	6.49	6.57	5.60
Base 90-s delay dry coil tests						
90-s Delay sensible cooling (Btu) [d]	1283	1553	3465	5598	11,285	4637
90-s Delay AC energy use (kWh) [e]	0.276	0.281	0.553	0.836	1.677	0.725
90-s delay sensible efficiency (EER*) [f = d/e/1000]	4.65	5.52	6.27	6.69	6.73	5.97
90-s delay fan energy (kWh) [g]	0.011	0.011	0.011	0.011	0.011	0.011
Base no delay dry coil (>30 min between tests)						
No delay dry coil sensible cooling (Btu) [h]	857	1109	2965	5094	10,791	4163
No delay dry coil energy use (kWh) [i]	0.264	0.270	0.541	0.825	1.666	0.713
No delay dry coil efficiency (EER*) [j = h/i/1000]	3.24	4.11	5.48	6.17	6.48	5.10
No delay dry coil v. no-delay wet coil [k = 1 − c/j]	−17.0%	−25.2%	−9.5%	−5.2%	−1.4%	−11.7%
EFC® tests	**1**	**2**	**3**	**4**	**5**	
EFC® sensible cooling (Btu) [l]	1602	1893	3837	6186	11,864	5076
EFC® AC energy use (kWh) [m]	0.287	0.293	0.564	0.855	1.696	0.739
EFC® sensible efficiency (EER*) [n = l/m/1000]	5.58	6.47	6.80	7.23	7.00	6.62
EFC® fan energy vs. no delay (kWh) [o]	0.023	0.023	0.023	0.030	0.030	0.026
EFC® savings v. no delay wet coil base [p = 1 − c/n]	**32%**	**20.5%**	**11.9%**	**10.2%**	**6.1%**	**16.1%**
EFC® fan energy vs. 90-s (kWh) [q = o − g]	0.011	0.011	0.011	0.019	0.019	0.014
EFC® savings v. 90-s base [r = 1 − f/n]	**16.6%**	**14.7%**	**7.8%**	**7.5%**	**3.8%**	**10.1%**

laboratory tests also measured total sensible cooling capacity for 60 min at the same conditions. Table 3 and Figs. 1 and 2 provide the following cooling tests for unit #3: (1) no delay wet coil base, (2) 90-s delay dry coil base, (3) no delay dry coil base, and (4) Smart EFC® variable delay.[4] Cooling savings vary from 6.1% to 32% compared to no delay base, and 3.8% to 16.6% compared to 90-s delay base. Average

[4] AC systems operating with no fan-off delay leave water on the evaporator coil for successive AC cycles, if the AC off cycle time is less than 15–30 min (hereafter referred to as initial "wet coil" conditions). AC systems with fixed 30–90-s fan-off delays leave less water on the coil for successive AC cycles (hereafter referred to as initial "dry coil" conditions). If the AC off cycle is greater than 30 min, then the AC system will generally operate with initial dry coil conditions with the exception of high humidity conditions where the AC system has not been previously operating.

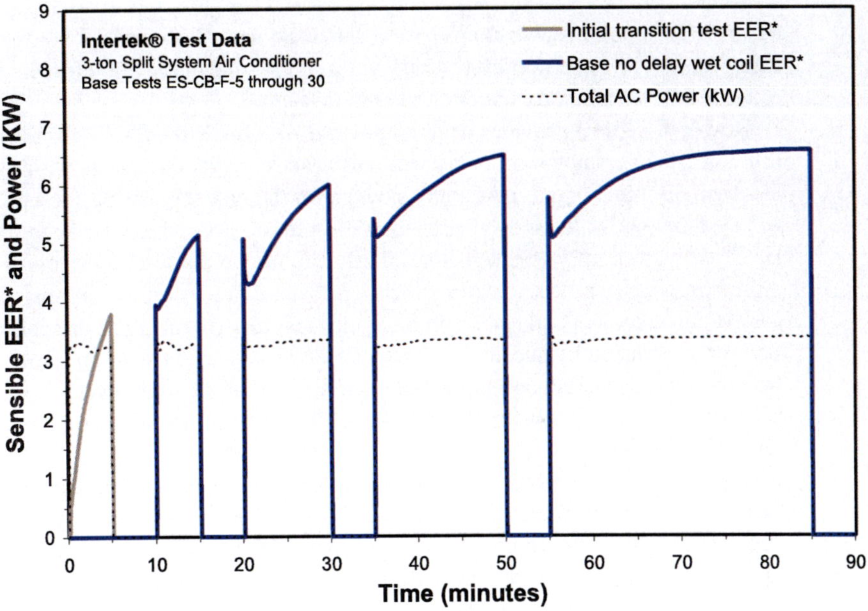

Fig. 1 Intertek® cooling tests 3-ton unit #3—no delay wet coil base

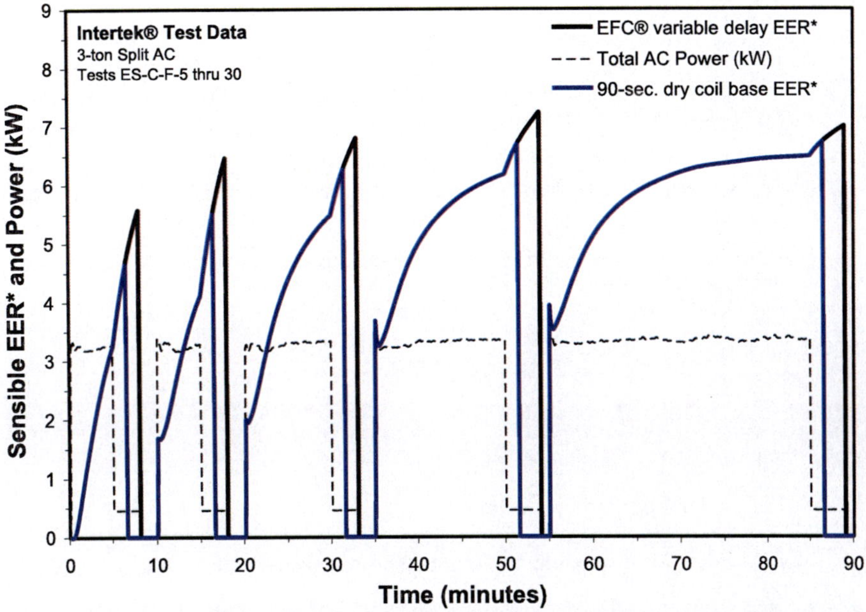

Fig. 2 Intertek® cooling tests 3-ton unit #3—Smart EFC® v. 90-s dry coil base delay

Smart EFC® cooling savings are 16.1% versus no delay base and 10.1% versus 90-s delay base. The base no-delay short cycles are significantly less efficient than the Smart EFC® which corrects for short cycling fault.

Row k of Table 3 provides negative savings of −1.4% to −25.2% for the no delay dry coil base versus the no delay wet coil base for successive tests with off cycles less than 30 min. Negative AC-only savings for the no delay dry coil tests show that AC systems with off times greater than 30 min or successive tests with the Smart EFC® deliver slightly less sensible initial evaporative cooling and require slightly more compressor power to satisfy the thermostat call for cooling compared to successive no delay wet coil tests with off cycles less than 30 min. The dry coil negative savings are caused by successive Smart EFC® tests starting with an initial dry coil which reduces sensible cooling at the beginning of the AC-only cycle. The negative EER* impact is −1.4% for the 30-min cycle and approaches zero for cycles of 60-min or longer.

Table 4 and Fig. 3 provide field tests of the 5-ton AC unit #2 with the Smart EFC® FDD. Tests were performed at average OAT of 93.2 ± 0.06 °F and 21% total system duct leakage @ 25 Pa. Table 4 shows the Smart EFC® providing normalized cooling savings ranging from 9.7% to 24.4% (row p) compared to no-delay wet coil base tests. Normalized savings are adjusted based on the EFC® AC-only sensible capacity (row i) required to match base no delay wet coil sensible capacity (row b) and Smart EFC® normalized energy (row n) to match base no delay wet coil normalized capacity. The Smart EFC® EER* negative impact is −4.4% to +1.7% (row k) for EFC® AC-only compared to no-delay wet coil base for successive tests. The +1.7% indicates an initial dry coil at 30 min off time. The average difference between field and laboratory test results is −0.2% ± 1.2% based on Eq. (2) (row q) indicating that the Smart EFC® FDD performs as good or better than the EFC® product tested at Intertek®.

The ratio of sensible cooling capacity for each test divided by the total sensible cooling capacity for 60 min tests is defined as the cooling Part Load Ratio (PLR) as shown in Eq. (1). The cooling PLR is used to normalize the cooling savings for each group of tests.[5]

$$PLR_c = \frac{Q_{c_o}}{Q_{c_r}} \tag{1}$$

where,

PLR_c = cooling Part Load Ratio,

Q_{c_o} = delivered sensible cooling capacity measured for each test (Btu or J), and

Q_{c_r} = total sensible capacity measured at same conditions for 60 min (Btu or J).

[5] Weighted average test results are based on tests performed at approximately the same PLR where the baseline is either zero or a fixed fan-off time delay for the same AC unit.

Table 4 Field tests 5-ton unit #2—Smart EFC® FDD v. no-delay base with 21% duct leakage

Compressor on time (min)	6	6	10	15	30	Ave.
Base no delay wet coil tests	**11**	**12**	**13**	**14**	**15**	
Base PLR	0.085	0.086	0.150	0.229	0.482	0.207
Base sensible cooling (Btu) [a]	2626	2654	4617	7051	14,810	6352
Base AC energy use (kWh) [b]	0.459	0.458	0.774	1.168	2.368	1.05
Base sensible efficiency (EER*) [c = a/b/1000]	5.72	5.80	5.97	6.04	6.25	5.96
Smart EFC® FDD tests	**16**	**17**	**18**	**19**	**20**	
EFC® FDD sensible cooling (Btu) [d]	3824	3812	5904	8544	16,566	7730
EFC® FDD AC energy use (kWh) [e]	0.497	0.497	0.806	1.235	2.392	1.09
EFC® FDD sensible efficiency (EER*) [f = d/e/1000]	7.69	7.67	7.32	6.92	6.93	7.31
EFC® FDD preliminary savings v. no delay [g = 1 – c/f]	25.6%	24.4%	18.6%	12.7%	9.7%	18.2%
EFC® extra fan energy (kWh) [h]	0.045	0.045	0.053	0.057	0.062	0.052
EFC® AC-only sensible cooling w/o delay [i]	2475	2508	4465	7017	14,816	6256
EFC® AC-only energy (kWh) [j = e – h]	0.452	0.451	0.753	1.178	2.330	1.033
EFC® AC-only sensible EER* (Btu/W) [k = i/j/1000]	5.48	5.56	5.93	5.96	6.36	5.86
EFC® AC on cycle EWV impact on EER* [k]	–	**–4.4%**	**–0.7%**	**–1.4%**	**1.7%**	**–1.2%**
EFC® extra cooling to match base [l = a – i]	150	146	152	34	–6	95
EFC® normalized sensible capacity (Btu) [m = d – i + b]	3974	3958	6057	8578	16,560	7825
EFC® normalized energy (kWh) [n = j*b/i + h]	0.525	0.523	0.832	1.240	2.391	1.102
EFC® normalized sensible EER* (Btu) [o = m/n/1000]	7.57	7.56	7.28	6.91	6.93	7.25
EFC® cooling savings [p = 1 – d/g]	**24.4%**	**23.3%**	**18.1%**	**12.7%**	**9.7%**	**17.6%**
Lab test Fig. 4 Eq. (2) savings [q = 0.0468*PLR$^{-0.6928}$]	25.7%	25.5%	17.4%	13.0%	7.8%	17.9%
Field minus lab test Fig. 4 Eq. (2) difference [r = p – q]	–1.3%	–2.2%	0.7%	–0.3%	1.9%	–0.2%

Figure 4 provides test data of the Smart EFC® cooling energy savings versus cooling PLR for the dry coil base and wet coil base. Figure 4 provides three regression equations for calculating EFC® cooling energy savings. Equation (2) is used to calculate EFC® cooling savings versus the dry coil base.

$$\Delta\eta_c = 0.0468\left(PLR_c\right)^{-0.6928} \tag{2}$$

where, $\Delta\eta_c$ = Smart EFC® cooling savings versus dry coil base.

Equation (3) is used to calculate EFC® cooling energy savings versus the no-delay wet coil base.

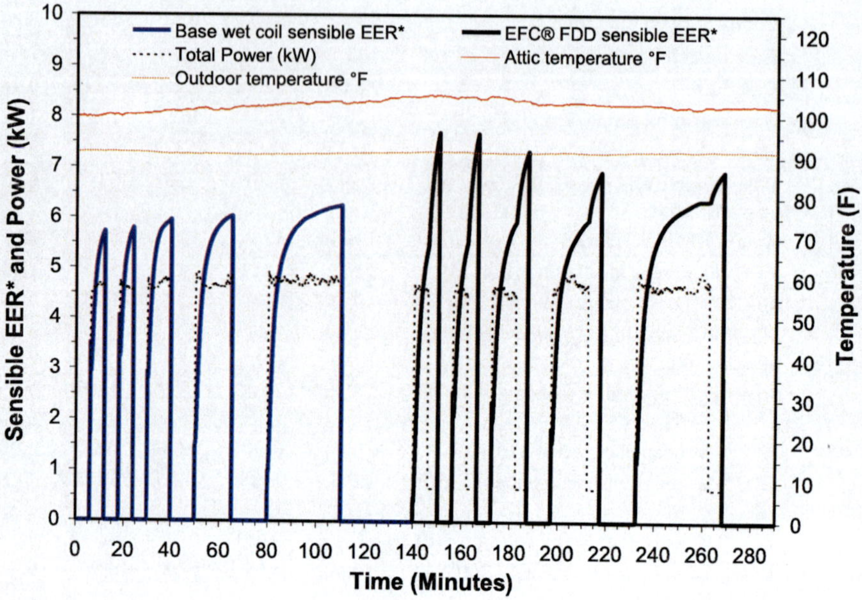

Fig. 3 Field tests 5-ton unit #2—Smart EFC® FDD v. no-delay base with 21% duct leakage

$$\Delta\eta_c = 0.0418(PLR_c)^{-0.5711} \tag{3}$$

where, $\Delta\eta_c$ = Smart EFC® cooling savings versus wet coil base.

The relationship between the initial no delay dry coil base EER* and PLR compared to initial wet coil no delay base conditions is provided in the following power function regression equation.

$$\Delta\eta_a = 0.009(PLR_s)^{-1.095} \tag{4}$$

where, $\Delta\eta_a$ = No delay dry coil base EER* compared to no delay wet coil base EER* from Table 3 (row k).

Equation (4) is used in the DOE2 post processor to adjust Smart EFC® savings in subsequent time steps to account for initial dry coil conditions during successive Smart EFC® tests which reduce sensible cooling at the beginning of the AC-only cycle. The mass and energy balance is determined using the above regression equations in the DOE2 post processor to adjust EFC® savings in subsequent time steps based on PLR.

Laboratory and field tests demonstrate that the Smart EFC® improves energy efficiency by delivering more cooling capacity and providing longer off times. The EFC® also prevents evaporator coil icing by evaporating water from the coil at the

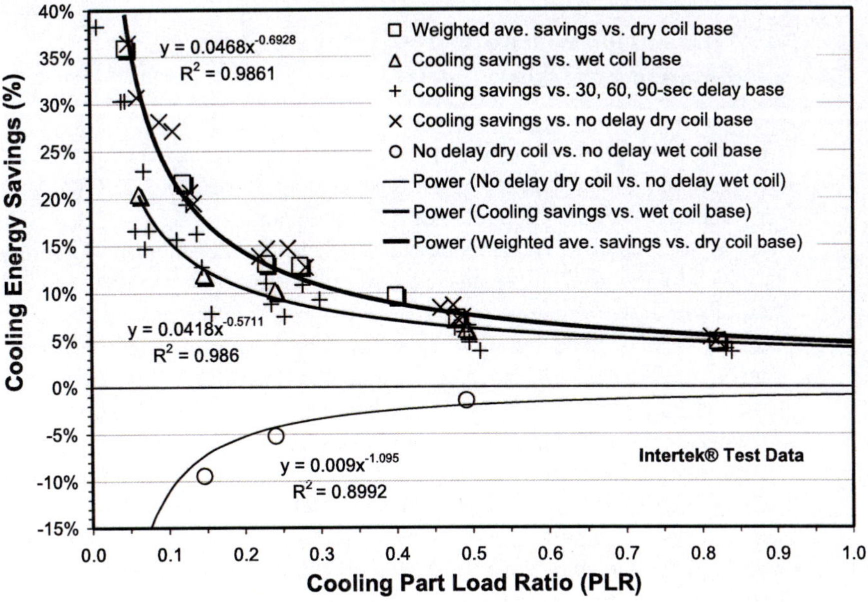

Fig. 4 Cooling energy savings versus part load ratio for Smart EFC®

end of each AC cycle. This prevents ice formation when the evaporator coil temperature is below freezing which can be caused by low airflow, dirty air filters, low refrigerant charge, low cooling setpoint, and refrigerant restrictions. Coil icing can reduce evaporator airflow by 17–37%, reduce efficiency by 4–12%, and cause continuous AC operation [3].

Data from a sample of 5582 AC units in California showing 77% of air conditioners with zero base fan-off delay, 11.9% with 30-s base delay, 7.8% with 60-s base delay, and 3.3% with 90-s base delay [4]. These values are used to determine weighted average cooling savings based on field and laboratory tests of the six AC units and different base fan-off delays.

Building energy simulation software, post processors, and the Database for Energy Efficiency Resources (DEER) residential single-family, multi-family, and mobile home building prototypes were used to evaluate the baseline HVAC energy use and peak demand for each building prototype and 16 California climate zones [10]. The average annual EFC® cooling energy savings are 10.9% ± 3.5% and the weighted average cooling PLR is 0.21 based on building simulations and housing stock weights for each climate zone from California housing stock data and US Census data [4, 11, 12].

4 Gas Furnace Heating Test Data and Energy Savings Analysis

The laboratory performed 48 split- and packaged gas furnace heating tests (24 baseline tests and 24 measure tests). The tests were performed at 72 °F (22.2 °C) return air DB and 53 °F (11.7 °C) return air WB temperatures and 47 °F (8.3 °C) DB outdoor air temperature. The laboratory tests measured the additional heating capacity provided by the EFC® using an extended fan-off time delay which varies as a function of the heat-source operational time compared to the baseline system with no time delay or a fixed fan-off time delay. The laboratory tests measured heating capacity output (Btu or J) with and without the Smart EFC® for furnace operational times varying from 5 to 30 min. The laboratory tests also measured total heating capacity for 60 min at the same conditions.

Table 5 provides the following Intertek® gas furnace heating tests for the 3-ton packaged unit #4: (1) 90-s base, (2) 120-s delay base, and (3) Smart EFC® FDD variable delay. Heating energy savings vary from 5.6% to 38.3% with 90-s delay base, and savings vary from 4.2% to 24.6% for 120-s delay base. Average EFC® FDD heating energy savings are 20.3% versus the 90-s delay base and 14.2% versus the 120-s delay base.

The ratio of heating capacity for each test divided by the total heating capacity for 60 min tests is defined as the heating Part Load Ratio (PLR) as shown in Eq. (5).

Table 5 Intertek® gas furnace heating tests unit #4—EFC® FDD v. 90- and 120-s delay

Furnace on time (min)	3	7	8	15	30	Ave.
Base 90-s delay tests	**109**	**111**	**113**	**115**	**117**	
Base 90-s heating energy (Btu) [a]	875	3461	4185	9559	21,619	7940
Base 90-s gas furnace energy use (kWh) [b]	3026	7774	8952	16,081	32,695	13,706
Base 90-s heating efficiency (EER*) [c = a/b]	28.9%	44.5%	46.8%	59.4%	66.1%	49.2%
Base 120-s delay tests	**51**	**53**	**55**	**57**	**59**	
120-s delay heating energy (Btu) [c]	1070	3755	4485	9887	21,952	4637
120-s delay gas furnace energy use (kWh) [d]	3026	7774	8952	16,081	32,695	0.725
120-s delay heating efficiency (EER*) [e = d/e]	35.3%	48.3%	50.1%	61.5%	67.1%	5.97
EFC® gas furnace heating tests	**52**	**54**	**56**	**58**	**60**	
EFC® heating energy (Btu) [f]	1419	4564	5339	10,826	22,907	5076
EFC® gas furnace energy use (kWh) [g]	3026	7774	8952	16,081	32,695	0.739
EFC® heating efficiency (EER*) [h = f/g]	46.9%	58.7%	59.6%	67.3%	70.1%	6.62
EFC® fan energy vs. 90-s delay (kWh)	0.011	0.017	0.018	0.018	0.018	0.026
EFC® savings v. 90-s delay [j = 1 – c/h]	**38.3%**	**24.2%**	**21.6%**	**11.7%**	**5.6%**	**20.3%**
EFC® fan energy vs. 120-s (kWh)	0.008	0.014	0.014	0.014	0.014	0.014
EFC® savings v. 120-s base [k = 1 – e/h]	**24.6%**	**17.7%**	**16.0%**	**8.7%**	**4.2%**	**14.2%**

The heating PLR is used to normalize the gas furnace heating energy savings for groups of tests.

$$PLR_h = \frac{Q_{h_o}}{Q_{h_r}} \tag{5}$$

where,

PLR_h = heating part load ratio of delivered heating capacity for each test divided by the total heating capacity of the equipment (dimensionless),

Q_{h_o} = delivered heating capacity measured for each test (Btu or J), and

Q_{h_r} = total heating capacity measured at same conditions for 60 min (Btu or J).

Laboratory test data of the gas furnace heating energy savings versus PLR are shown in Fig. 5. Gas furnace heating energy savings are calculated using regression Eq. (6) based on the PLR.

$$\Delta\eta_h = 0.0674\left(PLR_h\right)^{-0.4394} \tag{6}$$

where, $\Delta\eta_h$ = Smart EFC® FDD gas furnace heating savings compared to baseline.

Table 6 and Fig. 6 provide two sets of 10-h field tests of the gas furnace unit #2 controlled by a thermostat. Each set of tests was performed with 6% duct leakage at 25 Pa: The base test (black curve) includes a 120-s fixed fan-off delay and the

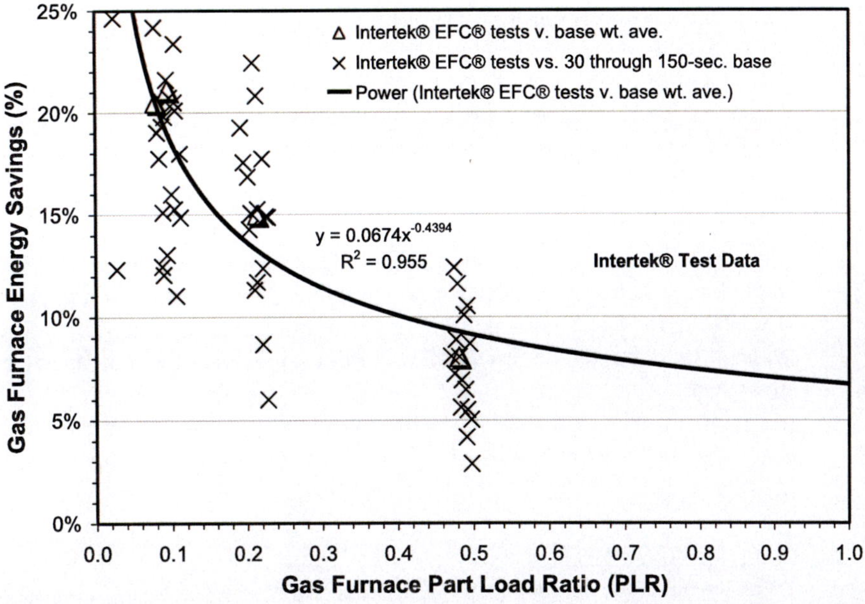

Fig. 5 Intertek® gas furnace heating tests—Smart EFC® savings versus part load ratio

Table 6 Gas furnace unit #2 10-h tests—Smart EFC® FDD HSF vs. 120-s base delay

Description	Row	Total
Base furnace on time (min)	a	233.3
EFC® furnace on time (min)	b	204.6
EFC® normalized furnace on time (min)	c = b*(s/p)	191.0
EFC® normalized savings based on furnace on time (%)	**d = 1 – (c/a)**	**18.1%**
Base furnace energy input (Btu)	e	382,982
EFC® furnace energy input (Btu)	f	336,585
Normalized base furnace energy input based on delta T (Btu)	g = e*[s/p]	314,238
Base furnace off time (min)	h	382.3
EFC® furnace off time (min)	i	411.0
Base heating (Btu)	j	222,811
Base heating efficiency (%)	k = j/e	58.2%
EFC® heating (Btu)	l	227,093
EFC® heating efficiency (%)	m = l/f	67.5%
EFC® normalized heating (Btu)	n = l*[s/p]	243,243
EFC® normalized heating efficiency (%)	o = n/f	72.3%
EFC® additional heating energy (Btu)	p = n – j	20,432
Base outdoor air temp. (°F)	o	40.9
Base indoor air temp. (°F)	p	72.4
Base average outdoor minus indoor air temp. delta T_{BASE} (°F)	q = p – o	31.5
EFC® ave. outdoor air temp. (°F)	r	38.9
EFC® average indoor air temp. (°F)	s	72.7
EFC® average outdoor minus indoor air temp. delta T_{EFC} (°F)	t = s – r	33.7
EFC® furnace savings unadjusted for delta T	**u = 1 – [f/e]**	**12.1%**
EFC® average Part Load Ratio (PLR)	v	0.183
EFC® heating savings based on Eq. (6) $\Delta\eta_h = 0.0674(PLR_h)^{-0.4394}$	w	14.2%
EFC® normalized heating savings based on delta T (%)	**x = 1 – [g/e]**	**17.9%**
EFC® normalized heating savings based on efficiency and delta T (%)	**y = 1 – [k/o]**	**19.5%**

default fan speed provides 1080 cfm (509.7 lps) airflow. The Smart EFC® FDD test includes a variable fan-off delay and approximately 4 min after each thermostat call for heating the EFC® energizes the fan relay to High Speed Fan (HSF) which provides 1154 cfm (544.6 lps) airflow.[6] Prior to 4 min, the fan operates at the default speed.

Table 6 (row x) indicates normalized gas savings of 17.9% for the Smart EFC® FDD HSF based on normalized EFC® gas usage of 314,328 Btu (row g) versus base

[6]The furnace factory default fan-off time delay is 90-s. The default heating medium fan speed delivers 1080 cfm (509.7 lps) and the High Speed Fan (HSF) normally used for cooling delvers 1154 cfm (544.6 lps). Approximately 93.3% of forced air units operate the fan at high speed when the fan relay is energized either by itself or with the furnace operating simultaneously, and only 6.7% operate the fan at a low speed when the fan relay is energized by itself.

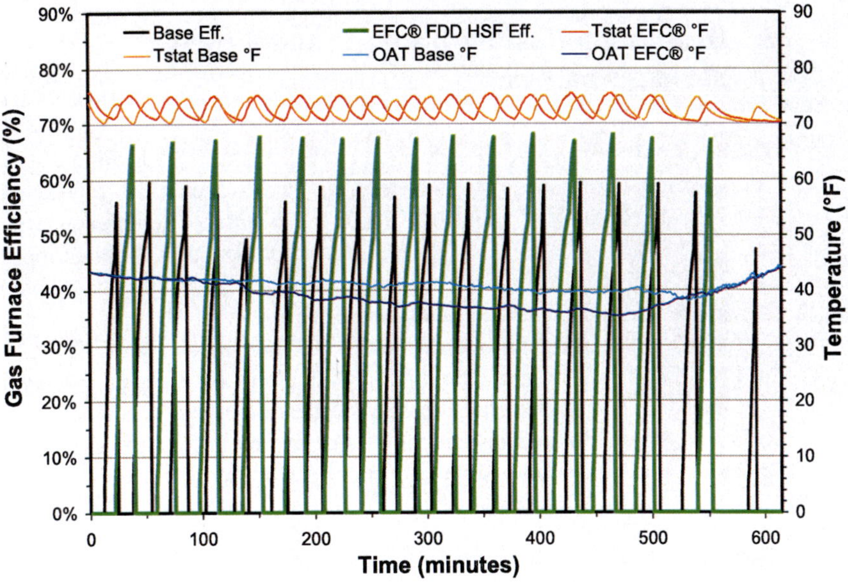

Fig. 6 Gas furnace unit #2 10-h tests—Smart EFC® FDD HSF vs. 120-s base delay

gas usage of 382,982 Btu (row e).[7] Normalized savings based on furnace on time are 18.1% (row d). The Smart EFC® provides longer off times and 15 heating cycles consuming 314,328 (row g) Btu of non-normalized gas usage with average furnace operation of 13.1 min and average thermostat temperature of 72.7 °F (row s) and 33.7 °F delta T (row t). The base has 18 heating cycles with average furnace operation of 12.5 min and base thermostat temperature of 72.4 °F (row p) and 31.5 °F delta T (row q). The 17.9% normalized gas savings (row x) are 23% greater than the 14.6% calculated savings based on Eq. (6) indicating that calculated savings are conservative. The EFC® FDD delivers 19.5% savings (row y) based on 72.3% efficiency (row o) versus 58.2% base efficiency (row k).

Data from a sample of 5582 gas furnace units in California indicate 2.6% of gas furnaces with 45-s fan-off delays, 8.3% with 60-s delays, 1.3% with 75-s delays, 57% with 90-s delays, 24% with 120-s delays, and 6.7% with 150-s delays [4]. Data from a sample of 5206 units indicate 93.3% have base high-speed fan control 6.7% do not [4]. Assuming weighting factors of 69.3% for 90-s delay and 30.7% for 120-s delay, the weighted average gas furnace heating savings are 18.4% based on data in Table 5. Based on eQuest simulations, the average annual heating PLR values range from 0.11 to 0.2 and the weighted average heating PLR 0.14 [4]. The average annual EFC® heating savings are 17.7% ± 2.1% based on Eq. (6), and housing stock weights for each climate zone from California housing data and US Census data [4, 11, 12].

[7] Smart EFC® HSF normalized gas usage of 314,328 Btu is based on non-normalized EFC® gas usage of 336,584 Btu times ratio of 31.5 °F delta T for base divided by 33.7 °F delta T for EFC®.

5 Heat Pump Heating Test Data and Energy Savings Analysis

The laboratory performed 48 split-system Heat Pump (HP) heating tests (24 baseline tests and 24 measure tests). The tests were performed at 17 °F (−8.3 °C), 35 °F (1.7 °C), 47 °F (8.3 °C), and 62 °F (16.7 °C) outdoor temperatures and 70 °F (21.1 °C) DB and 55 °F (12.8 °C) WB indoor temperatures. Laboratory tests measured the Smart EFC® additional heating capacity, and also measured total HP heating capacity for 60 min at the same conditions.[8]

Table 7 provides the following Intertek® HP heating tests at 47 °F (8.3 °C) OAT for the 1.5-ton split-system HP unit #5: (1) no delay base, (2) 65-s delay base, and (3) Smart EFC® variable delay. Smart EFC® heating energy savings vary from 3.7% to 71% compared to no delay or 65-s base delay and PLR values ranging from 0.02 to 0.83 based on 48 laboratory tests (not all test data are shown). EFC® heating

Table 7 Intertek® heat pump heating tests unit #5—Smart EFC® v. 0- and 65-s delay

Heat pump on time (min)	2	5	10	20	30	50	Ave.
Base 0-s delay tests	**125**	**126**	**127**	**128**	**129**	**130**	
Base no delay HP heating energy (Btu) [a]	36	256	953	2974	5268	9863	3225
Base no delay HP input (kWh) [b]	0.04	0.11	0.23	0.47	0.71	1.20	0.46
Base no delay HP efficiency [c = a/b/3412]	0.24	0.68	1.24	1.87	2.18	2.41	1.44
EFC® HP heating energy (Btu) [d]	91	437	1330	3531	5862	10,481	3622
EFC® HP input (kWh) [e]	0.05	0.12	0.23	0.48	0.72	1.21	0.47
EFC® HP heating efficiency [f = d/e/3412]	0.56	1.10	1.66	2.17	2.39	2.54	1.74
EFC® fan energy vs. no delay (kWh)	0.004	0.006	0.009	0.011	0.010	0.010	0.008
EFC® HP savings v. no delay [g = 1 − c/f]	**56.7%**	**38.0%**	**25.5%**	**13.9%**	**8.8%**	**5.1%**	**24.7%**
Base 65-s delay tests	**131**	**132**	**133**	**134**	**135**	**136**	
Base 65-s delay HP energy (Btu) [h]	95	366	1135	3212	5522	10,126	3410
Base 65-s delay HP input (kWh) [i]	0.05	0.11	0.23	0.47	0.71	1.20	0.46
Base 65-s delay HP efficiency [j = h/i/3412]	0.59	0.94	1.45	2.01	2.27	2.47	1.62
EFC® HP heating energy (Btu) [k]	148	513	1430	3642	5981	10,605	3720
EFC® HP input (kWh) [l]	0.05	0.12	0.24	0.48	0.72	1.21	0.47
EFC® HP heating efficiency [m = k/l/3412]	0.85	1.25	1.76	2.22	2.42	2.56	1.84
EFC® fan energy vs. 65-s (kWh)	0.008	0.014	0.014	0.014	0.014	0.014	0.013
EFC® HP savings v. 65-s base [m = 1 − j/m]	**30.6%**	**24.6%**	**17.6%**	**9.8%**	**6.3%**	**3.7%**	**15.4%**

[8] Heat pump input Btu values are based on measured kWh times 3412 Btu/h.

energy savings vary from 5.1% to 56.7% versus the no delay base, and EFC® savings vary from 3.7% to 30.6% versus the 65-s delay base. Average Smart EFC® HP heating energy savings are 24.7% versus the no delay base and 15.4% versus the 65-s delay base.

The ratio of heating capacity for each test divided by the total heating capacity for 60 min tests at the same test conditions is defined as the Part Load Ratio (PLR) as shown in Eq. (7). The PLR is used to normalize HP heating energy savings for each group of tests. Laboratory test data of the EFC® heating energy savings versus PLR are shown in Fig. 7. EFC® FDD HP heating energy savings are calculated using regression Eq. (7) based on the PLR (from Fig. 7).

$$\Delta\eta_h = 0.0526\left(PLR_h\right)^{-0.4499} \tag{7}$$

where, $\Delta\eta_h$ = Smart EFC® heat pump heating savings compared to base.

Figure 7 shows the Smart EFC® FDD HP heating savings varying from 3.1% to 29% compared to baseline fan-off delays of zero or 65 s and PLR values ranging from 0.05 to 0.83 based on 48 lab tests. Data from a sample of 3114 heat pump units in California indicate 78% of heat pump heating units have no delay and 22% have 65-s fan-off delays [4]. Based on the eQuest simulations, the average annual heating PLR values range from 0.09 to 0.27 and the weighted average heating PLR 0.13 [4]. The average annual EFC® heat pump heating energy savings are 10.7% ± 2% based on Eq. (7), and housing stock weights for each climate zone from California housing stock data and US Census data [4, 11, 12].

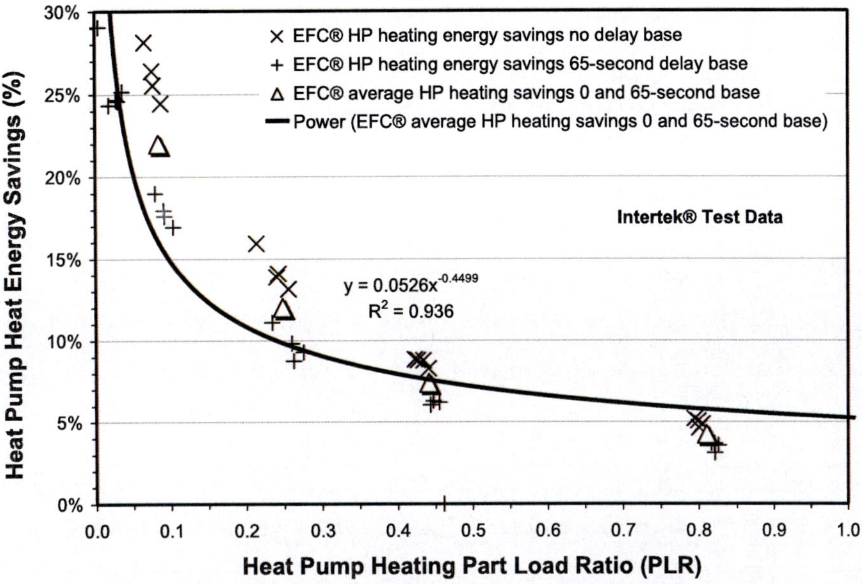

Fig. 7 Intertek® HP heating tests—Smart EFC® FDD heating savings versus PLR

6 Hydronic Heating Test Data and Energy Savings Analysis

The laboratory performed 20 split-system hydronic hot water heating tests. The tests were performed at 47 °F (8.3 °C) outdoor temperatures with 130 °F (54.4 °C) and 140 °F (60 °C) hot water temperature and 70 °F (21.1 °C) DB and 55 °FF (12.8 °C) WB indoor temperatures. The laboratory tests measured the additional heating capacity provided by the EFC® using an extended fan-off time delay which varies as a function of the hydronic heating operating time compared to the baseline system with no time delay or a fixed 60-s time delay. The laboratory tests also measured total hydronic heating capacity for 60 min at the same conditions.

Table 8 provides the following Intertek® hydronic heating tests at 130 °F for the 1.5-ton split-system unit #6: (1) no delay base, (2) 60-s delay base, and (3) EFC® variable delay. EFC® hydronic heating energy savings vary from 4.2% to 66% with no delay base, and savings vary from 2.3% to 28.7% for 60-s delay base. Average EFC® heating energy savings are 24.6% versus the no delay base and 12.1% versus the 60-s delay base. The PLR is used to normalize the EFC® hydronic heating energy savings for each group of tests. Laboratory test data of the EFC® hydronic heating energy savings versus PLR are shown in Fig. 8. EFC® hydronic heating savings are calculated using regression Eq. (8) based on the PLR (from Fig. 8).

Table 8 Intertek® hydronic heating tests unit #6 at 130 °F—EFC® v. 0- and 60-s delay

Hydronic heating on time (min)	2	5	10	20	30	50	Ave.
Base 0-s delay tests	**173**	**174**	**175**	**176**	**177**	**178**	
Base no delay HYD heating energy (Btu) [a]	122	512	1869	4260	6325	10,834	3987
Base no delay HYD input (Btu) [b]	970	2365	4584	9223	14,102	23,893	9189
Base no delay HYD heating efficiency [c = a/b]	12.6%	21.6%	40.8%	46.2%	44.9%	45.3%	35.2%
Base 60-s delay tests	**179**	**180**	**181**	**182**	**183**	**184**	
Base 60-s delay HYD energy (Btu) [d]	256	674	2099	4458	6540	11,040	4178
Base 60-s delay HYD input (Btu) [e]	970	2365	4584	9223	14,102	23,893	9189
Base 60-s delay heating efficiency [f = d/e]	26.4%	28.5%	45.8%	48.3%	46.4%	46.2%	0.40
EFC® HYD heating energy (Btu) [g]	360	839	2379	4709	6854	11,303	4407
EFC® HYD input (Btu) [h]	970	2365	4584	9223	14,102	23,893	9189
EFC® HYD heating efficiency [i = g/h]	37.1%	35.5%	51.9%	51.1%	48.6%	47.3%	0.45
EFC® fan energy vs no delay (kWh)	0.007	0.009	0.010	0.010	0.012	0.010	0.010
EFC® HYD savings v. no delay [j = 1 − c/i]	**66%**	**39%**	**21.4%**	**9.5%**	**7.7%**	**4.2%**	**24.6%**
EFC® fan energy vs. 60-s (kWh)	0.007	0.009	0.010	0.009	0.012	0.010	0.009
EFC® HYD savings v. 60-s base [k = 1 − f/i]	**28.7%**	**19.7%**	**11.8%**	**5.3%**	**4.6%**	**2.3%**	**12.1%**

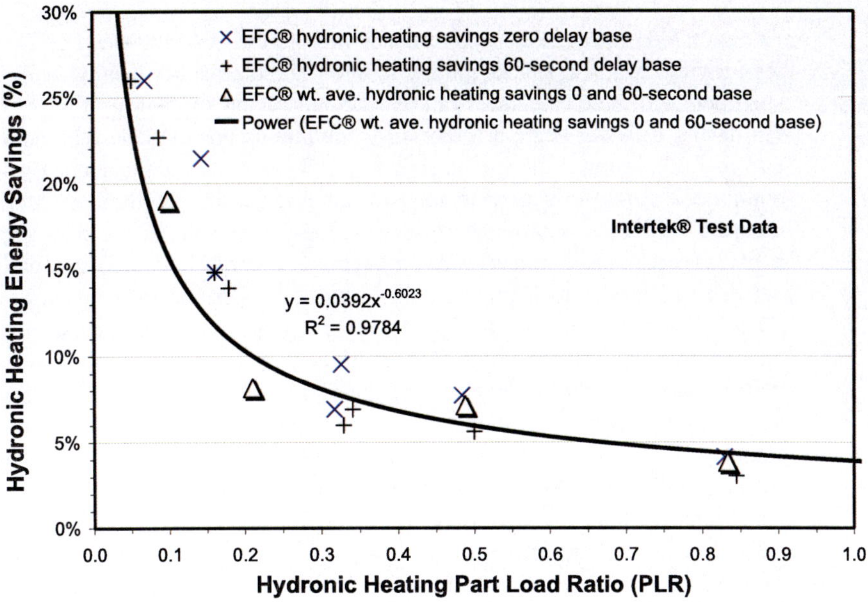

Fig. 8 Intertek® hydronic heating tests—Smart EFC® FDD savings versus part load ratio

$$\Delta\eta_h = 0.0392\left(PLR_h\right)^{-0.6023} \tag{8}$$

where, $\Delta\eta_h$ = EFC® hydronic heating savings versus base.

Figure 8 shows the Smart EFC® FDD hydronic heating savings varying from 2.3% to 26% compared to baseline fan-off delays of zero or 60 s and PLR values ranging from 0.056 to 0.85 based on 20 lab tests. Data from a sample of 1291 hydronic units in California indicate 72.3% have no delay, 17.6% have 30-s delays, and 10.1% have 60-s fan-off delays [4]. Based on the eQuest simulations, the average annual heating PLR values range from 0.09 to 0.20 and the weighted average heating PLR 0.12 [4]. The average annual EFC® heating energy savings are 18.3% ± 2.3% based on Eq. (8), and housing stock weights for each climate zone from California housing stock data and US Census data [4, 11, 12].

7 Fan-On Continuously Fault Detection Diagnostics and Energy Savings

The Smart EFC® reduces fan-on operation by 50–75% to save energy by providing a fraction of continuous or intermittent fan-on operation or turning off the fan during unoccupied periods. Fan-on correction energy savings are based on evaluation studies regarding continuous fan-on operation for new furnaces with an Electronically

Commutated Motor (ECM) and furnaces without the ECM [13–15, 18]. These studies provide fan-on operational practices for three seasons (heating, cooling, and shoulder) and three categories (auto, continuous, and sporadic). Auto operates the fan during calls for heating or cooling, continuous operates the fan continuously, and sporadic operates the fan intermittently. These studies indicate occupants select continuous fan-on operation to improve air circulation, air filtration, and thermal comfort [15]. Table 9 provides residential fan-on operational practices for participants who purchased a new ECM furnace and a control group of nonparticipants before and after the installation of new furnaces [15]. After installing a new furnace, 16.3% of participants and non-participants operate fans continuously during the heating season, 20.3% operate fans continuously during the cooling season, and 9.7% operate fans continuously during the shoulder season [15]. The weighted average estimate of participants and nonparticipants who operate fans continuously year round is 15.4% which is less than the average of 10–30% of people worldwide who suffer from allergic rhinitis [16]. This estimate represents the percentage of occupants who operate fans continuously for air filtration to reduce indoor pollution, aerosol, or virus concentrations.

Table 10 provides field data from five Multi Family (MF) sites and 1445 apartments where smart communicating thermostats were installed and 99 occupants or 6.9% set fan-on continuously and requested turning off the fan-on setting. An unknown number of occupants set fan-on continuously, but did not request turning off the fan. The Smart EFC® with FDD automatically corrects these faults.

Table 11 provides weighted average Smart EFC® savings for single family (SFM), multi family (MFM), and mobile homes (DMO) based on DOE2 simulations and 15.4% weighted average estimate of occupants who operate HVAC fans

Table 9 Residential fan-on operational practices [15]

Season	Fan operation practice	Participants with ECM furnace (n = 150) before (%)	Participants with ECM furnace (n = 150) after (%)	Nonparticipant furnace replacers (n = 82) before (%)	Nonparticipant furnace replacers (n = 82) after (%)	Average participant and nonparticipant furnace replacers after (%)
Heating	Auto	90.7	73.3	91.5	90.2	
Season	Continuous	9.3	25.3	6.1	7.3	16.3
	Sporadic	0.0	1.3	2.4	2.4	
Cooling	Auto	81.3	65.3	85.4	82.9	
Season	Continuous	14.7	30.7	7.3	9.8	20.3
	Sporadic	4.0	4.0	7.3	7.3	
Shoulder	Auto	86.7	81.3	91.5	87.8	
Periods	Continuous	7.3	13.3	2.4	6.1	9.7
	Sporadic	6.0	5.3	6.1	6.1	
Wt. ave.	**Continuous**					**15.4**

Table 10 Multi family apartments in California with continuous fan-on operation

Site	MF apartments [a]	HVAC fan-on continuously [b]	Fan-on [c = b/a] (%)
MF site #1	194	40	20.6
MF site #2	259	10	3.9
MF site #3	320	18	5.6
MF site #4	320	16	5.0
MF site #5	352	15	4.3
Total sites	1445	99	6.9

continuously. Base fan energy (col. a) assumes fan operation only during cooling and heating cycles (i.e., auto mode). Weighted average kWh savings are 20.7% ± 4.3% for fan-on correction, and total weighted average kWh savings are 25% ± 3%. Fan-on correction heating savings are 4.1% ± 2.1%, and total weighted average heating savings are 17.4% ± 2.6%. MFM fan-on correction heating savings are −2.4% due to extra fan heat and less duct leakage in apartments where ducts are located in conditioned space.

8 Discussion

The Smart EFC® recovers latent energy from the AC evaporator coil by operating the fan after the compressor turns off to evaporatively cool the conditioned space, over satisfy the cooling thermostat setpoint, and lengthen the off cycle. For heating systems, the Smart EFC® operates the fan after the heating system turns off to over satisfy the heating thermostat setpoint and lengthen the off cycle. Tests results indicate that over-sized HVAC systems with frequent on-off cycles can realize greater savings than properly-sized systems with longer cycles. Laboratory and field tests demonstrate that the EFC® can prevent evaporator coil icing by continuing to operate the fan and evaporate cold-water condensate from the coil at the end of each cooling cycle which prevents ice formation when the evaporator coil temperature is below freezing. This helps maintain thermal comfort, efficiency, and equipment life per the ACCA HVAC Quality Standards [17].

Intertek® tests of the AC unit #3 found Smart EFC® cooling savings ranging from 3.8% to 32% with average savings of 16.1% versus the no delay base and 10.1% versus the 90-s delay base. Field measurements of the 5-ton AC unit #2 with 21% duct leakage and the same unit with the Smart EFC® found normalized cooling energy savings of 18.2% based on 20 tests. Intertek laboratory tests of gas furnace unit #4 found Smart EFC® heating savings ranging from 4.2% to 38% and average savings of 17.3%. Field measurements of gas furnace unit #2 with 21% duct leakage and the same system with the Smart EFC® found average normalized heating savings of 19.5% based on 15 EFC® cycles over 10 h and 18 base cycles over 10-h. Intertek® laboratory tests of heat pump unit #5 found heating energy savings ranging from 3.7% to 56.7% with average savings of 20.1% based on 48 tests. Intertek®

Table 11 Smart EFC® annual savings for single, multi-family, and double-wide mobile homes

Building	Base fan energy kWh/year [a]	Fan-on cont. extra fan kWh/year [b]	Fan-on cont. extra cooling kWh/year [c]	Fan-on cont. extra heating MJ/year [d]	Fan-on cont. weight average [e] (%)	FDD fan-on correct heat savings MJ/year [f] (%)	EFC® cool savings kWh/year [g] (%)	FDD fan-on correct cool + fan savings kWh/year [h] (%)	Smart EFC® FDD total savings kWh/year [i] (%)
SFM	231.1	3121	659	4040	15.4	2.5	4.2	23.5	27.7
DMO	318.1	2365	425	10,588	15.4	8.3	4.2	13.7	17.9
MFM	115.9	1757	535	−1719	15.4	−2.4	5.5	29.3	34.8
Average	**253**	**2832**	**581**	**5840**	**15.4**	**4.1**	**4.3**	**20.7**	**25.0**

laboratory tests of hydronic unit #6 found heating energy savings ranging from 2.3% to 66% with average savings of 18.4% based on 20 tests. Weighted average savings for the Smart EFC® are 17.4% ± 2.6% for heating (therms) and 25% ± 3% for cooling plus fan (kWh). About 80% of electricity savings are from reducing continuous or intermittent fan-on operation by 50–75% or turning off the fan during unoccupied periods. According to the US EIA, California consumed 7.88 quadrillion Btu (quads) or 8.32 exajoules (EJ) in 2017, and the California residential sector consumed 1.42 quads or 1.49 EJ with 27% for space heating and 4% for space cooling [1]. Assuming the Smart EFC® can save 25% ± 3% on electricity (cooling plus fan) and 17.4% ± 2.6% on heating, the potential annual energy savings are 0.08 quads (0.09 EJ) or 1% ± 0.1% of total California energy use in 2017. The potential US energy savings for the EFC® are about 1.3 quads (1.4 EJ) or 1.3% ± 0.2% of total US energy use of 101 quads (106.5 EJ) based on 4.15 quads (4.38 EJ) for heating and 2.22 quads (2.34 EJ) for cooling.

9 Conclusions

Laboratory and field tests indicate that the Smart EFC® saves energy by providing variable fan-off delays to increase cooling or heating capacity and extend off-cycle durations and reducing continuous or intermittent fan-on operation by 50–75% or turning off the fan during unoccupied periods. Cooling and heating tests demonstrate improved thermal comfort by exceeding thermostat set points and providing longer off-cycle times from variable fan-off delays based on cooling or heating on/off cycles. Tests indicate that over-sized HVAC systems with frequent on-off cycles will realize greater savings than properly-sized systems with longer cycles. Laboratory tests, field tests, and occupant surveys indicate that the EFC® improves air filtration, air circulation, and thermal comfort. Based on laboratory and field tests, market share data, housing stock data, and building energy simulations, the Smart EFC® can save about 25% ± 3% on cooling plus fan energy and 17.4% ± 2.6% on heating. About 80% of the electricity savings are from overriding fan-on settings. According to the US EIA, in 2017 California's residential sector energy consumption was 1.42 quads or 1.49 EJ with 27% for residential space heating and 4% for residential air conditioning [1]. The Smart EFC® with fan-on correction can save about 1% ± 0.1% of the total annual energy use in California and 1.3% ± 0.2% of the total annual energy use in the US.

Acknowledgements Ean Jones managed the laboratory tests at Intertek in Plano, Texas, and performed field tests in Reno, Nevada, with funding from GreenFan® Inc. Site inspections were performed by Robert Eshom. Building energy simulations were performed by Pete Jacobs of Building Metrics Inc.

References

1. United States Energy Information Agency (USEIA). 2014. Residential Energy Consumption Survey. Washington, D.C.: https://www.eia.gov/consumption/residential/ and https://www.eia.gov/state/print.php?sid=CA.
2. Rufo, M. 2002. California's Secret Energy Surplus: The Potential for Energy Efficiency. http://www.p-2.com/PEERS/Hewlett-Foundation-Report-9-23-02.pdf.
3. Mowris, R., Eshom, R., Jones, E. 2015. Laboratory Measurements and Diagnostics of Residential HVAC Installation and Maintenance Faults. 8th EEDAL'15, Lucerne, Switzerland.
4. Mowris, R. 2016. Efficient Fan Controller® (EFC®) for Residential HVAC Systems. Work Paper EFC173PHVC138. Prepared by Verified® Inc.
5. ANSI/AHRI. 2017. Standard 210/240. Standard for Performance Rating of Unitary Air-Conditioning and Air-Source Heat Pump Equipment.
6. USGAO. 1975. S. 622—94th Congress: Energy Policy and Conservation Act. Energy Policy and Conservation Act of 1975 (EPCA) (Pub.L. 94–163, 89 Stat. 871, enacted December 22, 1975) Retrieved from https://www.govtrack.us/congress/bills/94/s622.
7. ANSI/ASHRAE STANDARD 37-2009. 2009. Methods of Testing for Rating Electrically Driven Unitary Air-Conditioning and Heat Pump Equipment.
8. ANSI Z21.47-5th Edition 2006/CSA 2.3-5th Edition 2006—Standard for Gas-Fired Central Furnaces. American National Standards Institute.
9. ANSI/ASHRAE. 1987. ASHRAE 41.2-1987 Standard Methods for Laboratory Airflow Measurement.
10. Database for Energy Efficiency Resources (DEER) at http://www.deeresources.com/.
11. US Census Bureau. 2010. Population and Housing Units Count 2010. https://www2.census.gov/library/publications/decennial/2010/cph-2/cph-2-1.pdf.
12. Posesta, A., et al. 2018. California's Housing Future: Challenges and Opportunities Final Statewide Housing Assessment 2025, Calif. Dept. of Housing and Community Development. http://www.hcd.ca.gov/policy-research/plans-reports/docs/SHA_Final_Combined.pdf.
13. Wisconsin Department of Administration. 2003. Electricity Use by New Furnaces: A Wisconsin Field Study. Technical Report 230-1. Madison, Wisconsin: Wisconsin DOA.
14. Wisconsin Department of Administration (DOA). 2004. ECM Furnace Fan Impact Assessment. Focus on Energy Statewide Evaluation Report. Madison, Wisconsin. https://library.cee1.org/system/files/library/1814/997.pdf.
15. Public Service Commission of Wisconsin (PSCW). 2009. Focus on Energy Evaluation ECM Furnace Impact Assessment Report Final Report. Prepared by T. Talerico and R. Winch, Glacier Consulting Group, LLC for PA Consulting Group. Wisconsin DOA.
16. Pawankar, R. et al. 2012. White Book on Allergy 2011–2012 Executive Summary. World Health Organization. https://www.worldallergy.org/UserFiles/file/ExecSummary-2013-v6-hires.pdf.
17. American National Standards Institute (ANSI)/Air Conditioning Contractors of America (ACCA). 2007. ANSI/ACCA. 2013. Standard 4: Maintenance of Residential HVAC Systems and ANSI/ACCA. 2015. Standard 5: HVAC Quality Installation Specification.
18. Pigg, S. Talerico, T. 2004. Electricity Savings from Variable-Speed Furnaces in Cold Climates. 2004 American Council for an Energy Efficient Economy (ACEEE) Summer Study on Energy Efficiency in Buildings, Asilomar, CA, Published by ACEEE, Washington DC.

Energy Consumption Scenarios and Policy Implications for Home Automation and IoT

Anson Wu and Paul Ryan

Abstract The additional energy use due to network-connected devices is projected to be 3.5% of total world electricity consumption in 2030. This paper draws on work done for the IEA 4E Electronic Devices and Networks Annex (EDNA) to provide a unique view of where energy is consumed and the potential scenarios of total worldwide energy consumption. A Total Energy Model developed as part of the project illustrates the main downstream categories of energy use as well as the upstream energy use of data centres (DCs) and wide area networks (WANs).

Domestic appliances are increasingly becoming automated and connected to home networks. They are part of a wider trend of Internet of Things (IoT). As devices connect, energy consumption will increase not only in the device itself but via the devices it connects through including home routers, broadband networks and upstream in the data centres and servers. Energy consumption varies by which type of network it connects through, the energy consumed by Wi-Fi routers and broadband will differ to mobile networks and base stations, as well as low-power options such as Bluetooth and Zigbee. Furthermore, the connectivity and processing routes between devices differ. They might process commands directly and connect via home networks, or they may depend on the cloud and data centres to process and communicate.

This paper analyses various energy consumption scenarios of both upstream and downstream equipment, based on three variables: sales, network types, and connectivity/processing route. It presents best and worst cases that may occur and discusses the policy implications for maximising energy efficiency.

A. Wu (✉)
Hansheng Ltd., Abingdon, UK
e-mail: anson.wu@hansheng.co.uk

P. Ryan
Energy Consult, Melbourne, VIC, Australia
e-mail: paul@energyconsult.com.au

P. Bertoldi (ed.), *Energy Efficiency in Domestic Appliances and Lighting*, Springer Proceedings in Energy, https://doi.org/10.1007/978-3-030-79124-7_17

1 Background

The additional energy use due to network-connected devices is projected to be 3.5% of total world electricity consumption in 2030 [1]. Domestic appliances are increasingly becoming automated and connected to home networks. They are part of a wider trend of Internet of Things (IoT). The number of voice assistant speakers is expected to grow over the next 5 year from 2.5 billion to 8 billion in 2023 [2].

As devices connect, energy consumption will increase not only in the device itself but via the devices it connects through, including home routers, broadband networks and upstream in data centres and servers (Fig. 1). Energy consumption varies by which type of network it connects through, the energy consumed by Wi-Fi routers and broadband will differ to mobile networks and base stations, as well as low-power options such as Bluetooth and Zigbee. Furthermore, the connectivity and processing routes between devices differ. They might process commands directly and connect via home networks, or they may depend on the cloud and data centres to process and communicate.

For many devices, the network connections only provide connected standby and other very basic functions, e.g. switching a bulb on/off. These require very little data which means high throughput and relatively high power Wi-Fi connections are an unnecessary use of energy. Instead, low power networks could be used in conjunction with Wi-Fi [3] or even zero network standby power [4]. However, the low power consumption means that the signal range is short and may result in many routers or signal repeaters being installed around the home to ensure reliable connection. This is especially true for battery operated sensors which have very low power and may be placed far away or on the edge of the building envelope, e.g. window/door sensors. The number of repeaters required depends on the house size and signal range, and the risk is that this may negate any energy savings in the devices.

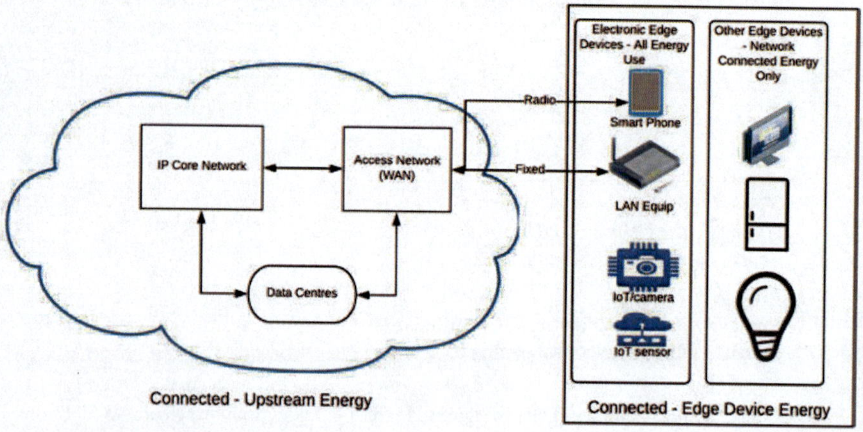

Fig. 1 Network connected energy consumption

Smart devices and IoT are already proven to provide net savings; smart thermostats are now more common and are covered under ENERGY STAR [5], while occupancy sensors for lighting are extremely common in commercial buildings and are covered under ENERGY STAR Smart Homes Energy Management Systems (SHEMS) [6]. Similar to smart meters, maximising potential savings may demand that homes are covered by Wi-Fi or low power networks and are ready for sensors and controllers to be installed. Policies may be aimed to address how this can be achieved with the smallest additional energy consumption.

This paper draws on work done for the IEA 4E Electronic Devices and Networks Annex (EDNA) to provide a unique view of where energy is consumed and the potential scenarios of total worldwide energy consumption [1]. It analyses various residential energy consumption scenarios and discusses the policy implications for maximising energy efficiency. The work covers both upstream equipment in the data centre and internet as well as downstream equipment in homes, businesses and other location. It is based on three variables: sales, network types, and connectivity/processing route and presents realistic best and worst cases that may occur.

2 Methodology

Six different scenarios were created to cover the envelope of possibilities for residential energy connected devices. These were then input into the TEM to calculate the main downstream categories of energy use as well as the upstream energy use of the data centres (DCs) and wide area networks (WANs). The scenarios ("Sc") are:

Sc1. Business as usual (BAU) scenario and is based on the TEM assumptions [1].

Sc2. A scenario which assumes that consumers will demand high speed internet throughout the house for streaming and VR etc., which can only be achieved through 5 GHz. To achieve this, Wi-Fi mesh or repeaters will be installed into all homes with broadband as required, increasing sales and energy consumption of these devices. While this is not expected under normal market conditions, this scenario provides an upper bound for additional energy consumption to provide reliable high speed Wi-Fi connectivity.

Sc3. A scenario as Sc2 but which assumes that low energy networks are also installed in all homes with repeaters installed as required. While this is not expected under normal market conditions, this scenario provides an upper bound for additional energy consumption to provide reliable LE network connectivity.

Sc4. A scenario as Sc3 but which assumes that all sensors and IoT devices such as smart bulbs use low energy networks instead of Wi-Fi where possible. All network connected devices such as TVs and game consoles also use low energy networks instead of Wi-Fi during standby. This scenario represents the maximum device savings possible with existing low energy network technology.

Sc5. As Sc4 but which assumes that the low energy hub can control all automation locally. While the hub power consumption increases to provide the functionality, it reduces some of the upstream power consumption.

Sc6. As Sc5 but which assumes zero network energy is achieved for all devices (including desktop PCs and set top boxes (STBs)) with the exception of smart lights (required to provide LE network repeater functionality), voice assistant speakers/hubs (e.g. Alexa, Google hub, Xiao Ai), and network hubs (routers, modems, repeaters, etc.). This represents the maximum savings with current technology and also assumes zero network energy is achievable.

The changes to the model are summarised in Table 1.

2.1 *Estimating Global Number of Repeaters*

The number of repeaters required in the home depends primarily on the size of the house, and the range of the network. The number of walls, construction and shape will also impact the coverage. Therefore, to calculate the global number of repeaters per house, a global distribution of house sizes was required and since this data was not available, it had to be calculated.

Table 1 Summary of scenarios

Scenario	Hubs	Devices	WAN
Sc1	BAU	BAU	BAU
Sc2	From 2021, Wi-Fi repeater stock increases linearly to 59% of all households with broadband by 2025	BAU	BAU
Sc3	As Sc2 From 2021 sales of LE gateways matches sales of integrated routers.	BAU NB smart bulbs sales under BAU already provide sufficient repeaters for full house coverage.	BAU
Sc4	As Sc3	All sales of smart lights and sensors switch to LE from 2021. All new devices have LE network standby instead of Wi-Fi, reducing standby power by 0.6 W, except voice assistants, from 2021. Existing LE devices standby power unchanged.	BAU
Sc5	As Sc3 Power consumption of low energy network hub increases 0.4 W from 2021.	As Sc4	WAN energy for IoT reduced 75% from 2021
Sc6	As Sc5	All devices except voice assistants, smart lights have zero network standby power from 2021.	As Sc5

There is very limited research on global house sizes and no data was found for the majority of regions. Only two sources were identified which showed regional average house sizes [7, 8] and the accuracy of the data was not verified. The data available showed that average new house size varied significantly from 45 m^2 in Hong Kong, China to 214 m^2 in Australia. Given the uncertainty and the wide variation in size, it was therefore important to consider the impact this might have on the number of repeaters. Of the 23 regions, most of the data was from Western Europe, North America, Australia but also included China, Russia, and Japan. This is therefore likely to skew towards larger homes but is also where broadband internet is most common. In addition to this, the variation in house sizes from the United States (US) Housing Census showed even greater variation from 50 m^2 to over 500 m^2 [9].

The global distribution was created by applying the US housing distribution pattern to the average house size of each of the 23 regions with available data to get a distribution for each region. These were then summed in 25 m^2 bins and weighted by the number of households in each region. The global average house is estimated to be approximately 85 m^2.

The average number of 5 GHz Wi-Fi and LE hubs (router and repeaters) could then be calculated based on the estimated range in Table 2. The hub range estimates are relatively conservative compared to the line of sight range. This is to account for the high signal attenuation of walls as well as the assumption that coverage must be sufficient to the edge of the house. Even with conservative range estimates, the number of hubs is relatively low, although significantly higher than Sc1.

3 Results

The TEM calculates the additional network energy consumed for all devices in both active and standby modes including the upstream energy in the DC and internet (Table 3). This does not include the active energy consumed by a device providing its main function, e.g. the TV active mode network energy excludes the screen, video inputs and video signal processing, unless the device's primary function is network connectivity, e.g. routers.

Figure 2 shows the energy consumption for Scenario 1 (BAU) from 2015 to 2030. For clarity, there are six device categories (Table 4). Total energy consumption is expected to increase steadily from 517 TWh in 2019 to 726 TWh in 2030, an increase of 40%. This is primarily as a result of increasing standby energy

Table 2 Hub range and global average number of hubs per house

Hub type	Hub range (m^2)	Global average number per house for full coverage
5 GHz Wi-Fi	75	1.59
LE hub	50	2.11

Table 3 Energy consumption modes

Mode	Description
Upstream	Energy consumed in the data centre and internet providing functionality to the end device, e.g. streaming video to the TV
Standby	Energy consumed by device in network standby
Active	Additional energy consumed by the device to maintain the active network

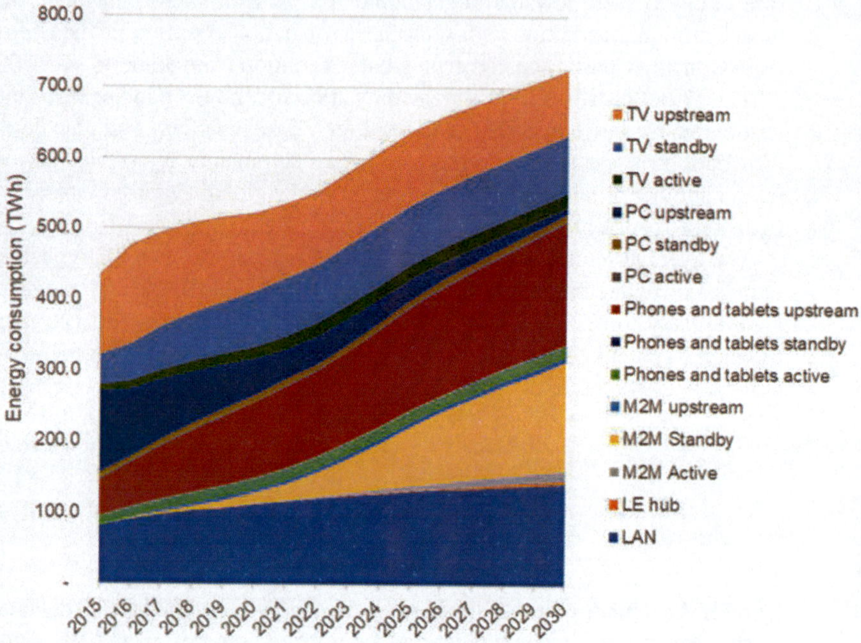

Fig. 2 Growth of additional network energy consumed for residential devices under scenario 1 (BAU) from 2015 to 2030

Table 4 Device categories

Category	Devices
TV	TV, audio equipment, STBs and game consoles
PC	Desktop, laptop computers, network attached storage, printers
Phones and tablets	Smartphones, non-smartphones and tablets
M2M	All other smart devices, Internet of Things, connected appliances etc.
LE Hub	Low energy network hub
LAN	Integrated modem routers and all other LAN networking devices

consumed by the growing number of internet connected devices in the M2M category causing an increase in energy from 10 to 152 TWh in 2030.

The data centre and internet energy consumption resulting from the M2M category (M2M upstream in Fig. 2) is a very small proportion of the total energy. This contrasts with the much more significant 75 TWh growth in the upstream energy consumed by the apps and streaming video on phones and tablets. However, TEM calculates upstream energy based only on the assumed amount of data being consumed and therefore has a higher level of uncertainty.

While M2M standby energy is high, there are also other large contributors to the overall energy consumption including TVs (upstream and standby) and the LAN.

Using the same approach, energy consumption was calculated over the same time period for each scenario. Figure 3 gives a comparison of the scenarios in 2030, which most clearly shows the impact of each change. The contribution of each device to the total energy consumed is also more clearly presented.

Scenarios 2 and 3 show the increase in energy consumed to provide complete network coverage in the house. Wi-Fi mesh networks and repeaters have a much larger energy consumption, a 36 TWh increase, compared to LE network hub (11.5 TWh). This is because the LE hub has much lower power consumption, and the repeaters are already integrated into the devices under the business as usual scenario.

The increase in energy consumed by the LE hubs is significantly outweighed by the savings if all devices adopted LE network standby mode (Sc4) where 50 TWh of savings are estimated. This is approximately 35% of the projected energy increase

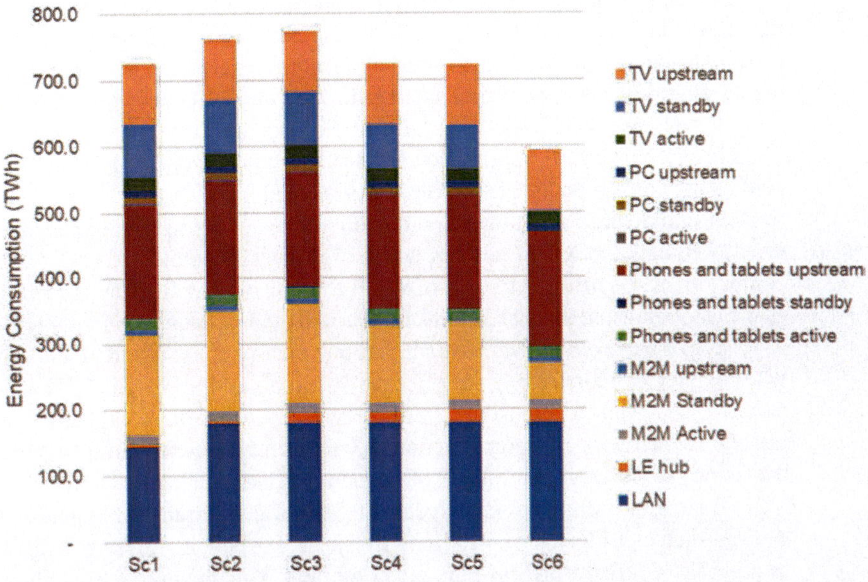

Fig. 3 Scenarios compared for additional energy consumed for residential network devices in 2030

from M2M devices in standby. However, there is no significant net energy savings compared to Sc1 since complete 5 GHz Wi-Fi is also included.

Moving automation from the cloud to local automation on the LE hub in Scenario 5 causes active energy to increase by less than 5 TWh, but there is only a savings of around 6 TWh from reduced upstream energy. This cannot be considered significant given the level of uncertainty in the input data, and no significant net savings are achieved compared to Scenario 1.

Only Scenario 6 can achieve a net reduction in energy consumption compared to Scenario 1, with total energy consumption dropping to 591 TWh. However, this is still greater than the estimated energy consumption in 2019 (517 TWh). This is primarily due to voice assistant speakers but also the number of smart lights which are needed as repeaters, which consume 56 TWh combined.

4 Discussion and Conclusions

Energy consumption for network connected devices is expected to increase, and from all the scenarios modelled, only scenario 6, zero network standby power, is projected to reduce total energy below BAU. However, scenario 6 is only likely to be possible with widespread use of low energy networks. Based on current technology, Wi-Fi standby power consumption is too high. From this perspective Scenarios 3–5 provide some useful guidance about how policies could prepare households to expand the use of LE networks.

By estimating a global distribution of house sizes it was also possible to estimate the total energy required to achieve sufficient coverage of LE networks, which are often considered unreliable. An average of only 1.1 repeaters are needed per home, which should be met by existing smart bulb sales and BAU energy consumption. As a result, the energy increase is small, and the potential benefits will outweigh this. Therefore, considering policies to increase the number of LE hubs is useful and could steer consumers to LE smart devices instead of Wi-Fi.

This could then be followed with policies encouraging the expansion of LE standby to new models of devices which are typically connected such as TVs and computers, while limiting the use of Wi-Fi standby on new IoT devices and sensors, as well as connected devices such as fridges and other appliances similar to Scenario 4. Research has even shown that smartphone energy consumption can be reduced when connected to low energy ZigBee networks rather than Wi-Fi [3]. Furthermore, policies may want to consider limiting the types of mains-powered devices that include LE network repeater functionality to increase savings further. One aspect that is not explored is the timing of introducing such changes. Currently sales of M2M devices are small but are expected to increase very rapidly. In addition, these devices such as fridges have very long lifespans. Introducing these policies as early as possible will therefore have the greatest effect. This may require additional support for research and standards to ensure devices can switch seamlessly between low energy standby and higher energy active Wi-Fi.

While there is concern about the DC and WAN (upstream) energy consumed for smart devices, based on the TEM modelling, it appears to be very small for IoT devices since they generate a relatively small amount of data and is therefore not significant enough to warrant policies to address this, especially compared to the impact of streaming. Cloud connected devices may also benefit from greater flexibility to create more innovative automation and control patterns to save energy in other systems such as residential heating. However, local automation may be more reliable, especially when there are internet connection problems. This additional reliability may increase consumer trust and using both approaches may be warranted to increase adoption rates and total savings.

Only very aggressive reductions in network standby to near zero Watts will be effective in limiting future standby energy consumption (Scenario 6). Because the difference in energy consumed between the scenarios, excluding scenario 6, is only in the region of 10%, the policy focus should be expanded and complemented beyond network standby where greater savings may be available:

- Reductions in network energy consumption should not come at the cost of larger energy savings available through better control of lighting, heating and other systems within the home [10]. If an increase in network energy is inevitable, policies may want to place more emphasis on maximising the potential savings. ENERGY STAR SHEMS is an example of a potential policy option.
- Achieving net reduction in network connected energy compared to 2019 may not be possible, but it is essential that other research and policies try to tackle the upstream energy consumption used for TVs, phones and tablets. Another specific area of focus is voice assistant speakers and other voice assistant devices [11].

These scenarios are based on the TEM model and BAU inputs. However, sales and sales projections change frequently and new devices are being developed which all impact the energy projections. Additional research is recommended to improve our understanding of the energy impacts:

- Updating the TEM to more accurately assess the upstream impacts of different types of internet activities, e.g. streaming, IoT, AI.
- Modelling the energy impacts of delaying the introduction of new policies to devices, especially those with long lifetimes.
- Collaborating with IoT platforms to test new ideas for maximising potential energy savings that are easily deployed such as software, UI and other options.
- Expanding research into industrial and commercial IoT. IoT in industry could be significantly more complex, generating more data and consuming more energy. They are more likely to use different sensors and networks, e.g. local 5G to cover larger areas. More research in how IoT is being applied by industry/manufacturing and what technology is used to estimate the energy costs and potential savings could help to make informed decisions about the value of policy intervention.

References

1. Ryan, P et al *Total Energy Model for Connected Devices*, IEA EDNA 4E, 2019 https://www.iea-4e.org/document/429/total-energy-model-for-connected-devices
2. https://techcrunch.com/2019/02/12/report-voice-assistants-in-use-to-triple-to-8-billion-by-2023/
3. Zhang, Y and Li, Q. *HoWiES: A Holistic Approach to Zigbee Assisted WiFi Energy savings in Mobile Devices*, Proc. IEEE INFOCOM 2013
4. Xergy Consulting, *Getting to Zero: An Evaluation of Zero Network Standby Power*, IEA 4E EDNA https://www.iea-4e.org/document/430/getting-to-zero-an-evaluation-of-zero-network-standby-power
5. https://www.energystar.gov/products/connected_thermostats
6. https://www.energystar.gov/products/spec/smart_home_energy_management_systems_pd
7. http://demographia.com/db-intlhouse.htm
8. http://shrinkthatfootprint.com/how-big-is-a-house
9. https://www.census.gov/programs-surveys/ahs/data/interactive/ahstablecreator.html#?s_areas=a00000&s_year=n2017&s_tableName=Table1&s_byGroup1=a6&s_byGroup2=a1&s_filterGroup1=t1&s_filterGroup2=g1&s_show=S
10. Urban, B et al *Energy Savings from Five Home Automation Technologies: A Scoping Study of Technical Potential*, Fraunhofer USA, 2016 https://www.cta.tech/CTA/media/policyImages/Energy-Savings-from-Five-Home-Automation-Technologies.pdf
11. Horowitz, N et al *The energy impacts of smart speakers and video streaming devices* NRDC, 2019 https://www.nrdc.org/resources/energy-impacts-smart-speakers-and-video-streaming-devices

Consumer Demand Side Flexibility in Europe

Paolo Falcioni

Abstract The energy system in Europe is undergoing a transformation driven by decentralisation, decarbonisation and digitalisation. This brings along new market actors, empowers the role of consumers and opens up to new business opportunities. Consumers will play a vital role in this process through adapting their power demand if incentivised and willing to change their consumption patterns. Renewable energy resources contribute largely towards this transition, but their availability is limited. Therefore, new technologies need to be put in place to respond to this opportunity. Demand side flexibility that is materialised by consumers via home appliances represents a considerable potential. To ensure that smart appliances, a central energy manager and a grid can communicate effectively, interoperability is essential. Industry has been working towards this end by developing a Smart Appliances REFerence (SAREF) ontology that was standardised in December 2015. Taking in consideration that households account for more than 25% of final energy consumption in the European Union (EU), the potential is enormous. The right regulatory framework could enhance the innovation, drive the uptake of smart appliances and technologies and incentivise consumers to actively participate in the energy market.

1 Introduction

The energy market is facing unprecedent challenges. New market actors emerge, and new business models are being developed to respond to these changes. There is a rising pressure for decentralization and decarbonization of energy production and consumption. From regulatory point of view, the European Commission presented the most ambitious package so far, The Clean Energy for all Europeans, tackling the transformation of the energy system. Also, a long-term vision of a decarbonized and prosperous Europe by 2050 was set in an outlook "A Clean Planet for all", providing guidance for future political initiatives. The new set of legislation highlights the role

P. Falcioni (✉)
APPLiA - Home Appliance Europe, Brussels, Belgium
e-mail: paolo.falcioni@applia-europe.eu

P. Bertoldi (ed.), *Energy Efficiency in Domestic Appliances and Lighting*, Springer Proceedings in Energy, https://doi.org/10.1007/978-3-030-79124-7_18

of energy efficiency and promotes the active role consumers in the changing energy system.

2 Demand Side Flexibility and a Cleaner Energy

Demand side flexibility (DSF) implies consumers reacting to variable energy prices or other market incentives, resulting in a change of their energy consumption. This could mean in practical terms postponing a dishwasher cycle during a peak time or scheduling a washing machine cycle for when solar energy is available or simply adjusting a temperature of an air-conditioner. These cases might differ in an impact that this change might have on consumer and therefore might predict the level of consumer willingness to adapt its behaviour.

The role of DSF becomes more prominent in a current transition towards a clean energy system where consumers prefer purchasing energy coming from renewables, such as solar panel or wind turbine. Since the energy from renewables is not available on demand, unlike in case of conventional energy production, new solutions have to be found to fully grasp its benefits. To ensure a proper integration of renewable energy sources in the system, the market needs to be further decentralized and smart technologies needs to be deployed to balance demand and supply in real time.

Provided that variable prices are available, the consumer can be rewarded for changing his energy pattern. Furthermore, consumers producing their own energy can sell the surplus to a utility or other consumer through a local energy community, for example. The potential of such communities lies in cost-effectiveness as well as bringing positive social, economic and environmental impact. At system level, DSF can alleviate strain from the grid, better manage increased demand during peak periods as well as balance energy needs against real-time supply.

Currently, there are several obstacles to use effectively of real-time price tariffs such as lacking knowledge among user on how to benefit from those tariffs as well as lack of building automation and control systems in place [1]. To conclude, the demand side flexibility has a great potential to provide economic, environmental, and societal benefits once those obstacles are overcome.

3 Changing Regulatory Environment

Increasing digitalization, expansion of Internet of things and rise of artificial intelligence applications are changing our perception of energy demand. Together with a higher presence of smart connected appliances, this contributes to empowering consumers and their patterns. They no longer play exclusively the role of end-consumers but, on the contrary, we encounter new generation of so called "prosumers". End user equipped with a solar panel can produce their own energy and sell the surplus.

When looking at latest trends that shape both the energy and smart appliances market, smart homes are predicted to increase tenfold by 2021 [2], while the number of users of smart appliances has been growing annually by 30.9% between 2017 and 2023. With 80% Europeans considering the idea of smart home appealing, there is a clear potential. This trend is accompanied by constantly decreasing cost of appliances, amounting to 43% between 2008 and 2016 [3].

Consumers are directly empowered via their smart appliances. A smart appliance can be defined as being "capable of automatically changing and optimising its consumption patterns in response to external stimuli", according to Regulation (EU) 2017/1369. Other benefits include increased energy efficiency or enhanced use of renewable energy. As our homes become more connected, new risks emerge, especially when it comes to cyber threats and privacy concerns. This leads increased regulation in sectors that have not been traditionally considered. This can be demonstrated with the new Cyber Security Act that entered in force in June 2019 and upcoming Regulation on Privacy and Electronic Communications that is currently negotiated. This comes on the top of the already existing General Data Protection Regulation, known as GDPR.

To further enhance uptake of smart technologies and achieve more efficient use of energy, the Commission is about to establish a new tool to better reflect a potential of smart appliances, devices and systems in buildings. The so-called Smart Readiness Indicator (SRI) will indicate which smart solutions are present in a building and how "smart ready" the building is or has a potential to be. The smartness would include the ability to adapt to occupant's needs, participate in demand response or to adjust energy consumption through the use of renewables [4]. This has potential to further enhance innovation and penetration of smart technologies for both residential and commercial buildings.

4 Demand Side Management Techniques

Demand side management (DSM) focuses on minimizing peak load requirement that is put on the grid. The algorithm used depends both on a two-way communication between an energy supplier and a customer, and therefore also on his/her willingness to reduce or postpone the electricity consumption [5]. The most frequently used techniques include peak reduction techniques, filling valleys, moving tips, conservation strategy, strategic growth, and flexible modelling techniques. These techniques differ in both scale and the way the electricity is used. Peak clipping (1) refers to reduction of a load during peak demand and this could be done directly via consumers directly controlling their energy smart appliances. This comes hand in hand with so called valley filling (2) where consumption is encouraged during off-peak periods. Since the cost of production decreases, the consumer benefits from more favourable price and this contributes to better energy efficiency of the whole system. Various incentives can be put in place to motivate consumer to change their energy patterns. Conservation technique (3) involves a reduction in both energy demand and consumption by consumers, which can be done mostly via implementation of new technologies and uptake of energy efficient home appliances. Strategic load (4) growth seeks to increase the load level, going beyond the valley filling. This might include electrification or deployment of alternative solution to primary fuels. Load shifting (5), on the other hand, aims to shift loads from peak to off-peak period. Space heating and cooling storage as well as domestic hot water storage have a considerable potential to enable load shifting without changing the overall consumption. When it comes to flexible load shape (6), a consumer agrees to limit his energy demand at certain times, depending on the real-time needs and system conditions and includes variability in reliability of energy. This might include the possibility for utility to interrupt loads if necessary (Fig. 1).

The load management and demand response are believed to become of particular interest and potential in residential scenario [6], which is mostly driven by smart grid implementation.

There are two main approaches to demand side management: price-based and incentive-based. In the first case, consumers are encouraged to actively participate in demand response based on price information, where a real time pricing, time of use, critical and peak time pricing are used [7]. In the latter, consumers agree to accept certain conditions by energy supplier and adjust their energy consumption when asked to do so. This includes direct load control, interruptible service and emergency demand response. The impact on consumer comfort is higher than in the first case. In residential context, direct load control and price-based programs are suitable. The consumer can either shift program manually or via home energy

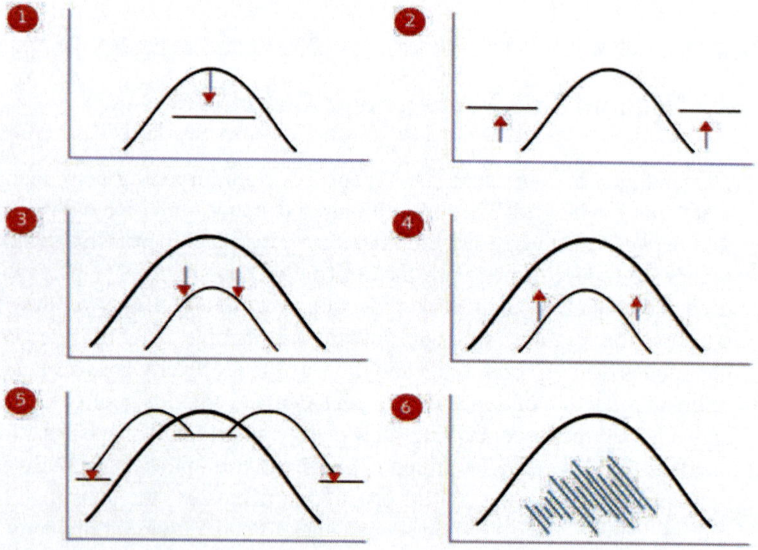

Fig. 1 Demand side flexibility techniques figure

manager that is capable of controlling the consumption of several smart appliances. This is mostly used by electric water heaters, air conditioners, refrigerators, washing machines, washer dryers and dishwashers [6].

5 Use Cases

Connected homes equipped with smart appliances able to communicate with a grid through a home energy manager represent a great opportunity towards optimising energy consumption and minimizing the electricity cost for users [8]. To ensure that connected appliances, most likely produced by different manufacturers, are capable to communicate with home energy manager and being able to receive the messages from home energy manager and grid, they need to "understand" each other. Furthermore, they need to translate messages they receive into actions so that they adapt their behaviour, based of grid conditions, for example. Interoperability, according to European Telecommunications Standards Institute (ETSI), can be described as "ability of two or more systems or applications to exchange information and to mutually use the information that has been exchanged". Further, two use cases are mentioned, to demonstrate how interoperability can be achieved using the same ontology. To that end, the Smart Appliances REFerence ontology (SAREF) provides a common language that can be used by differently branded home appliances to exchange messages with home energy manager.

Our use cases include an EV charging station, a PV panel, a heat pump, a dishwasher and a washing machine that will exchange messages and adjust their consumption based on signals received from a grid. To optimize energy streams and improve grid stability, a three-step methodology [9] is applied: use of prediction, planning and real time control of appliances.

The first use case describes a situation of overload protection. In this case, a charging station is charging an electric vehicle. When grid condition changes and a peak occurs, a signal is sent to the station to pause. Once good grid condition improves, the station can resume its operation (Fig. 2).

The second case demonstrates a usage of surplus energy. If a PV panel receives a prediction of a cloudy day ahead, energy production increases. The energy surplus that becomes available can be used for a heat pump which therefore receives a signal to start loading. Once the PV production drops, the heat pump pauses the load (Fig. 3).

These use cases were part of a demonstration organised at APPLiA premises, developed by EEBUS and Energy@home initiatives. Interoperability was achieved using the Smart Premises Interoperable Neutral-message Exchange (SPINE), a neutral language based in SAREF.

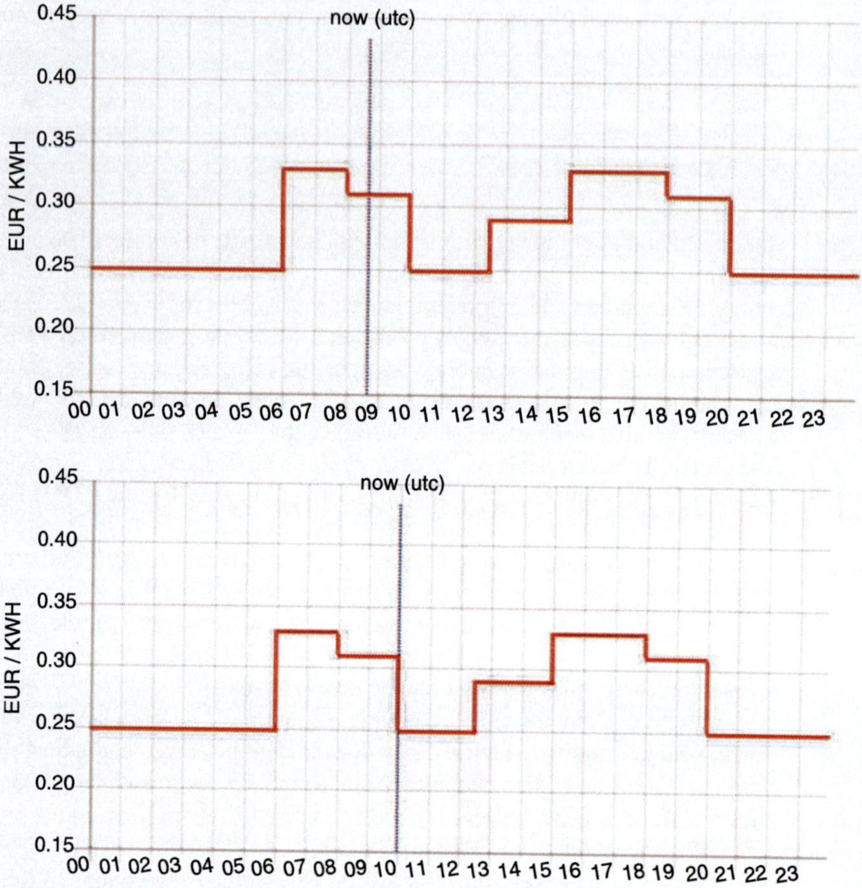

Fig. 2 Overload protection scenario

6 Flexibility Potential Benefits

The flexibility potential also depends on the type of appliances. Looking at the overall energy consumption of households in Europe, space heating represents 64.7% of the total consumption, followed by water heating (14.5%) and lighting & appliances (13.8%). The table also suggest the share of different energy sources whereas gas (36.9%) and electricity (24.4%) prevail. Renewables amount to 16% of household energy consumption [10] (Fig. 4).

The following categories are evaluated: periodical, energy storing appliances, behavioural and heating, ventilation and air conditioning (HVAC) appliances. The flexibility potential also depends on the way appliance is controlled and in it can receive external signals from a grid.

When it comes to periodical appliances, a cycle is set by a user and no interaction is needed during the execution. There are several options to adjust the energy

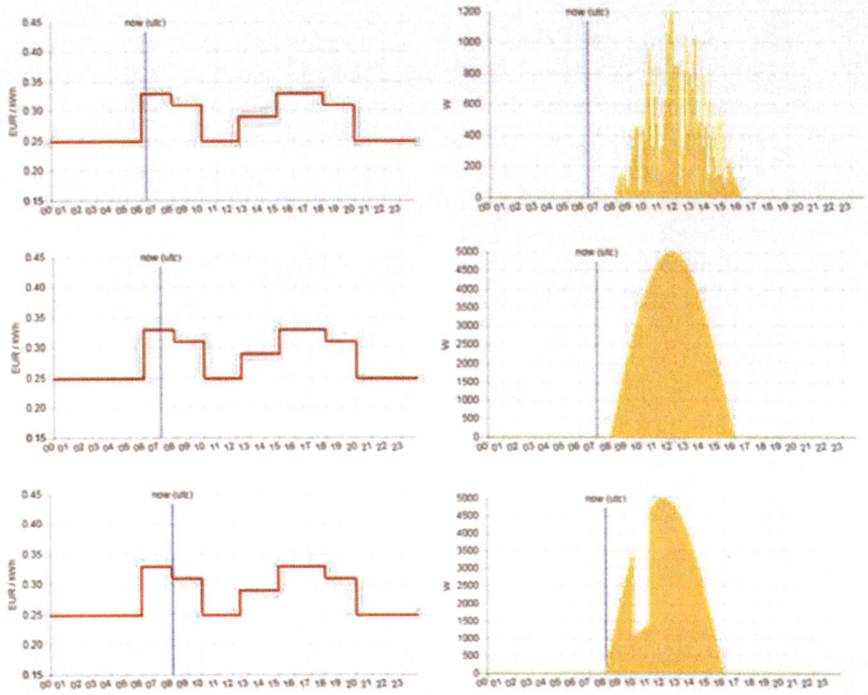

Fig. 3 Surplus energy use: Incentive table on the left side and PV production on the right side

Electricity	24.4	3.6	0.3	2.8	2.7	13.8	1.2
Derived Heat	7.6	6.0	0.0	1.6	0.0	0.0	0.0
Gas	36.9	28.1	0.0	7.0	1.8	0.0	0.0
Solid Fuels	3.4	3.1	0.0	0.2	0.0	0.0	0.0
Oil & Petroleum Products	11.8	9.6	0.0	1.5	0.7	0.0	0.0
Renewables and Waste	16.0	14.3	0.0	1.4	0.2	0.0	0.1
Total	**100.0**	**64.7**	**0.3**	**14.5**	**5.4**	**13.8**	**1.3**

Fig. 4 Energy consumption in households for the main energy products in Europe, 2017

consumption, such as anticipating or postponing power demand in time, remote activation with a set deadline for program execution or alteration of energy consumption pattern. The overall energy consumption is relatively small compared to other categories. When taking into consideration the number of households in Europe, the energy shifting potential was estimated as follows: 4.86 GWh for washing machines, 3.03–9.47 GWh for tumble driers and 8.17 GWh for dishwashers per day [11]. For energy storage appliances, it is possible, for example, to launch the compressor of a fridge or postpone water heater operation based on demand response signals. Energy shifting potential of 1.56 GWh [11] for refrigerators was estimated. HVAC appliances represent the biggest flexibility potential. Depending on product and season, the energy load can be significantly reduced. When it comes to electric heating (electric radiators, electric and hybrid heat pumps), they usually already include thermostat control and more recently, consumers are being offered a smart controller as usual practice. Flexibility potential could amount to 100 GWh per day during the winter period [11]. When it comes to ventilation, shifting potential lies mostly in non-residential sector. The potential amounts to 65 GWh per day on average during the summer period [11].

When having a closer look at flexibility potential of a thermal appliances, Fig. 5 shows the presence of flexibility. On the left side, usual appliance operation without using flexibility is shown and on the right side, an operation when flexibility is enabled is portraited [12].

The controller works with an upper and lower temperature limit, so called comfort limits that are usually set by the end user. When the room temperature reaches the lower limit, the heating is automatically switched on until the temperature reaches the upper limit. Once the flexibility is activated, an external command can switch the appliance back on at time *t1*, or in the case of *t2*, when heating would be normally switched off if the flexibility was not enabled.

Demand response implementation brings along several benefits for both utilities and end-users. These include not only better system operation and overall efficiency but also direct savings on energy bills. It empowers consumers and encourages them to participate in the energy market. Indirectly, this also brings benefits to other consumers since the overall wholesale market prices are lower when shifting loads to off-peak periods. Although it needs to be noted that materialization of the flexibility potential depends on the participation of consumers in shifting or reducing their consumption during peak hours [13]. It should be ensured that consumers are equipped with information tools to make informed choices when managing their loads. Also, they should be fairly informed on charges that may apply when using a

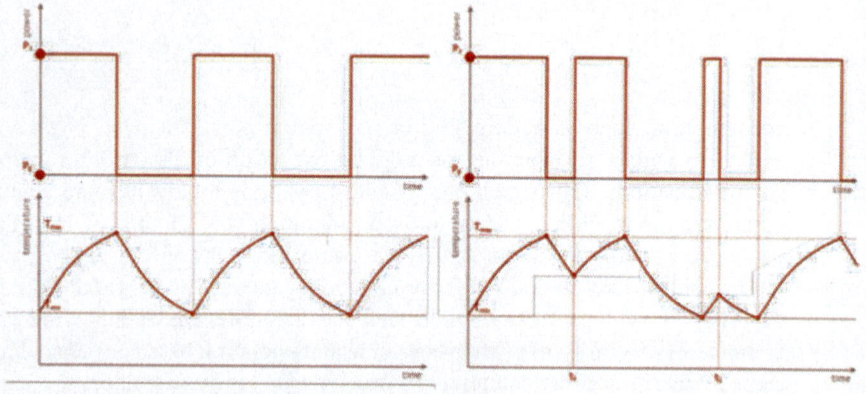

Fig. 5 The presence of flexibility in a thermal appliance that supports power modulation and a more advanced controller

demand response program. This will be not only beneficial to consumers, motivating them to take part in demand response, but also for the optimal functioning at system level.

7 Conclusions

The changing energy system is not only bringing new opportunities but is also facing new challenges. From consumer perspective, active participation in the energy market implies lower energy prices, savings and possible financial compensation for self-production and consumption. From the technological point of view, the new generation of smart appliances brings unprecedent possibilities. Thanks to connection to the grid, this can contribute to balancing of demand and supply in real time and therefore avoid black-outs and better manage peak times. This balance will benefit the stability of the whole electricity system. Furthermore, a greater integration of renewable energy sources into energy mix, for example via local energy communities, would lead to environmental benefits and societal benefits. On the other hand, there is a lot of room for improvement. Consumers incentivization is believed to remain the biggest obstacle.

The article demonstrated that demand side flexibility has potential to further optimize the energy consumption and facilitate the integration of renewable sources into energy system. Consumers become more active and have more possibilities to engage on the energy market. They should be incentivized to change their behaviour and consider providing flexibility to the system. Also, the different demand response programs should be clear and easy to understand. Last but not least, end users should be equipped with information tools to be able to easily track their consumption and make informed decisions.

In order to fully materialise the potential of demand side flexibility in the residential scenario, the role of the home energy manager, coordinating the home environment, is crucial. First, the energy manager has to receive a signal from the grid and be able to translate it to a concrete action that would result in altered consumption of an appliance. Secondly, the communication has to happen seamlessly and for that, interoperability has to be ensured at all levels. It needs to be noted that a lack of building and automation system is an obstacle to a greater uptake of smart technologies.

Looking forward to regulatory initiatives, if set correctly, they might create a right framework that would further enhance the new opportunities that the changing energy market offers. The new business models that are entering the market and enabling active consumer participation and sharing of aggregated flexible demand will unlock the flexibility and contribute to optimal functioning of the market as whole.

References

1. A.-H. Mohsenian Rad, A. Leon-Garcia, "Optimal Residential Load Control with Price Prediction in Real-Time Electricity Pricing Environments", *IEEE Transaction on Smart Grid, Vol. 1, No. 2,* September 2010
2. European Political Strategy Centre. (2018), "10 Trends Reshaping Climate and Energy", 2018, p. 13
3. APPLiA. "By the numbers: The Home Appliance Industry in Europe 2017-2018". Appeal of Smart Home products. 2018
4. Energy Ville, Vito, Waide Strategic Efficiency, "Second technical study to support the establishment of a common European scheme for rating the Smart Readiness of Buildings"; 26 March 2019, Brussels
5. Anjana S. P., Angel T. S. "Intelligent Demand Side Management for Residential Users in a Smart Micro-Grid", 2017 IEEE International Conference on Technological Advancements in Power and Energy (TAP Energy).
6. Nikolaos G. Paterakis, Ozan Erdinç, João P.S. Catalão, "An overview of Demand Response: Key elements and international experience", Renewable and Sustainable Energy Reviews 69 (2017) 871–891.
7. O. Ayan, B. Turkay, "Domestic Electrical Load Management in Smart Grids and Classification of Residential Loads", 5th International Conference on Electrical and Electronics Engineering. 2018
8. P. Siano, "Demand response and smart grids survey", Renewable and Sustainable Energy Reviews, vol. 30, pp. 461-478, 2014.
9. A. Molderink, V. Bakker, M. G. C. Bosman, J. L. Hurink, G. J. M. Smit, "Management and Control of Domestic Smart Grid Technology", IEEE Transactions on Smart Grid, Vol. 1, No. 2. September 2010
10. APPLiA, "By the numbers: The Home Appliance Industry in Europe 2017-2018", Energy consumption in households for the main energy products in Europe in 2017.
11. Vito, Viegand & Maagøe, Armines and Bonn University. (19 January 2019). Ecodesign Preparatory Study on Smart Appliances (lot33) MEErP Tasks 1-6, Final report. Annex 1.
12. Vito, Viegand Maagoe, Armines, Wuppertal Institut. (2018, October). Preparatory study on Smart Appliances. Task 7 Report—Policy and Scenario Analysis. Figure 22; p. 73
13. J. S. Vardakas, Ch.V. Verikoukis, "A survey on Demand Response Programs in Smart Grids: Pricing Methods and Optimization Algorithms", IEEE Communications Surveys and Tutorials. 2015

Using Feedback on Computer States to Improve Power Management Behaviors

Joy E. Pixley, Sabine Kunrath, Raquel Fallman, Sergio Gago-Masague, and G. P. Li

Abstract Computer power management settings can potentially save substantial energy by putting computers into low-power modes when not being used. However, previous research shows that sleep settings are often disabled in office desktops. The Power Management User Interface (PMUI) feedback app was designed to give users feedback on their computer idle states and to encourage enabling of sleep settings. Unlike current energy feedback devices, PMUI is a standalone free software application that does not require installing additional equipment. PMUI was field tested in 407 computers (303 treatment subjects and 104 control subjects), with a minimum 1-month baseline period and 2-month treatment period for each subject. At baseline, only 13% of computers had computer sleep settings enabled, but 56% of subjects reported the settings were enabled. Findings suggests user confusion about settings that is correlated to lack of use and lack of knowledge. Subjects exposed to the PMUI application were significantly more likely than control subjects to enable their computer sleep settings and to reduce the delay time. Treatment subjects' computers subsequently spent less time idle and more time in sleep mode than control subjects' computers. Overall, these results provide strong evidence that feedback on computer states can effectively induce desktop users to improve their power management settings and thus save energy, without the need for separate plug meter devices to measure energy usage.

1 Introduction

Many feedback interventions have been designed to change users' behaviors with the goal of reducing energy demand at home and at the workplace, with varying degrees of success [1–5]. The current study focused on behaviors toward one specific device: the desktop computer. Desktop computers are less prevalent than portable computers, but their contribution to energy consumption continues to be high.

J. E. Pixley (✉) · S. Kunrath · R. Fallman · S. Gago-Masague · G. P. Li
University of California, Irvine, CA, USA
e-mail: jpixley@uci.edu; skunrath@uci.edu; rfallman@uci.edu; sgagomas@uci.edu; gpli@calit2.uci.edu

P. Bertoldi (ed.), *Energy Efficiency in Domestic Appliances and Lighting*, Springer Proceedings in Energy, https://doi.org/10.1007/978-3-030-79124-7_19

A recent study estimates the installed base in the US of desktop computers to be 72 million, compared to 122 million for portable computers. However, the higher energy consumption for desktop computers results in an estimated annual electricity consumption of 18 TWh compared to only 5.1 TWh for portable computers [6]. Desktop computers are major contributors of wasted "idle" energy use in homes and businesses [7]. Current ENERGY STAR® guidelines estimate average power consumption for desktops at 2.3 W in sleep mode compared to 48.1 W in idle mode [8]. In one case study, desktops with sleep settings enabled (regardless of delay time) spent an average of 12% of the week in idle mode, compared to 68% for those without sleep enabled [9]. The solution is simple to conceive but difficult to implement: ensure that desktops spend as much time in sleep mode and as little time idle as possible.

Unfortunately, despite having power management options available, most desktop users are not applying them. Physical audits and monitoring studies in commercial and university buildings have found that in practice a high percentage of computers were left on unnecessarily when not being actively used, and that substantial energy savings could be possible with better power management practices [e.g., 10–13].

In contrast to research observing computers directly, most surveys show high rates of users reporting that their computer sleep settings are engaged [14]. Some of the discrepancy between self-reports of enabled sleep settings and observations of idling computers appears to be due to user confusion about sleep settings [15]. One study that linked self-report to research observations for the same subjects found that although 86% of subjects had reported their computer sleep settings enabled, only 30% of those subjects actually had their settings enabled [16]. Furthermore, those who rated themselves as more knowledgeable about computers were more likely to be accurate about their settings. This suggests that educating users about sleep settings and giving them feedback about their current power management settings and the computer's sleep behavior may be an important step to saving energy.

Previous studies have tried encouraging office workers to save energy with their computers and other equipment by providing advice and feedback to employees about the power consumption of their workstations, covering a range of plug load devices [17–19] or the whole office [20]. A few feedback studies focused on computer states and energy use [21, 22] or provided the user or researchers with usage information disaggregated for specific devices [23, 24]. Few feedback studies have examined user behavior in connection with computer states and power modes [21, 22]. However, most studies suffer from small sample sizes or were not able to utilize an experimental randomized control trial design. More studies are needed to assess the potential of feedback on influencing computer users' behaviors.

One approach to improving computer energy efficiency in commercial enterprises is centralized IT control, in which a company's IT department uses a software application or service to remotely control its employees' computers, including their power management settings. This works well for many companies and can save substantial energy. However, it is not appropriate in many situations, such as residential computer use and small companies with limited IT resources. Also, even where centralized IT control is already used to facilitate updates and backups, a policy of mandatory sleep settings can face resistance from IT personnel and employees [22, 25]. Thus, a method for encouraging voluntary improvements in energy-saving computer behaviors could be useful in many circumstances.

To address this problem, the research team at University of California, Irvine developed a software application, the Power Management User Interface (PMUI), designed to encourage voluntary use of automatic sleep settings. PMUI was developed based on design principles established by prior research on using feedback to motivate behavior, which describe the importance of giving timely feedback [26–28], offering specific, actionable tips [28, 29], presenting data in multiple, simple to understand graphs and figures [26, 30, 31], comparing users' current outcomes to previous outcomes [32, 33] or to an ideal standard or goal [5, 34], rewarding the desired behavior with happy face emoticons [35], appealing to common norms by using pro-environmental messages [36, 37], and engaging the user with interactive features [27, 30].

This paper presents the results of a field test of the PMUI interface, with 407 staff members of a major university. The overall research question asks whether the PMUI interface encourages subjects to improve their sleep setting behaviors, relative to the baseline period, and whether this behavior change is significantly better than for the control group, who are only reminded how to access their standard sleep settings.

The primary research question is whether use of the PMUI application can positively affect subjects' behavior: specifically, encouraging them to improve their automatic sleep settings by either enabling them (if they were disabled) or reducing the delay time (if they were already enabled). This behavior is assessed at two time points. First, it is hypothesized that initial exposure to the PMUI application will result in more treatment subjects improving their computer sleep settings than control subjects, who are exposed only to their standard sleep settings interface. If this were the only effect of PMUI, it would indicate that PMUI is more effective as an informational tool than simply reminding subjects to check their standard sleep settings. Second, it is hypothesized that the PMUI application has an ongoing and persistent effect past the initial effect, as indicated by larger improvement in sleep settings between initial exposure and the end of the study (2 or more months later) compared to the control group. If PMUI is shown to have an effect over time in addition to the initial exposure effect, it would indicate that ongoing feedback and encouragement had an important role.

Although the PMUI application does not offer feedback to users on display settings, possible improvements in display settings are also examined.

2 Methods

2.1 *The Power Management User Interface Feedback Application*

The Power Management User Interface (PMUI) application was designed to encourage users to better utilize their computers' existing sleep settings by (1) providing a clear, simple sleep settings interface for computer and monitor/display, (2) providing consistent cues about the importance and desirability of reducing idle time; (3) providing clear advice about how to reduce idle time; (4) providing multiple types of feedback reports on how much time the computer spends idle, including comparisons to previous time periods and to a "target" profile; and (5) using multiple types of encouragement, including environmental impact measures and smiling versus sad emoticons. To simplify the interface, the PMUI application did not provide feedback to users on the idle time of their monitors. PMUI sends a weekly pop-up reminder for users to check their usage report for the week (see Fig. 1). Users can access the application using a tray icon, which opens onto the sleep settings page (see Fig. 2). A laboratory pretest walked 22 subjects through the use of every feature and asked open-ended questions to assess their comprehension and interpretation. Pretesting was staggered so that later subjects were presented with revisions based on earlier subjects' responses to confirm that the changes were effective. Other details about the software are presented elsewhere [38, 39].

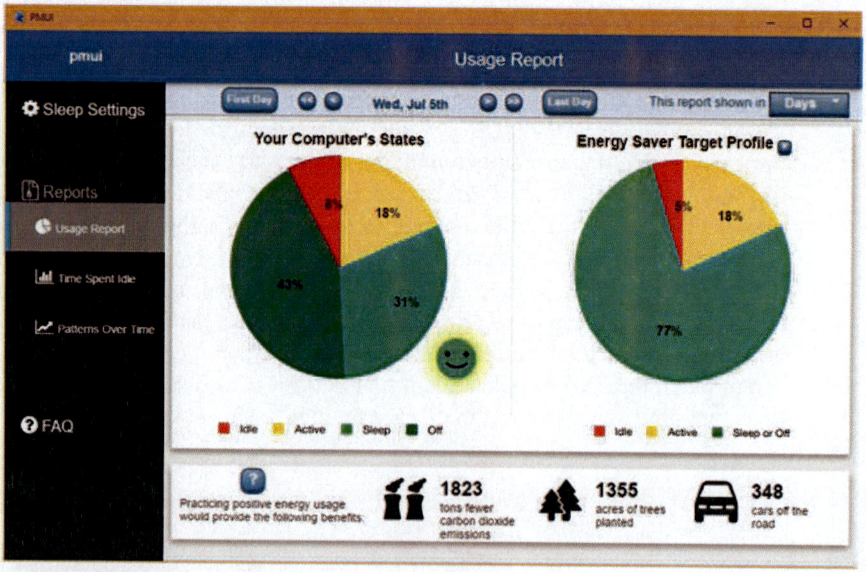

Fig. 1 PMUI usage report page

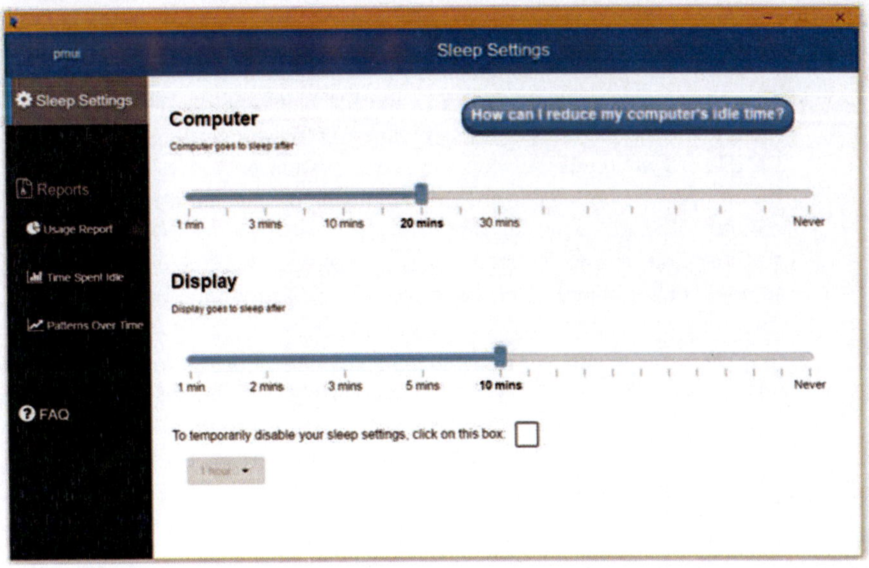

Fig. 2 PMUI sleep settings page

2.2 *Data Collection*

All schools and other units on campus who might be eligible for inclusion were identified. Recruitment began with the administrative units whose computers were managed by the Office of Information Technology. Other units were added in waves, focusing on those with substantial numbers of staff, and limited to those where the unit did not have a sleep setting policy in place and where the IT manager, director, dean, or other representative approved the use of the software. The participating units represent a wide range of staff on campus, including humanities, physical sciences, administration, and facilities management. Data collection lasted from March 2017 to May 2018.

Once a unit was chosen, staff contact information was obtained from the university directory or from the unit itself. The IT manager sent an email alerting staff that the recruitment email was not a scam and that participation was voluntary. Potential subjects were then emailed a recruitment letter detailing the study, including a link to the consent form. If interested, they filled out an online form verifying their eligibility and asking for potential times to schedule their first appointment. The public name of the study shown to subjects was "[University Name] Computer Energy Study" to reduce introducing bias by mentioning power management or sleep settings in the initial materials.

Individuals were eligible if they were university staff, aged 18 or older, who were the sole users of a desktop computer on campus and had the ability to change their own sleep settings.

Student research assistants (RAs) conducted three research visits per subject. Subjects were given a $25 Amazon gift card at each of the three visits, for a total of $75. At the first research visit (T1), the RA went over the informed consent document and obtained the subject's signature before proceeding. The RA installed the PMUI software on the desktop, set to observation-only mode. The RA also plugged the computer and its monitor(s) into a power strip, which was plugged into a power monitor. The specially programmed power monitors transmitted real-time energy consumption data to the study's secure servers. Baseline data on computer sleep settings and on energy use was collected for a minimum of 4 weeks, subtracting vacations, university holidays, and other leaves.

After T1, subjects were assigned to the control or experimental condition, at a one-to-three ratio. This process was not entirely random, but depended on the order in which the Access database sorted new cases. In addition, a few cases were reassigned at or before T2 (e.g., cases where a researcher accidentally mentioned the PMUI app to a control subject).

At the second research visit (T2), subjects filled out an online survey on the RA's tablet. For the treatment group, RAs activated the PMUI GUI and showed the subject how to access the app. For the control group, RAs showed the subject how to access their computer's standard computer settings. Included in the extensive training the RAs received was how to manage this last step without implying that the subject didn't already know.

The final research visit (T3) was conducted a minimum of 8 weeks after T2. Subjects filled out another online survey, including questions asking them to evaluate their standard computer sleep settings and, for the treatment group, to evaluate the PMUI app. At this visit, the RA removed all hardware and software.

2.3 Subjects

The final sample consisted of 407 subjects, all university staff. This included research and administrative staff and post-doctoral scholars, but not faculty or students. The majority were women (68%). Almost half (45%) of subjects identified as white, with 28% identifying as Asian or Pacific Islander, 11% identifying as Hispanic or Latino, 8% identifying as more than one race or ethnicity, and less than 5% identifying with any other group. Compared to Census estimates for California, the sample is less white (versus 72.4%), less Hispanic (versus 37.2%) and more Asian (versus 15.2%). However, the state race and ethnicity data is for people of all ages and education, whereas staff members who use computers at a university (and at other similar enterprises) are likely to be of working age and more highly educated than average. Indeed, all but 11% of the sample are college-educated: 53% have a bachelor's degree, 28% have a master's degree, 1% have a different professional degree, and 6% have a doctorate. By contrast, the Census shows that for the overall population in the state, only 32% of persons 25 and older have a bachelor's degree or higher. Subjects were asked to identify which official Census occupation category fit them best. The majority were professionals, administrative support, or higher-level administrators. Almost all (98%) reported working full-time, defined as at least 30 h/week, or a 75% appointment. These demographic differences from the overall state averages are to be expected with a sample focused on office workers in administrative and professional settings, who comprise most users of office desktop computers.

The majority of desktops in the study (92%) used the Windows operating system, most using Windows 7 (77% of the sample) or Windows 10 (15% of the sample). The other 8% used Mac operating systems, mostly the most recent build at the time, macOS Sierra (19 of the 31 Mac computers, and 5% of the sample).

Multiple monitors were common: at the beginning of the study, 37% of subjects had one monitor, 62% had two monitors, and six subjects (less than 2%) had three or four. Seven subjects changed the number of monitors they had over the course of the study, but the percentages in each group stayed the same. Twenty-five desktops (6%) were all-in-one computers; these were all Apple computers, although two of them were running Windows operating systems. All-in-one computers were coded as having one monitor; nine of them were also connected to a second monitor.

As planned, three in four subjects were in the treatment group (303, or 74.5%). The treatment and control groups did not differ significantly on any of the demographic characteristics measured, nor on the operating system of their computer or the number of monitors.

2.4 *Measures*

The sleep settings for the PMUI app are the same as those in the standard sleep settings, that is, those that come with the Windows or Mac OSX operating system. Only sleep settings for the computer and monitor/display are assessed (not, for instance, hard drive sleep, hybrid sleep, or hibernation). Sleep settings that are set to "never" are considered "disabled" while sleep settings set for a delay time (e.g., 20 min until sleep) are considered "enabled." A shorter delay time indicates that the subject reduced the number of minutes of idle before the computer or monitor goes into sleep mode, while a longer delay time indicates the opposite.

The hypotheses are tested by looking at changes in the sleep settings over the course of the subject's study period, particularly whether the sleep settings at the end of the study period (T3, about 2 months after the intervention at T2) differ from the settings observed at the end of the baseline period (T2). The current analysis thus does not capture subjects who made multiple changes and ended with the same setting. For instance, those who initially enabled their sleep settings but did not persist—that is, went back to disabling their settings—are considered as "no change" in the end.

3 Results

3.1 Initial Computer Sleep Settings

Consistent with prior research, only a minority of subjects (14%) had their computer sleep settings enabled at the first research visit. Of those computers, the most common sleep delay time was 30 min (the default), with almost equal proportions of others exhibiting higher and lower delay times. Far more computers had display sleep settings enabled at baseline (83%). Almost all of them had delay times set at 30 min or less, with a substantial minority at 15 min or less. Only nine subjects changed their settings during the baseline period, suggesting the lack of a Hawthorne effect. Four subjects enabled computer sleep and two disabled it, while three subjects enabled display sleep and two disabled it (two subjects changed both computer and display sleep).

At the end of the baseline period, the control group had a slightly higher proportion of computers with computer sleep enabled (17% vs. 14% in the treatment group) and a slightly lower proportion with display sleep enabled (81% vs. 84%), but neither difference was statistically significant.

3.2 Changes to Sleep Settings

As the primary goal of PMUI is to encourage users to enable their sleep settings, the first analysis is whether the group using the PMUI application were more likely to enable their sleep settings than the control group. As shown in Fig. 3, the main effect of PMUI on behavior is clear and substantial. Among subjects who had their computer sleep settings enabled prior to the intervention at T2, slightly more treatment subjects retained them, but the difference is not significant. However, among those with disabled computer sleep settings, treatment subjects were much more likely to enable them than control subjects (59% versus 15%, $p < 0.0001$).[1] The results are similar, if not as strong, for those who began with their display settings disabled, with 56% of treatment subjects versus 30% of control subjects enabling them ($p = 0.0484$).

Clearly, exposure to the PMUI application is more effective at encouraging energy-saving settings than reminding subjects of their normal sleep settings. The next set of analyses assess whether this effect occurs mainly upon first exposure to the information, or if it is amplified by more extended, longer-term feedback and encouragement.

[1] All the analyses in this paper use bivariate chi-square tests to compare results for treatment versus control groups; no other variables were controlled for.

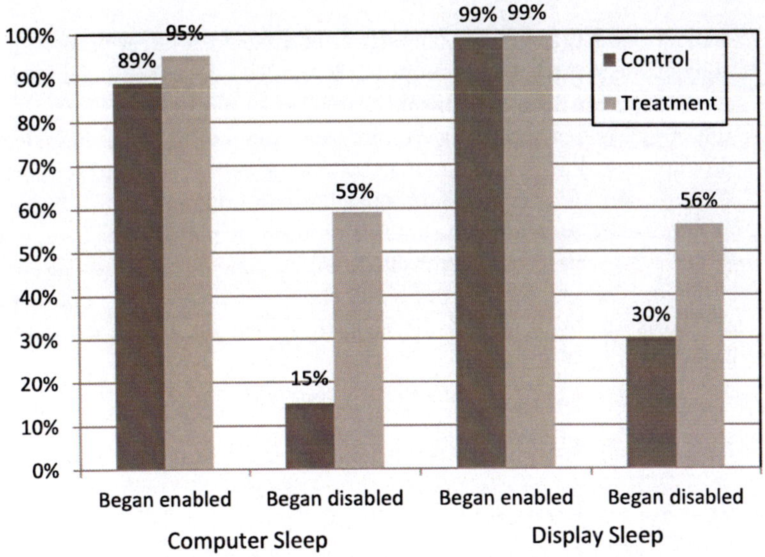

Fig. 3 Percentage of computers with sleep settings enabled at T3 by experimental group and initial settings at T2

Table 1 Change in computer sleep settings after initial intervention (T2), by condition

		Control group		Treatment group	
Status at T2	Change after T2	#	Percent	#	Percent
Disabled	No change	71	83	186	71
	Enabled sleep	15	17	76	29
	N	86		262	
Enabled	No change	16	89	31	76
	Disabled sleep	0	0	1	2
	Increased sleep delay	1	6	0	0
	Decreased sleep delay	1	6	9	22
	N	18		41	
Total		104		303	

3.2.1 Immediate Response to Intervention

As it is possible that the subject experimented with changing during or shortly after the research visit at T2 and then reverted to the previous settings, these analyses examine the change in status (if any) from the end of the day prior to T2 to the end of the day after T2 occurred. The possible subject actions following the intervention at T2 depend on whether the computer sleep settings were already enabled; they are listed in Table 1. Among subjects who had computer sleep disabled prior to the intervention, subjects in the treatment group were significantly more likely to enable sleep than those in the control group (29% versus 17%, $p = 0.0342$). Among subjects who already had computer sleep enabled, a larger proportion of treatment subjects decreased (that is, improved) the sleep delay (22% versus 6%); however, this is not statistically significant, likely because of the small sample size. Overall, this suggests that subjects were more motivated to improve their sleep settings by looking at the PMUI program than by looking at their standard settings, even though the two interfaces displayed the same basic information.

Immediate effects of the intervention on display sleep settings are shown in Table 2. The observed relationship is in the same direction as with computer sleep settings. More subjects in the treatment group than in the control group enabled display sleep or decreased the delay time, but the difference is not statistically significant.

3.2.2 Long-Term Changes over the Experimental Period

The next analysis assesses the relationship between the subjects' initial response after T2 and the eventual outcome, that is, whether continued use of PMUI had an additional effect on the outcomes shown at the end of the study. For the control subjects who began with disabled computer sleep settings, those who did not change them during T2 were highly unlikely to change them later in the study (only 1%), while 20% of those who enabled their sleep settings during T2 had disabled them again by the end of the study (see Table 3). By contrast, of the treatment subjects who did not enable their settings during T2, almost half (45%) enabled their computer settings later in the intervention period and kept them enabled until T3. Also, almost all of the treatment subjects who enabled their settings at T2 still had them enabled at the end of the study (95%). All the treatment subjects with enabled sleep settings who did not change them at T2 still had them enabled at T3; the other cells in the table are too small to produce reliable percentages. In short, the continued positive changes for the treatment group after the change at T2 suggest the effectiveness of the ongoing intervention of the PMUI feedback.

Table 2 Change in display sleep settings after initial intervention (T2), by condition

Status at T2	Change after T2	Control group #	Control group Percent	Treatment group #	Treatment group Percent
Disabled	No change	16	80	30	63
	Enabled sleep	4	20	18	38
	N	20		48	
Enabled	No change	74	88	208	82
	Disabled sleep	0	0	1	0
	Increased sleep delay	0	0	1	0
	Decreased sleep delay	10	12	45	18
	N	84		255	
Total		104		303	

Table 3 Relationship of change in computer sleep settings after initial intervention (T2) to outcome at end of study (T3), by condition

Status at T2	Change at T2	Control group #	Sleep disabled at T3 (%)	Sleep enabled at T3 (%)	Treatment group #	Sleep disabled at T3 (%)	Sleep enabled at T3 (%)
Disabled	No change	71	99	1	186	55	45
	Enabled sleep	15	20	80	76	5	95
Enabled	No change	16	13	88	31	0	100
	Disabled sleep	0			1	0	100
	Increased sleep delay	1	0	100	0		
	Decreased sleep delay	1	0	100	9	22	78
Total		104			303		

Table 4 Relationship of change in display sleep settings after initial intervention (T2) to outcome at end of study (T3), by condition

Status at T2	Change at T2	Control group #	Sleep disabled at T3 (%)	Sleep enabled at T3 (%)	Treatment group #	Sleep disabled at T3 (%)	Sleep enabled at T3 (%)
Disabled	No change	16	88	13	30	67	33
	Enabled sleep	4	0	100	18	6	94
Enabled	No change	74	0	100	208	0	100
	Disabled sleep	0			1	0	100
	Increased sleep delay	0			1	0	100
	Decreased sleep delay	10	10	90	45	2	98
Totals ($N = 407$)		104			303		

The same analysis for display sleep settings is shown in Table 4. As most subjects had their display sleep settings enabled for the entire study, there are not many differences by experimental condition. In both control and treatment groups, all of the subjects who already had display sleep enabled and made no change at T2 still had display sleep enabled at the end of the study, as did most of the subjects who enabled display sleep at T2. Among subjects with disabled settings who did not enable them at T2, treatment subjects were more likely than control subjects to later enable them, but this relationship is not statistically significant. A small number of subjects who decreased the display delay time at T2 later disabled display sleep.

3.2.3 Changes in Delay Times for Computer and Display

Although enabling sleep settings has the largest impact on how much time the computer or monitor spends idle, reducing the delay time for sleep settings also saves energy. Table 5 shows the proportion of subjects in each group who already had sleep settings enabled before the intervention at T2 who either increased or decreased their delay time. Subjects in the treatment group were somewhat more likely than control subjects to reduce the delay time for their computer sleep settings, but this relationship only approached the level of significance (39% versus 11%, $p = 0.0532$). The difference for the larger number of subjects who began with display sleep settings is more pronounced, with 47% of treatment subjects reducing their display delay time compared to 13% of control subjects ($p < 0.0001$).

4 Discussion

The field test results presented here suggest that users can be encouraged to save substantial amounts of energy on desktop computers, if given clear feedback on how their decisions affect the energy consumption of the devices. A substantial minority of control group subjects (15%) enabled their sleep settings after being shown the standard settings page, and left those settings enabled for at least another 2 months. This is an interesting result in itself, as it suggests that many computer users are honestly unaware that their sleep settings are disabled, and are willing to enable them without any additional feedback or encouragement.

Table 5 Change in sleep setting delay times over intervention period, by condition

	Control group		Treatment group	
	Number	Percent	Number	Percent
Computer sleep settings				
Delay time at T3 versus T2				
Same	13	72	23	56
Shorter	2	11	16	39
Longer	1	6	0	0
Disabled	2	11	2	5
N	18		41	
Display sleep settings				
Delay time at T3 versus T2				
Same	71	85	122	48
Shorter	11	13	120	47
Longer	1	1	11	4
Disabled	1	1	2	1
N	84		255	

Outcomes for the treatment subjects are even more dramatic: 59% of those with disabled sleep settings enabled them after using the PMUI app. This is more than three times higher than the effect for the control group. Also, those subjects who already had their settings enabled were more likely to reduce the delay time for their computer sleep and for their monitor sleep if they were exposed to the PMUI app.

Subjects might be more likely to pay attention to the information relayed by PMUI than by their standard sleep settings, given that PMUI is new to them and includes engaging, colorful reports. If so, this could result in a stronger response for the treatment group at the T2 research visit, but no discernable additional long-term effects. However, additional analyses indicate that PMUI had additional effects over the course of the experimental period, suggesting that its effect is not solely due to information provided at the intervention visit.

One limitation is the length of the study period, as treatment subjects were only observed about 2 months after first encountering the PMUI app at T2. For behavioral change interventions, persistence is a serious concern. For in-home energy feedback devices or behavioral modification devices (e.g., Fitbit step counters), if subjects stop accessing the feedback after a few weeks or months, or otherwise appear to lose interest, their behaviors are likely to revert to their previous level. It is worth stepping back to consider the types of energy-saving behaviors most programs encourage, which are usually divided into investment decisions versus curtailment behaviors. For investment decisions, the focus is on convincing users to buy more energy-efficient appliances. Once that purchase is made, the energy savings (for that device, at least) are locked in, and the user need not persist in thinking about it. By contrast, curtailment behaviors (such as remembering to turn lights off, choosing the "no dry" option on the dishwasher, or hanging up laundry instead of using clothes dryers) must be repeated on a regular basis, and thus require longer-term commitment and persistence.

A third type of energy-saving behavior gets less attention, which is changing settings. This behavior change is similar to investment behavior, in that the user may adopt more energy-efficient settings for, say, their thermostat, after which they do not need to persist in thinking about it. However, this behavior resembles curtailment to some extent also. As the user's experience is altered by the setting, it may be noticeable and even onerous enough to feel like curtailment, possibly leading to the user reverting the setting to the earlier, less energy-efficient level. Enabling sleep settings on a computer is conceptually much the same as thermostat settings. Simply enabling the settings once does not guarantee that sleep will remain enabled. Subjects who experience computer delays or other problems after enabling sleep (for example, complications when trying to remote into their work desktops from home), whether or not actually related to sleep, may decide that the cost of dealing with the problem is not worth the benefits, and disable sleep again. This may have happened to some proportion of the treatment subjects, who ended up with disabled sleep settings at T3. However, 2 months may be a reasonable period for measuring persistence in sleep settings, especially for subjects who changed their settings early on in that period. These subjects work full-time and most use their campus computers every day. If they have their sleep delay set for 30 min (the default) or less, they would return to a sleeping computer several times per day. This provides many repeated experiences of computer sleep over time, and it seems plausible that anyone who would be annoyed enough to disable sleep after 3 months would already be annoyed enough to do so after a few weeks. Later in-depth analyses of the timing of changes and the number of subjects who changed their minds about enabling sleep will hopefully shed more light on this, as will analyses of the reasons given for not using sleep settings, and the qualitative comments subjects made about problems with engaging sleep.

Another limitation is that the study focused on a specific population: staff members at a university. However, compared to many other organizations, a university provides a wide and diverse range of subjects, working in varied fields and with varied occupations and tasks. They are more highly educated than the average worker, but perhaps no more so than the average worker whose job involves regular use of a desktop computer. Still, the results cannot be assumed to apply to employees of small businesses or to home users of desktops. Additional field tests would be needed to assess the utility of the PMUI app in these other settings.

Overall, these results are strong, and bode well for future use of feedback interventions for saving energy on office computers and other plug load devices by encouraging more pro-environmental behavior among individual users.

In summary, these results are promising, and offer strong support for the idea that giving users engaging feedback and actionable information can encourage meaningful pro-environmental behavior change. The current results are specific to office desktop computers, but could reasonably be applied to other home or office devices for which power management settings and the energy impacts of user behavior may be misunderstood.

References

1. Darby, Sarah. (2006). *The Effectiveness of Feedback on Energy Consumption*. Oxford, UK: Environmental Change Institute University of Oxford.
2. Delmas, Magali A., Miriam Fischlein, & Omar Isaac Asensio. (2013). Information strategies and energy conservation behavior: A meta-analysis of experimental studies from 1975 to 2012. *Energy Policy, 61*, 729-739. doi:https://doi.org/10.1016/j.enpol.2013.05.109
3. Ehrhardt-Martinez, Karen, Kat A. Donnelly, & John A. Skip Laitner. (2010). *Advanced Metering Initiatives and Residential Feedback Programs: A Meta-Review for Household Electricity-Saving Opportunities*. Washington, D. C.: American Council for an Energy-Efficient Economy.
4. Electric Power Research Institute. (2009). *Residential Electricity Use Feedback: A Research Synthesis and Economic Framework*. Palo Alto: Electric Power Research Institute. Retrieved from https://www.epri.com/#/pages/product/1016844/?lang=en.
5. Karlin, Beth, Joanne F. Zinger, & Rebecca Ford. (2015). The effects of feedback on energy conservation: A meta-analysis. *Psychological Bulletin, 141*(6), 1205-1227. doi:https://doi.org/10.1037/a0039650
6. Urban, Bryan, Kurt Roth, Mahendra Singh, & Duncan Howes. (2017). *Energy Consumption of Consumer Electronics in U.S. Homes in 2017*. Boston, MA: Fraunhofer USA Center for Sustainable Energy Systems.

7. Delforge, Pierre, Lisa Schmidt, & Steve Schmidt. (2015). *Home Idle Load: Devices Wasting Huge Amounts of Electricity When Not in Active Use* (NRDC Issue Paper 15-03-A). San Francisco, CA: National Resources Defense Council. Retrieved from https://www.nrdc.org/sites/default/files/home-idle-load-IP.pdf.
8. ENERGY STAR. (2018). ENERGY STAR Computer Power Management Savings Calculator. Retrieved from www.energystar.gov/products/power_mgt/LowCarbonITSavingsCalc.xlsx
9. Pixley, Joy E., & Stuart A. Ross. (2015). *The Effects of Using Power Management in Office Desktops: A Case Study*. Paper presented at the Behavior, Environment, and Climate Change Conference, Sacramento, California.
10. Acker, Brad, Carlos Duarte, & Kevin Van Den Wymelenberg. (2012, August 12-17). *Office Plug Load Savings: Profiles and Energy-Saving Interventions.* Paper presented at the ACEEE Summer Study on Energy Efficiency in Buildings, Asilomar, California.
11. Bensch, Ingo, Scott Pigg, Karen Koski, & Rana Belshe. (2010). *Electricity Savings Opportunities for Home Electronics and Other Plug-In Devices in Minnesota Homes. A technical and behavioral field assessment.* Madison: Energy Center of Wisconsin. Retrieved from https://www.seventhwave.org/sites/default/files/257-1.pdf.
12. Cabrera, David F. Motta, & Hamidreza Zareipour. (2011). *A Review of Energy Efficiency Initiatives in Post-Secondary Educational Institutes.* Paper presented at the Energy 2011: The First International Conference on Smart Grids, Green Communications, and IT Energy-Aware Technologies.
13. Mercier, Catherine, & Laura Moorefield. (2011). *Commercial Office Plug Load Savings Assessment* (CEC-500-2011-010). Sacramento, CA: California Energy Commission.
14. Pixley, Joy E., Stuart A. Ross, Ankita Raturi, & Alan C. Downs. (2014). *A Survey of Computer Power Modes Usage in a University Population* (CEC-500-2014-093). Sacramento, CA: California Energy Commission.
15. 1E & Alliance to Save Energy. (2009). *PC Energy Report 2009: United States, United Kingdom, Germany.*
16. Pixley, Joy E., & Stuart A. Ross. (2014). *Monitoring Computer Power Modes Usage in a University Population* (CEC-500-2014-092). Sacramento, CA: California Energy Commission.
17. Gulbinas, Rimas, & John E. Taylor. (2014). Effects of real-time eco-feedback and organizational network dynamics on energy efficient behavior in commercial buildings. *Energy & Buildings, 84*, 493-500. doi:https://doi.org/10.1016/j.enbuild.2014.08.017
18. Mulville, M., K. Jones, G. Huebner, & J. Powell-Greig. (2017). Energy-saving occupant behaviours in offices: change strategies. *Building Research and Information, 45*(8), 861-874. doi:https://doi.org/10.1080/09613218.2016.1212299
19. Murtagh, Niamh, Michele Nati, William R. Headley, Birgitta Gatersleben, Alexander Gluhak, Muhammad Ali Imran, & David Uzzell. (2013). Individual energy use and feedback in an office setting: A field trial. *Energy Policy, 62*, 717-728. doi:https://doi.org/10.1016/j.enpol.2013.07.090
20. Nilsson, Andreas, Kristin Andersson, & Cecilia Jakobsson Bergstad. (2015). Energy behaviors at the office: An intervention study on the use of equipment. *Applied Energy, 146*, 434-441. doi:https://doi.org/10.1016/j.apenergy.2015.02.045
21. Kamilaris, Andreas, Jodi Neovino, Sekhar Kondepudi, & Balaji Kalluri. (2015). A case study on the individual energy use of personal computers in an office setting and assessment of various feedback types toward energy savings. *Energy and Buildings, 104*, 73-86. doi:https://doi.org/10.1016/j.enbuild.2015.07.010
22. Pollard, Carol Elaine. (2016). Up close and personal. *International Journal of Sustainability in Higher Education, 17*(1), 68-85. doi:https://doi.org/10.1108/IJSHE-03-2014-0043
23. Gandhi, Priya, & Gail S. Brager. (2016). Commercial office plug load energy consumption trends and the role of occupant behavior. *Energy & Buildings, 125*, 1-8. doi:https://doi.org/10.1016/j.enbuild.2016.04.057

24. Orland, B., N. Ram, D. Lang, K. Houser, N. Kling, & M. Coccia. (2014). Saving energy in an office environment: A serious game intervention. *Energy and Buildings, 74*, 43-52. doi:https://doi.org/10.1016/j.enbuild.2014.01.036
25. Hackel, Scott, Chris Plum, Maureen Colburn, Greg Marsicek, Thea Rozenbergs, Nicole Kessler, Rick Carter, & Lindsey Kieffaber. (2016). *Impacts of Office Plug Load Reduction Strategies. Quantifying plug load usage, the potential for reduction, and the impact on users (COMM-20140512-87091): Prepared by Seventhwave, Center for Energy and Environment (cee), and LHB for the Minnesota Department of Commerce, Division of Energy Resources. Retrieved from* https://www.cards.commerce.state.mn.us/CARDS/security/search.do?method=showPoup&documentId=%7B5A402E71-6933-4A8A-BFC4-D9BE445B4FD7%7D&documentTitle=358673&documentType=6
26. Abrahamse, Wokje, Linda Steg, Charles Vlek, & Talib Rothengatter. (2005). A review of intervention studies aimed at household energy conservation. *Journal of Environmental Psychology, 25*(3), 273-291. doi:https://doi.org/10.1016/j.jenvp.2005.08.002
27. Roberts, Simon, & William Baker. (2003). *Towards Effective Energy Information: Improving Consumer Feedback on Energy Consumption*. Bristol, UK: Centre for Sustainable Energy.
28. Vine, D., L. Buys, & P. Morris. (2013). The effectiveness of energy feedback for conservation and peak demand: A literature review. *Open Journal of Energy Efficiency, 2*(1), 7-15.
29. Armel, K.Carrie, Abhay Gupta, Gireesh Shrimali, & Adrian Albert. (2013). Is disaggregation the holy grail of energy efficiency? The case of electricity. *Energy Policy, 52*, 213-234. doi:https://doi.org/10.1016/j.enpol.2012.08.062
30. Gupta, Abhay, & Prateek Chakravarty. (2013). *Impact of Energy Disaggregation on Consumer Behavior* (White Paper). Sunnyvale, CA: Bidgely.
31. Fischer, Corinna. (2008). Feedback on household electricity consumption: a tool for saving energy? *Energy Efficiency, 1*(1), 79-104. doi:https://doi.org/10.1007/s12053-008-9009-7
32. Murugesan, Latha Karthigaa, Rashina Hoda, & Zoran Salcic. (2015). Design criteria for visualization of energy consumption: A systematic literature review. *Sustainable Cities and Society, 18*, 1-12. doi:https://doi.org/10.1016/j.scs.2015.04.009
33. Canfield, Casey, Wändi Bruine de Bruin, & Gabrielle Wong-Parodi. (2017). Perceptions of electricity-use communications: Effects of information, format, and individual differences. *Journal of Risk Research, 20*(9), 1132-1153. doi:https://doi.org/10.1080/13669877.2015.1121909
34. Karjalainen, Sami. (2011). Consumer preferences for feedback on household electricity consumption. *Energy and Buildings, 43*(2-3), 458-467. doi:https://doi.org/10.1016/j.enbuild.2010.10.010
35. Kluger, Avraham N., & Angelo DeNisi. (1996). The effects of feedback interventions on performance: A historical review, a meta-analysis, and a preliminary feedback intervention theory. *Psychological Bulletin, 119*(2), 254-284. doi:https://doi.org/10.1037/0033-2909.119.2.254
36. Schultz, P. Wesley, Jessica M. Nolan, Robert B. Cialdini, Noah J. Goldstein, & Vladas Griskevicius. (2007). The Constructive, Destructive, and Reconstructive Power of Social Norms. *Psychological Science, 18*(5), 429-434. doi:https://doi.org/10.1111/j.1467-9280.2007.01917.x
37. Asensio, Omar Isaac, & Magali A. Delmas. (2015). Nonprice incentives and energy conservation. *Proceedings of the National Academy of Sciences, 112*(6), E510-515. doi:https://doi.org/10.1073/pnas.1401880112
38. Steinhorst, Julia, & Christian A. Klöckner. (2018). Effects of Monetary Versus Environmental Information Framing: Implications for Long-Term Pro-Environmental Behavior and Intrinsic Motivation. *Environment and Behavior, 50*(9), 977-1031. https://doi.org/10.1177/0013916517725371
39. Pixley, Joy E., Sergio Gago-Masague, & Raquel Fallman. (2019). *Software Development Report for the Power Management User Interface Study*. Sacramento, CA: California Energy Commission.
40. Pixley, Joy E., Sabine Kunrath, Sergio Gago-Masague, Raquel Fallman, & G.P. Li. (2021). *The Power Management User Interface Study*. Sacramento, CA: California Energy Commission.

The Energy Impacts of Video Streaming Devices and Smart Speakers

Noah Horowitz and Gregg Hardy

Abstract Approximately 100 million smart speakers and video streaming devices have been installed in the US and many more around the world. NRDC and its consultant Pacific Crest Labs measured the energy use of these devices both in on and standby modes. We found that both types of devices were very energy efficient, and that they were able to achieve low standby power levels, around 1–3 W of power, even though they were continuously connected to the internet and had fast wake times.

Many of the newest TV models include the capability to wirelessly connect an adjacent smart speaker and to wake and control the TV through voice commands, negating the need to use the remote control. For a few of the models we tested, the TVs standby power use increased from its baseline of <1 W of power to roughly 20 W of power when linked to the smart speaker. As TVs spend most of its time in standby mode this could cause a TV's overall energy use to more than double. If one quarter to one half of TVs in the US operated this way, national annual TV energy use would increase by 1.3–2.6 billion dollars. Per conversations we had with the TV manufacturers and additional testing, we were able to confirm that TVs could be wakened and controlled by voice command for only 1–2 W.

We hope our paper serves as a wake-up call to the industry to focus on this issue and to optimize their products in order to avoid a world-wide increase of several billion dollars of annual electricity use and related pollution.

1 Introduction

Sales for new internet connected smart speakers and video streaming devices have skyrocketed in the past few years. However, very limited data is available on the power consumption of these devices when in on or standby power modes. To gain

N. Horowitz (✉)
Natural Resources Defense Council, San Francisco, CA, USA
e-mail: nhorowitz@nrdc.org

G. Hardy
Pacific Crest Laboratory, Portland, OR, USA

P. Bertoldi (ed.), *Energy Efficiency in Domestic Appliances and Lighting*, Springer Proceedings in Energy, https://doi.org/10.1007/978-3-030-79124-7_20

further information on the energy use of these products, the Natural Resources Defense Council (NRDC), a U.S. based environmental advocacy group and its consultant Pacific Crest Labs purchased the leading models for laboratory testing.

Smart speakers are a relatively new product category that was initially popularized by Amazon with the launch of its popular Echo product line. This product and similar models from Google, Apple and other manufacturers allow the user to say a command in the vicinity of the device such as "play songs by the Rolling Stones" or "what is the weather going to be tomorrow in New York City?" These devices are connected wirelessly to the internet and have built in speakers to play back music and other audio content.

Video streaming devices allow users to view an almost unlimited range of content on their television, both live TV and video from numerous "apps" such as Netflix or YouTube. These devices are physically connected to a TV and come in one of two forms, a small box or a thumb drive shaped "dongle". These devices are connected to the internet wirelessly or via an Ethernet connection and come with their own remote control.

Our study attempted to answer the following questions:

- How much power do some of the most popular models of smart speakers consume, both when they are playing music and when they are only in standby mode, awaiting a voice command? Are some models more efficient than others in standby mode? What are the national impacts of smart speakers in terms of total annual energy use and operating costs?
- How much power do some of the most popular models of video streaming devices consume, both when streaming video content accessed from the internet via a wireless connection and when in standby mode? Are some models more efficient than others in standby mode? What are the national impacts of video streaming devices in terms of total annual energy use and operating costs?
- What impact does linking smart speakers to new televisions have on energy use? For example, how does linking an Amazon Echo or Google Home smart speaker to a 2018 model-year television have on the television's standby power and overall annual energy consumption? If this feature is deployed widely with new televisions, what are the national environmental and economic impacts? And are there design options that could deliver this functionality with minimal incremental standby power use?

All of the models tested were purchased directly from retail outlets to ensure testing was done on the same models that consumers buy. Our study did not investigate or take into account the energy consumed by the modem and router equipment in a home that allows the above equipment to connect and receive information from the internet. We based this decision on the fact that users already had this equipment in their home and that it was already always on to access the internet for other purposes such as for reading, sending e-mails, and obtaining other content.

While this study focused on models available in the U.S., similar products are sold world-wide and we anticipate the general findings and recommendations to apply worldwide.

2 Smart Speaker Testing

For this study we tested the:

- Amazon Echo—2nd generation
- Apple HomePod
- Google Home
- Google Home Mini
- Harman Kardon Invoke

The first four represent some of the top selling models and represent the vast majority of sales, whereas the Harmon Kardon Invoke was selected as it represents Microsoft's entry in the market with its Cortana voice assistant software. Each of the products has a "voice assistant" feature which listens for user commands started by a key word. For example, the Amazon products are listening for the term Alexa, Apple products respond to Siri and Google products to the phrase OK Google. All of these devices remain in a standby state when they are not being actively used whereby they are always "listening" for a command and maintaining their network connection.

2.1 Smart Speaker Test Method

Each of these devices was purchased new and set up in its default configuration. If there were forced menu choices (e.g., options to select crash reporting or location services), we chose to opt in. That resulted in selecting the options most likely to consume the most power. Once the unit was set up and connected to a local wide area network (e.g., a live internet signal via a wireless router), we conducted the following measurements:

On mode testing—With a voice command, we asked each smart speaker to play news, classical music, and popular music. We selected a volume level representative of the level one would be likely to use while listening to music in the background (typically 3 on a scale of 10). News and music were played in equal proportions for several minutes, and the average power value for each of the three content types was calculated in watts. We used a simple average of these three power values to approximate the typical power used by each smart speaker in the "on" mode. Note: The power values reported were meant to estimate on mode power use. We recognize that measured power may vary depending on the specific content being played and the volume level selected.

Standby mode testing—While music or news was playing, we told the device to stop playing and then measured the device's standby power over an 18-h period. We monitored the variations in standby power use during this time and measured average standby power use after power use stabilized and remained essentially unchanged over time. (Note: Most, but not all, devices stabilized within a few minutes.)

Wake-time testing—We measured the time it took each speaker to respond to a voice command when the device was in standby mode. For this test, we asked the sleeping device, "What time is it?" and then started a timer, stopping when the speaker began to provide an audio response. This wake time is an indication of how quickly a device is able to power up from its low-power state.

Our intention when performing these tests was to understand how these products behave in typical real-world conditions rather than to strictly follow existing industry or government test methods that might deliver slightly greater precision during on mode testing but would require much more sophisticated procedures.

All testing was performed using a Tektronix PA1000 power meter calibrated in accordance with the International Electrotechnical Commission (IEC) standard IEC 17025 [1].

To calculate each smart speaker model's annual energy use (kWh/year), we applied a daily duty cycle of 1241 h/year in on mode and 7519 h/year in standby [2]. These values represent roughly 3.5 h/day of active usage and 20.5 h in standby. For video streaming devices we applied a duty cycle of 7 h/day in on mode and 17 h/day in standby, the same duty cycle used by the U.S. Environmental Protection Agency (EPA) in its ENERGY STAR program for set-top boxes (STB), to determine each model's annual energy use.

2.2 Smart Speaker Testing Results

The on and standby mode power levels observed during audio testing for each smart speaker are provided in Table 1.

All of the devices used little power to stream audio content. These values are quite impressive given that the devices are both maintaining an internet connection and delivering sound. While the Apple HomePod and Harman Kardon Invoke had slightly higher on-mode power levels, these devices may deliver louder bass or greater fidelity, which typically require some additional power.

Other than the Harman Kardon model, all of the units achieved less than 2 W of power in standby mode. The wake times that we observed were all within 1–2 s. These standby levels are quite acceptable given the devices' ability to always be listening for the wake command, such as Alexa, Google, or Siri; to maintain a connection to the home wireless network; and to have very fast resume times. We are unaware of a technical reason why the Harman Kardon Invoke and other smart speakers shouldn't be able to achieve a standby level of less than 2 W. It should be noted that the Harman Kardon speaker was the only device we tested that ran on Microsoft's Cortana voice assistant software, which might be the source of the slightly elevated standby power level.

Table 1 Average power use by smart speakers during on and standby modes

Product	On-mode power (W)	Standby power (W)
Google Home Mini	1.7	1.4
Amazon Echo (2nd Gen.)	2.4	1.6
Google Home	2.2	1.9
Apple HomePod	5.9	1.9
Harman Kardon Invoke	4.2	3.8

Energy impacts of smart speakers and video streaming devices, NRDC, August 2019 [4]

Table 2 Annual energy use, U.S. stock, and national energy use of smart speakers

Product	Annual energy use (kWh/year) per model	U.S. stock (millions)	National energy use (GWh)
Google Home Mini	12.3	4	49
Amazon Echo (2nd Gen.)	15.2	35	532
Google Home	17.1	8	137
Apple HomePod	21.6	3	65
Harman Kardon Invoke	33.4	Insig.	Insig.
Total		50	783

Energy impacts of smart speakers and video streaming devices, NRDC, August 2019

Table 2 provides each model's annual energy use, as well as the estimated number of units installed in the United States as of fall 2018. The installation numbers were obtained from Consumer Intelligence Research Partners (CIRP), and we assumed that the energy of the product we tested represents roughly the average of the product family reported by CIRP [3]. The annual energy use for these models ranged from 12 to 33 kWh/year, which is lower than the typical bookshelf/mini-speaker systems that were previously used and consumed around 75 kWh/year [4].

Per our modeling, we estimate that the smart speakers currently installed in America's homes consume a total of 783 gigawatt hours per year (GWh/year), meaning U.S. households are consuming $102 million worth of electricity per year to power these devices.

3 Video Streaming Device Testing

We tested these video streaming devices:

- Amazon Fire TV 4K Ultra HD HDMI Dongle
- Apple TV 4K set-top box
- Google Chromecast 4K Ultra HDMI Dongle
- Roku Ultra 4K/HDR set-top box

All of these devices except Google Chromecast are controlled by a separate remote control that comes with the streaming device. Google Chromecast enables users to stream (or "cast") video and music from laptops, tablets, and smartphones to a television without the use of a separate remote control. Casting involves starting a video on a mobile device and then clicking on a "casting" button that asks you which device you would like to cast to. Once the user selects the display device, the video appears there but can still be controlled from the mobile device. Given the rich user interface of mobile devices, this enables a user to more quickly and easily select and control content to play on his or her television. Audio casting offers similar capability. Chromecast dongles (roughly the size of thumb drive) host no apps; they only support media casting.

3.1 Video Streaming Device Test Method

For the on-mode power measurement, we recorded for 2 min the power levels while an SDR show from Netflix was playing.[1] We then stopped the show via the remote control, turned off the television via its remote control, and continued to measure the streaming device's power for approximately 8 h. We analyzed the data, determined when stabilization occurred, and reported the average standby power level after stabilization.

We also measured the wake time from a video streaming device remote control command to the appearance of its home screen on a connected TV.

3.2 Video Streaming Device Test Results

Table 3 shows the on and standby power level testing results and time it took to display the home screen.

The on-mode power levels ranged from 2.3 to 3.3 W. The standby power levels ranged from 0.7 to 2.7 W. The Google and Roku products had higher standby levels (more than 2 W) than the other devices, which were less than 1 W. Although not a huge difference, the extra watt of power that is consumed whenever the user is not streaming content adds up at the national level, given that more than 100 million streaming devices are in use in the U.S.

Using the duty cycle of 7 h on, 17 h standby, the annual energy use for these devices ranged between 11 and 25 kWh/year, which is quite similar to the values we saw for the smart speakers we tested.

[1] Most of today's content is formatted as SDR. High Dynamic Range (HDR) formats, such as Dolby Vision, offer new, enhanced contrast.

Table 3 Annual power use by video streaming devices during on and standby modes, start-up latency, and annual energy use

Product	On-mode power (W)	Standby power (W)	Time to home screen (s)	Annual energy use (kWh/year)
Amazon Fire TV	2.3	0.9	1.8	11
Apple TV	2.9	0.7	6.7	12
Google Chromecast Ultra	2.6	2.2	1.0	20
Roku Ultra	3.3	2.7	6.2	25

Energy impacts of smart speakers and video streaming devices, NRDC, August 2019

Table 4 Annual energy use, U.S. stock, and national energy use of leading video streaming devices

Product	Per unit annual energy use (kWh/year)	U.S. stock (millions)	National energy use (GWh/year)
Amazon Fire TV	11	11	122
Apple TV	12	9	101
Google Chromecast Ultra	20	6	128
Roku Ultra	25	15	376
Total		41	727

Energy impacts of smart speakers and video streaming devices, NRDC, August 2019

To determine national impacts, we gathered data on the number of these devices installed in the United States as of the first quarter of 2018 [5]. The total amount of electricity consumed by these devices nationally is approximately 727 gigawatt hours per year (GWh/year). That translates to around $94 million in annual consumer electricity costs (Table 4).

If all of these devices in the U.S. were able to achieve 1 W in standby, it would save 204 GWh/year (worth $26 million annually) on consumers' utility bills and prevent 143,948 metric tons of carbon dioxide equivalent (CO2e) emissions per year.

4 Linking a Television to a Smart Speaker and Operating It Through Voice Commands

As we began our study we became aware of a new and exciting feature that enables the user to link a smart speaker to a television. Once the initial setup is completed, the user can, with certain new televisions from the 2018 model year and later, control the television via his or her voice rather than a remote control. In a few limited cases, the user can also wake the television from standby mode simply with a voice command.

Table 5 Television models tested

Product	Model	Size	Technology	Operating system
2018 LG	OLED55C8PUA	55″	OLED	WebOS
2018 LG	SK8000PUA	49″	Edge-lit with local dimming	WebOS
2018 Samsung	QN55Q8FN	55″	QLED full Array with local dimming	Tizen OS
2018 Samsung	55NU8000	55″	Edge-lit with local dimming	Tizen OS
2018 Sony	XBR55X900F	55″	Full Array with local dimming	AndroidTV
2018 Vizio	P55-F1	55″	Full Array with local dimming	SmartCast
2017 TCL	55S405	55″	Full Array, no local dimming	Roku TV
2017 Westinghouse	WA50UFA1001	50″	Edge-lit, no local dimming	Amazon Fire

Energy impacts of smart speakers and video streaming devices, NRDC, August 2019

As part of our research we decided to also study what impact, if any, this new capability might have on a television's standby power use and resulting annual energy use. Although today's televisions typically have low standby power levels (less than 0.5 W), it was possible the numbers could go up considerably if a television had to keep additional circuits "awake" in order to listen for a voice command.

To perform this testing, we accessed eight different 2017 and 2018 television models previously purchased by the Northwest Energy Efficiency Alliance (NEEA) for its own energy testing. The models are listed in Table 5. They represent a range of brands and corresponding operating systems, which are potential determinants of the television's standby power level when linked to a smart speaker.

4.1 Test Method for Measuring TV Standby Power When Linked to a Smart Speaker

We attempted to link the Amazon Echo 2nd Generation Speaker and the Google Home Speaker to each of the above televisions. (Neither the Apple HomePod nor the Harman Kardon Invoke supported this capability.) It took time to navigate the user menus and, in some cases, required calling customer support to link the television and smart speakers together. We made sure the televisions and speakers were connected to a wide area network and a live internet signal before conducting testing. Otherwise the televisions were in default configuration per the U.S. Department of Energy TV power measurement test method.

When setting up each television for the first time, we accepted any software updates if prompted. We then elected to disable future software updates, so we could repeat our testing without the variability that new updates could cause. All testing was done in the fall of 2018.

For each television, we performed a standby power test before we linked it to a smart speaker. We then turned off the television by pressing the power button on the

TV's remote control. We measured the television's standby power overnight in the same way that we tested smart speakers and OTT boxes.

We then linked each television to a smart speaker and measured its standby power over a period of 2 h. We chose this shorter testing interval after preliminary tests suggested that television power levels were fairly stable in this network-standby mode.

4.2 Results and Potential Impacts

Table 6 summarizes the results for standby power tests conducted on the Amazon Echo smart speaker where in some cases the user can both wake and control (e.g., change channel, adjust volume, etc.) the TV via voice commands, while in other instances the user could control the TV but not wake it by voice command.

We also tested televisions with the Google Home speaker. Table 7 summarizes results for models we were able to successfully connect to and control with this smart speaker.

The key findings from this testing:

- The televisions that could be controlled, but not wakened, by voice command were able to maintain low standby levels of less than 0.5 W while preserving their normal wake times (less than 10 s).

Table 6 Results from interactions tests with TVs and Amazon Echo smart speaker (SS)

Product	Echo 2nd generation
Samsung 8000	Control only TV: 0.4 W SS: 1.6 W
Samsung Q8	Control only TV: 0.3 W SS: 1.6 W
LG OLED	Control only TV: 0.2 W SS: 1.6 W
TCL	Control only TV: 0.3 W SS: 1.6 W
Sony	Wake & control TV: 21.2 W SS: 1.6 W
Westinghouse	Wake & control TV: 22.9 W SS: 1.6 W
Vizio	Wake & control TV: 19.8 W SS: 1.6 W

Energy impacts of smart speakers and video streaming devices, NRDC, August 2019

Table 7 Results from interactions tests with TVs and Google smart speaker (SS)

Product	Google Home
LG OLED	Control only TV: 0.3 W SS: 1.9 W
LG SK	Control only TV: 0.3 W SS: 1.9 W
Sony	Wake & control TV: 21.1 W SS: 1.9 W
Vizio	Wake & control TV: 18.8 W SS: 1.9 W

Energy impacts of smart speakers and video streaming devices, NRDC, August 2019

- Each of the televisions that supported both wake and control capabilities resulted in standby power levels between 18.8 and 22.9 W on average. This represents a roughly 20-W increase in standby power from its less than 1 W level and more than doubles the annual energy consumption of many TVs.

Given that a typical television spends around 19 h/day in standby mode and only 5 h/day in on mode, a significant jump in standby power mode can dramatically affect a television's overall annual energy use [6]. In Fig. 1 we show how linking a television that can be wakened by voice control to a smart speaker can produce high television standby power levels. Here the television's annual energy use jumps from 106 kWh/year all the way to 248 kWh/year, more than doubling. To put that into perspective, two of these TVs would consume as much energy per year as a new U.S. household refrigerator which consumes around 500 kWh/year.

4.3 *Potential National Impact of Poor Wake by Voice Implementation in TVs*

Although the combined feature of waking and controlling a television by voice is just beginning to come on the market, we understand that it will become much more common in the 2019 models that will be widely available in the second half of the year. If this trend continues and manufacturers fail to optimize their televisions for low standby power levels when linked to smart speakers, we could see large increases in national energy consumption. If one-fourth to one-half of all televisions in the United States have this feature and a 20.8-W increase of standby power, we would see an increase in national annual energy consumption of 9709–19,418 GWh, which is equivalent to the annual electricity generation of three to six large (500-MW) power plants and more than the annual electricity consumption of all the households in the city of Houston. This extra electricity use would add $1.3 to $2.5 billion to national household electric bills.

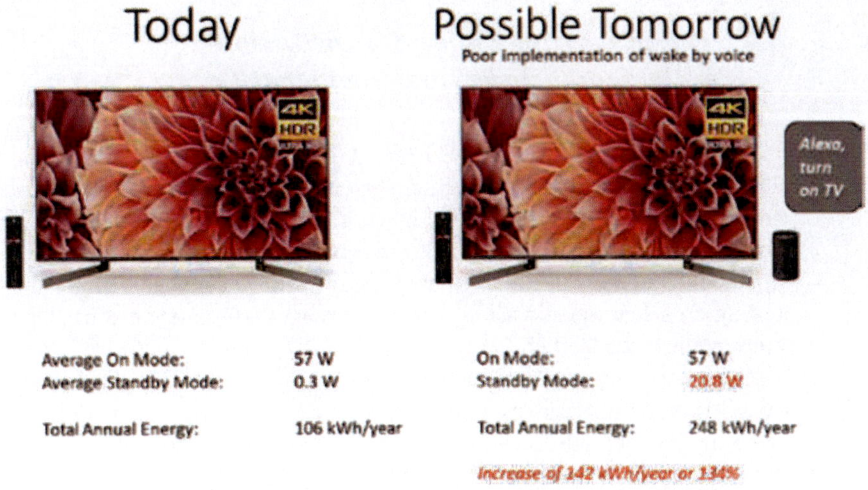

Fig. 1 Energy impact of wake by voice feature in TVs linked to a smart speaker

We have presented this information in informal individual discussions with some of the leading television manufacturers and hope it serves as a wake-up call, drawing proper attention to this issue as the industry increasingly incorporates the wake-and-control feature and designs future televisions. We believe television manufacturers should be able to deliver this great new feature for only 1–2 W of television standby power. This could be accomplished in a number of ways, including the addition of a low-power network proxy chip or similar capability in a system-on-chip block with power islanding, whereby all other components/features such as the television tuners are turned off.

We are confident that a low-standby power engineering solution exists because the Amazon Fire TV dongle can act as a network proxy that wakes the television from low-power sleep via HDMI Consumer Electronics Control (CEC). If one uses the Fire TV dongle with, for instance, the Sony television, the two products use a total of 1.4 W in standby—0.5 W for the television and 0.9 W for the Fire TV dongle.

We understand that adding components increases cost. Our hope is that manufacturers add this capability for little cost in the system on a chip as the smart speaker use case gains market penetration.

Another implementation we might see in the future are Internet-connected televisions that come with full smart speaker–type capabilities built into them, eliminating the need to purchase and install a smart speaker near the television to achieve hands-free operation. We believe that these implementations can achieve low-power standby as well. Amazon provides many low-power reference designs that manufacturers can leverage to develop Alexa-enabled products with low standby power. Other voice assistant developers may provide low-power reference designs as well. The ball is in the court of the TV manufacturers.

4.4 Learnings from Follow-Up Discussions with Manufacturers and Additional Testing

In March and April 2019, NRDC reached out to several television manufacturers to share our testing results and preliminary analysis. During these discussions we also sought to gain feedback from the manufacturers, as well as any information they could share about the future of wake and control features in new televisions and any potential reductions in standby power. Afterward we conducted targeted additional testing on a few models using each one's most up-to-date software. During these conversations we learned:

Sony has developed and pushed a software update to its 2018 model year televisions. After our discussions with Sony, we accepted the software update and retested the Sony model XBR55X900F. The standby power level for this television, when connected to the Amazon Echo and with the wake and control command feature enabled, was 8.2 W, down from the 21.2 W observed during our initial testing prior to the software update. We were unable to connect this television to our Google Home smart speaker.

Vizio informed us that the model we had tested, the P Series P55-F1, is one of the company's high-end models, and the high standby power level of 18–19 W when connected to Amazon Echo or Google Home was due to the high-power chip set used in the P Series televisions. The company said we should see much lower standby power levels for its more mainstream, higher-selling units. To confirm this, we purchased an entry-level model (Vizio V505-G9 SmartCast) and performed testing after connecting it to the Amazon Echo and Google Home Mini. The results of our testing are shown in Table 8.

Vizio advertises that its SmartCast televisions, which include the new model we purchased, can be turned on and off and controlled with both Amazon Echo and Google Home smart speakers. However, we were unable to achieve full functionality with the Echo speaker. We could wake the television with the Echo but could not control it or turn it off. To determine if we could control the TV, we attempted to change the volume and to start YouTube and Netflix. We were surprised to find that we could wake the television by voice from a 1-W sleep with the Quick Start feature turned off, even though the setup instructions stated that we would have to turn on Quick Start to achieve wake-by-voice functionality. This is evidence that a television can be turned on by voice command for less than 1 W from sleep. The sleep power level should have nothing to do with whether television control works when it's on. We believe that with software bug fixes to its on-mode control, Vizio could fully enable Echo wake and control with a standby power level of less than 1 W.

Voice control with the Google Home Mini smart speaker behaved as expected. With Quick Start enabled we could wake, control, and turn off the television by voice. With Quick Start disabled we could still control and turn off the television by voice command, but we could not wake it by voice. We are optimistic that a software upgrade could enable wake and control of this television with both Amazon Echo and Google Home smart speakers.

Table 8 Wake-by-voice characteristics of Vizio P55-FI TV with Echo and Google smart speakers

	Echo 2nd generation		Google Mini	
Characteristics	QS on	QS off	QS on	QS off
Television standby power (W)	8.3	0.9	8.3	1.0
Wake by voice time (s)	7	10	4	N/A
Wake television by voice?	Yes	Yes	Yes	No
Change volume by voice?	No	No	Yes	Yes
Turn off television by voice?	No	No	Yes	Yes

Energy impacts of smart speakers and video streaming devices, NRDC, August 2019

Table 9 Wake-by-voice characteristics of Lg SK8000PUA TV with Amazon Echo smart speaker

	Echo 2nd generation
Characteristics	QS on
Television standby power (W)	0.5
Wake by voice time (s)	7
Wake television by voice?	Yes
Change volume by voice?	Yes
Turn off television by voice?	No

Energy impacts of smart speakers and video streaming devices, NRDC, August 2019

LG 2018 models that we tested in the fall of 2018 with June 2018 software provided only voice control to operate the television once it was turned on manually. But with April 2019 software installed, the LG SK8000PUA was able to wake by voice command after being in a 0.5-W sleep state overnight, as shown in Table 9.

These results were extremely encouraging, as the Vizio and LG televisions were able to achieve less than the 1-W standby level we are recommending to the industry while providing user-acceptable wake times. In addition, in each case this was achievable through a software update alone.

Since we performed this testing, LG has removed the wake-by-voice capability for its models introduced in 2018. We were unable to learn from LG why this decision was made. We could not perform additional follow-up testing on the 2019 models from LG or other manufacturers as they were not available prior to the initial writing of this report.

5 Conclusion and Recommendations

Since new smart speaker and entertainment-related devices studied in this report will only continue to grow in popularity both in the United States and internationally, it's essential for manufacturers to ensure their products use as little electricity as possible, especially when they are waiting to be used and in standby mode. The greatest priority should be placed on new television designs to ensure that television standby power levels remain low even when linked to a smart speaker and when capable of being wakened and controlled through voice commands. Doing so will avoid unnecessary increases in consumer energy bills and electricity generation, which will decrease the likelihood of more power plant pollution that leads to health problems and warms our climate.

5.1 Recommendations

- TV manufacturers should design their models to achieve 1 W in standby mode when configured to wake and be controlled through a linked smart speaker.
- Smart speaker manufacturers should achieve standby power levels of less than 2 W, like the vast majority of today's models already achieve.
- Video streaming device manufacturers should achieve standby power levels of no more than 1 W, like the levels already attained by Apple's and Amazon's recent products.
- Consumers who have streaming devices and subscribe to pay TV from their cable or satellite company should look to download the app from their service provider and return their set top box. Via the apps, consumer will receive the same content and experience, and avoid the extra energy use and associated utility bill costs.
- Regulators and standards bodies should review and update their test procedures and policies (labels and efficiency standards). Test methods need to better reflect real world usage and capture typical on and standby mode power levels. (Today's test methods fail to connect the tested device to a live internet signal, and do not require devices like smart speakers to be connected to the TV before testing.)

References

1. ANSI Webstore, "ISO/IEC 17025:2017," https://webstore.ansi.org/standards/iso/isoiec170252017?gclid=Cj0KCQjw6cHoBRDdARIsADiTTzZO5w7Ms-V1eVGwQMkQXqhVBnybU2JcJA0pxMnbwnQx_-I3m0Xa3JwaAjsGEALw_wcB.
2. Duty cycle figures are derived from assumptions for shelf speakers in Bryan Urban, "Consumer Electronics in U.S. Homes: Energy Use in 2017," Fraunhofer USA presentation at Energy Efficiency in Domestic Appliances and Lighting (EEDAL) 2017 annual meeting, Irvine California, September 13–15, 2017, https://www.cse.fraunhofer.org/hubfs/2017%20-%20Urban%20-%20EEDAL%20-%20CTA%20Energy%20in%20Homes.pdf.
3. John Koetsier, "Amazon Echo, Google Home Installed Base Hits 50 Million." Forbes.com, August 2, 2018. https://www.forbes.com/sites/johnkoetsier/2018/08/02/amazon-echo-google-home-installed-base-hits-50-million-apple-has-6-market-share-report-says/#2f35c4e4769c.
4. Natural Resources Defense Council, "Energy Impacts of Smart Speakers and Video Streaming Devices," August 2019. https://www.nrdc.org/sites/default/files/gadget_report_r_19-07-b_13_locked.pdf.

5. "Parks Associates: Nearly 40% of U.S. Broadband Households Own a Streaming Media Player," Cision PR Newswire, May 31, 2018, https://www.prnewswire.com/news-releases/parks-associates-nearly-40-of-us-broadband-households-own-a-streaming-media-player-300657338.html.
6. Per the duty cycle assumptions in the DOE test method. See "Energy Conservation Program: Test Procedures for Television Sets," Federal Register, October 25, 2013, https://www.federalregister.gov/documents/2013/10/25/2013-24346/energy-conservation-program-test-procedures-for-television-sets.

Impact of New Approaches to Address Energy Management Gaps on Total Energy Use in Computer Workstations

Michael J. Klopfer, Joy E. Pixley, Saniya Syed, and G. P. Li

Abstract Power management in computer workstations commonly relies on user behavior, typically involving the continual assessment of keyboard and mouse usage combined with timers to initiate sleep/standby power management actions. Specifically, if the user does not provide input on the keyboard or mouse within the timer period, then the user is not considered to be present, continued computer use is not expected, and the device automatically initiates entering sleep/standby mode. In this indirect manner, the presence of the user is determined, and accordingly, the intent of the user to continue employing the device is inferred. In this study, we investigated the effectiveness of power management in desktops and the impact of alternative sensing and control approaches for workstation power management. This study used a hybrid model approach where the dataset from a 115 subject real-world, observational study was used to seed the behavioral model to evaluate savings potential of traditional power management with different modeled strategies in addition to modeled energy savings obtained by using an independent, USB-based power management motion sensor device. Such a device provides sleep triggers based on a motion sensing and an independent power management timer to determine the presence of the user and trigger initiation of sleep as opposed to using a keyboard/mouse. CalPlug modeled and compared energy management capabilities for both systems. The USB motion sensor device produced between 12% and 67% energy savings with action on two specific mechanisms: (1) elimination of sleep blocking events that prevent normal entrance into sleep states, and (2) prevention of users unintentionally disabling or setting sleep to extended periods and leaving them this way permanently. This sleep blocking effect was observed during 13.5% of all idle periods where sleep would have occurred sooner. Findings from this work highlight the continued concern with sleep blocking affecting the operation of

M. Klopfer (✉) · J. Pixley · G. P. Li
California Plug Load Research Center, University of California, Irvine, CA, USA
e-mail: mklopfer@uci.edu; jpixley@uci.edu; gpli@calit2.uci.edu

S. Syed
California Plug Load Research Center, University of California, Irvine, CA, USA

D+R International, Silver Spring, MD, USA
e-mail: ssyed@drintl.com

P. Bertoldi (ed.), *Energy Efficiency in Domestic Appliances and Lighting*, Springer Proceedings in Energy, https://doi.org/10.1007/978-3-030-79124-7_21

computer power management as well as the need for judicious sleep management settings and the potential for independent systems to improve computer energy management by reducing the impact of sleep blocking events.

1 Introduction

Computer workstations are plug loads that consume significant amounts of energy in both residential and commercial settings. Prior investigations have shown that power management is often enabled by default for users, but these settings can become disabled inadvertently, contributing to a substantial increase in energy usage [1, 2].

Several studies have performed physical setting audits or energy measurements of computers and other electronic office equipment in situ, in commercial and university buildings [2–5]. These audits and monitoring studies have found that, in practice, a high percentage of computers were left on unnecessarily when not being actively used. With better power management practices, substantial energy savings could be possible (e.g., [6–9]). Multiple studies have shown that in the absence of office policies or IT control of computers, the majority of office desktops have their computer sleep settings disabled [1, 10, 11]. Similarly, it has been shown that in a number of cases, workstations can experience power management that fails to act as required when it is in an unmanaged state. This may be due to legitimate processes such as updates, backups, virus scans, and actively playing audio/video. In other, illegitimate cases, desktop and similar types of un-required notifications to users from the operating system, un-observed playing media, site contents in open browsers, etc. may cause the countdown timer responsible for activating power management sleep states to be reset. Power management launch issues can also be due to the unwanted action of programs, peripherals or utilities operating on the workstation which prevents sleep modes from being activated.

In this study, we sought to look deeper into the prevalence of sleep blocking and external means that could mediate power management for workstations to augment onboard power management capabilities. In typical implementation, the sleep-state based power management system on workstations involves the use of countdown timers (often referred to as a power management or sleep timer) which are reset via an action, typically a keyboard or mouse movement indicating user input. Timers can be suspended either legitimately or illegitimately to delay entering sleep mode. An example of legitimate sleep blocking would be a video display application that prevents sleep from occurring when a user is watching video. Contrarily, illegitimate sleep blocking would be a hidden tab or peripheral preventing sleep from occurring. This illegitimate sleep blocking also can encompass low-frequency actions of programs or peripherals causing unwanted system wakeups, e.g. frozen or looped tasks, active windows, etc.

Several products exist in the marketplace that use either centrally managed or independent workstation managed approaches to power management. For centrally controlled approaches, an onboard utility daemon and a supervisory system independently tracks activity and mediates the workstation entering sleep in non-activity scenarios. An alternative approach uses a locally placed sensor to identify motion in the workspace to indicate user presence in front of the workstation and, accordingly, manage entering sleep in non-activity scenarios. An example of this type of device is a USB-based motion sensor. Both systems can be used in conjunction with an advanced power strip (APS) to provide behavior mediated shutdown of workstation peripherals (Tier-2 APS type control).

In office settings, promoting power management best practices either by policy or centralized power management control is a component of the problem. Two general classes of approaches have been used: empowering the user to make better decisions or using tools to help improve or retain power management effectiveness without user involvement. Both approaches have been demonstrated as effective [2, 12].

2 Materials and Methods

2.1 *Idle Period Determination and Power Management Classification*

The 115 subject 2014 dataset for the CalPlug monitoring study [1] was reanalyzed to identify patterns in regard to the duration of inactivity periods. This dataset was originally collected using Verdiem Surveyor software (Seattle, WA, USA) and, in segments of 15-min-long reporting periods, the data captures computer status including: On (active or idle), Sleep, or Off states. The CalPlug developed the Marginal Intervention Savings of Energy Reporter (MISER) (available at https://github.com/CalPlug/MISER) tool which was used to tabulate the individual contiguous idle periods for each day for all subjects whose computer was active for any period during the day [13]. Making the assumption that power management prevents idle periods longer than the sleep timer duration except in limited cases, the duration of periods longer than the sleep timer duration is subset out of the total idle periods. Cross-referencing this against the workstations with power management enabled (20 subjects) allows the calculation of the following metrics:

1. *Incidence percent (%)*: Ratio of the number of idle periods greater than the sleep timer setting divided by the total number of idle periods for a single subject across all study days. This metric provides the frequency of incidences where idle periods are greater than the timer duration compared to all idle periods. In normal usage this ratio should typically be low.
2. *Period average ratio*: Ratio of the average period length between the idle periods longer than the timer and the average of all idle periods for a single subject across all study days. This metric provides the comparative average duration for

the two sets. This ratio provides a qualitative metric for how distinct the subset of extended idle periods is compared to the full dataset.

3. *Average power management overage period (min/day)*: Average period of time per day for all study days where the workstation was in an idle state longer than the sleep timer duration. This metric can be used to categorize the general effectiveness of the power management settings.

2.2 Marginal Interventional Savings Calculation

The CalPlug MISER tool was used with the same dataset to calculate potential energy savings based on a strictly applied power management timer duration. By varying the "interventional" power management setting, the difference in savings performance between different interventional settings, in addition to onboard power management performance, can be evaluated. This approach allows the calculation of marginal savings performance. Savings is expressed at this state in minutes per period. To calculate energy savings, state energy usage on a per-state (On [Active or Idle], Off, or Sleep) basis provides baseline energy use values (per Eq. 1). Change in states of operation can be considered adding or subtracting time spent in each of these states. Conversion of time spent in On [Idle] to Sleep is considered savings.

$$Energy\ Baseline\left[\frac{Wh}{period}\right] = T_{On}\left(P_{On}\right) + T_{Off}\left(P_{Off}\right) + T_{Sleep}\left(P_{Sleep}\right)$$

$$Yearly\ Savings\ Change\left[\frac{kWh}{year}\right] = \left(S_{idle} * \frac{365.25\ days}{year * 60\ \min} * \frac{P_{On}}{1000}\right) - \left(S_{idle} * \frac{365.25\ days}{year * 60\ \min} * \frac{P_{Sleep}}{1000}\right) \quad (1)$$

The MISER program provides daily usage information and an Excel calculator is used to determine yearly energy usage or savings. As a result of startup delays and limited observed use, only activation of sleep mode was considered; Off was not considered a state valid for savings conversion in this estimation.

The use of a Tier 1 Advanced Power Strip in combination using power sensing to turn off peripherals when an attached computer enters sleep mode can provide Tier-2 type control for device peripherals by eliminating both standby load along with active primary device load. An estimation of peripheral savings can also be calculated by modeling a Tier 2 approach. Estimation of savings requires knowledge of time spent in active versus off/standby for each peripheral. As limited information is available in modeling, an estimation factor was used in calculation to provide the average On versus Off state power.

2.3 *Sensing and Occupancy Detection Comparison*

Evaluating occupancy sensing as a factor was based on the operational sensing provided by an Onset HOBO UX90-006 motion/light sensor and a HOBO UX120-018 plugload meter (Onset Computer Corp., Bourne, MA). This workstation was using CalPlug's PMUI power management software to provide comparative state transition information [2]. The study was performed in a 10′ × 10′ isolated office with no person traffic except for the test user.

3 Results

3.1 *State Usage Summary*

As reported in the original study, only 20 of 115 (17.4%) evaluated university workstations in the study were determined to have power management active by evaluation. The remainder of the evaluated workstations were found without power management enabled. MISER was used to tabulate the average percentage of study time and corresponding standard deviation for each workstation in the study, providing additional granularity beyond previously published results for the monitoring study from which the dataset was sourced. Periods of time where the reporting utility could not determine a state or sub-state are marked as "Unknown". Typically these periods exist in the dataset near transitions or during partially observed study days. During weekdays, the active period of 13.2% is substantially lower than the 33.3% the 24 h day that the 8 h work day corresponds to. Workers are, on average, using their computer actively for a cumulative period much shorter than the 8 h work day (Fig. 1). The time in the Sleep state on average is reduced during the weekend compared to the weekday likely due to workstations being turned off by users during this period. Similarly, the drop in active usage for weekends compared to weekdays does not strongly convert to extended idle or sleep during weekends, but likely is encapsulated in increased workstation presence in the Off state during the weekends. The idle periods are generally short in length with large numbers of small periods and a few large periods corresponding to the length of 1 day (see [1]). Due to the nature of the analysis, contiguous idle periods between the end of one workday and the start of another are broken into subcomponents each of approximately 7–8 h in length (420–480 min). This can be observed in Fig. 2. As this length is longer than any common power management timer setting, this artifact does not impact calculations. Consequently, day breaks of idle periods in calculation can contribute approximately 1.0% error per event (Table 1).

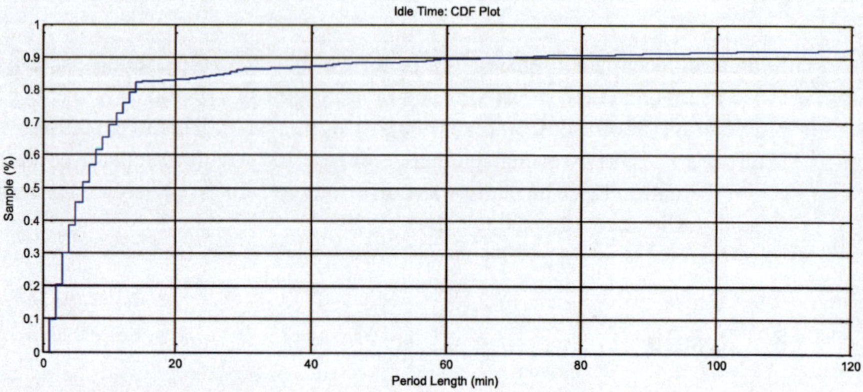

Fig. 1 Cumulative distribution function showing idle periods up to 120 min in length for all study workstations (n = 115). The vast majority of periods are under 20 min in length. At 120 min, only 7% of all samples are still unaccounted

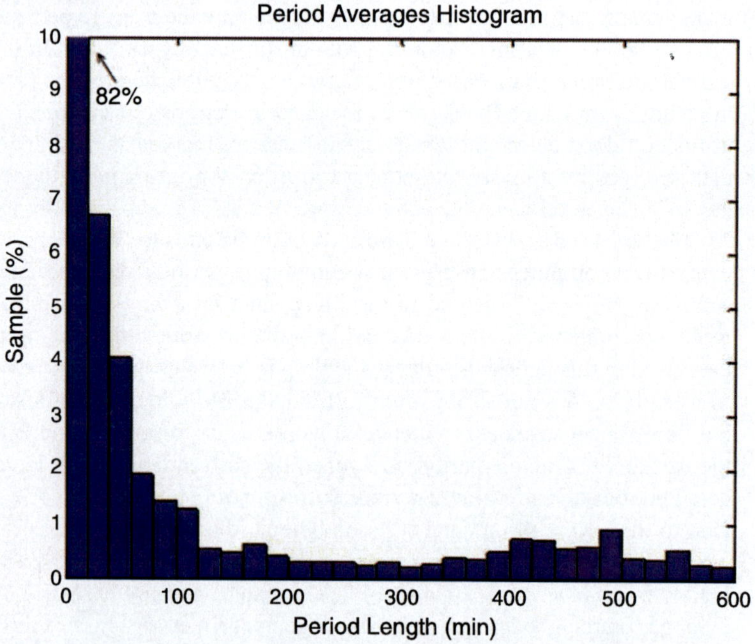

Fig. 2 Idle period length across all days and subjects from lengths 5 to 600 min in 20 min histogram bins. A local maximum exists near 480 min which corresponds to 8 h (evocative of the period of a full work day) but also coincidentally the approximate difference between the start of a day and the start of the workday and the end of the workday and the end of a day, creating an artifact

Table 1 The time spent in each state for the 115 observed office desktops in the monitoring study with sub-states shown

	Weekday average		Weekend average		Overall average	
Computer state	Percent	s.d. (%)	Percent	s.d. (%)	Percent	s.d. (%)
On	77.7	31.0	68.7	41.7	75.1	34.6
User active	*13.2*	*7.4*	*1.0*	*3.3*	*9.7*	*8.5*
User idle	*64.0*	*31.3*	*66.3*	*42.4*	*64.6*	*34.9*
User unknown	*0.5*	*3.6*	*1.5*	*2.3*	*0.8*	*1.8*
Sleep	8.2	20.0	6.9	21.8	7.8	20.6
Off	11.8	22.3	21.0	36.1	14.4	27.3
Unknown	2.2	10.1	3.4	14.0	2.64	11.4

3.2 Observed Power Management Performance

Of the 20 workstations with power management enabled, a large disparity exists in observable performance. The averaged sleep power management setting is 14 min while the average overage period in minutes per day for each workstation beyond the power management setting was 258.4 ± 118.4 min (95% confidence interval-CI) (see Table 2). For a hypothetical workstation that consumes 30 W in active mode and 1 W in sleep, this corresponds to a yearly added consumption of 45.6 ± 20.9 kWh (95% CI). A total of 4 of the 20 evaluated workstations in the study (20%) were operational with greater than 700 min (approximately 11.5 h) per day of energy usage beyond the power management setting. As the average number of days evaluated with operation (observed On state) was 54.8, this is clearly an incidence of power management improperly functioning. As extended idle periods were observed throughout the study period for these subjects, this was likely not due to the user changing the power management period mid-study (Fig. 3).

Excluding these four grossly underperforming workstations, the updated average per day overage is 26.0 ± 20.4 min (95% CI). A large standard deviation can be observed as performance within this group was divergent and overage periods ranged from 0 to 555 min/day. Using the same computer state values as before in calculation, this corresponds to 4.6 ± 3.6 kWh (95% CI) per year. The disparity shows energy usage has a strong correlation with power management performance. Furthermore, substantial failures, when they do occur, can be severe and contribute to substantially high energy usage. Omitting the systems with observed substantial overages due to catastrophic power management dysfunction, sleep blocking (both legitimate and illegitimate) contributes to (on average) 1.8% of added daily operation time due to delays in sleep. The set time for the power management sleep timer did not appear to strongly correlate with general performance of overage period average number per day or length.

Table 2 Numerical comparison of a subset of 20 subjects with power management enabled in the 2014 monitoring study

	Subject 42	Subject 83	Subject 52	Subject 71	Subject 43	Subject 29	Subject 79	Subject 11	Subject 66	Subject 3	Subject 107
Computer PM setting	9	9	9	9	10	10	11	11	11	25	30
Overage average period (min)	310.06	11.15	2.11	20.55	2.39	4.53	46.62	365.12	318.20	336.41	6.00
Number of overages	161.00	125.00	53.00	295.00	18.00	19.00	455.00	214.00	152.00	148.00	1.00
Incidence%	36.10%	21.89%	15.32%	21.63%	5.90%	18.63%	21.18%	42.71%	27.59%	8.88%	1.09%
Average period ratio	2.64	3.72	0.38	2.14	0.49	0.75	2.90	2.23	3.38	9.07	0.65
Overage per day (min)	1062.13	58.08	4.31	110.22	1.87	6.62	275.45	1240.24	1179.68	732.18	0.43
Idle periods per day	9.49	23.79	12.81	24.80	12.71	7.85	27.54	7.95	13.44	0.01	6.57
Overage periods per day	3.43	5.21	1.96	5.36	0.75	1.46	5.83	3.40	3.71	2.18	0.07
On-state days	47	24	27	55	24	13	78	63	41	68	14
Total study days	51	37	39	78	68	71	98	63	41	70	38
Workstation type	Win.	Win.	Win.	Win.	Win.	Win.	Mac.	Win.	Win.	Win	Win.

Subject 30	Subject 61	Subject 101	Subject 21	Subject 97	Subject 86	Subject 90	Subject 35	Subject 46	Average	Std. dev.	Median
30	30	30	60	60	120	180	180	240	53.70	69.37	27.50
48.12	20.20	28.79	26.33	176.63	6.67	0.00	16.00	0.00	87.29	131.79	20.37
78.00	5.00	29.00	12.00	160.00	3.00	1.00	2.00	1.00	96.60	120.52	41.00
4.19%	0.93%	2.98%	2.34%	16.16%	0.37%	0.18%	0.38%	0.19%	12.43%	12.94%	7.39%
4.57	2.38	2.26	2.85	3.42	1.11	0.00	0.51	0.00	2.27	2.10	2.25
68.24	3.37	15.46	15.80	392.50	0.67	0.00	0.76	0.00	258.40	429.77	15.63
33.82	17.83	18.00	25.60	13.75	26.83	15.49	12.67	18.52	16.47	8.49	14.62
1.42	0.17	0.54	0.60	2.22	0.10	0.00	0.05	0.00	1.92	1.95	1.44
55	30	54	20	72	30	35	42	29	41.05	19.59	38.00
55	39	76	34	89	43	49	68	52	57.95	18.71	53.50
Win.	Win.	Win.	Win.	Win.	Win.	Win.	Win.	Win.	N/A	N/A	N/A

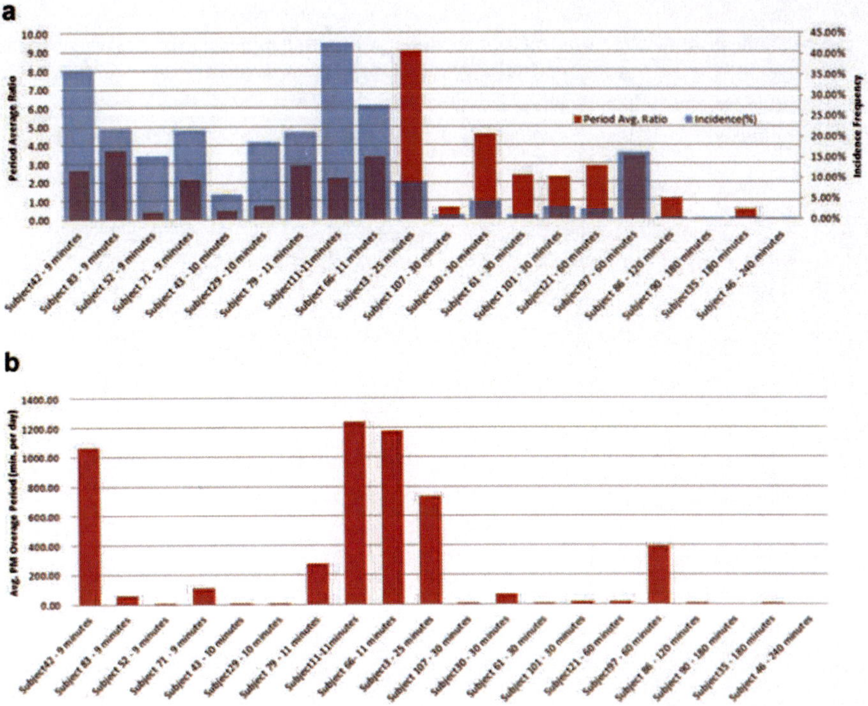

Fig. 3 (**a**, **b**) Graphical comparison of a subset of 20 subjects with power management enabled in the 2014 monitoring study

3.3 *Interventional Savings in All Study and PM Subset Workstations*

Considering all workstations (with power management and without) in a single set, a marginal savings calculation was performed for simulated sleep settings ranging between 5 and 300 min. With a sleep setting of 5 min, a savings of 880.8 ± 84.7 min (95% CI) could be realized (Table 3 and Fig. 4). Because of the total length of a day and the potential for carry-over to the next day, the potential for power management with even relatively long delay periods can produce substantial savings. In Fig. 5 this is further illustrated as even consistent simulated long power management sleep timer settings lead to savings greater than 40% as compared to the average of all study workstations. This fact highlights the general importance of power management even if with longer settings.

As the majority of observed idle periods are short in duration, increased savings is available due to short timer lengths. This can be observed visually in the representing figure as an inflection between approximately 5 and 60 min modeled intervention period length for weekdays but not for weekends where short idle periods are generally nonexistent due to lack of user presence at the workstation.

Table 3 Summary of calculated minutes per day period estimated power savings (for a subset of all study days) due to the action of simulated power management

Intervention period (min)	Average savings (min/day)	Standard deviation (min/day)	Average energy savings (kWh/year)
5	880.8	463.6	156.8
30	782.6	446.1	138.1
60	717.9	424.2	126.7
120	623.4	384.2	110.0
300	414.2	274.5	73.1

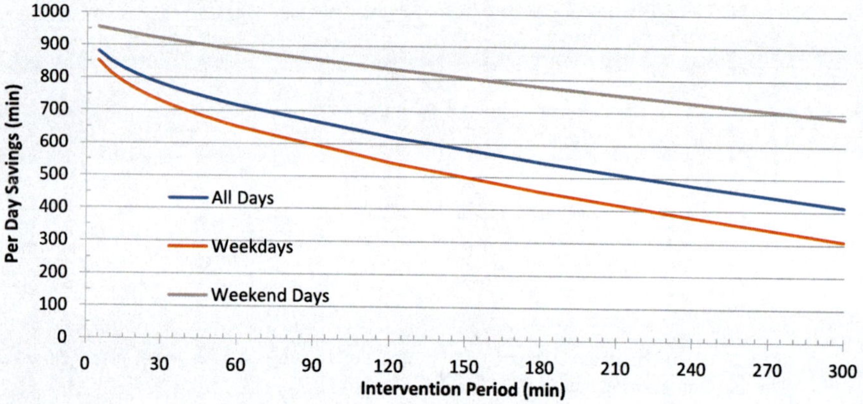

Fig. 4 Study-average estimated savings due to simulated change in power management sleep timer setting considering weekday versus weekend evaluation periods

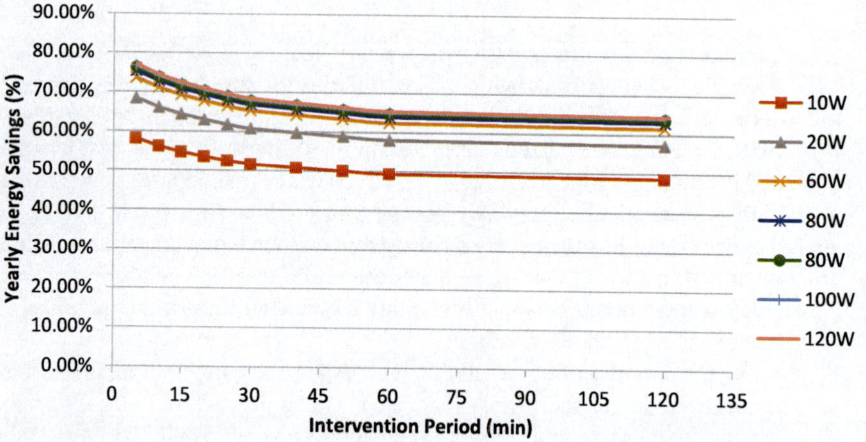

Fig. 5 Baselined yearly savings for different modeled On/Active state power loads with a high modeled Sleep state power load of 2.5 W for multiple intervention period settings for all study computers (n = 115)

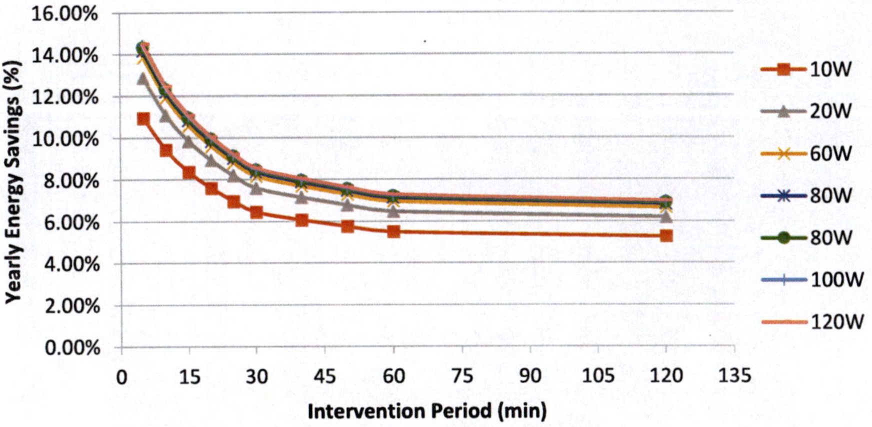

Fig. 6 Baselined yearly savings for different modeled On/Active state power loads with a modeled sleep state power load of 2.5 W for multiple intervention period settings for study computers as found with PM enabled excluding workstations with gross power management operational issues (n = 17)

Considering the study subset with power management enabled (17 subjects, excluding the 4 specified cases) the energy savings potential is substantially less than presented in Fig. 5 as no extended events contribute to multi-day idle periods (Fig. 6).

3.4 Marginal Interventional Savings

The distribution of idle periods can be used to seed a simulation to estimate the difference in savings between two ideal case operations. This is essentially the On time that can be saved considering an ideal operation between two intervention period settings. In this manner the marginal benefit can be estimated and compared against the potential marginal cost related to potential user interruption. The details of how different settings can produce savings is presented in Fig. 7.

3.5 Sensing and Occupancy Detection Comparison

A weak correlation between active usage versus idle for energy usage was observed. This is indirectly observed in Fig. 7 where extended period of use (as evidenced by a time correlated occupied signal from the motion detector) is not strongly correlated to a unique increase in energy usage (Fig. 8).

An extended live evaluation was used to observe the effect of specific activities on the impact of motion versus keyboard/mouse used as an indicator for

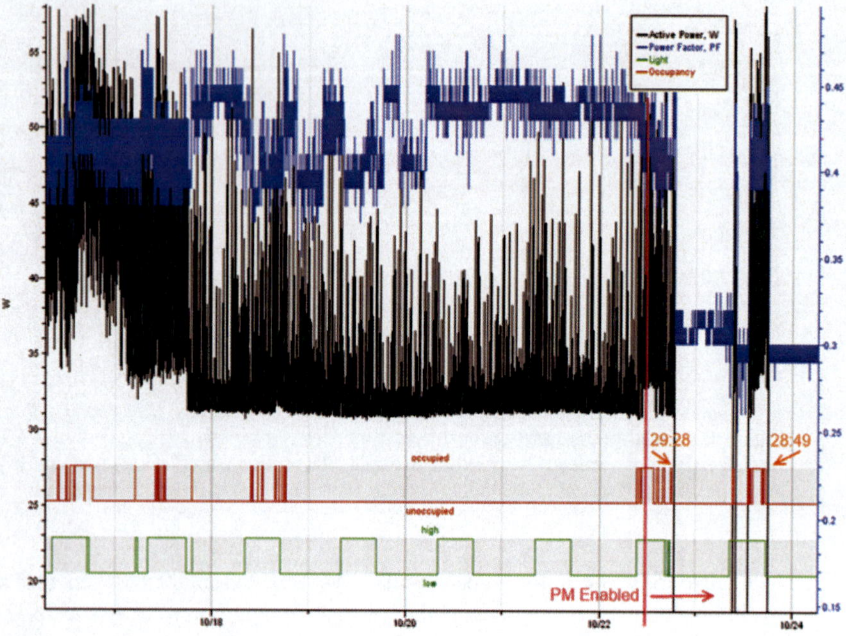

Fig. 7 Evaluation of motion as an indicator for power management

engagement and to assess the impact of daily activities on energy savings. In this test, as in previous evaluations, no background motion was permitted to be observed other than the computer user based on the configuration of the evaluation space. Using the same setup configuration as the prior live test during a 5 day period, the user logged activities that were performed. During this test, the computer PM timer was set to 30 min. Actuation of sleep was compared to time points of measured motion. The time of the last motion event is compared to the timer duration (listed in Table 4 as sensor delta period). The duration between when sleep was expected to happen and the last motion event is presented. In all cases a motion shutdown would have occurred yet a short timer delay occurred likely due to residual motion following the last keystroke.

4 Discussion

Improving the effectiveness of computer power management can lead to massive savings. Within just the context of the monitoring study, a majority of computers were shown to operate without effective power management set up or in cases where power management was active, catastrophic failures were observed, resulting in loss of substantial savings opportunities. Clearly getting power management to

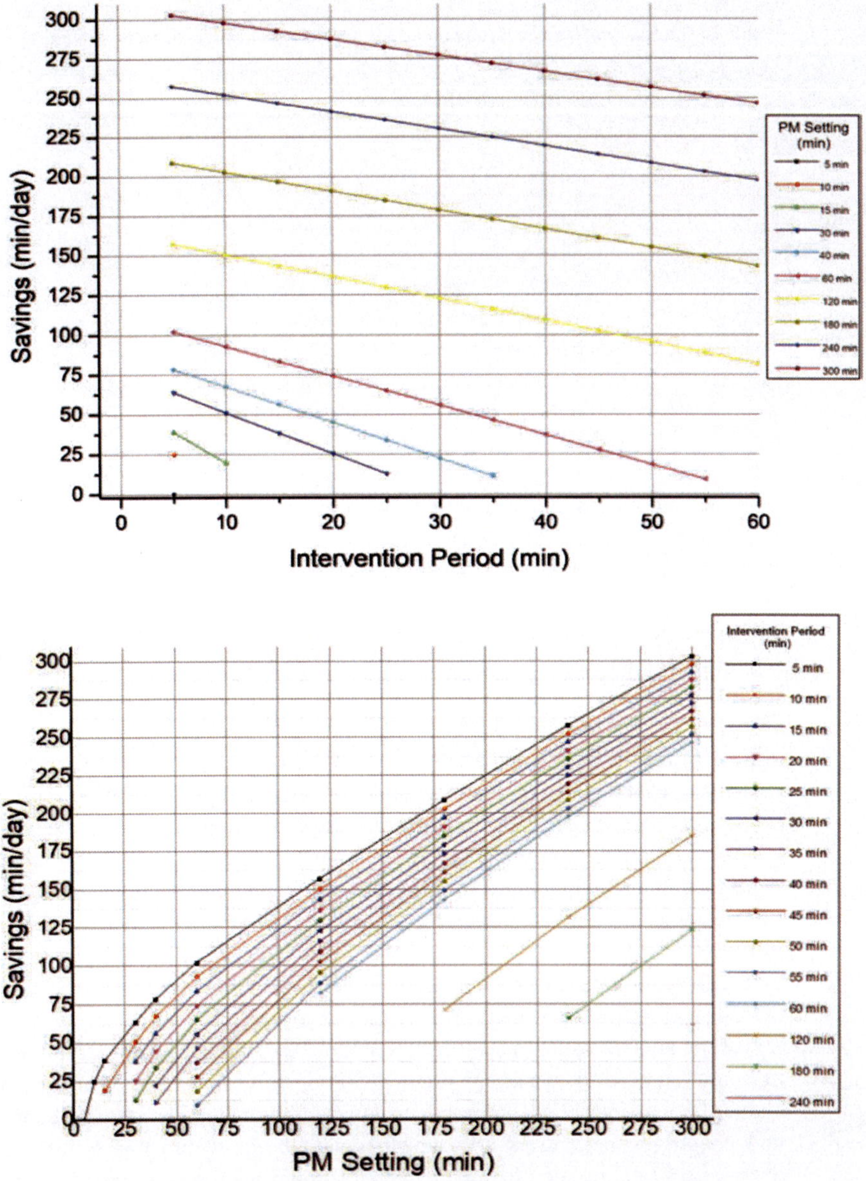

Fig. 8 (**a, b**) Savings for all subjects (with and without PM enabled) as a function of simulated PM period compared to the action of an alternative choice of time power management sleep timer duration (denoted as Intervention Period (min))

Table 4 Summary of sensor common periods and time deltas for the evaluation test period

Sensor delta period (min:s)	User activity (comments)
2:02	User left workspace area
1:50	User left workspace area
0:39	User left workspace area
0:18	User left workspace area
6:09	User left, workspace area, someone entered the workspace to leave a note on the desk in front of sensor
4:38	User cleaned up then left workspace area
1:35	User left workspace area
14:38	User conducted meeting near computer area but not using the computer during the period. User continued to trigger resets of the motion timer but not the PM timer

operate (and remain operational based on retained settings), then operate correctly to provide consistent transitions into sleep are two major considerations. The results of the 2014 monitoring study highlighted the low incidence of power management present on university workstations while simultaneously confirming screen power management as prevalent. This suggests users may be confused between screen blanking (screen power management) and computer power management. As none of these systems were centrally managed for power management nor did a centralized IT infrastructure or policies exist, the findings of this study are likely applicable to similar office desktop deployment scenarios. Based on presented observations, it is clear that even when enabled, sleep blocking, background tasks, and wakeups negate some power management effectiveness, both legitimate and illegitimate. As a large variability exists in the length of recorded idle periods, this suggests multiple contributing pathways in some or all of these factors contributing to abnormal, likely illegitimate power management override and excessive energy usage. The use of presence sensing may provide a means to negate illegitimate energy usage; however, application usage must take into consideration that such a device will cause a system to go into sleep even if a legitimate process is continuing when the user leaves the area. Best practices for browser code execution and media playing as well as improved general operating systems API controls to override power management should be considered to improve continued judicious power management controls and encourage both users and developers to build within constraints to reduce energy use. Luckily the continued shift of development toward mobile applications inherently requires judicious power management. The mobile-inspired design of current operating systems considers energy usage as a function of battery life. While inherent design trends may implicitly draw attention to improved energy management, the authors have independently observed illegitimate sleep blocking occurring on both mobile and desktop workstations at the date of publication of this work. Browsers with open video tabs often prevent sleep along with cloud service sync in addition to backups and updates. While design is improving, sleep blocking may potentially be a growing problem due to more content-rich websites and advertisements that can prevent computer power management from taking effect. The current incidence of sleep blocking is a worthy investigation. Although the 2014 CalPlug monitoring study dataset was used for the current research study, the CalPlug's 2018 PMUI study dataset (finalized immediately following the submission of this manuscript) [2] may provide deeper and more up-to-date insight into the prevalence of sleep blocking and the real impact on energy usage. Further evaluation of more modern datasets can help determine the rate of sleep blocking and common root causes. From here a plan can be developed with potential mitigation actions for stakeholders.

The use of alternative detection of occupancy other than keyboard/mouse activity, especially employed as an external device to allow independent triggering, has the potential to reduce the effects of extended sleep blocking and users with misconfigured settings. Motion events in lighting and HVAC controls have shown the potential to reduce usage by improved occupancy determination. Improved integration of different systems using motion from various sources as well as alternative passive sensing methods are a major potential focus of continued investigation with wide potential applications to energy management. From this study, motion events specifically were well correlated with power management triggering in a known working configuration. Compared to scenarios where power management is not configured or is operating incorrectly, even limited power management provides substantial energy savings. An example of this is a 1 h sleep timer: while a long period itself, this can prevent potentially 7 h of wasteful overnight operation, leading to nearly 30% savings directly preventing just overnight operation. As observed with Tier 2 APS devices for entertainment applications, extraneous or irrelevant background motion can lead to unwanted resetting of the countdown timer. When compared to no power management, using motion sensing is likely an improvement. Such a system can ensure consistent power management implementation when plugged in, as the operation of this device cannot be as easily changed as operating system settings. However, in a situation in which power management is functioning efficiently, the probability of chance motion canceling sleep requests is potentially high, limiting the ability to eke out additional savings. As power management was largely misconfigured across evaluated systems, such external devices would substantially improve savings as they would provide a stopgap measure. There are specific challenges to using external signaling including not precluding legitimate actions that may suspend sleep such as backup processes and video viewing. Continued development of such devices to reduce interference in legitimate tasks by Operating System (OS) communication is a step to improving the applicability of these devices to a wider audience. Proving the effectiveness of such devices paves the way for integration of motion sensors into monitors and keyboards to reduce total user peripherals while adding consistent energy management functionality. Continued evaluation of specific implementations of solutions based on this approach can be used to further develop a model to advance practical alternative sensing for computer power management.

Acknowledgements Funding for the work detailed in this report and tools developed to support the work was provided by Southern California Edison Company (Edison International) and TrickleStar, Inc. The authors greatly appreciate the assistance of Genoveva Paz with copyediting this report.

References

1. Pixley, J. and S. Ross, Monitoring Computer Power Modes Usage in a University Population. California Plug Load Research Center, University of California, Irvine, 2014.
2. Pixley, J., *Power Management User Interface (PMUI)*, California Energy Commission (CEC-500-2019-017), 2019.
3. Mercier, C. and L. Moorefield, Commercial Office Plug Load Savings and Assessment. 2011, California Energy Commission. PIER Energy-Related Environmental Research Program.
4. Pixley, J.E., et al., *A Survey of Computer Power Modes Usage in a University Population.* 2014, California Energy Commission.
5. Hafer, M., Inventorying Plug Load Equipment and Assessing Plug Load Reduction Solutions on a University Campus. Sustainable Stanford, 2015.
6. Acker, B., C. Duarte, and K. Van Den Wymelenberg. *Office Plug Load Savings: Profiles and Energy-Saving Interventions.* in *ACEEE Summer Study on Energy Efficiency in Buildings.* 2012. Asilomar, California.
7. Bensch, I. and S. Pigg, *Unplug for Savings*, in *Home Energy Magazine*. 2010. p. 18-20.
8. Motta Cabrera, D.F. and H. Zareipour. *A Review of Energy Efficiency Initiatives in Post-Secondary Educational Institutes.* in *Energy* 2011*: The First International Conference on Smart Grids, Green Communications, and IT Energy-Aware Technologies*. 2011.
9. Mercier, C. and L. Moorefield, Commercial Office Plug Load Savings Assessment. 2011, California Energy Commission: Sacramento, CA.
10. Barr, M., C. Harty, and J. Nero, Thin Client Investigation Including PC and Imaging State Data (Tasks 1, 2 & 3). San Francisco, CA, 2010.
11. Mercier, C. and L. Moorefield, *Commercial Office Plug Load Savings and Assessment: Executive Summary.* California Energy Commission, 2011.
12. Zeng, K. and M. Smith, Tier 2 Advanced Power Strips in Residential and Commercial Applications. 2015.
13. Klopfer, M., et al., *USB Motion Sensor Computer Energy Management System: Technology Evaluation Report*. 2019, California Plug Load Research Center (Prepared for Tricklestar).

State Based Approximation of Behavior Contribution to Energy Usage

Michael J. Klopfer, Joy E. Pixley, Saniya Syed, and G. P. Li

Abstract Understanding how behavior affects consumer plug load device energy usage provides improved guidance on the development of testing standards, voluntary agreements, and incentive programs. In this study the authors demonstrated an approach for developing and analyzing device use profiles for common residential devices, to determine the range of energy usage bounds in plausible real-world usage scenarios and the relative impact of usage types. Three aspects of how devices are used were taken into consideration: the total amount of active use per day, the pattern of that use over the course of the day (e.g., how many periods of use), and power management settings and behaviors. For each aspect, at least three levels were specified (low, moderate, high). The bounds of each category were justified based on observed or reported device usage and the features of the specific device. Device use profiles were constructed using all possible combinations of the levels of the three aspects such that the impact of each aspect was evaluated while the other aspects were held constant in simulated operation. For each device, power consumption measurements were taken for each steady-state and transitional state observed in normal usage, and operational state chains were determined. A simulation tool, the Plug Load Simulator Suite 1.2 (PLSim), was developed for inputting device-state test results and modeling energy consumption across device use profiles. Multivariate regression models were used to identify the proportion of variance in energy consumption across profiles that was explained by each of the three behavior aspects. One device—the satellite set-top box—showed almost no variation across profiles: energy consumption was not responsive to either the amount of active use or to power management settings. Power management was a significant factor predicting energy consumption for all ten remaining devices. Devices varied in the effects of the active use and pattern of use aspects. Four patterns are exhibited:

M. J. Klopfer (✉) · J. E. Pixley · G. P. Li
California Plug Load Research Center, University of California, Irvine, CA, USA
e-mail: mklopfer@uci.edu; jpixley@uci.edu; gpli@calit2.uci.edu

S. Syed
California Plug Load Research Center, University of California, Irvine, CA, USA

D+R International, Silver Spring, MD, USA
e-mail: ssyed@drintl.com

P. Bertoldi (ed.), *Energy Efficiency in Domestic Appliances and Lighting*, Springer Proceedings in Energy, https://doi.org/10.1007/978-3-030-79124-7_22

(1) strong impacts of both active use and power management aspects, with active almost as high as power management (4K and HD televisions); (2) significant impact of the active use aspect but much lower than the impact for power management (streaming device, video game console, desktop computer, and laptop computer); (3) significant impact for power management only (sound bar and both pod coffee makers); and (4) significant impact of pattern of use that exceeds that of power management (rice cooker). Assessments of the implications for each device class, and extensions for other similar devices, are discussed at length in the report; overall conclusions are summarized here. This approach provides an objective means to classify the impact of aspects of device usage between users and across devices within a specific category and shows promise for future evaluations.

1 Introduction

Utility companies have become increasingly more interested in understanding the real-world energy usage of devices in homes in their efforts toward a more sustainable grid. A particularly difficult area of this investigation is varying behavioral energy usages of miscellaneous electric loads (MEL) and plug/process loads (henceforth collectively referred to as plug loads). While many households contain the same plug load devices, there is a large range of usages from the standard usage estimates of how these devices are being used in the home which leads to the question of how to prepare the grid for these highly variable loads. With the total number of consumer electronics devices expected to rise in each household [1, 2], finding strategies to manage energy consumption from plug loads in current and new categories of devices has become very important.

Standard energy testing, reporting, and modeling protocols provide manufacturers, regulatory authorities, and consumers a way to evaluate energy use and to compare energy efficiency options. However, the single point evaluation tests on individual devices are not able to capture how these devices are used in real-life situations, when they are often connected to other related devices as a part of a network, and subject to various users' behavior. These issues lead to a significant roadblock when trying to assess the true saving potential of emerging technologies or to effectively communicate results to utilities and consumers [3].

Plug loads in general continue to be a growing source of residential and commercial total building loads in part also due to efficiency gains for space heating, water heating and lighting and due to new categories of devices and increased numbers of current device categories in use (see Fig. 1) [1, 4]. Considering macro changes, both the average living space and total numbers of devices have increased over time. Cisco estimates 13 devices per person by 2021 [5]. The average home boasts now more than 7 screens and 60% of a nationally representative sample of survey respondents use devices more than 3 h per day [6]. The 2016 EIA energy outlook to 2040 predicts miscellaneous loads (in which plug loads are included) will increase by with an adjusted average growth rate of 1.4% per year to 2040 with

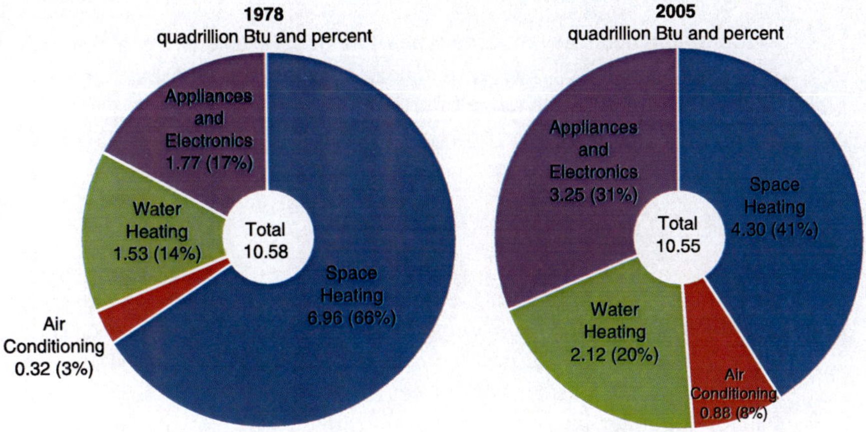

Fig. 1 The changing nature of residential energy consumption in the U.S. (Source: 1978 and 2005 Residential Energy Consumption Survey [1])

a commercial growth of 11.5% [7]. Audio systems and game consoles are major annual consumers [8], with electric grills, coffee machines acting as major active load devices [9, 10]. Considering both commercial and residential applications, plug-in devices are responsible for approximately two-third of California's residential electricity consumption, including 20% for TVs and office equipment and another 11% for miscellaneous devices [11]. It is estimated that plug-in equipment and miscellaneous loads will be responsible for 70% of electricity demand growth from 2015 to 2024 [12]. Therefore, while household plug load devices are individually low in energy demand, they collectively pose major challenges to future sustainability plans, such as California's Zero Net Energy (ZNE) initiative for all new homes set for 2020.

Assessing potential intervention strategies for both utility energy efficiency and demand response programs within multiple sub-classes of plug loads requires understanding the root cause source of waste. For analysis, by comparing variations in usage against controlled parameters, changes in energy use can be localized to specific, energy efficiency efforts can be applied in two main modes, passive and active. The next point is the boosting of the capability of power management features, this includes expanding the capability of features present or adding additional features.

2 Methods

In this study, we demonstrate a new modeling approach for estimating how energy usage can be impacted by behavior, across multiple devices and categories. This is essentially the inverse of standardized testing, in which one test protocol is applied to multiple models of the same type of device. Instead, we examine one model of each type of device, holding the device parameters constant, and vary the test parameters to reflect the range of ways it could be used in real-world conditions.

2.1 Device Selection and Testing

First, a set of residential plug loads was identified, prioritizing those where usage logically affects energy consumptions; for instance, devices that must always stay on were omitted. The 10 device types selected include: 4K (UHD) television, HD television, game console, satellite set-top box, audio sound bar, streaming device, desktop computer, laptop computer, pod coffee maker (two models), and rice cooker.

Second, each device was tested to determine all the states observed during use (including warming up, idle, and powering down periods) and the power consumption during each state. This step required a comprehensive examination of the operating conditions and mechanisms of each device, particularly as they affected the timing and stages of shifting from off or idle into active use, and from various active modalities into low-power and soft-off states. At this stage, all the power management functions were also examined and assessed for their implications for energy use. For details on the extensive testing performed in the larger project, see [13].

2.2 Device Usage Profiles

Third, device use profiles were constructed that vary on three main aspects of how the device is used in real-life situations:

1. How much the device is used per day (Active)
2. The timing or pattern of that usage (Pattern)
3. What power management settings or user behaviors affect efficient use (PM)

This approach was developed in an earlier project conducted using the SIM Home testing lab [3]. For each aspect, at least three levels are defined: usually low, moderate, and high. Whenever available, ENERGY STAR testing estimates for amount of active use were used to establish the moderate or standard level. Survey data on how devices were used in homes was analyzed to determine the reasonable range: the median was used for "moderate" active use, while the 10th percentile was considered "low" usage and the 90th percentile was considered "high" usage. See the first SIM Home report for more detail [3]. In the absence of any quantitative data, assumptions were made about realistic use cases.

For a given duration of use, the number of periods throughout the day could potentially affect energy usage. Many devices incur transition costs, such as warm up or cool down times, or experience idle periods prior to entering a sleep or standby mode. The pattern aspect reflect this, with "low" indicating that all the usage occurred in a single period, "moderate" being two periods, and "high" being more periods. For instance, 4 h of computer use may occur all in one sitting or be divided into two periods of 2 h, or even more periods of shorter duration. For devices with automatic standby settings, additional pattern levels were created to reflect varied periods between usage periods: for instance, if a device idles for an hour before transitioning to a low-power mode, whether the next use is in 30 min versus 4 h affects the amount of time spent idling.

The power management aspect affects what state the device is in when it is not being used. This involves two factors: whether automatic power management settings are enabled (and if so, at what delay setting) and whether the user turns the device off when finished or not. Here, the "moderate" level is set where the user does not act but the device is set at the factory-set automatic settings for transitioning off or to sleep mode when not used. "High" PM is where the user turns off the device regularly when it is not used and/or engages any automatic PM settings at high levels. "Low" PM is when where any automatic PM features are disabled (if possible) and the user never turns the device off. Given the great variation possible with PM behaviors, many devices were given more than three levels of PM to better reflect real-life conditions.

A set of device use profiles was created for each device by combining all levels of all three aspects. The profiles are represented in the form "active-pattern-PM," so low-high-mod means Active-low, Pattern-high, PM-mod. A device that has three levels for each aspect which are all compatible would thus have 27 device use profiles, ranging from low-low-low to high-high-high. A device with multiple levels of any aspect would have more profiles. Some levels of active and pattern cannot be logically combined, resulting in fewer profiles. For instance, 13 h of television watching (Active-high) could not be divided into four periods with 5 h between (Pattern-high1) and still be contained within the 24 h period. Other combinations were not plausible: for instance, a half hour of television watching (Active-low) would not reasonably be broken up into four periods over the course of the day (Pattern-high1 and high2).

For each device, a "standard profile" was determined which matched the ENERGY STAR or other standardized testing protocol or, in the absence of such, approximated a standardized protocol. The standard profile represented Active-mod (the median amount of active use), Pattern-low (usage all in one period), and PM-mod (factory PM settings, no user intervention).

A full description of the device usage profiles used for all devices can be found in the second SIM Home report [13].

2.3 Energy Use Modeling

The Plug Load Simulator Suite 1.2 (PLSim)[1] is an open-source simulation tool that was developed in support of this project. PLSim is used to rapidly tabulate energy usage based on device states as modeled through the profiles. Device testing was used to verify state and general usage along with total energy consumption. The state model was developed and verified against the actual device using a modeled usage plan to verify the developed state mode. This tool is universal for all modeled devices and was the primary tool used in this work for energy modeling. A per-device, state-wise energy usage XML database is generated by testing device operation per aforementioned testing approaches. This database provides a list of states. A developed time-based (temporal) profile maps the time the device spends in a given period for a given action to energy usage. Daily energy usage is calculated as a temporal combination of all event states during a 24 h period (see Eq. 1). In this relationship PS_l is used to model the lowest energy usage state (assumed to be the lowest power modeled state) while Nx, Px, Tx, are used to model the average power consumption and number of periods a state exists during a 24 h period.

Calculation of daily energy usage based on generic event frequency and classification:

Shown explicitly with two terms:

$$EC = \frac{P_l\left(24-\left(\left(N_1T_1\right)+\left(N_2T_2\right)\right)\right)+\left(P_1\left(N_1T_1\right)+P_2\left(N_2T_2\right)\right)}{1000} \tag{1}$$

Where:

Variable	Value
EC	Daily energy consumption in kWh for the modeled system.
$N_{X(shown:\ 1,2)}$	Average number of events of a particular event of a given duration that occurs in a 24 h period: the value for x increments for each event.
$T_{X(shown:\ 1,2)}$	Average duration (in hours) for a particular event which occurs in the 24 h period: the value for x increments for each event.
$P_{X(shown:\ 1,2)}$	Average power consumption (in watts) for a particular event which occurs in the 24 h period: the value for x increments for each event.
P_l	Average power consumption (in watts) for the lowest power operational mode, such as soft-off, sleep mode, or standby mode.

The set of device use profiles were then programmed into the PLSim tool that CalPlug developed. Adding power consumption data from in-house testing of the device in all possible states into PLSim produces the total simulated energy consumption for each profile. For comparison, lower and upper boundaries are also

[1] See: https://github.com/CalPlug/PlugLoadSimulator

modeled, showing the energy consumption if the device remained in the lowest usage state possible or the highest usage state possible for 24 h.

2.4 Energy Use Analyses

The PLSim calculations produced estimates of total 24-h energy consumption for each device use profile for each device. These were assessed in three steps.

For each device, a graph showing the energy consumption across profiles was produced, ordered so that the overall patterns of different consumption results could be visually attributed to changes in active use, pattern, or PM. This descriptive step helped give insights into the data, and to verify and make sense of the summary results shown later. To illustrate, the graph of energy use estimates across device use profiles is shown below for the desktop computer that was tested (see Fig. 2). The first three profiles all have low active use (30 min) and low pattern (that is, all at one time), and differ only on PM. The pattern shows a very high energy consumption for profile #1, with PM-low (disabled sleep settings) and much lower for those with PM-mod or PM-high. That same pattern repeats through the rest of the graph. The energy consumption rises with the amount of active use, especially for profiles with moderate or high PM. However, the primary distinguishing pattern continues to be the much higher energy usage for profiles with PM-low versus all others. As these graphs are illustrative rather than analytical, they are presented illustratively at this point in the discussion (Sect. 3) here (see [13]).

Next, for each device, simulated energy consumption for the standard device use profile was compared those with the median, highest, and lowest energy consumption estimates, and the range. These provide quantitative assessment of how much

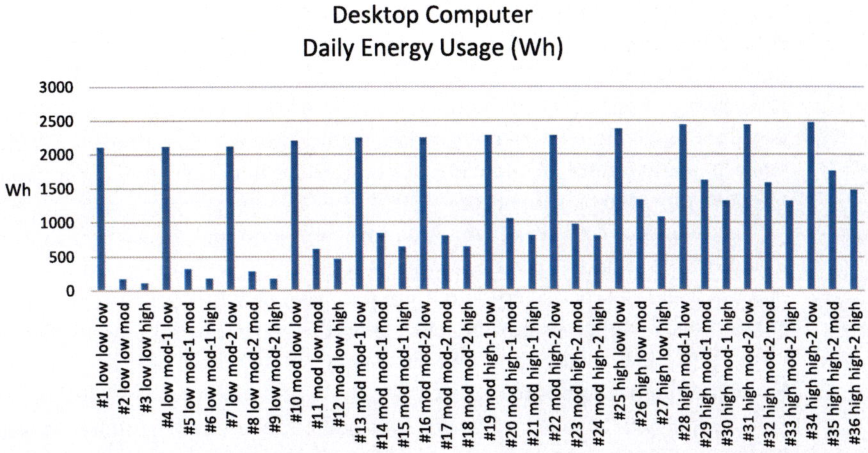

Fig. 2 Desktop computer energy usage profiles

higher or lower energy consumption could be for a particular device in at least some homes, compared to estimates based on the standard testing protocol. Although the simulated results do not allow for comparisons across real households, they do help identify potential problem areas that deserve additional scrutiny.

Finally, multivariate regression analyses were run. For each devicc, thc number of device use profiles was the sample, the estimated energy consumption for each profile was the dependent variable, and the predictor (independent) variables were the levels of active, pattern, and PM. The standard device use profile is the default in all models. This means Active-mod is the omitted variable and the regression tests the effect of Active-low versus Active-mod, and the effect of Active-high versus Active-mod. Likewise, Pattern-low and PM-mod are the omitted variables, with the effects of the other levels of each aspect compared against them. This allows the model to show which levels of which aspect significantly differ from each other. The regression analyses also indicate how much of the variance in energy consumption is explained by the three different aspects (e.g., whether variation in power management matters more for energy consumption than amount of active use).

3 Results

3.1 Range of Energy Consumption Across Profiles, by Device

Given a reasonable range of usage behaviors, what energy consumption results would we observe in these tests? That is, if we assumed that all devices in all households were operated according to the standard device use profile, how inaccurate would our estimates be about the highest- and lowest-usage households? If the range for a device is relatively small, this suggests that standard tests would give good estimates across a range of households. However, if the device's energy consumption is high across all usage patterns, perhaps additional development of low-power states could help lower consumption during non-active periods. Alternately, if the range of energy consumption across profiles is very large, especially in terms of values much higher than the standard testing profile, this points to possible intervention points either in reducing active use consumption or in promoting more effective power management.

The ranges of energy consumption for the highest and lowest profile, compared to that of the standard profile, are presented in Fig. 3. Three general patterns are observed: devices with very small ranges; devices with low or moderate ranges, either mostly higher or mostly lower than the standard profile; and devices with large ranges that span in both directions from the standard profile but err more on the side of higher values than lower.

For instance, the standard profiles for pod coffee makers exhibit almost as high of energy consumption as for the 4K television, and higher than the HD television. However, almost all the variations in how pod coffee makers are used result in lower

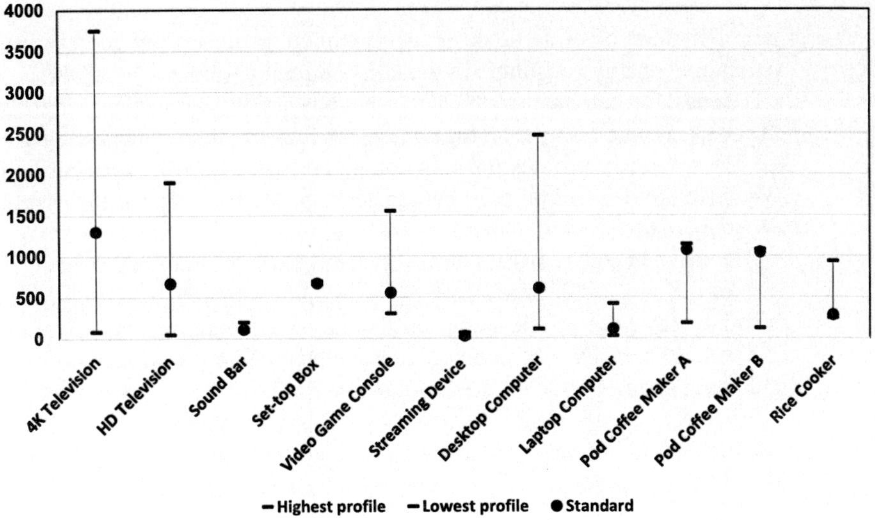

Fig. 3 Range of daily energy consumption of profiles, by device

consumption, whereas the top range for televisions is substantially higher. The rice cooker shows the opposite pattern, with other ways of using the device resulting in higher energy consumption than the standard profile. The same is true for the desktop computer, game console and set-top box: the standard profiles for these devices show similar energy consumption. For both the desktop computer and game console, variation in usage can lead to higher consumption, whereas there is almost no variation by usage for the set-top box.

The range between the minimum and maximum should be considered in the context of the median energy consumption for that device. For instance, a range of 50 Wh would be small if the device's median energy consumption was 1000 Wh but substantial if it was 100 Wh. Logically, it is possible for the maximum profile to be more than 100% higher than the standard profile (that is, use more than twice as much energy) but the difference between the standard and minimum profiles must be less than 100% than the standard profile, probably much less (as 100% lower would mean zero energy consumption for the minimum profile).

Among the entertainment devices, the two televisions show the largest energy consumption ranges, but the range for the video game console is also high, especially compared to the small ranges for the sound bar and set-top box. The streaming device shows a large range, but the energy consumption of the standard profile is so low that this is not substantively important. The television profiles' energy consumption estimates range by almost 300% of the standard device use profile: most of that represents how much higher the maximum profile is, but the minimum profiles save an impressive 93–94% energy compared to the standard profile. By contrast, the satellite set-top box shows the lowest ranges for all the device shown here. For most of these devices, the maximum estimates are more impactful on variation in energy consumption than the minimum estimates.

The desktop computer tested here uses substantially more energy in the standard device use profile than the laptop computer. However, their relative ranges are similar. The minimum profiles for both computers save a similar proportion of energy relative to the standard profiles, but the desktop's maximum profile uses proportionally more than that of the laptop.

The two pod coffee makers have very similar results. For both, the range is about 90% of the standard profile's energy consumption, and almost all of that is due to the minimum usage profile being substantially lower than the standard profile. Note that pod coffee makers have power management features available but are shipped with those settings disabled, so the moderate PM level has no low-power mode, and all PM levels are more energy saving than the standard profile.

By contrast, the rice cooker shows a very large range of energy consumption, with the maximum being much higher than the standard profile.

3.2 *Effects of Usage Aspects*

The team used regression models to evaluate what proportion of the variation in modeled energy consumption across device use profiles can be attributed to the three aspects tested here, given the specific definitions of each aspect used. Particularly for devices with a large range of estimated values, is the deviation from the standard default device use profile largely due to differences in the amount of active use, in the timing or pattern of that use, or in the power management settings or behaviors?

For each device, four models were run: one for each aspect alone and one full model including predictor variables for all three aspects. The sample for each model is the set of device use profiles for that device, and the dependent variable is the energy consumption calculated for each profile. Each regression model produces an R^2 statistic indicating the proportion of the variance in the dependent variable explained by the parameters in that specific model. For example, if the R^2 statistic for the Active Model were 0.50, that would indicate that 50% of the variance in energy consumption across the device use profiles was due to whether they had high, moderate, or low active use.

The results of the regression models are summarized in Fig. 4. The asterisks indicate which models were statistically significant. These results reveal major differences in the relative importance of active use, pattern, and power management for energy consumption across these devices.

These effects should also be considered within the context of how much energy consumption varied across profiles for that device. For example, for the set-top box, power management explains 72% of the variance in energy consumption. However, compared to the standard profile for the set-top box, energy consumption only ranges from 2% lower to 4% higher across other profiles (see Table 1), so there is

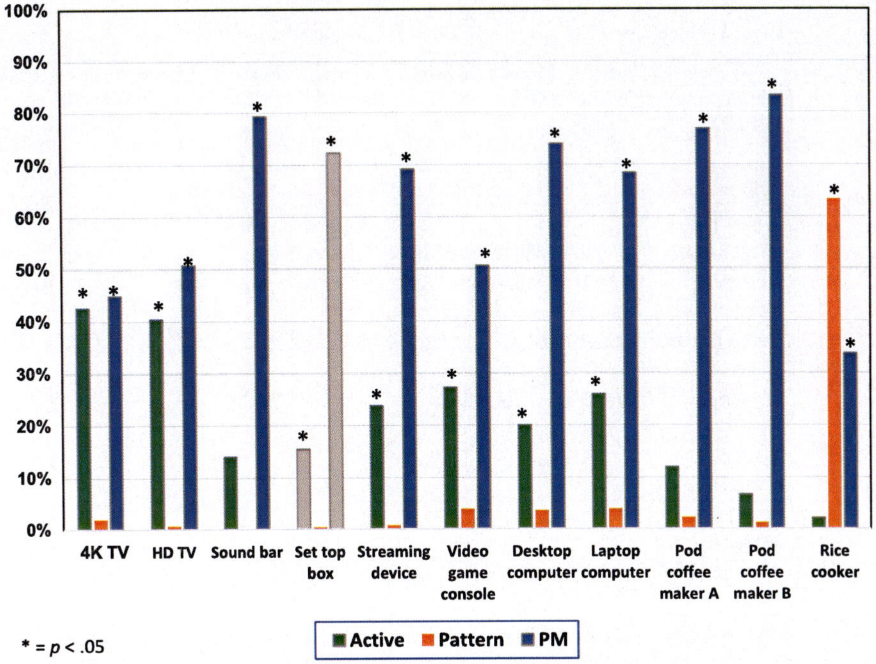

Fig. 4 Proportion of variance in energy consumption explained by each aspect

essentially no variance to explain. For that reason, the set-top box is not discussed here. For the sound bar and the streaming device, the variation is higher as a percentage of the fairly low standard energy consumption although the range still represents a small increase or decrease in energy consumption.

4 Discussion

The impact of behavior on energy use is a major consideration to produce accurate evaluation of device energy consumption for both modeling and mitigation efforts. In this project, the authors evaluated the effects on energy consumption of three aspects of user behavior toward devices: the amount of active use, the pattern of that use, and power management. This method was applied to common residential plug load devices. Using carefully defined device use profiles, comparisons across profiles were used to identify the impact of each aspect, to prioritize and focus efforts to address improving energy usage in specific devices. Future testing and evaluation methods could expand upon this method to continue to refine procedures to match devices as they evolve, considering behavior and usage. The analyses in this report address two main questions. The first question is how large a range of energy consumption outcomes the device use profiles generate for each device, based on

Table 1 Summary of ranges and differences from standard profile

	Std (Wh)	Median (Wh)	Min (Wh)	Min % from Std	Max (Wh)	Max % from Std	Range (Wh)	Range % of Std
4K Television	1305.1	1981.5	82.4	94	3746.7	187	3664.3	281
HD Television	667.4	743.3	48.5	93	1903.2	185	1854.7	278
Sound Bar	111.7	148.1	90.4	19	198.0	77	107.6	96
Set-top Box	669.9	684.6	654.1	2	699.7	4	45.7	7
Streaming Device	28.1	35.4	6.9	76	68.9	145	62.0	220
Video Game Console	556.9	644.2	303.5	46	1557.8	180	1254.3	225
Desktop Computer	609.5	1310.3	110.0	82	2472.5	306	2362.5	388
Laptop Computer	112.3	243.1	28.9	74	423.3	277	394.4	351
Pod Coffee Maker A	1076.9	639.9	189.0	82	1147.6	7	958.6	89
Pod Coffee Maker B	1046.4	535.7	123.3	88	1081.4	3	958.1	92
Rice Cooker	282.2	529.4	249.0	12	937.9	232	688.9	244

reasonable assumptions about the range of real-life usage. A follow-up to this question is whether the range is primarily higher or lower than the standard profile produced by standardized testing protocols. If the range is fairly small or if it is evenly distributed, standardized tests are more likely to produce accurate estimates of real-life outcomes if averaged over a large number of households. However, if the range is very small, this raises a new concern: that the energy use of the device is not responding to the amount of time the device is actively being used and that any power management features are ineffective at saving energy.

The second question is how much of the variation in that range of energy consumption outcomes across profiles is explained by differences in active use, versus by differences in pattern of use or differences in power management behaviors. Energy consumption is expected to vary by active use: if that is the main driver of energy use, then energy-saving strategies would logically focus on reducing the operational costs of active use for that device. If engaging power management options fails to save energy, that suggests those options are ineffective. However, if much of the variation attributable to power management behaviors results in higher energy usage, that suggests that power management options may not be effectively engaged by users, which prompts additional research and development into modes and user interfaces that will work in everyday usage. Although the device usage profiles are based on observed or self-reported behaviors as much as possible, several assumptions had to be made when defining the levels of each aspect. Thus, any conclusions are limited by the extent to which the assumptions about average and also extreme behaviors (that is, behaviors at the 10th and 90th percentile) are accepted as reasonable.

The answers to both of these questions differed greatly across the plug load devices tested here. For that reason the discussion below focuses on specific devices, with the ordering adjusted to help compare similar device results.

5 Specific Devices

5.1 *Televisions*

The 4K television tested here uses almost twice as much energy in its standard profile than the HD television (1305 Wh versus 667 Wh). However, the pattern of results for the two devices are otherwise similar. Both televisions produce a large range of energy consumption estimates across the profiles (3664 Wh or 281% of standard for 4K and 1855 Wh or 278% of standard for HD). In both cases, the upper range is twice as large as the lower range. For instance, for the 4K television, the lowest-use profile is only 82 Wh (lower by 94% of the standard profile) while the highest-use profile is 3747 Wh (higher by 187% of the standard use profile). This suggests that, to the extent that these device use profiles reflect real-life behavior patterns, estimates based on standard usage would underestimate total usage across households.

The high energy consumption for the standard profiles for these devices, especially for the 4K television, and the very high range in energy consumptions shown by the profiles (especially, higher than the standard profile), motivates a close look at the relative impact of the three device use aspects. The two televisions show a pattern not seen in other devices, in which the active use aspect is almost as strong of a predictor of variation in energy consumption as the power management aspect.

The pattern aspect does not significantly impact energy consumption for televisions, which makes sense, given the lack of a substantial boot-up period.

As active use and power management are both important contributors to energy use for televisions, they provide avenues for potential energy savings. Reducing energy consumption during active use is already a main consideration in energy efficiency regulations aimed toward manufacturers. These results could encourage stricter regulations for devices such as televisions. Options that adjust the screen brightness can potentially save energy during active use, although not enough is known about how these features are used (or misused) in real households, and whether this behavior negates any possible savings.

On the other hand, improving power management options and their use could potentially save as much energy with less extensive modifications to the devices themselves. Both of these televisions have very low-power standby modes; the challenge is to transition the device into standby mode whenever feasible. Both televisions have a feature that transitions the device to sleep after a delay period with no signal from the connected source. The feature for the 4K television has three delay settings from 15 min (default) to 60 min, while the HD television has only one delay setting, 10 min. This feature can potentially save substantial energy, but only in specific circumstances: when the television is receiving content from an external device (rather than through apps in a smart TV), and when the user either turns that device off when done or has that device set to sleep after a short period of inactivity. Otherwise, the sleep transition due to lack of signal will not activate.

The 4K television also has an auto-off feature that transitions to sleep mode in the absence of user input (i.e., through the remote control), but the lowest possible delay period is 4 h, which is the default setting. The HD television has no such auto-off feature. One relatively simple improvement would be providing such a feature in all televisions, and offering shorter delay period options, as short as 1 h, and making a 2-h delay the default setting. Informative user interfaces are essential for encouraging users to enable (or not disable) PM settings and to understand how to use them effectively, such as motivating explanations on the PM setting screen, and a signal that warns users of an impending sleep transition, so that they can easily avoid it by pressing a key on their remote control.

5.2 *Video Game Console*

The video game console tested here showed results closest to that of the HD television in terms of its standard profile energy consumption (557 Wh) and its range of consumption across profiles (225%). However, the total range was not as large as for the HD television (1254 Wh versus 1855 Wh). Compared to televisions, the lower range for the video game console is smaller (lower by 46% of the standard profile) while the upper range is almost as large (higher by 180% of the standard profile). In other words, the video game console uses less energy than the HD television (and much less than the 4K television), but a larger proportion of its estimated use is higher than the standard profile compared to lower than it.

Like several of the other devices, the video game console showed a large impact of power management on variation in energy consumption across devices, and a smaller but significant impact of active use. The difference between the two is less pronounced compared to most other devices (that is, a relatively larger impact of active), making the video game console more similar to the televisions in this respect. As such, the results support a similar approach to that described above for televisions, focusing on reducing consumption during active use and improving power management.

The video game console tested here has a standby state (called "rest") which uses 10.7 W compared to active game play at 69.5 W and a main system menu page which uses almost as much energy, at 63.7 W. When the user stops playing, the system automatically switches to the main system menu page and stays there until the user turns off the device or the automatic standby delay is activated. The standby delay period can be set from a minimum of 20 min to a maximum of 5 h. As with any other device, effective user interface instructions may help motivate users to enable the standby option and use a shorter delay setting. Users may be especially reluctant to turn off gaming devices or let them sleep because of fears that their game progress will be lost, even if this concern is unfounded; added security and reassuring communication may be helpful here. Another point to raise is that the main system menu page, if not interacted with, is functioning similarly to a standby mode, and yet consumes almost the same power as actively gaming and continues to do so no matter how long the device goes unused. This suggests that exploring a "deep idle" mode, similar to that for computers, could save additional energy by pausing certain processes after some shorter period of inactivity.

5.3 Set-Top Box

The set-top box provided a unique pattern of all the devices tested here, in that the energy usage of the standard profile (678 Wh) was higher than many others, while the range of energy use estimates across profiles was negligible. The total range across profiles is only 45.7 Wh, or 4% above standard and 2% below standard. In fact, this mirrors the maximum boundary conditions range for this device, which consumes 699.7 Wh at the highest possible use state (active video) for 24 h versus 654.0 Wh at the lowest possible use state (standby). This quantifies the extent to which the energy use for this device is not responsive to any variation in behavior: the device uses essentially the same energy while idle as while active. Indeed, the minimum and maximum device use profile results were the same as the minimum and maximum boundary conditions—that is, the least the device could possibly use (if on standby all day) and the most it could use (if actively used all day). This reflects the fact that set-top boxes must maintain continuous connections for program and encryption services; thus, even in its lowest-power standby mode, it uses substantial power. As a result, users are limited in how much they can affect energy savings on this device.

Given the lack of variation in energy consumption across profiles, the multivariate analyses explaining which use aspect caused that variation is moot. A closer examination of the operations of this device shows that the power management features of this device are completely ineffective, due to the high power usage of the idle and standby states relative to the active state. The device uses 27.25 W while in standby mode compared to 29.16 W while being actively used. The only power management option is a delay time of 4 h; this leaves the device in idle mode, which uses the same power as the active use. Ideally, shorter delay times would also be available and used as the default, but without an effective lower-power mode to transition into, this is of secondary concern.

If it is not possible to reduce relative power consumption during standby because too many of the same functions must operate even when the device is not in use, then the only avenue for saving energy with this device is to reduce consumption during the active use mode.

5.4 Streaming Device

The streaming device uses the lowest energy for its standard profile of all the devices tested here, and shows one of the narrowest absolute range of energy consumption estimates, at 62.04 Wh between the lowest and highest device use profiles. This is only somewhat larger than the range for the set-top box (at 45.7 Wh). Because the streaming device has such low standard energy consumption, the proportional range is moderate compared to the other devices (76% lower and 145% higher than the standard profile). That said, relative to other devices tested here, there is not a substantial absolute amount of energy variation to explain, or to save.

The streaming device benefits from having an aggressive power management as a default setting, with an elaborate and engaging screensaver. The applications that run on this streaming device do not appear to contribute substantially to sleep blocking for idle states (device sitting at a paused video or menu). However the device will continue to play active video wastefully even if not being viewed, which suggests one possible avenue of saving energy.

5.5 Sound Bar

When presented in comparison to other devices, the sound bar most closely resembles the streaming device, in that the standard profile energy use is low compared to most others tested here (112 Wh), and the range between the lowest and highest device use profiles is relatively narrow (108 Wh, or 96% around the standard profile). However, the absolute range of energy consumption is over twice as large as that for the streaming device, indicating more potential room for substantive energy savings.

The sound bar is one of the three devices where the majority of variation in energy consumption across profiles is explained by power management, and neither active use nor pattern are significant factors. Indeed, if all profiles using PM-low were removed, the maximum profile usage would drop by 48 Wh, cutting the range by almost half. This illustrates the importance of doing more research on how these devices are used in actual households, to help establish which assumptions are reasonable for high and low behavioral usage.

5.6 Desktop and Laptop Computers

Desktop and laptop computers show a similar range of energy consumption estimates (388% around the standard profile for the desktop and 351% for the laptop) and the pattern is similar as well, in that the upper range is much larger than the lower range. However, as the desktop computer uses so much more energy in its standard profile than the laptop computer (609.5 Wh versus 112.3 Wh), the absolute range for the desktop is much larger, and the substantive effects of the higher-use profiles are even greater. Put another way, the highest-use profile for the desktop uses 1862.9 Wh more than the standard profile (306% more) whereas the highest-use profile for the laptop uses 311.0 Wh more than the standard profile (277% more). In terms of how much the standard profile potentially underestimates real-life usage, the desktop is second only to the 4K television. Depending on how many households exhibit higher-use profile behaviors compared to those who exhibit lower-use profile behaviors, average estimates assuming the standard use profile could be off by enough to negate variation in any other household plug load device.

Like the video game console, streaming device, and set-top box, active use and power management aspects are significant contributors for desktops, but power management has a much greater impact. Pattern of use shows more impact for computers than for other devices covered so far, but not enough to achieve significance. In theory, pattern should make a difference for computers in that they automatically transition to a "long idle" state after being in a "short idle" state for 10 min, although the difference in power consumption is not large and may be overshadowed by other factors. Pattern may also interact with sleep settings in ways that are not represented by the average state use estimates utilized here, and which are beyond the scope of this report to explore.

Power management options and low-power states are well-developed in both desktops and laptops. As the PLSim results confirm, enabling sleep settings is a highly effective way to reduce energy consumption in computers, especially during long periods of user inactivity (such as overnight or during work hours for residential computers). The challenge not currently addressed by regulations or voluntary agreements is how to get more users to enable (or not disable) their computer sleep settings. As there are valid reasons why some users would need to prevent their computers from entering a low-power mode, either permanently or occasionally, it would be infeasible to remove the option of disabling sleep settings and make them involuntary. Instead, efforts to reduce energy in computers would be more fruitfully turned toward research into how users behave toward computer power management. Specific tasks include designing more effective and convincing user interfaces, and understanding and addressing the barriers that lead to users disabling or otherwise underutilizing computer power management options.

5.7 Pod Coffee Makers

The two pod coffee makers showed very similar standard profile energy usage estimates (1076.9 and 1046.4 Wh), which are higher than any device other than the 4K television, and similar ranges (89% of the standard profile versus 92%). Both pod coffee makers showed a unique pattern in this set of devices, in that most of the range was lower-energy compared to the standard profile: the highest-use profile was only 7% above the standard profile for Model A and 3% above the standard profile for Model B. In other words, most of the variation in usage predicted by the current model results in lower energy consumption than the standard profile.

The pod coffee makers are also unique among this set of devices in that the active aspect is not a significant factor explaining variation in energy use across profiles. Although the heating and brewing cycle during the active period is quite energy-intensive, it requires only 2 min to heat the water cache from cold state, and 1 min to dispense each cup. Instead, power management accounts for the majority of the energy consumption variation across profiles. Energy consumption is very similar across profiles using PM high-1 and high-2 (in which auto-off is enabled and set at 2 h and the user turns the device off after use, respectively) but much higher for profiles using PM-mod, in which auto-off is disabled. There is no low level of power management, as the standard PM aspect—the factory default—is already as inefficient as possible.

The first solution to saving energy with pod coffee makers thus seems straightforward: change the factory default so that auto-off is enabled. This assumes that users are less likely to disable the setting if it is enabled by default than they are to enable the setting if it is disabled. This would not change the potential range of the device use profiles, but it would shift the standard profile down considerably, and make the higher-use profiles less likely to occur in actual households. Offering shorter delay periods—such is already done for the more advanced Model A device—would also save energy, and may be considered as a default setting. This could work well for households where only a few cups of coffee (or tea) are brewed within a short time frame every morning.

Unlike many other devices discussed (especially in the entertainment category) these devices are not intended to run for extended duration, as their utility comes from producing a product (a cup of coffee) quickly rather than providing screen time. Accordingly, leaving the device on for extended periods beyond producing coffee is wasteful. At the same time, users may become frustrated if the pod coffee maker takes what they perceive as “too long” to warm up from a standby state when they want a cup of coffee, especially as one expected benefit of pod coffee makers is their speed and convenience. One possible reason why users disable sleep settings is that they get annoyed at waiting for the device to resume from sleep mode. Speeding up the warm-up period and providing a user interface showing the progress of the device in warming up may help prevent this annoyance, allowing a shorter sleep delay time to be effective without reducing user satisfaction.

5.8 *Rice Cooker*

Although the rice cooker, like the pod coffee makers, also involves heating and keep-warm states, the results here show a drastically different pattern of effects. The standard profile for the rice cooker produces a fairly low energy consumption estimate compared to other devices in this set (282.2 Wh), and although the range is somewhat smaller in absolute terms than that for the pod coffee makers, almost all the other profiles showed higher consumption than the standard profile. That is, the standard profile is almost as low as the lowest-use profile, and most of the other device use profiles result in higher energy consumption.

The rice cooker is unique among the devices tested here in that pattern of use explains a large and significant proportion of variance in energy consumption across device use profiles. The amount of active use—in this case, how much rice is cooked in total that day—has little effect in this analysis, but the number of times the rice cooker is used does. A closer look reveals that this is because the additional amount of energy used to cook, say, three cups of rice is incrementally small compared to the amount of energy used to cook one cup of rice. This comes down to timing: with the white rice used for testing here, it takes 32.5 min to cook one cup of rice, and only an additional 8 min to cook three cups of rice. However, changing the pattern and cooking that total of three cups of rice in two or three fresh batches over the course of the day (say, for lunch and dinner separately), requires a new baseline level of cooking time. In other words, with a pattern of use spread out over multiple periods per day, it takes more total time to provide the same amount of rice. This differentiates cooking appliances from experiential devices such as a television or computer, where the amount of time actively watched or used is synonymous with the amount of service received. As such, although the pattern aspect reveals the additional energy consumption, it is the consumption during active use that would need to be reduced in order to save energy. The rice cooker is similar to other category devices not tested here that involve heating water and/or keeping food or liquids warm, such as drip coffeemakers, under-sink or table-top water heaters, hot pots, and electric pressure cookers, and some conclusions can be cross applicable.

While power management is also significant, it is less impactful than the pattern of use over the course of the day. Other things being equal, profiles using low power management—where the user keeps the rice warm most of the day—use much more energy than others, whereas turning the rice cooker off as soon as it's done saves only a small amount of energy compared to leaving it on for another hour (say, until the meal is over). The rice cooker is unique among devices tested here, in that users deliberately leave the device on in the keep-warm state. An online search reveals many people who prefer to make a large pot of rice and keep it warm all day, despite warnings about food safety. According to the current results, it uses more energy to make a new, smaller pot of rice three times a day (and turn off the warmer after 1 h) than to make one large pot and keep it warm all day. So if a user perceives these as the competing options, the "worse" power management strategy would actually use less energy.

6 Overall

6.1 *Range of Energy Consumption*

The range of energy consumption across profiles for each device is shown to identify the highest and lowest energy usage that would be seen in real-life usage, given the assumptions in the profile definitions. The ranges are compared against the "standard" profile that represents or approximates the standard testing procedure. This indicates not only the percentage difference from the standard usage but also whether more of the range is above the standard or below it.

A device exhibiting a moderate range in energy consumption across profiles is not necessarily a bad sign, either for energy efficiency or for the accuracy of standard testing protocols. It is reasonable that devices would use more energy if actively used more hours, and that devices would save more energy if more aggressive power management features were used. Likewise, a very small range is not necessarily a good sign, as it indicates that the device does not effectively reduce energy use for shorter active periods or in response to power management.

Ideally, the range of device use behavior—and thus profile energy usage—would be normally distributed around the standard profile, in which case using standard testing methods would produce accurate and reliable estimates of the population. The current study cannot speak to whether this is the case, as it depends on how common the device use behaviors discussed here are in the population, which would require consumer behavior research that the field is sorely lacking.

The larger the range in possible energy consumption outcomes, the more likely it is that the real-life pattern of outcomes are skewed, which is especially concerning when results show energy consumption levels much higher than the standard profile. For most of devices tested here (both televisions, video game console, desktop and laptop computers, and rice cooker), the upper range was much larger than the lower range, indicating that deviations resulting in higher use would be more extreme than deviations resulting in lower use. Only the two pod coffee makers exhibited more profiles with energy consumption below the standard profile than above, which is due to the power management settings being disabled by default for those devices.

6.1.1 Active Use

The duration or frequency of active usage is a significant factor influencing energy consumption for many of the evaluated devices. Indeed, were power management not being tested, the effects of active use would be more pronounced for most devices.

Even considering the weight of power management, active use explained 40–43% of variation in energy consumption for the 4K and HD televisions, and 20–27% of variation for the desktop, laptop, video game console, and streaming device. Of these devices, the televisions, video game console, and desktop computer use a relatively large amount of energy in their standard profile compared to others. It is especially troubling how much more energy is used by the newer 4K television than the HD television. These results thus add weight to efforts to reduce power draw during the active state for these devices.

It is important to distinguish between active use (when the user directly benefits from the device being active) and the active state itself, which may continue long after active use has ended, if power management fails (that is, automatic low-power settings are disabled and the user neglects to manually turn off the device). Thus, the high power draw of the active state contributes to the energy waste attributed to the power management aspect in these results. More aggressive improvements to energy efficiency during the active state would thus also save energy during user-idle time, when the devices are left on and unused either prior to or in the absence of automatic transitions to a low-power state.

Reducing energy usage during operation typically requires comparable device utility: that is, to modify the device so that it uses less power without sacrificing functionality or features. For computers, this would be improving the way energy is used during idle periods. When not required, the device will self-regulate to passively save energy. After testing multiple generations of computers for this project, improvement in idle energy usage was easily observed. Promotion of alternate solutions when possible helps. For example a substantial energy penalty is paid to stream online content on a video game console versus a dedicated streaming player.

6.1.2 Pattern of Use

The pattern of use for this investigation was defined as the number of times or periods the device was used, and the amount of time between those uses, given a specific amount of total active use. One way pattern of use can affect overall energy consumption is if the device requires an energy-intensive warm-up or boot-up period at the beginning of each use or if it has a long or otherwise wasteful cool-down period after each use. Some devices tested here, such as the video game console, do have a separate boot-up and/or shutting down process with a relatively high power draw. However, as these processes are quite short in duration, the resulting contribution to overall energy consumption by restarting the device multiple times during the day is not substantial.

Another significant issue for pattern of use is the idle time due to automatic sleep or auto-off settings with long delay times, which accumulates every time the device is used and left idle again. In this case, pattern of use can be seen as an example of a power management problem, in which the solutions are to reduce the amount of energy used when the device is on but idle and reduce the amount of time the device spends idle. Some devices exhibited a small effect of pattern due to long delay times, such as the streaming device. All other things being equal, pattern does matter in such situations. However, this effect was overwhelmed by variation in active use and in power management behaviors.

The rice cooker provided a third way in which pattern of use matters for energy consumption: when the device requires a baseline amount of energy for a single use, with fairly small distinctions between a small versus large amount of product or service provided. Here, the effect of pattern can be interpreted as an effect of active use, in that the only solution would be to reduce the baseline energy consumption for the active cooking state. The lesson should also apply to other types of kitchen appliances that cook food or heat water. Although it would seem that pod coffee makers would suffer from this effect, the design has largely addressed the problem already: instead of heating the entire reservoir of water, the pod coffee makers only heat enough water for a single cup at a time, greatly reducing the impact of a long keep-warm period even when power management settings are disabled.

In sum, pattern of use can affect energy consumption of plug-load devices, but the effects for most plug-load devices are small compared to effects of power management and active use.

6.1.3 Power Management

The power management definitions used for the device use profiles combined two factors: settings that automatically transition the device into a sleep or soft-off state after a specified delay time of inactivity, and whether or not the user turns off the device immediately after using it. For every device, a moderate level of power management is defined with whatever automatic setting is the factory default (if any) along with the most likely user reaction at the end of use. Most devices have at least one low level, in which any power management setting is disabled and the user leaves the device on, and at least one high level, in which the user turns the device off after each use, negating any effect of automatic power management setting. Given this wide range of behaviors, it is not surprising that power management had a significant impact on energy consumption for all devices, and was the primary factor in variation across profiles for most devices. Still, while few would question the general idea that power management is important, this study helps show the importance of systematically examining when and how much specific power management behaviors (both settings and manual shut-downs) affect energy consumption.

The devices studied in the current project revealed three main failure points for power management: when automatic settings are disabled or otherwise ineffectively utilized, when low-power modes do not save much energy, and when devices remain in a fully functional active state during long periods of idle. A potentially missed opportunity for reducing energy consumption was also identified: automatic transitions to a low-power state based on the status of connected devices was shown to be very effective in one device, and could be effective in others.

The most pressing problem is how to get more devices to automatically transition into sleep or other low-power modes. Unlike those of earlier generations, all of these devices offered at least one low-power mode and an automatic power management setting for transitioning to it. However, if automatic sleep or auto-off settings are disabled, they do not save any energy. Worse, they result in devices remaining on for long periods, even all day long, every day. CalPlug's field study shows that many office desktop computers are left idle at all times, but little research is available to indicate how often users leave other devices on all the time. However, the effect of not using power management and leaving devices in the active state all day long is so large that even if only a small proportion of households do this, it would take a much larger proportion of households consistently enacting stringent power management behaviors to counteract all the wasted energy.

The simplest step to getting more devices into low-power states is to enable the energy-saving settings by default. To their credit, most devices already do this. The pod coffee makers are the one exception: for both models, the user would have to realize that the setting existed, realize that it was not enabled, and figure out how to enable it. For some devices, it may be possible to take away the users' ability to disable power management settings without reducing user satisfaction; this is already done with smart phones, and users have broadly accepted that limitation. However, this approach could be problematic for other devices where users are accustomed to having more control, especially for those where users may have valid reasons for leaving the device on and idle for long periods (for instance, computer users who cannot remotely access their work desktops if they are in sleep mode). More research into when and why users disable their sleep settings would be needed before the effects of enforcing settings could be estimated.

A more complicated issue is how to design the power management settings and the associated user interface to best encourage users to keep them enabled. Although little research has been done on this topic, anecdotal evidence—including countless tech forums answering users' questions about why their devices are mysteriously turning off—indicates two problems. First, users are confused about sleep settings. Second, the most common response to being annoyed by even a few undesired sleep and shutoff events is to disable all automatic PM settings. Once settings are disabled, users may forget they even exist. Most manuals and settings pages, including for the devices tested here, do little to explain the reasoning behind the settings, or encourage users to change the settings to a longer delay period rather than disable them, or to try to motivate users with energy-saving or "green" messages, which have worked in other applications. Furthermore, almost nothing is known about exactly what settings users would ideally want to use, what signals might work to help prevent unwanted shut-down events, or what their annoyance threshold is for how long they're willing to wait for a device to restart from sleep. For instance, if a television gave users a certain signal that it was going to switch off in 5 min, so that users could easily forestall a false auto-off, it could be possible to set the default auto-off delay time to 1 h instead of 4 h without any decrease in user satisfaction. Much more research and development is needed to fully address these issues.

The second main failure point was illustrated in the current research by the set-top box, for which the stand-by mode uses almost as much energy as the fully functional active state. As a result, the device showed almost no variation in energy consumption across device use profiles, even when the power management settings were enabled. The solution here is simple, at least in concept: reduce the energy consumption of the supposedly low-energy state. The overarching goal is to get devices to spend more time in the low-power state, and the more that goal is met, the more important it is to incrementally reduce power draw in the sleep or standby state.

The third failure point identified here is when devices spend considerable time at full power during periods of inactivity when they could conceivably enter a lower-power idle state. Computers lead by example here, by shifting into "short idle" and then into "long idle" states in the absence of user input; these states pause certain processes to save energy, yet leave the device ready to quickly resume full activity when the user returns. While worthwhile efforts to further reduce energy in idle state for computers continue, efforts to do the same for other devices are warranted. For example, when the video game console is not running a game, it switches to a main menu state that uses almost as much energy, where it remains indefinitely unless it transitions to sleep or is turned off. Reducing the consumption of the main menu state would be a substantial improvement.

Finally, use of linked devices for guiding power management was an effective approach for the sound bar. Specifically, an input-specific power management option switches the sound bar off when the device sending audio input to the sound bar, such as a television, sends no input for 5 min. This feature could function similarly in other connected devices that offered no such option, indicating a missed opportunity for savings. For example, any device that requires the television to display content could be set to transition to standby or soft-off if the television is turned off or transitions to sleep mode. A related alternate approach is to use a Tier-2 Advanced Power Strip (Tier-2 APS) to turn off devices, as well as reduce the burden of standby loads from these devices.

In sum, the evaluation of the effects of power management and potential improvements in its use and effectiveness, especially if combined with variation in active use and use patterns, is a rich area for continued investigation.

6.1.4 Evaluation Method Limitations

The current analysis shows the promise of the device use profile approach for assessing potential energy consumption across various users. However, the approach is inherently limited by the quality and reliability of the information used to define the model's parameters. In the current format, each aspect—active use, pattern of use, and power management—was defined with at least three levels of behavior: low, moderate, and high. Unfortunately, solid data on how devices are used in the field is sorely lacking. As such, most of the definitions used here were constructed by the research team based on assumptions and anecdotal observations, and for power management, by the options offered by the device. Self-reports of the amount of active use per day are available for a few of the devices here, but even for those, many survey questions use categories (e.g., a range of hours of TV use) rather than point estimates. No reliable data could be found on how people actually use power management in other devices, or on patterns of usage over the day. For this reason, no attempt was made to further differentiate weekday versus weekend use or to extrapolate to estimated annual energy consumption, which would require additional levels of assumptions that could not be warranted.

In short, as with any research of this sort, the results of the device use profile analysis are only as good as the assumptions that underlie its measures. Even assuming that the definitions are accepted as representing *some* users, the lack of data means no conclusions can be drawn about *how many* users fit each profile. The findings provide useful boundary conditions and a range of energy consumption results based on reasonable behaviors. If most of those behaviors produce much higher energy consumption estimates than the standard or "moderate" device use profile, this reduces the chance that natural variation in device use will average to the standard testing's mean and increases the chance that users with higher energy use profiles will outweigh those with lower-use profiles. This illustrates the importance of conducting more research on how devices are actually used in real-life situations, if accurate estimates of annual energy consumption for device types are desired.

The findings demonstrate the utility and potential of the PLSim tool that was developed and described here. In the present version, each parameter was input into a template usage file. This file was used to create a usage schedule in PLSim. This manual process serves as an effective demonstration, but can be tedious in usage and warrants additional development to add automation.

References

1. U.S. Energy Information Administration. (2011, March 28). Share of energy used by appliances and consumer electronics increases in U.S. homes. Retrieved from https://www.eia.gov/consumption/residential/reports/2009/electronics.php.
2. U.S. Energy Information Administration. (2019). *Annual Energy Outlook 2019 with projections to 2050*. Washington, D.C.: U.S. Energy Information Administration (EIA). Retrieved from https://www.eia.gov/outlooks/aeo/pdf/aeo2019.pdf.
3. Xia, L., Pixley, J. E., & Gago-Masague, S. (2017). *SIM Home Final Report*. Rosemead, CA: Prepared by California Plug Load Research Center (CalPlug) for Southern California Edison. Retrieved from https://www.etcc-ca.com/reports/sim-home?dl=1565392230.
4. Nordman, B., & Sanchez, M. (2006). Electronics Come of Age: A Taxonomy for Miscellaneous and Low Power Products. In *Proceedings of the 2006 ACEEE Summer Study on Energy Efficiency in Buildings*. Washington, D.C.: American Council for an Energy-Efficient Economy. Retrieved from https://aceee.org/files/proceedings/2006/data/papers/SS06_Panel9_Paper22.pdf.
5. Cisco. (2016). VNI Complete Forecast Highlights. North America 2021 Forecast Highlights. Retrieved from https://www.cisco.com/c/dam/m/en_us/solutions/service-provider/vni-forecast-highlights/pdf/North_America_2021_Forecast_Highlights.pdf?dtid=osscdc000283.
6. ReportLinker. (2017, April 27). When You Give a Kid a Tablet, He'll Ask for More Time [Results from a Representative Online Survey]. Retrieved from https://www.reportlinker.com/insight/give-kid-tablet-ask-time.html.
7. U.S. Energy Information Administration. (2016). *Annual Energy Outlook 2016 with projections to 2040*. Washington, D.C.: U.S. Energy Information Administration (EIA). Retrieved from https://www.eia.gov/outlooks/aeo/pdf/0383(2016).pdf.

8. Delforge, P., & Horowitz, N. (2014). *The Latest-Generation Video Game Consoles. How Much Energy Do They Waste When You're Not Playing?* [Online]: Natural Resources Defense Council (NRDC). Retrieved from https://www.nrdc.org/sites/default/files/video-game-consoles-IP.pdf.
9. Gelber, K. (2017). *The Powerful Pull of Plug Loads and the Hidden Costs of Plugging In*. Boulder, CO: E Source. Retrieved from https://www.esource.com/tas-1700025-001/powerful-pull-plug-loads.
10. Navigant Consulting, & SAIC (now Leidos). (2017). *Analysis and Representation of Miscellaneous Electric Loads in NEMS*. Washington, D.C.: Prepared for the U.S. Energy Information Administration (EIA). Retrieved from https://www.eia.gov/analysis/studies/demand/miscelectric/.
11. Palmgren, C., Stevens, N., Goldberg, M., Barnes, R., & Rothkin, K. (2009). *2009 California Residential Appliance Saturation Study*. [Online]: Prepared by KEMA, Inc. for California Energy Commission. Retrieved from https://ww2.energy.ca.gov/appliances/rass/previous_rass.html.
12. Delforge, P. (2015). *Plug-In Equipment Efficiency: A Key Strategy to Help Achieve California's Carbon Reduction and Clean Energy Goals*. [Online]: Natural Resources Defense Council (NRDC). Retrieved from https://www.nrdc.org/sites/default/files/home-idle-load-plug-in-efficiency-IB.pdf.
13. Klopfer, M., Pixley, J. E., Syed, S., & Li, G. P. (2019). *Advancing Plug Load Efficiency with Behavioral-Focused Usage Evaluation* (ET17SCE1190). Rosemead, CA: Prepared by California Plug Load Research Center (CalPlug) for Southern California Edison. Retrieved from https://www.etcc-ca.com/reports/advancing-plug-load-efficiency-behavioral-focused-usage-evaluation?dl=1586231844.

Regulating On- and Off-Grid Lighting and Appliances in Eastern and Southern Africa

Naomi Wagura and Elisa Lai

Abstract With increased electrification and economic growth, the demand for appliances and lighting is starting to grow in eastern and southern Africa. However, some countries that still have significant populations that are living completely without electricity have realized that only using the traditional grid-extension model is insufficient to reach the most remote rural population quickly and cost-effectively, and thus missing the Sustainable Development Goal 7 of 'universal energy access by 2030' target. As a result, these countries have adopted renewable off-grid electrification models, utilizing decentralized energy systems such as mini-grids and solar home systems to fully electrify the hard-to-reach populations. This trend of hybrid on and off-grid electrification plans is likely to continue in the eastern and southern African countries with significant but sparse off-grid populations.

In lock step with the expansion of energy access across this region, the demand for energy services like lighting, cooling, cooking, heating and water supply is increasing for households electrified using both on- and off-grid solutions. While one of the differences between the two is that grid-connected households will use alternating-current (AC) appliances while off-grid households often use direct-current (DC) appliances, the differences between the appliances used goes beyond their power source. In some cases, some of the performance metrics for on- and off-grid appliances are similar. These differences and similarities have to be considered when regulating these appliances.

Most countries in the region have not yet regulated the performance and quality of both on- and off-grid lighting and appliances. This paper focuses on lighting and refrigerators, two product groups that have seen rapidly growing consumer demand in the region. It explores the differences and similarities between the on- and off-grid products, the current state of the market for these products, and the current policy landscape for each of these product types, and provides recommendations for governments that are considering regulating the performance and energy efficiency of these products.

N. Wagura · E. Lai (✉)
CLASP, Washington, DC, USA
e-mail: elai@clasp.ngo

P. Bertoldi (ed.), *Energy Efficiency in Domestic Appliances and Lighting*, Springer Proceedings in Energy, https://doi.org/10.1007/978-3-030-79124-7_23

1 Introduction

In some countries in east and southern Africa like Kenya, Malawi, Rwanda, South Sudan and Tanzania, the electrification rates have increased by more than 50% in the past 5 years (Fig. 1). On the other hand, the access to electricity in 14 out of 22 countries in the region is still below 50%.

In order to achieve universal energy access, more and more countries in the region now recognize that off-grid solutions like solar home systems can help increase access to electricity in hard-to-reach rural areas and are integrating these solutions in their national electrification planning. For example, Madagascar's energy policy lays out an electrification strategy that combines the use of traditional grid, mini-grids and solar home systems to achieve its electrification target of 70% access by 2030.[1] The electrification strategies in Kenya and Ethiopia also expressly include off-grid technologies like solar home systems. These strategies identify approximately 2 million households in Kenya[2] and approximately 5.7 million households in Ethiopia[3] that could be electrified using solar home systems and mini-grids.

Increased energy access using both traditional grid (and AC mini-grids) and DC power sources like solar lighting kits, solar home systems and DC mini-grids will translate to increased uptake of lighting and appliances. Government standards and labelling (S&L) programs would help pushing these countries' markets toward

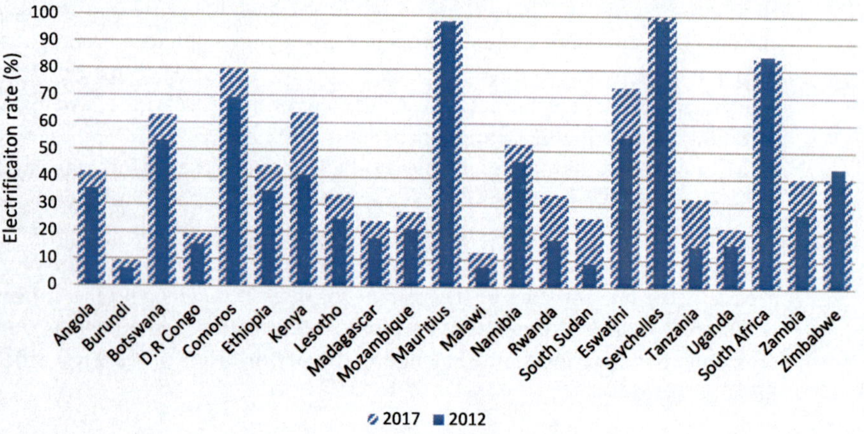

Fig. 1 Change in electrification rates in east and southern Africa between 2012 and 2017. (Source: World Bank)

[1] The policy can be accessed here http://www.ore.mg/Publication/Rapports/LettreDePolitique.pdf

[2] The Kenya National Electrification Strategy 2018.

[3] http://ethioenergybuz.com/energy-investment-climate/policy-regulation/13-ethiopia-launched-national-electrification-program-implementation-road-map-and-financing-prospectus

higher quality and efficient products, and would save the consumers energy and money.

Minimum energy performance standards (MEPS) specify the minimum efficiency requirements for products manufactured in or imported into a country. For most products, the MEPS also give specifications on other performance features. For example, in addition to efficacy, lighting MEPS may also specify requirements on the lifetime, colour temperature and mercury content (for fluorescent lamps). Energy labels affixed on products provide consumers with information on the key performance metrics of the appliance. Labels can either be endorsement or comparative labels. Endorsement labels communicate to the consumer that the labelled appliance is the best of its kind in terms of performance while comparative labels provide consumers with performance information to compare similar products.

Given the anticipated increased uptake of on- and off-grid appliances, introducing standards and labelling programs—a proven low-cost method of increasing the quality and performance of AC appliances—would have the following advantages for governments and consumers in Africa:

- Prevent product dumping from more developed appliance markets that have already instituted S&L programs. This includes importation of old inefficient appliances and appliances that do not meet the performance standards in source countries. Standards provide consumer protection while ensuring consumers are getting high-quality products and energy services.
- Help consumers differentiate the quality appliances when making purchasing decisions. This is especially important for off-grid consumers who are typically cash-poor and therefore purchasing an appliance is a significant investment. Labels can provide benchmarks and data for consumers to compare products based on their expected energy performance.

This paper dives into the similarities and differences between on- and off-grid lighting and refrigeration technologies, and gives an overview of the markets for these products to provide an understanding of the differences and similarities between the technologies and markets for these products. It then compares the performance of these two technologies based on similar metrics to determine if they are comparable and the current policy landscape in the region. Finally, we give some recommendations to governments on how to go about regulating these two types of products in the region.

2 Defining the Technologies: Differences and Similarities Between On- and Off-Grid Lighting and Refrigerators

Product definition is a key element in appliance policy development that determines physical characteristics and scope of the regulations. To pave the way for the policy development discussions, this section aims to establish a common understanding for both "on-grid" and "off-grid" appliances—including whom these technologies are designed for and how they are different from and similar to each other.

2.1 *Understanding End-Users*

The first step in understanding the differences between on-grid and off-grid technologies is to understand the needs of the end-users of these technologies.

2.1.1 Similarities Between On- and Off-Grid Appliance Users

Constrained incomes are a concern for both on- and off-grid appliances users in the region. Additionally, the power supply in most grid connected rural villages in the region is often unreliable and prone to poor grid conditions such as brownouts and blackouts. Similarly, although to a larger extent, off-grid users have energy supply constraints.

2.1.2 Differences Between Off- and On-Grid Appliance Users

Compared to households in urban and peri-urban areas connected to a reliable grid, energy supply is more of a constraint for off-grid users reliant on solar home systems.

Many people in un- and under-electrified "off-grid" villages rely on fuels like kerosene and diesel for their daily energy needs. For many of these villages, grid connectivity is an unrealistic solution due to technical and economic limitations. The introduction of decentralized solar technologies such as pico-PV off-grid lighting products and solar home systems provides an opportunity for off-grid consumers to obtain modern energy services.

2.2 *Design Considerations for On- and Off-Grid Appliances*

Due to the similarities and differences between the appliance users, on- and off-grid appliances have some similar design considerations but also some differences.

2.2.1 Similarities Between the Design Considerations for On- and Off-Grid Appliances

The income constraints experienced by both types of users require that appliances sold to on- and off-grid consumers are affordable. However, the appliances also have to be good quality in order for the users to derive maximum value from their appliances.

The appliances for both applications in the region also have to able to withstand energy supply constraints. This is especially true for users in grid-connected rural villages that experience frequent power quality issues and users of off-grid power systems [1].

2.2.2 Differences Between the Design Considerations for On- and Off-Grid Appliances

On-grid appliances are conventional appliances designed for use in AC grid systems. These are available in global markets and can be purchased from ordinary retail stores in the east and southern African region.

"Off-grid appliances" are products that are designed for and/or marketed for use with off-grid energy systems, such as a solar home system. As illustrated in Fig. 2, off-grid solar home system (SHS) typically includes a photovoltaic (PV) module, a low-voltage direct current (DC) battery, a charge controller, wiring, and sockets [2, 3]. A typical SHS includes a PV module with power rating ranging from 10 to 350 W and can power a number of lights and small DC appliances, such as a radio, cellphone charging, TVs, and fans.

Off-grid appliances are usually low-voltage (e.g. 12 V) DC appliances, but some AC appliances can be used with off-grid systems by converting the DC energy supplied by off-grid systems to AC supply using inverters.

In terms of technology, off-grid and on-grid appliances are quite similar since they deliver similar energy services e.g. lighting and cooling. But because off-grid appliances are often used in relatively harsh conditions, quality, affordability and durability are key design considerations.

Some of the distinct design considerations for off-grid appliances are listed below, though the list is far from comprehensive.

Fig. 2 An Illustration of a solar home system kit

- **Energy consumption**: Like on-grid consumers, off-grid consumers want to enjoy sufficient lighting service (e.g., provided by a solar lantern) to read at night, cooling provided by a refrigerator to enjoy a cold soda, and having a television to get access to information and entertainment. However, off-grid appliances need to deliver the level of energy services that off-grid consumers desire while operating under the environmental, i.e. limited and unpredictable solar hours, and infrastructure constrains, i.e. limited PV panel and battery capacity. Energy efficiency thus become a particularly important consideration in off-grid appliances design. For example, a 60 Wp solar module and a 30 Ah battery can power a 25 W incandescent bulb for 6 h per day, but the same solar system can power a highly-efficient 50 L DC refrigerator for 24 h.
- **Affordability**: Because their users typically have very little disposable income, off-grid appliances need to be designed in a way that balances performance and cost. Off-grid consumers are among the poorest, most under-served people in the world. In East Africa—Tanzania, Kenya, and Uganda alone—over 55 million people are living in extreme poverty, living on less than $1.90 a day [4]. According to a World Resources Institute study, cash-poor people living in off-grid villages spend up to $0.40 a day, or roughly 30% of their earnings on energy [5]. Highly-efficient appliances require much less energy and therefore reduce the size of off-grid energy systems needed to provide the same service, lowering costs considerably and making these technologies affordable for more off-grid consumers. Research indicates that by using the most-efficient lights, TV, fan, radio and mobile phone charging, the upfront purchase cost of a solar home system and appliances can be reduced by as much as 50% [6]. The introduction of pay-as-you-go (PAYG) technology, which enables off-grid consumers to use basic mobile phones to pay off the entire SHS and appliances in small installments on a weekly or monthly basis, greatly improve the affordability of off-grid appliances.
- **Durability**: Durability is also particularly important in the case of off-grid appliances. Many off-grid consumers live in remote areas with almost no access to repair technicians or replacement components. Once an appliance stops working, most off-grid households cannot afford to replace faulty appliances. Making off-grid appliance durable, low-maintenance, or easily repairable by local technicians are essential design considerations for high-cost appliances such as refrigerators.
- **Interoperability and electrical resilience**: The off-grid appliance market is largely unorganized and there is no standardization around connectors and voltage ranges between solar systems and appliances. Off-grid appliances need to be tolerant to constant voltage fluctuations from solar and battery systems. The voltage being too high or too low can cause electrical failure due to overheating, causing safety hazards for off-grid consumers.

An off-grid appliance manufacturer needs to take all these factors into consideration, likely make some trade-offs between these factors and design appropriate products based on the characteristics most demanded by consumers and market conditions.

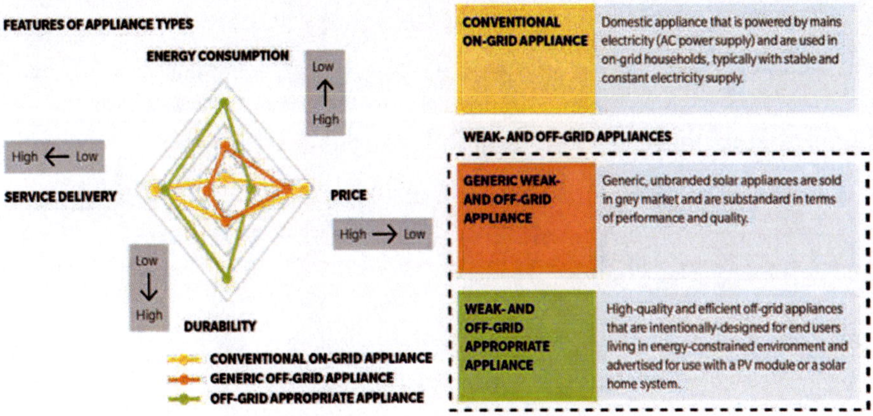

Fig. 3 Categorization of appliance types

While many manufacturers can produce low-voltage DC appliances, only a few design and produce appliances that are tailored for use in the off-grid context and environment. Given the newness of this market, and limited information about it, many appliance manufacturers are not familiar enough with this market to design and market off-grid products (Fig. 3).

2.3 *On- and Off-Grid Lighting and Refrigerator Technology Types*

This section explores the similarities and differences between on- and off-grid lighting and refrigerator technology types.

2.3.1 On- and Off-Grid Lighting

General service lamps (GSLs) are typically used in on-grid residential sectors for general lighting purposes. Based on the US Department of Energy's definition, GSLs include the following types: compact fluorescent lamps (CFLs), general service light-emitting diode (LED) lamps, organic light-emitting diode (OLED) lamps, and any other lamps that are used to satisfy lighting applications traditionally served by general service incandescent lamps (GSILs) [7].

The light sources or light emitting components in off-grid solar lighting products are typically same as those of on-grid lamps. But there are additional components in off-grid solar lighting systems, including: (1) a solar photovoltaic module integrated or separate from the lighting product as the energy source, (2) lighting sources which can be LEDs, CFLs, or other light emitting components, (3) a power control unit, and (4) batteries [8].

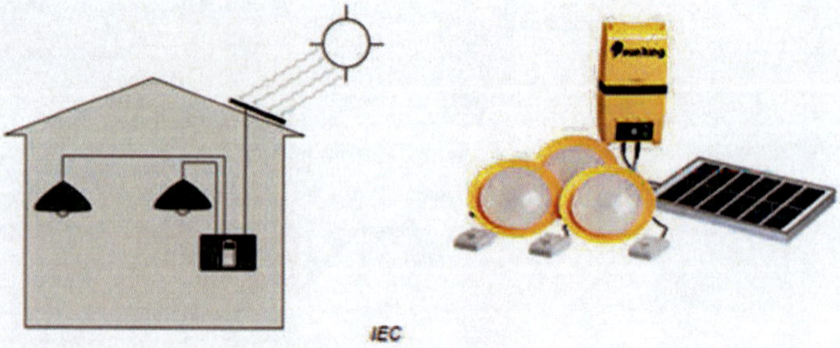

Fig. 4 An illustration of fixed components mounted indoors (left) and an example of a pico solar PV kit (Sun King Home 60). (Source: IEC TS 62257-9-5)

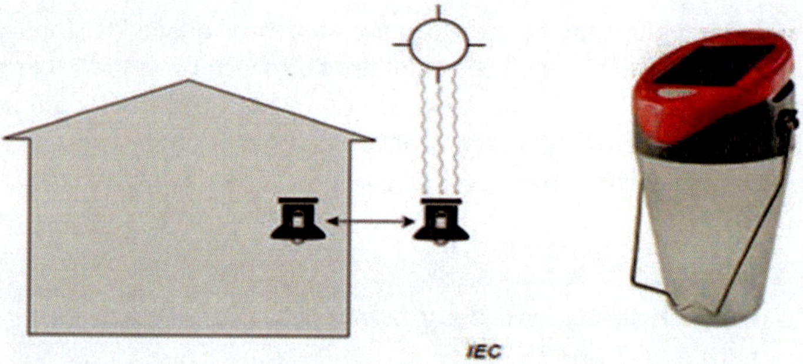

Fig. 5 An illustration of portable components that can be used either indoors or outdoors (left) (Source: IEC TS 62257-9-5). A photo of solar lantern (d.light S30) as an example of portable components (right)

The off-grid solar lighting system can have multiple different configurations, including the following types of components, for various use cases—indoor, outdoor, and portable.

- Fixed components: Light sources are designed for permanent or semi-permanent mounting and use in place (Fig. 4).
- Portable components: Light sources are portable and typically contain an internal energy source. For example, a solar lantern can include a portable light with a battery that can be charged from the solar module or from another battery in a fixed enclosure (Fig. 5).

2.3.2 On- and Off-Grid Household Refrigerators

According to the definition included in the internally-accepted refrigerator testing standard, IEC 62252, a refrigerating appliance, is an "insulated cabinet with one or more compartments that are controlled at specific temperatures and are of suitable size and equipped for household use, cooled by natural convection or a forced convection system whereby the cooling is obtained by one or more energy-consuming means [9].

The physical characteristics of off-grid refrigerators are similar to the definition in the IEC 62552. However, off-grid household refrigerators often have dual uses—for both household and small commercial uses—and can be powered by any of a wide range of power input options including AC, DC and solar direct drive.

There are different technology and power supply types of off-grid refrigerators, as defined in the Global LEAP[4] Off-Grid Refrigerator Test Method [10]:

- Continuous power supply refrigerator: Refrigerator designed for continuous (24/7) AC or DC power supply, generally without any integrated thermal or electrical battery. The design of this refrigerator is similar to that of a conventional refrigerator.
- Solar battery-based refrigerator: Refrigerator designed for intermittent power supply, generally containing an integrated thermal and/or electrical battery allowing autonomous operation during periods when power supply is lacking (Fig. 6).
- Solar direct drive refrigerator: DC supply refrigerator designed for direct connection with a photovoltaic solar panel, generally containing an integrated thermal and/or electric battery to allow autonomous operation during the night (Fig. 6).

Off-grid refrigerators are designed to operate without constant power. Thus, the autonomy of the refrigerators, i.e., how many hours the refrigerator can maintain cooling without power input, becomes an important design parameter. It is common

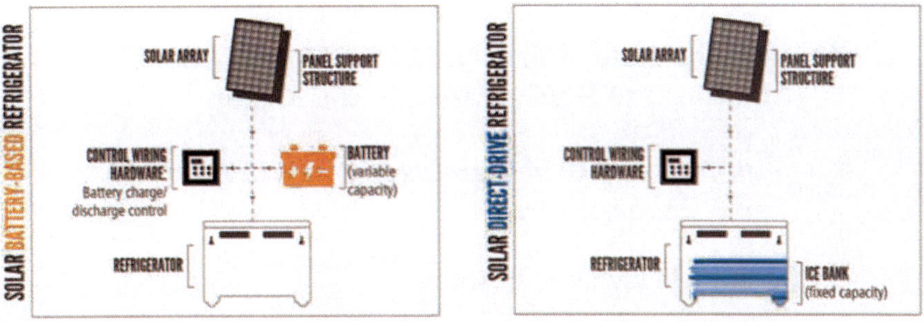

Fig. 6 Illustration of battery-powered refrigerator and solar direct drive refrigerators

[4]The Global LEAP Awards is a global competition that aims to provide the off-grid industry with information on the world's best and most energy-efficient off-grid appliances.

to see off-grid refrigerators built with thicker insulation in the cabinet and thermal storage devices such as phase-change material to improve temperature management when there is no power being supplied to the unit.

2.4 Conclusion on the Similarities and Differences Between On- and Off-Grid Lighting Products and Refrigerators

As elaborated in the sub-sections above, there are several similarities between energy-consuming products designed for on- and off-grid use. The users have similar economic constraints and in some cases, energy supply constraints. This means that products targeted at both user types in the region should be affordable while also able to adapt to the energy constraints.

However, the energy constraints of off-grid consumers are higher than those of their on-grid counterparts. Additionally, off-grid products are usually used in harsher conditions with limited access to technicians and replacement parts. Manufacturers of these off-grid products therefore have to balance the energy consumption, durability, interoperability and electrical resilience qualities required for these appliances with affordability.

For lighting products, while the light emitting components of on- and off-grid products are similar, off-grid lighting products have additional power system components either integrated into or separate from the light emitting components. For refrigerators, the general physical characteristics of both on- and off-grid products are similar. However, off-grid refrigerators can be powered by a range of power supply options which means there are refrigerators designed for use with a constant power supply, those with integrated batteries that are designed for use with intermittent power sources and solar direct drive refrigerators which are connected directly to a solar photovoltaic panel.

3 Current State of the Market for On- and Off-Grid Lighting and Refrigerators in the Region

3.1 On- and Off-Grid Lighting Markets in East and Southern Africa

3.1.1 On-Grid Lighting Market

Figure 7 shows that the size of the on-grid lamps market varies across the region. The general service lamp market in South Africa was approximately 130 million lamps in 2017 while that in Uganda was approximately 700,000 lamps [11].

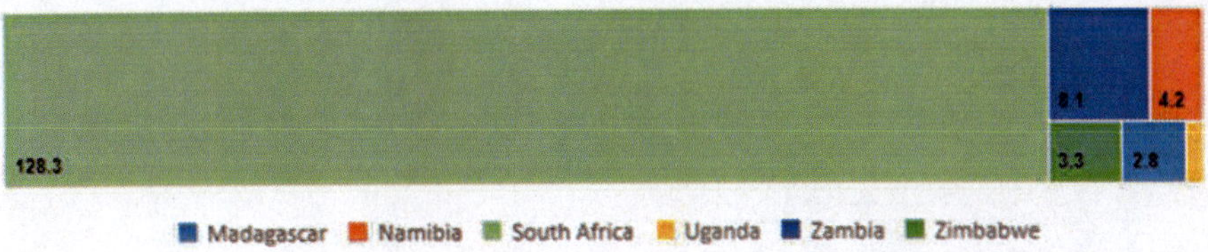

Fig. 7 On-grid lamps market—in millions—for select markets in East and Southern Africa. (Based on data from the UN COMTRADE database)

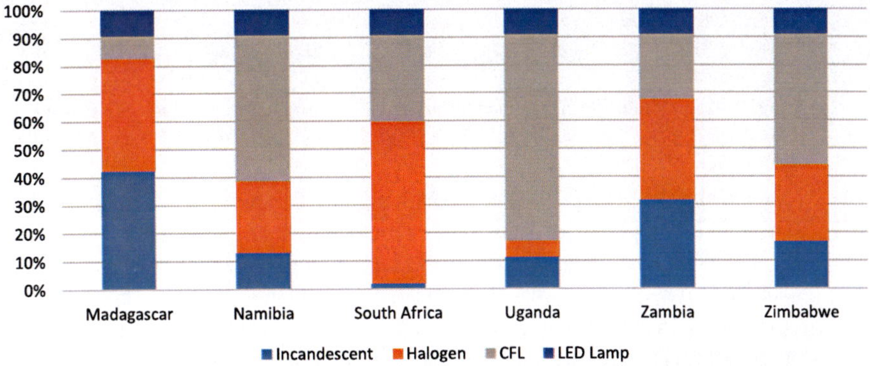

Fig. 8 Proportion of various lamp technology types in select countries in East and Southern Africa. (Source: [11])

In the on-grid market in sub-Saharan Africa, LED lighting is becoming increasingly popular, with the price of LED lamps now competitive with conventional lighting products such as compact fluorescent lamps. As shown in Fig. 8, LED lamps now account for approximately 10% of the total general service lighting market in these representative countries in the region. In some markets like Zambia and Madagascar, incandescent lamps still take up more than 30% of the market.

Most markets in the region are purely import-based although there is some local assembly of lamps. For example, LEDlite in Kenya is collaborating with an Indian LED manufacturer to set up a local LED lamp and luminaire assembly plant and Lesedi Electricals in Botswana assemble lamps in-country.

3.1.2 Off-Grid Lighting Market

Off-grid solar (OGS) devices, including solar lighting kits and solar home systems, currently provide electricity access to an estimated 73 million households globally. In sub-Saharan Africa, the potential market for OGS products is estimated to be 124 million households that are off-grid and 47 million households that are connected to

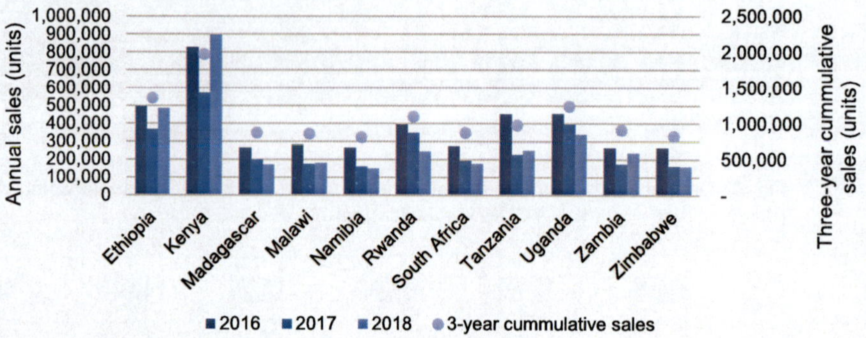

Fig. 9 Annual and 3-year cumulative off-grid solar product sales in select countries in East and Southern Africa. (Source: [13])

an unreliable grid [12]. The global market for pico-solar products[5] lanterns is expected to grow by a CAGR of 22% between 2017 and 2022.

From GOGLA[6] market data, the largest off-grid solar markets in eastern and southern Africa are primarily in eastern Africa (Fig. 9). In the last 3 years, Kenya, Ethiopia, Rwanda and Uganda recorded the largest number of units sold. In 2018, pico-solar products made up approximately half of the total volume of off-grid solar device sales in sub-Saharan Africa [13]. Factors such as increased penetration of pay-as-you-go (PAYG)[7] product financing mechanisms and continued technology innovations to bring down costs and improve service delivery will mean that the off-grid solar products will reach more and more of the region's off-grid population.

The luminous efficacy of cool white LED packages[8] has increased more than fivefold in the past 15 years from 25 to 160 lumens/W [14]. This increased efficacy has transformed both the on- and off-grid lighting market. Almost all lighting products in the off-grid market are now LED-based. For example, in Kenya, the share of LED-based products in off-grid lighting solutions increased from 87% to 99% between 2009 and 2014 with the use of solar PV as the primary power source of the lighting products increasing from 1% to 35% within that period [15]. Efficacy is especially important in the off-grid context since the power draw of the load is directly proportional to the size and consequently the cost of the solar PV system.

[5] Lanterns and simple multi-light systems (which may enable mobile charging) of 0–10.999 Wp. These enable partial or full Tier 1 electricity access to a person or household (GOGLA).

[6] GOGLA (Global Off-grid Lighting Association) is a global association for the off-grid solar energy industry.

[7] Pay-as-you-go is a financing mechanism for off-grid solar products where the customers put down a fraction of the product cost as a deposit and pays back the remaining amount over time through daily, weekly or monthly payments over a fixed period. The products are equipped with a lock-out mechanism where the product automatically turns off when the payment date passes without payment. The customers especially in Eastern Africa make their payments through mobile money platforms which are ubiquitous in the region.

[8] LED packages produce and mix the various colours of light to form white light.

For pico-solar lighting products, switching from a CFL lighting source to LED leads to a 33% decrease in cost [15].

The off-grid solar market in the region is primarily import-based with manufacturing done in China, India, and other Asian countries. However, there is some local assembly in Southern Africa—e.g., Maxlite and Specialized Solar Systems manufacture solar home systems in South Africa.

3.2 *On- and Off-Grid Refrigerator Markets in East and Southern Africa*

In rural areas in east and southern Africa where smallholder farming is the key economic activity, greater access to off-grid refrigeration could reduce losses and potentially increase earnings from farm produce. However, refrigerator penetration—both on-and off-grid—in rural East and Southern African countries remains extremely below. In Uganda, Tanzania, and Kenya, the penetration rate is as low as 1–2%—due to the high costs and limited product availability [2, 3].

3.2.1 On-Grid Refrigerator Market

The market for AC-powered refrigerating appliances in the region is small but growing. In 2018, the number of units sold varied across countries from approximately 2 million units in South Africa (not shown in the chart) to a few hundred in the Seychelles (Fig. 10).

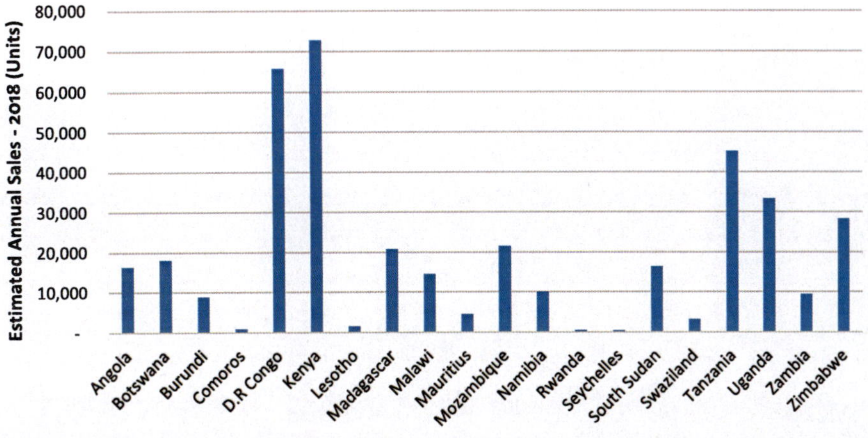

Fig. 10 Estimated annual refrigerator sales in 20 countries in East and Southern Africa. (Source: Euromonitor)

While most refrigerators in the region are imported, some countries like South Africa, have some local refrigerator assembly plants that assemble international brands [16].

3.2.2 Off-Grid Refrigerator Market

According to the 2018 Efficiency for Access Appliance Market Survey, refrigerators are one of the appliances with the highest consumer demand [17]. The off-grid refrigerator market is estimated to have immense market growth potential. It is estimated that by 2030, the global market size for off-grid refrigerators could grow to 14.3 billion USD if efficient, appropriately priced products become accessible to all households with the purchasing power to buy a refrigerator [18].

Due to the nascent nature of the market and limited market penetration, an approximation on the potential size of this market in east and southern Africa is difficult to make. A forthcoming Efficiency for Access report estimates that the addressable market for solar off-grid refrigerators among small shops, bars and restaurants in Kenya that are either off-grid or are connected to an unreliable grid is US$ 20 million [18].

The current prices of off-grid refrigerators are well above the affordability for most off-grid consumers. Figure 11 shows the retail prices for a sample of off-grid refrigerators with varying capacities and technology types. The prices of these units would therefore have to drop in order for the market to expand. Several initiatives of the Efficiency for Access Coalition are supporting the development of this market

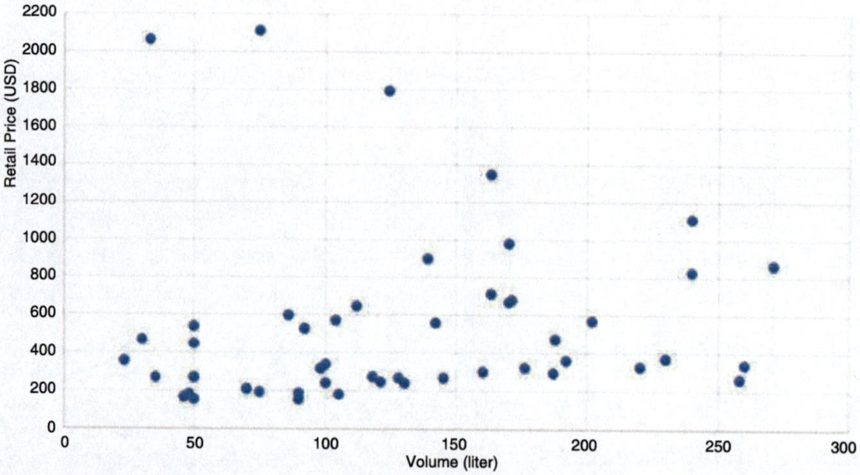

Fig. 11 Price vs. capacity of select off-grid refrigerators—N = 53. (Source: Equip Data)

by running product awards, financial incentives programs and providing research and development grants.[9]

There is some local manufacturing of off-grid refrigerators in the region. For example, Palfridge—with two models recognized by Global LEAP Awards in 2017 as the best-performing products in the medium and large refrigerator categories—manufacture their refrigerators in Swaziland.[10]

3.3 *Conclusions of the On- and Off-Grid Lighting and Refrigerator Markets in East and Southern Africa*

The on-grid lighting market in the region varies greatly between South Africa at the higher end and smaller economies like Uganda. The off-grid solar products market, which includes off-grid lighting products is most vibrant in eastern Africa with smaller markets in southern Africa. In terms of lighting technology types for both markets, in the on-grid market, CFLs, halogen and in some countries incandescent lamps still dominate the market with LED lamps taking up approximately 10% of the market. In the off-grid market, LED lighting products dominate the market due to the superior efficacy and as a results reduction in power system requirements. There is some local assembly of LED lamps in the region but both the on- and off-grid market is dominated by imports.

The market for on-grid refrigerators is small but growing in the region while that of off-grid refrigerators is expected to grow particularly if the prices of these units decrease. While the markets for both on- and off-grid refrigerators is dominated by imports, some countries like South Africa assemble on-grid refrigerators locally while Swaziland manufacture off-grid refrigerators.

4 Comparisons Between the Performance of On- Versus Off-Grid Lighting Products and Refrigerators

While being conscious of the differences highlighted in the previous sections, this section compares the performance of on- and off-grid lighting products and refrigerators based on two key metrics: lumens/watt for lighting and energy consumption for refrigerators. The rationale behind this comparison is to determine if on- and off-grid lighting products and refrigerators can be subjected to the same performance requirements when regulating them.

[9] More information on the coalition's initiatives can be found on their website here.

[10] Global LEAP Awards: 2017 Buyer's Guide for Outstanding Off-grid refrigerators. https://storage.googleapis.com/e4a-website-assets/Global-LEAP-Buyers-Guide-Refrigerators.pdf

4.1 Energy Consumption of Lighting Products

The efficiency and performance of on- and off-grid lighting products are defined and measured differently. For on-grid lighting products, the efficiency metric often used is lumens/watt—the amount of visible light you get for the amount of power drawn. For off-grid lighting products, the efficiency of light points is part of the system efficiency but does not tell the whole story. The Lighting Global Quality Standards focus on the total solar system performance rather than the efficiency of only the light points. The Lighting Global Quality Standards require product manufacturers to accurately report total light output in lumens and run time per solar charging day (in hours) based on the brightest light setting. The rationale behind these off-grid lighting performance metrics is informed by how off-grid consumers make purchase decisions. They typically want to know how bright the lamp is and how many hours of lighting services they can get per day.

Nevertheless, a comparison of the lighting products included in pico-solar products could help compare the two technologies using the same metric. This could help evaluate if pico-solar lighting products—and the lighting products included in larger off-grid solar systems—should be subjected to the same efficacy requirements to which on-grid lamps are subjected. It could also help evaluate the effect such a requirement would have on the off-grid solar market.

To make the comparison, the authors acquired data on the performance of the pico-solar PV lighting products evaluated by Lighting Global Quality Assurance (LGQA). Additionally, as part of CLASP's support to the Kenyan and South African lighting standards' development, CLASP procured 29 LED lamps from retailers in Kenya and South Africa in 2018 and 2019. Further data was also acquired from the UNDP program supporting the standards and labelling program in South Africa, which procured 32 lamps from South African retail stores.

LGQA tests the performance of the pico-solar products according to IEC/TS 62257-9-5. This test method covers pico-solar products and solar home system kits. The tests on the light output (Luminous flux) reference international test methods like CIE084 if the tests are carried out in an integrating sphere or goniophotometer.[11]

For the on-grid lighting products, an accredited laboratory at the Swedish Energy Agency conducted the tests on 29 LED lamps procured in Kenya while an accredited lighting laboratory at Eskom in South Africa conducted tests on 32 lamps.

Figure 12 shows the comparison between the lighting efficacy of the pico-solar products (N = 170) tested under LGQA and the LED lamps procured from retailers in Kenya and South Africa (N = 61). Pico-solar products that were previously certified by LGQA (in grey) have been included in the comparison to illustrate the change in the performance of pico-solar lighting products.

[11] IEC TS: 62257-9-5:2018 includes an alternative test method for light output using the multi-plane method.

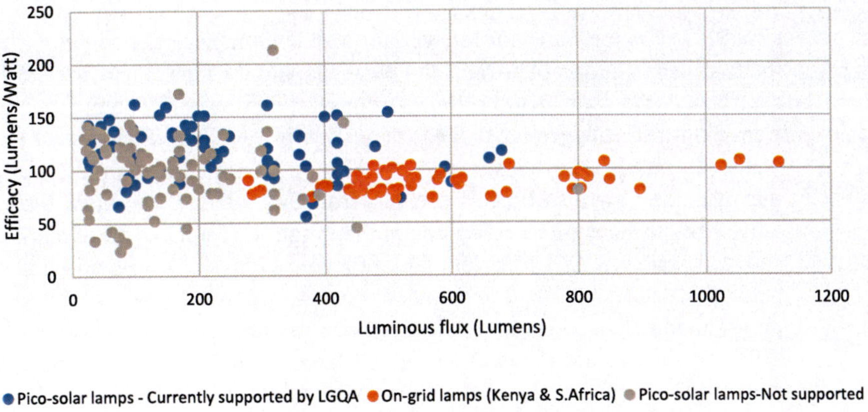

Fig. 12 Efficacy comparison between the pico-solar lighting products tested by LGQA and on-grid LED lamps procured in South Africa and Kenya

The median efficacy of the currently supported[12] pico-solar products is 114 lumens/W, which is higher than that of previously supported products, which is 101 lumens/W. The minimum efficacy of currently supported lamps is 56 lumens/W while that of previously supported products is 23 lumens/W. The median efficacy of the on-grid lamps is 89 lumens/W while the lamp while the lowest efficacy is at 73 lumens/W.

If the minimum efficacy level for both pico-solar lighting products and on-grid lamps was set at the level of the on-grid LED lamp with the lowest efficacy i.e. 73 lumens/W then 97% of the currently supported pico-solar lighting products and 79% of the previously supported products would meet the standard.

4.2 *Refrigerators Energy Consumption*

One of the key performance metrics for refrigerators is the daily energy (kWh/day) used to cool products stored to the target temperatures in the refrigerator compartments. As discussed in the section above, the IEC 62552:2015 test method can be used to measure the energy consumption and other performance metrics of refrigerators. This test method is the basis of the Global LEAP test method developed by

[12] Currently supported refers to the pico-PV lighting products whose certification credentials provided by the LGQA program have not expired. The LGQA certificates are valid for 2 years.

CLASP and is the basis for refrigerator standards in several economies including the European Union,[13] Australia and New Zealand,[14] China,[15] Korea[16] and Japan.[17]

To make the comparison on energy consumption, two data sources are used: the Kenya on-grid refrigerator database and Equip Data. The Energy and Petroleum Authority publishes a register that contains performance data on the refrigerators that meet the Kenya's MEPS.[18] The data from this database was used as a representative sample for on-grid refrigerators that might be found in other markets in the east and southern Africa region. The Efficiency for Access' Equip Data platform publishes data on off-grid appliances tested as part of the Global LEAP Awards program and also those procured directly from markets across the world. The off-grid refrigerators are tested at accredited laboratories using the Global LEAP test method.[19]

Since the current Kenyan refrigerator test method was based on the previous Australian/New Zealand standard (AS/NZS 4474.2:2009) the energy consumption measured using this test method has been normalised to the estimated energy consumption if consumption had been measured using IEC 62552:2015 at 32 °C ambient temperature [19]. The data from Equip Data comprises 29 refrigerator-only units: 10 models procured directly from various markets and 19 models that participated in the 2016–2017 Global LEAP Awards. For comparison, the on-grid refrigerators included in the analysis are only those that are classified as refrigerator-only models (N = 42). While the comparison is limited by the sample size and datasets available, Fig. 13 provide an indication on the energy consumption comparison of a small number of on- and off-grid refrigerators.

The annual energy consumption of off-grid refrigerators is between 43 and 532 kWh while the on-grid fridges range between 158 and 354 kWh. These on-grid refrigerators are all 1 or 2-star on the Kenya efficiency label scale, which goes up to five (5) stars. There are nine off-grid models, including those procured directly from the market, whose consumption is below the lowest energy consuming on-grid refrigerator. These would probably achieve a higher star rating on the Kenyan

[13] The standard can be accessed using the following link https://shop.bsigroup.com/ProductDetail/?pid=000000000030176266

[14] The standard can be accessed using the following link https://shop.standards.govt.nz/catalog/4474%3A2018%28AS%7CNZS%29/view

[15] The standard can be accessed using the following link https://www.chinesestandard.net/PDF/English.aspx/GBT8059-2016

[16] The standard can be accessed using the following link https://www.kssn.net/eng/webstore/ksinfo.asp?idx=130319

[17] The standard can be accessed using the following link http://kikakurui.com/c9/C9801-1-2015-01.html

[18] The register can be accessed here https://www.erc.go.ke/services/renewable-energy-2/energy-audit-firm-register/

[19] In the Global LEAP Off-Grid Refrigerator Test Methods, all off-grid refrigerator samples are required to be test in both 32 and 43 °C ambient temperatures given off-grid households often live in warmer climates. The off-grid refrigerator energy consumption data used in the analysis is measured in 32 °C ambient temperature.

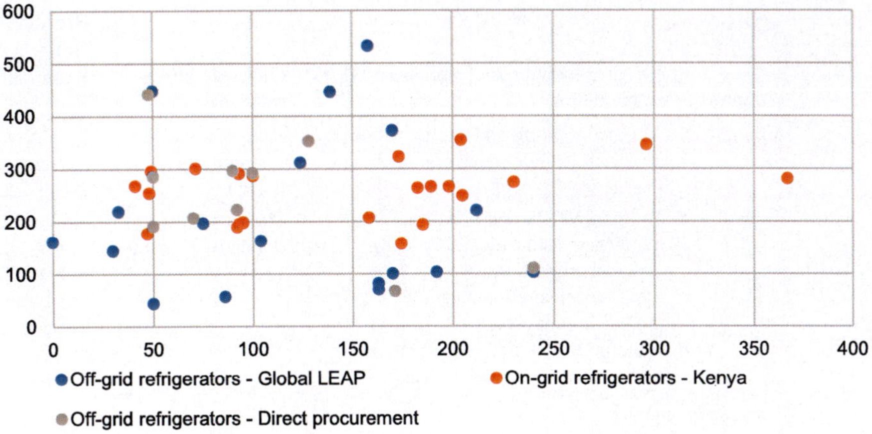

Fig. 13 Comparison between off- and on-grid refrigerator daily energy consumption

label.[20] There are also five off-grid refrigerators that consume more than the highest consuming on-grid refrigerator which would most likely not meet the Kenyan MEPS.

4.3 *Conclusions on the Efficacy and Energy Consumption of On- and Off-Grid Lighting Products and Appliances*

While a large percentage of the currently supported pico-solar lighting products would meet a 73 lumens/W efficacy level, it is important to note that the products certified by Lighting Global are designed to meet high quality standards in order to gain certification. These are therefore probably better quality than the products found in most markets in east and Southern Africa which are unregulated. However, as more countries in the region adopt Lighting Global (soon to be IEC) quality standards, these high quality off-grid lighting products will become more readily available.

Additionally, for on-grid lamps, as shown in the Sect. 3.1.1, markets in the region are still dominated by less efficient technologies such as CFL and incandescent which might not meet the efficacy level set. However, LED technology is increasing in efficacy and reducing in costs at an accelerated rate. This means that the difference between the cost of an LED lamp and that of a CFL lamp is decreasing. Given the superiority of LED lamps in terms of lifetime and efficacy, consumers in the region could benefit greatly from life cycle costs savings from switching from CFLs and incandescent lamps to LED lamps.

---

[20] Calculating the star rating requires more information on the off-grid refrigerators which was not readily available.

Table 1 Status of lighting and appliance policies in 11 countries in east and southern Africa

Country	On-grid lighting and appliances MEPS and labeling	Off-grid lighting standards	Broader Energy Efficiency and/or Renewable Energy Policy
Ethiopia	On-grid lighting and appliance S&L program is under development.[a]	Pico-solar quality standards adopted (2015)	Ethiopian National Energy Policy (2012)
Kenya	On-grid MEPS and labelling requirements developed in 2013. Implementation begun in 2018. Products covered include CFLs and domestic refrigerators. MEPS for all general service lighting products are currently under development.	Pico-solar quality standards adopted	National Energy And Petroleum Policy (2015) Draft Energy Efficiency Strategy (Under development)
Madagascar	None.	None	The 2015–2030 Nouvelle Politique de l'Energie
Namibia	Voluntary[b] MEPS.		National Energy Policy (2017)
Rwanda	On-grid MEPS & labelling program under development.[c]	Pico-solar quality standards adopted	Rwanda Energy Policy (2015)
Seychelles	Voluntary efficiency standards for appliances.		Energy Policy For The Republic Of Seychelles 2010–2030
South Africa	Implementation of on-grid MEPS and labelling program begun in 2017 and 2018. Products covered include refrigerators and CFLs. MEPS for all general service lighting products are currently under development.		Energy Efficiency Strategy (2005)
Tanzania	On-grid lighting and appliance MEPS & labelling under consideration for development.[d]	Pico-solar quality standards adopted (2016)	National Energy Policy (2015)
Uganda	Developed in 2011 but implementation has not begun. Includes MEPS and labelling requirements for CFLs.	Pico-solar and solar home system standards under consideration for adoption	Energy Policy (2002)
Zambia	None.	None	National Energy Policy (Revised 2008)
Zimbabwe	Regulations passed in 2017. They set MEPS maximum mercury content of lighting products and ban manufacture and sale of incandescent lamps.[e]		National Energy Policy, 2012

[a]CLASP correspondence with Ethiopian Energy Authority
[b]Namibia encourages use of the South African standards but these are not mandatory
[c]https://united4efficiency.org/wp-content/uploads/2017/12/UN-Environment-K-CEP-Rwanda-Cooling-Project-overview-20171108.pdf
[d]Source: CLASP correspondence
[e]The regulations can be accessed here: https://www.zera.co.zw/images/Inefficient%20Ligthing%20Products%20Ban%20Regulations.pdf

On refrigerators, if off-grid refrigerators were subjected to same MEPS requirement as on-grid refrigerators in Kenya, 76% of the models would most likely meet the requirements with a few units achieving star ratings of more than 2-stars on the 5-star Kenyan label. However, as discussed in Sect. 2, there are additional considerations beyond energy efficiency for off-grid refrigerators such as durability and affordability. This means that before subjecting off-grid refrigerators to similar MEPS requirements as on-grid refrigerators, regulators should consider the implications of increased efficiency on cost and other key design aspects of off-grid refrigerators.

5 Existing Lighting and Appliance Policies in the Region

Majority of the countries in East and Southern Africa have energy policies that include energy efficiency as shown in the last column of Table 1. However, only nine countries in the region have appliance energy efficiency regulations either under development or implemented. None of the countries in the region currently regulate refrigerators meant for off-grid use only i.e. those that cannot connect to the main grid.[21] In the off-grid lighting sector, only four countries in the region—Ethiopia, Kenya, Rwanda and Tanzania—have adopted pico-solar lighting quality standards while other countries in the region like Uganda are in the process of developing or adopting quality standards for pico-solar and solar home systems.

6 Recommendations

Based on these findings, the following conclusions and recommendations are aimed at providing the governments in these regions with a starting point when deciding how and when to regulate on- and off-grid appliances. These recommendations build on the broader recommendations provided by the Efficiency for Access Coalition's policy brief on this subject.[22]

---

[21] Refrigerators meant for off-grid use with the option to connect to the grid are covered by the Kenya MEPS requirement.

[22] The policy brief can be accessed here: https://clasp.ngo/publications/promoting-high-performing-off-grid-appliances

6.1 *Recommendation 1: Identify the Most Appropriate Regulatory Intervention for the Market and Develop a Roadmap*

There are three possible regulatory interventions that governments can use, individually or in combination; voluntary standards, mandatory standards and labelling schemes. A market assessment would enable a government to choose the right combination of interventions based on factors such as market penetration of efficient products, a cost-benefit analysis of implementing an intervention, consumer awareness and market size. Voluntary standards where the market players opt into an efficiency or quality standards scheme can help governments pave the way for stricter mandatory standards. Additionally, voluntary standards can be used to select products that can benefit from market development programs. For example, the Seychelles Energy-Efficiency and Renewable Energy Programme (SEEREP)[23] uses voluntary standards to determine which products qualify for a loan subsidy scheme offered to households that want to purchase high efficiency appliances. Voluntary standards could be implemented for off-grid refrigerators as part of market development programs.

Two types of mandatory schemes can be considered: mandatory test standards or mandatory performance standards. A mandatory testing and registration requirement that does not set a performance threshold would ensure that products get tested and gives the regulator insight into what products are being offered without setting performance or other requirements. Mandatory performance standards on the other hand could be introduced in more mature markets to eliminate the least efficient and poorest quality products from a market. Mandatory standards would be ideal for on-grid lighting products like general service lamps and refrigerators since more efficient, high quality products are already widely available globally at prices comparable to the less efficient alternatives like compact fluorescent lamps.

Labelling schemes could either be endorsement or comparative. Endorsement labelling schemes would help consumers differentiate the best performing lighting products and appliances based on a common evaluation metric. An endorsement labelling scheme could be implemented for off-grid lighting products and refrigerators to help consumers easily identify more efficient products. Comparative labels would help consumers make comparisons across products with similar uses.

As part of the implementation process for the regulatory intervention, governments should consider developing a road map. A roadmap would set out the goals and rationale for the regulatory intervention and when and how the interventions will evolve over time. For example when performance standards will move from being voluntary to mandatory.

---

[23] http://www.sec.sc/index.php/energy-efficency/promoting-renewable-energy

6.2 *Recommendation 2: Align with or Adopt Existing Test Methods*

For countries in the region that have not yet regulated on-grid products or those reviewing their existing regulations, aligning or adopting well-established and widely-used test methods for lamps and refrigerators would be recommended. These test methods are already widely adopted globally in large product markets like Europe, China, India and Japan. Since most countries in the region are purely import markets, aligning with the test methods used in the countries most of these products come from would be ideal. For example, most on-grid refrigerators in the region are imported from China and so countries in the region could adopt the IEC 62552:2015 test method. This ensures that importers can access accredited laboratories in the product source countries which lowers the barriers to complying with standards and labelling programs in the region. For on-grid general service lamps, IEC test methods such as IEC 62612 for measuring luminous flus of LED lamps are also widely adopted.

For off-grid lighting products, IEC TS 62257-9-5:2018 is an internationally recognised test method that was developed primarily by the Lighting Global program, which has experience testing numerous off-grid solar products. For off-grid refrigerators, the Global LEAP program has developed a test method as discussed in the sections above. Adopting or aligning with these test methods would reduce the burden of developing bespoke national test methods.

If a country decides to place mandatory efficiency requirements on off-grid refrigerators, then it would be recommended that they consider aligning with the IEC refrigerator standards and adding an Annex with the additional metrics evaluated as part of the Global LEAP test method for refrigerators marked as off-grid.

6.3 *Recommendation 3: Set Performance Requirements That Are Appropriate for the Market and Technology Types*

As discussed in Sect. 4.3 above, some off-grid lighting products and refrigerators would meet the energy consumption requirements if the level was set at energy consumption of the lowest-performing on-grid lamp or refrigerator. However, the regulator would first need to understand the implications on cost and other key design considerations that are unique to off-grid products.

If performance requirements are set for off-grid refrigerators, then the modified test method i.e. the one including Global LEAP test procedures, could be used to evaluate these products on the basis of qualities that are necessary for off-grid consumers such as durability and electrical resilience.

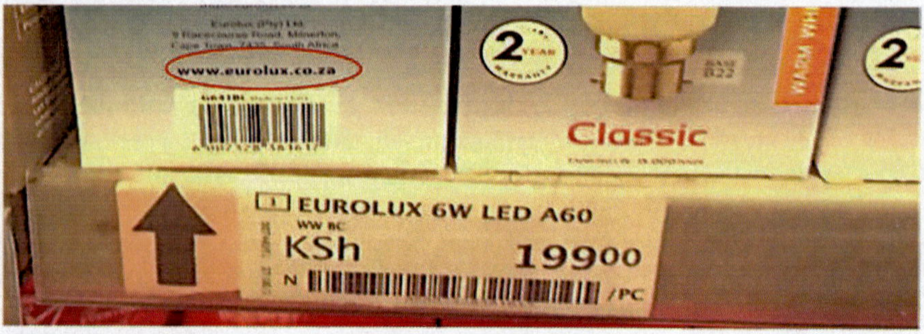

Fig. 14 A lamp distributed by a South African company on sale in a Kenyan retail shop

6.4 *Recommendation 4: Consider Regional Harmonization of Performance Standards*

There are three main trading blocs in the region: the Common Market for Eastern and Southern Africa (COMESA), East African Community (EAC) and Southern African Development Community (SADC). Within these trading blocs, products already move easily from one market to the other using both legal and illegal routes. Figure 14 illustrates this point. Regional harmonization eases the burden of compliance once standards and labelling requirements are put in place since countries can share data on products that are found to be non-compliant in their markets with their trading partners.

7 Overall Conclusion

There are some variations in off- and on-grid products. Additionally, the markets for these products vary across the products and across the countries. The countries in the region are also in different stages of developing policies including standards and labelling programs for these products. Ensuring product quality and performance is a critical step to help governments achieve their energy access targets. Low quality and inefficient appliances will most likely flood markets in the region since there are limited policies to help regulate the market and protect consumers.

The recommendations made on when and how to regulate the markets for on- and off-grid lighting products and refrigerators are based on these analyses on technologies, markets and policy and regulatory environment in the east and southern Africa region. The recommendations include: (1) identifying the most appropriate regulatory intervention and developing a roadmap for market players; (2) aligning with existing test methods for both on- and off-grid lighting products and refrigerators; (3) setting performance requirements that are appropriate for the market and (4) considering regional harmonization of performance standards.

References

1. Cosgrove-Davies, M., & Blimpo, M. (2019). *Electricity Access in Sub-Saharan Africa: Uptake, Reliability, and Complementary Factors for Economic Impact.* Washington DC: World Bank. doi:https://doi.org/10.1596/978-1-4648-1361-0.
2. Global LEAP. (2016, October). *Global LEAP Awards Solar Home System Guide.* Retrieved from Efficiency for Access: https://efficiencyforaccess.org/publications/global-leap-awards-solar-home-system-guide.
3. Global LEAP. (2016). *The State of the Off-grid Appliance Market.*
4. The East African. (2019, February 6). Sub-Saharan Africa is likely to miss poverty targets: WB. Nairobi, Kenya. Retrieved from https://www.theeastafrican.co.ke/news/ea/Sub-Saharan-Africa-is-likely-to-miss-poverty-targets/4552908-4969880-qoqvhxz/index.html.
5. WRI. (2007). *The Energy Market.* WRI.
6. Amol, P., Jacobson, A., Park, Y. W., Lee, R. G., Alstone, P., & Khare, A. (2015). *Sun-Connect news.* Retrieved from https://www.sun-connect-news.org/fileadmin/DATEIEN/Dateien/New/GlobalLEAP-PoweringAHome__1_.pdf.
7. Department of Energy. (2013). *10 CFR Parts 429 and 430—Energy Conservation Program: Energy Conservation Standards for General.* Retrieved from U.S Department of Energy: https://www.energy.gov/sites/prod/files/2016/02/f29/General%20Service%20Lamp%20NOPR_1.pdf.
8. IEC. (2018). IEC TS 62257-9-5:2018. *Recommendations for renewable energy and hybrid systems for rural electrification—Part 9-5: Integrated systems—Laboratory evaluation of stand-alone renewable energy products for rural electrification.*
9. IEC. (2015). IEC 62552-1:2015. *Household refrigerating appliances—Characteristics and test methods—Part 1: General requirements.* IEC.
10. Global LEAP. (2018). *Global LEAP Off-Grid Refrigerator Test Method.* Retrieved from Global LEAP: Global-LEAP-Off-Grid-Refrigerator-Test-Method-
11. CLASP. (2019). *SADC and EAC Lighting Market Assessment Report.* CLASP.
12. IFC, I. F. (2018). *Off-Grid Solar market Trends Report 2018.* IFC.
13. GOGLA. (2018). *Global Off-Grid Solar Market Report.* GOGLA.
14. Pattison, P. M., Hansen, M., & Tsao, J. Y. (2018). LED lighting efficacy: Status and directions. *Comptes Rendus Physique*, 134-145.
15. Alstone, P., & Jacobson, A. (2018). LED advances accelerate universal access to electric lighting. *Comptes Rendus Physique*, 146-158.
16. dti. (2012). *Technology Study on Consumer Energy Devices.*
17. Efficiency for Access. (2018). *Off-Grid Appliance Market Survey.* Efficiency for Access.
18. Efficiency for Access. (2019). *State of the Off-grid Appliance Market.* Efficiency for Access.
19. Harrington, L. (2018). *Household Refrigerators: Energy Modelling Methodology for MEPS 2021 Compliance.* Energy Efficient Strategies (EES).

Policy Measures to Prevent Dumping of Environmentally Harmful and Low Efficiency Cooling Appliances in African Countries: Kenya as a Case Study

Naomi Wagura and Ana María Carreño

Abstract African countries import most, if not all, all of their cooling appliances. Many developed and emerging economies around the world—that also import cooling appliances—have implemented policy measures, like standards and labelling, and refrigerant transition programs to remove inefficient cooling appliances that contain environmentally harmful refrigerants from their markets. In Africa, only eight countries have energy efficiency standards and labelling programs for cooling appliances (CLASP, Policy Database. Retrieved from CLASP: https://clasp.ngo/policie, 2019). In the absence of appropriate policies, inefficient and environmentally harmful products banned in other countries are dumped in unregulated markets, with detrimental outcomes for consumers and governments. This is the case in Africa, where products that cannot be sold in their country of origin (regulated markets) find their way to African countries that have not yet implemented policy measures. For instance, most room air conditioners found in African markets have energy efficiency ratios of between 2.4 and 3.4 W/W and yet the Chinese and Korean brands popular in these markets have high efficiency models sold in other markets (CLASP, Africa Air Conditioner Market Scoping Study. Retrieved from https://clasp.ngo/publications/scoping-study-of-african-air-conditioner-markets, 2018).

This paper illustrates how products that do not meet performance standards in export countries end up in unregulated markets like Kenya. It also shows how implementation of performance standards for room air conditioners (RACs) will help the Kenyan government control the importation of these low efficiency RACs. Additionally, it illustrates how performance standards can act as a de facto ban on the importation of second-hand products with untested performance. Before implementation of a standards and labelling program for cooling products, RACs in the Kenyan market had low energy efficiency levels and used environmentally harmful refrigerants like R-22. In 2018, Kenya implemented standards and labels for a broad

N. Wagura · A. M. Carreño (✉)
CLASP, Washington, DC, USA
e-mail: acarreno@clasp.ngo

P. Bertoldi (ed.), *Energy Efficiency in Domestic Appliances and Lighting*, Springer Proceedings in Energy, https://doi.org/10.1007/978-3-030-79124-7_24

range of cooling products including room air conditioners and refrigerators. In addition to the minimum energy performance standards (MEPS), Kenya is currently working on revisions to their refrigerant regulations to control the importation of equipment that use environmentally harmful refrigerants. This Kenyan case study illustrates the effectiveness of energy efficiency standards as a policy mechanism to prevent dumping of low efficiency products, and gives recommendations on how African countries can curb the importation of environmentally harmful RACs.

1 Introduction

Kenya is currently on an ambitious path to universal electrification [1]. Furthermore, and similar to other countries in Sub-Saharan Africa, the Kenyan economy is poised to grow [2] and temperatures are also expected to rise due to climate change. These three factors mean that more people will seek comfort cooling and they will have the economic capacity and access to electricity necessary to purchase and run room air conditioners. In addition to energy access, energy efficiency is a key priority in the Kenyan energy sector, with the Ministry of Energy implementing energy management policies for large electricity consumers [3] and efficiency standards and labelling for appliances [4]. Kenya is also in the process of ratifying the Kigali Amendment to the Montreal Protocol and working towards a 30% reduction of its business as usual greenhouse gas (GHG) emissions—143 MT CO_2—by 2030 [5].

Kenya's S&L program currently covers five products: room air conditioners, domestic refrigerators, compact fluorescent lamps, ballasts and three-phase caged motors. The standards and labelling requirements were developed in 2013 but the legal framework (regulations) supporting implementation and enforcement of the program was finalized in 2016. The *Energy (Appliances' Energy Performance and Labelling) Regulations* passed in 2016 made MEPS and labelling requirements mandatory. The regulations lay out the guidelines for compliance with the standards and labelling program. Kenya begun implementation of the standards and labelling program in July 2018.

Immediately after the implementation phase for the RAC standard begun in 2018, there were calls for revision from the industry. The standard's testing requirements were not well suited to the Kenyan climate.[1] As a result, this standard was revised and published in April 2019. The revisions involved changing the testing conditions to better reflect the Kenyan climate, increasing the MEPS level by 11% and aligning with the current international standard, ISO 5151. Enforcement of this revised standard has already begun with all RACs imported into the country

[1] The standard required testing RACs for import and sale in Kenya at T3 ambient temperature conditions. These temperatures are much higher than the maximum temperatures in most part of the country. The industry's complaint was that this condition characterized the performance of RACs for the Kenyan market incorrectly.

expected to meet this MEPS. The enforcement of the refrigerator MEPS also begun in 2018 with all units imported into and sold in the country required to meet the MEPS and to affix the energy label on the product at the point of sale.

CLASP carried out a baseline market assessment in 2018 to characterize the room air conditioner and refrigerator market in Kenya. The data collected was then used to evaluate the potential energy savings, avoided emissions and market transformation potential of revising the existing minimum energy performance standards.

This paper aims to demonstrate how implementation of the revised minimum energy performance standards for RACs will contribute to market transformation by eliminating the least efficient models from the market while leading to significant lifecycle cost savings for consumers in Kenya. It will also discuss how MEPS implementation has led to a de facto ban on imported second-hand domestic refrigerators in the Kenyan market.

2 Data Collection Methodology

The type of data collected includes:

1. Key stakeholders, sources of RACs, and supply chain;
2. RAC market size and characteristics;
3. RAC types, cooling capacities, and performance;
4. Costs associated with RAC purchase, installation, operation and maintenance; and
5. Energy sector and other economic data for evaluating energy, environmental, and economic impacts of RAC utilization in Kenya

The approach used to gather quantitative and qualitative data and information on RACs included desk research, stakeholder interviews, and field surveys:

1. Desk research—This involved a literature review of government reports and importation statistics from Kenya Revenue Authority (KRA).
2. Stakeholder interviews—Since Kenya imports all RACs, the team[2] surveyed the local representative offices of manufacturers and brand owners, importers, and dealers. These players provided data on approximate annual RAC sales, product prices, installation and maintenance costs and after sales services provided to consumers.
3. Field data collection—The data collection team visited 15 retail stores in four cities in Kenya. They collected data on the price and RAC technical characteristics from these stores. For products that did not have detailed technical characteristics on the nameplate, the team searched online manufacturers' websites and online shops.

[2] CLASP engaged a local partner, RenCon, to conduct on-the-ground data collection activities.

CLASP gathered product information for 103 models, out of which 99 were unique models.

3 Market Characteristics

Compared to other markets with hotter climates like North or West African countries, the Kenyan RAC market is small. From import data obtained from the customs body, the RAC imports range from 23,000 to 43,000 units per year (Fig. 1).

Kenya is an import-only RAC market and more than 30% of the products come from China with the rest coming from Thailand, South Korea, and Malaysia among others [6–8]. Majority (87%) of the RAC products are single-split units and 94% of the units have a cooling capacity of less than 36,000 btu/h. The most popular brands in the market are Daikin, LG, General, Samsung, Hotpoint, Ramtons, and Unionaire among others.

Since the implementation of the RAC MEPS and labelling program had not begun during the data collection period, none of the RACs had the Kenyan energy label. The products' energy efficiency information was therefore recorded from the nameplates and was only available for 71 out of the 103 products. All the products had an EER below 3.4 W/W as shown in Fig. 2.

Most (64%) of the RAC models had fixed speed compressors. As shown in Fig. 2, all the models with EER of less than 2.8 W/W were fixed speed RACs. The lack of differentiation between the efficiency of fixed and inverter type RACs could be due to the use of the EER metric which does not take into account how inverter RACs achieve energy savings.

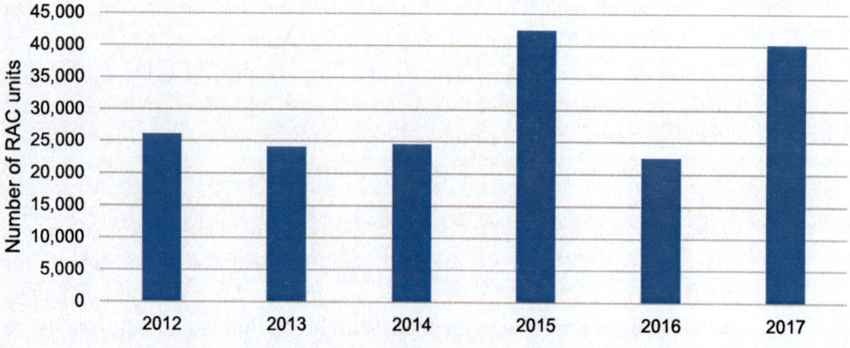

Fig. 1 Annual RAC imports into Kenya. (Source: Kenya Revenue Authority)

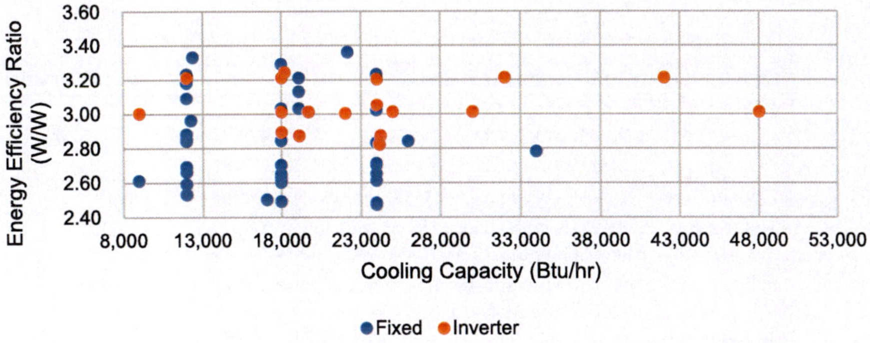

Fig. 2 Efficiency levels of RACs in Kenyan market in 2018. (Source: [7])

4 Comparison of the Efficiency Levels of Products in the Kenyan Market to the MEPS in Source Countries

As indicated in the section above, at least 30% of RACs in the Kenya are imported from China each year. For the models whose data was collected during CLASP's market assessment, 55% of the RACs were imported from China, 12% from Malaysia, 13% from Thailand, 8% from India and 5% from Korea with the rest from Egypt and Belgium [7].

Table 1 shows the minimum energy performance standard for fixed speed RACs in these main export countries in 2018. We use this information to evaluate if RAC units that do not meet the export country's MEPS are exported into Kenya.

Figure 3 shows that a number of the RAC models imported from China and Thailand would not meet the MEPS in their source countries. For the units imported from China, 71% of the units would not meet Chinese MEPS while for those imported from Thailand 43% would not meet the Thailand MEPS.

5 Using Regulation to Avoid Dumping of Low Efficiency RACs

Figure 3 compares the efficiency levels of RACs from the main export countries: China, Thailand, Malaysia, India and Korea with the previous and revised Kenyan MEPS. All the RACs that would not have met the previous Kenyan MEPS level of 2.8 W/W were imported from China and Thailand. With the proposed MEPS, 78% of the RAC models imported from China, 70% from Thailand, 50% from Malaysia and 60% from Korea would not meet the revised Kenyan MEPS (Fig. 4).

The revised MEPS has a significant effect in the market as it eliminates 73% of the least efficient models from the market [7]. This will push the efficiency levels in

Table 1 Fixed speed RAC MEPS in the countries that export RACs to Kenya

Country	MEPS for fixed split RACs	Implementation year
China[a]	For RACs with CC ≤ 4500 W, EER 3.20 W/W with 4500 W < CC ≤ 7100 W, EER 3.10 W/W with 7100 W < CC ≤ 14,000 W, EER 3.00 W/W	2010
Thailand	For RACs with CC ≤ 12,000 W, EER 2.82 W/W [8]	2010
Malaysia[b]	For RACs with CC < 4.5 kW, EER 2.9 W/W For RACs with CC ≤ 7.1 kW, EER 2.7 W/W	2012
India	For RACs manufactured between 2016 and 2017[c] with CC ≤ 7.1 kW, EER 2.7 W/W	2016/2017

[a]The Chinese standard was reviewed in July 2020 to increase the MEPS
[b]The Malaysian MEPS levels can be accessed here https://www.st.gov.my/contents/files/download/95/20180314-Guide_MEPS_For_AC.pdf
[c]Assumption is that RACs found in the Kenyan market in 2018 were manufactured in this period. The Indian RAC MEPS was revised to change the efficiency metric to ISEER starting Jan 2018

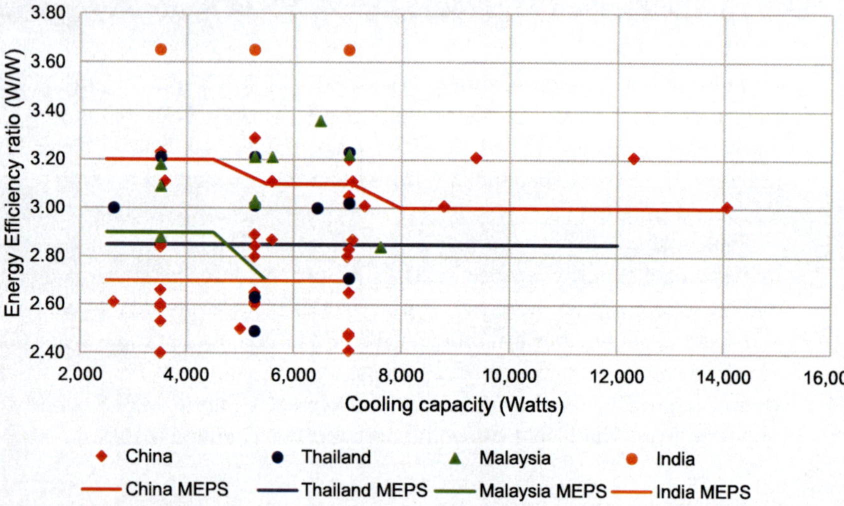

Fig. 3 Comparison of efficiencies of RACs found in the Kenyan market to the MEPS levels in the source countries

the Kenyan market up leading to lifecycle costs savings of between USD 62 and USD 96 for consumers.

6 Using Certification to Prevent Dumping of Second-Hand Products in Kenya

The market for second-hand refrigerators with untested performance in Kenya was thriving before implementation of the refrigerator standards. Second-hand refrigerators were imported from more mature appliance markets like Europe and were

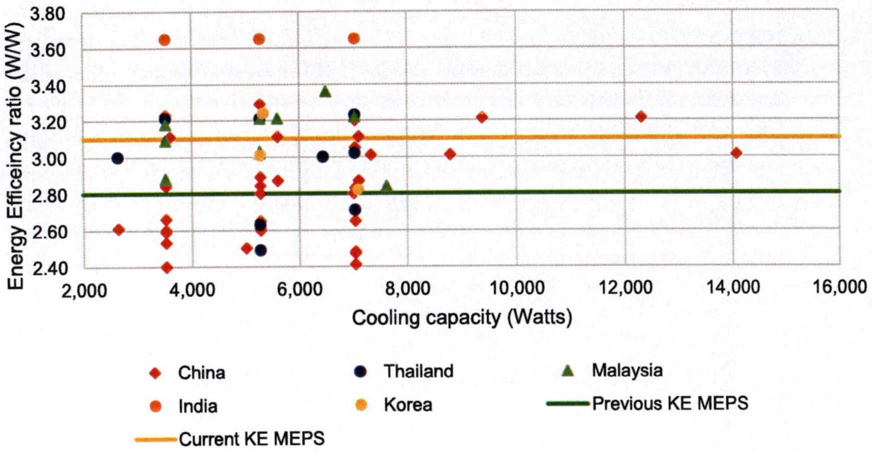

Fig. 4 Efficiency levels of RACs in Kenya with source countries indicated compared to previous and current MEPS levels

preferred by some consumers because they believed appliances that had been sold in these markets must be of superior quality.

The implementation of the Kenyan MEPS for refrigerators begun in 2018. In order to manufacture, import or sell refrigerators in the Kenyan market, the importer or manufacturer has to obtain certification from the Energy and Petroleum Regulatory Authority (EPRA) as proof that the product meets the performance standards. The 2016 Energy (Appliances' Energy Performance and Labelling) Regulations that made the MEPS mandatory also lay out the requirements that need to be met in order for EPRA to issue a registration certificate [4]. One of the requirements is that all refrigerator models intended for import or sale in Kenya should be tested in an accredited laboratory. The test report from this laboratory is then submitted to EPRA for inspection and to confirm that the model meets the Kenyan MEPS. Once confirmed, EPRA then issues a registration certificate that the importer can then use during importation. As an additional layer of checks on products entering the country, Kenya employs a Pre-export Verification of Conformity (PVoC) model. This involves sub-contracting firms that inspect goods destined for the Kenyan market at the country of export. In order to prove that the refrigerators meant for the Kenyan market meet Kenyan standards, the importer has to present the PVoC company with the registration certificate from EPRA. The PVoC company can verify the registration certificate again the register that is maintained on EPRA's website.

According to EPRA, the market has shrank since the implementation of the refrigerator MEPS. One of the main reasons is that the testing costs associated with obtaining a registration certificate for importation of these second-hand refrigerators significantly increases the costs of these appliances. The average testing costs for a refrigerator in Asia, where most refrigerators in the Kenyan market are imported from, ranges from US $ 885 and US $ 2500 [9]. The price increase in second-hand refrigerators due to the high testing costs reduces the advantage the untested appliances enjoyed before, as they were priced lower than new refrigerators with similar characteristics. Without the test report and the registration certificate from EPRA the refrigerators cannot be imported into the Kenyan market. This has made the importation of second-hand refrigerators difficult. As such, the refrigerator MEPS acts as a de facto ban on importation of second-hand refrigerators with untested performance.

7 Refrigerants Contained in RACs in the Kenyan Market

Most of the RAC units found in the Kenyan market—69%—contain R 410A which is a hydrofluorocarbon (HFC). However, 27% of the RAC market contain R-22, which is a hydrofluorocarbon (HCFC) (Fig. 5).

CLASP's RAC market assessment in Thailand found that the market share of R-22 room air conditioners shrank significantly from 79% in 2013 to just 4% in 2018 [8]. This is due to the Thai's government ban on the manufacture, sale and import of RACs containing R-22 by January 2018 [10]. Given that 4 out of the 13 RAC models imported from Thailand to the Kenyan market contained R-22, it is likely that the R-22 units that were still available in the Thai market by the end of 2017 were shipped to countries like Kenya that do not have regulations prohibiting importation of appliances containing HCFC.

China produces 90% of the HCFC used globally [11]. Under its Montreal Protocol obligations, China's production and consumption in 2030 should be only 2.5% of the baseline and intended only for use in servicing existing equipment using HCFCs. However, by 2020, production and consumption will still be at 65% of baseline. This explains why 28% of the units imported from China to the Kenyan market still contain R-22 (Fig. 6). However, with the assistance of the multilateral

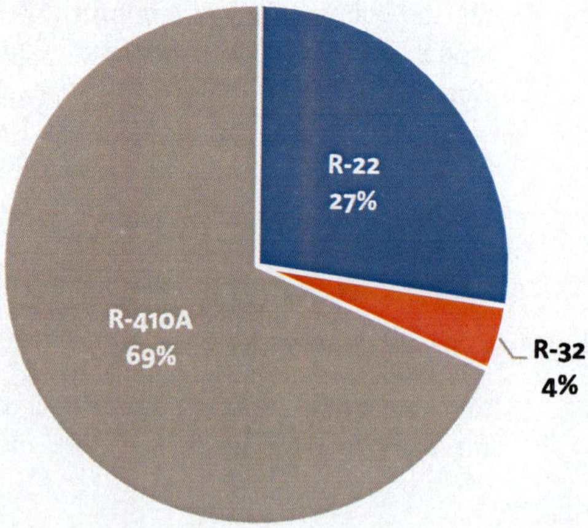

Fig. 5 Refrigerants contained in RACs sold in Kenya

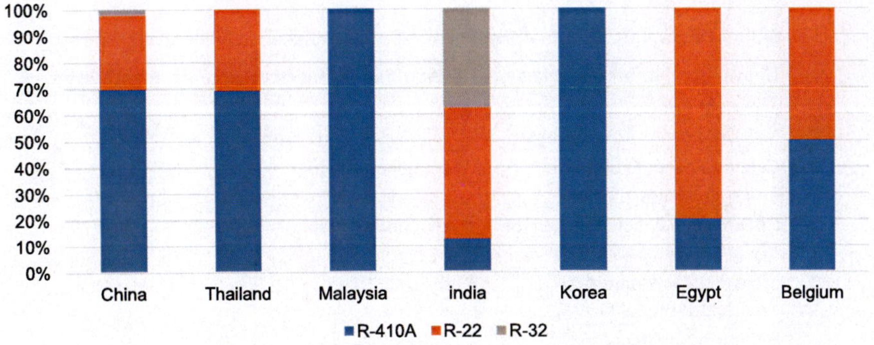

Fig. 6 Proportion of RACs with different refrigerants by source country

fund, UNIDO is supporting the conversion of HCFC refrigeration and air conditioning manufacturing lines to HFCs and natural refrigerants [12].

8 Current Regulations Controlling the Use of Environmentally Harmful Refrigerants

HCFCs and HFCs are substances controlled by the Montreal Protocol which was originally adopted in 1987 and subsequent amendments. In 2007, the countries party to the protocol decided to accelerate the HCFC phase-out schedule. Developed countries agreed to phase out HCFCs completely by 2020 and 2030 for developing countries [13]. In 2016, the parties to the Montreal Protocol agreed under the Kigali Amendment to begin phase-down of HFCs in 2019 for developed countries and a freeze in consumption in 2024 for Article 5, Group 1 countries like Kenya [14].

As shown in the section above, about a third of RACs in the Kenyan market contain R-22, a HCFC. HCFCs are ozone-depleting substances, while HFCs like R-410A and R-32 have zero ozone depleting potential but have global warming potential. For example, R-410A, which is popular in the Kenyan RAC market, has a global warming potential of 2008 while R-290, which is a natural refrigerant, has a GWP of 3 [15].

Kenya has ratified four out of the five amendments to the Montreal Protocol (Table 2) and is in the process of ratifying the Kigali Amendment.

As a signatory to the Montreal Protocol, Kenya is implementing the HCFC-phase down schedule as shown in Table 3. To provide the legal framework to support the phase-down of ozone-depleting substances, Kenya passed the Environmental Management and Co-ordination (Controlled Substances) Regulations in 2007. The regulations control the importation, manufacture and use of ODS by requiring that any person/entity that wishes to import or manufacture these substances gets a license from the National Energy Management Authority (NEMA). However, the regulations do not cover ozone depleting substances (ODS) contained in appliances like RACs, enabling the importation of RACs containing R-22 (a HCFC) without the licensing requirement. NEMA is currently in the process of revising these regulations [7].

Table 2 Kenya's status of ratification of the ozone-related multilateral agreements

Convention/protocol/amendments	Date of ratification/accession/acceptance/approval
Montreal Protocol on Substances that Deplete the Ozone Layer	November 1988
London Amendment to the Montreal Protocol (1990)	September 1994
Copenhagen Amendment to the Montreal Protocol (1993)	September 1994
Montreal Amendment to the Montreal Protocol (1997)	July 2000
Beijing Amendment to the Montreal Protocol (1999)	October 2013
Kigali Amendment to the Montreal Protocol (2016)	The process of ratification commenced in January 2017. Currently the Attorney General's office is reviewing the Cabinet Memo

Source: [6]

Table 3 Kenya's HCFC phase-down management plan

Base level average	2009–2010
Freeze both production and consumption	January 1, 2013
10% reduction	January 1, 2015
35% reduction	January 1, 2020
67.5% reduction	January 1, 2025
100% reduction	January 1, 2030

To control the importation and use of RACs containing HCFCs and eventually HFCs, Kenya will need to regulate these appliances under the Environmental Management and Co-ordination (Controlled Substances) Regulations. As a potential solution, Kenya could follow the South African approach—South Africa banned the importation of RACs containing R-22 in 2015 [16]. Additionally, ratification of the Kigali Amendment will give the ministry of environment the mandate to start implementation of the HFC phase-out schedule similar to the HCFC schedule.

9 Recommendations for African Countries on Policies to Avoid Dumping of Environmentally Harmful and Low Efficiency RACs

9.1 Recommendation 1: Adopt and Implement Standards and Labelling Policies for RACs

Products from the large Asian manufacturing countries including China dominate most RAC markets in Africa. Asian countries have S&L programs that will likely continue to increase the efficiency levels of products allowed for sale into their markets, but still allow the export of sub-standard products to other markets. For African countries without S&L programs for RACs, the current state of their market is likely similar to the Kenyan market in 2018 before implementation of the Kenyan MEPS, where a number of the RAC models imported from China and Thailand would not meet the MEPS in their source countries. Only ten countries in the region have implemented S&L programs for RACs [17]. Implementing S&L programs for RACs in African countries will help reduce the flow of low efficiency RACs, as shown by the standards and labeling program in Kenya, which will lead to significant market transformation by eliminating the least efficient RACs in the market.

9.2 Recommendation 2: Utilize Performance Standards as a Way to Prevent Dumping of Second-Hand Appliances

As seen from the Kenyan example, minimum performance standards can help prevent dumping of second-hand appliances in African markets by requiring that all appliances are tested to confirm that they meet performance standards. This ensures that consumers purchasing any appliances, whether new or used, are assured of the performance of the appliance. The PVoC model used in Kenya also protects the Kenyan consumers from sub-standard products. This model could be utilized in other African markets that are small import-only markets like Kenya to ensure compliance with the performance standards put in place.

9.3 Recommendation 3: Integrate the Refrigerant Transition Plans for Cooling Appliances into Environmental Regulations

The Montreal Protocol is one of the most successful international treaties and all 54 African countries that are UN member countries have ratified it. However, loopholes in the regulations like the one in Kenya where they do not regulate HCFCs contained in appliances means that the HCFC phase-out schedule might be challenging to implement. In addition, for countries like Kenya that have not yet ratified the Kigali Amendment, ratification should be high on the agenda. Once the phase-out begins in developed countries in 2019, RACs containing HFCs will end up in countries that have not yet developed HFC phase-out plans. Moreover, once these appliances enter the markets, it will be difficult and costly for the governments to implement HFC phase-out plans.

10 Conclusion

Like most African countries, the Kenya room air conditioner market is import-based with majority of the units coming from Asian countries. In 2018, before implementation of the Kenyan RAC MEPS begun, all models in the market had an energy efficiency ratio of less than 3.4 W/W. Most of these RAC models would not meet the MEPS in their source countries. The proposed MEPS in Kenya would eliminate majority of the low efficiency units imported from China, Thailand, Malaysia and Korea.

Before implementation of the MEPS for refrigerators begun, second-hand refrigerators with untested performance could be imported into the country. The Kenyan standards and labelling program requires testing and registration of all refrigerators—including second-hand refrigerators—imported into the country. This requirement has significantly reduced the importation of second-hand refrigerators with untested performance.

Room air conditioners containing HCFCs are still found in most markets in Africa, including Kenya. While Kenya is a signatory to the Montreal Protocol and has regulations that control the importation of HCFCs, these regulations do not control the importation of cooling appliances that contain HCFCs. Additionally, Kenya has not yet ratified the Kigali Amendment, which would begin the phase out of HFCs that have high global warming potential.

The evidence from Kenya shows that MEPS can be an effective tool for preventing dumping of low efficiency RACs and second-hand refrigerators with untested performance for other countries in Africa. Finally, integrating the refrigerant transition plans for cooling appliances into environmental protection regulations would help African countries phase out cooling appliances containing HCFCs and eventually HFCs.

References

1. Government of Kenya. (2018). *Kenya National Electrification Strategy*. GoK.
2. World Bank. (2018). *Analysis of Issues Shaping Africa's Economic Future*. Washington, DC.: World Bank.
3. ERPA. (20 de March de 2019). *The Energy (Energy Management) Regulations, 2012*. Obtenido de EPRA: https://renewableenergy.go.ke/downloads/policy-docs/the_energy_management_regulations_2012.pdf
4. EPRA. (2019, July 17). *Current Renewable Energy Regulations: ERC*. Retrieved from ERC: https://www.erc.go.ke/services/renewable-energy-2/global-fuel-economy-intiative-2-3/
5. Government of Kenya. (2015). *Nationally Determined Contribution*. Retrieved March 20, 2019, from UNFCCC: https://www4.unfccc.int/sites/ndcstaging/PublishedDocuments/Kenya%20First/Kenya_NDC_20150723.pdf
6. CLASP. (2019, July 17). *Policy Database*. Retrieved from CLASP: https://clasp.ngo/policies
7. CLASP. (2019). *Kenya Room Air Conditioner Market Assessment and Policy Options Analysis*. CLASP.

8. CLASP. (2019). *Thailand Room Air Conditioner Market Assessment and Policy Options Analysis*. CLASP. Retrieved from https://clasp.ngo/publications/thailand-rac-market-assessment-and-policy-options-analysis-2019
9. Super-efficienct Equipment and Appliance Deployment Inititative (SEAD). (2019). *Global Appliance Testing Costs Catalogue*. SEAD. Retrieved July 17, 2019, from https://storage.googleapis.com/sead-siteassets/2019-SEAD-Global-Appliance-Testing-Costs-Catalogue.pdf
10. JARN. (2017, 9 24). Thailand Accelerating Shift to R32. Retrieved from http://ejarn.com/detail.php?id=47149
11. World Bank. (2018). *China HCFC Phaseout Project Stage II: Combined Project Information Documents/Integrated Safeguards Datasheet*. Washington, DC: World Bank.
12. United Nations. (2016). *Executive Committee of the Multilateral Fund for the Implementation of the Montreal Protocol_ Project Proposals: China*. Retrieved July 2019, from http://www.multilateralfund.org/76/English/1/7625.pdf
13. UN Environment. (2019, July 15). *OzonAction*. Retrieved from UN Environment: https://www.unenvironment.org/ozonaction/who-we-are/about-montreal-protocol
14. International Institute of Refrigeration. (2019, July 17). *Summary Sheet | The Kigali Amendment to the Montreal Protocol*. Retrieved from International Institute of Refrigeration Website: http://www.iifiir.org/userfiles/file/webfiles/regulation_files/Synth%C3%A8se_Kigali_EN.pdf
15. UN Environment. (n.d.). *OzonAction Kigali Fact Sheet 3*. UN Environment. Retrieved from http://www.unep.fr/ozonaction/information/mmcfiles/7878-e-Kigali_FS03_Understanding_GWP&CO2e.pdf
16. CLASP. (2018). *Africa Air Conditioner Market Scoping Study*. Retrieved from https://clasp.ngo/publications/scoping-study-of-african-air-conditioner-markets
17. Carreño, A., & Taylor, C. (2019). Boosting African regional markets of highly. Washington, DC. Retrieved from https://storage.googleapis.com/clasp-siteattachments/2019-eceee-Summer-Study-Boosting-African-regional-markets-of-highly-efficient-appliances-through-effective-energy-efficiency-and-trade-policy.pdf

Experiences of Electric Pressure Cookers in East Africa?

S. Batchelor, E. Brown, N. Scott, and J. Leary

Abstract This paper seeks to highlight the emerging opportunity for manufacturers to enter the largely untapped market for efficient electric cooking appliances such as the Electric Pressure Cooker (EPC) in East and Southern Africa. The paper is an output of the UK Aid programme Modern Energy Cooking Services, a 5 year programme of work (2018–2023) led by Loughborough University. In East Africa, electricity networks are growing stronger and broader, opening up electric cooking to an almost entirely untapped market particularly in urban areas that are still dominated by charcoal. In each country, approximately ten million people pay for polluting cooking fuels, yet they have a grid connection that is not used for cooking. Historically this has been due to the pricing and unreliability of the grids. As Grids get stronger and appliances more efficient the affordability and convenience of electric cooking is becoming more realistic. In Southern Africa, electric cooking has been and is more popular, however inefficient appliances are placing a heavy strain on national utilities, many of whom are now looking to manage demand more sustainably. Again, the advent of energy efficient appliances changes the dynamic for the household.

S. Batchelor (✉) · N. Scott
Gamos Ltd., Reading, UK
e-mail: Simon@gamos.org

E. Brown
Loughborough University, Loughborough, UK
e-mail: E.D.Brown@lboro.ac.uk

J. Leary
Gamos Ltd., Reading, UK

Loughborough University, Loughborough, UK
e-mail: J.Leary@lboro.ac.uk

P. Bertoldi (ed.), *Energy Efficiency in Domestic Appliances and Lighting*, Springer Proceedings in Energy, https://doi.org/10.1007/978-3-030-79124-7_25

Cooking is deeply cultural and any new energy efficient cooking devices must be compatible with local foods and cooking practices. This paper presents insights from cooking diaries, focus groups and 'kitchen laboratory' experiments carried out in Kenya, Tanzania and Zambia. The results show that EPCs are not only acceptable, but highly desirable. Over 90% of the menu can be cooked in an EPC and certain foods require just one fifth of the energy of a hotplate. In real homes, participants with EPCs, rice cookers and hotplates chose the efficient appliances for approximately half their menu and for these dishes, they used roughly half the energy of the hotplate. Without training and with limited experience of the new devices, the trial participants in Kenya who cooked solely on electricity had a median daily consumption of 1.4 kWh/household/day, and the cooking of 50% of the menu on an EPC utilised 0.47 kWh/household/day of that total. Given that EPCs could have cooked 90% of the desired menu, with appropriate training and broader experience, the median could have been reduced to less than 1 kwh/day/household. This research feeds into a new UK Aid programme, *Modern Energy Cooking Services* and concludes with recommended design modifications that could enable users to do more cooking with EPCs and open up sizeable new market segments including strengthening weak-grid and off-grid.

1 Introduction

This paper seeks to highlight the emerging opportunity for manufacturers to enter the largely untapped market for efficient electric cooking appliances such as the Electric Pressure Cooker (EPC) in East and Southern Africa. The paper is an output of the UK Aid[1] programme Modern Energy Cooking Services, a 5 year programme of work (2018–2023) led by Loughborough University.

IMARC [1] estimate the annual EPC (or multicooker) sales to be worth $578 million (USD), however Africa accounts for just 5% of this. Globally, three billion people still cook with biomass, yet two billion of these now have access to electricity [2]. Cooking with biomass is estimated to cause in excess of four million deaths every year, due to respiratory illnesses from breathing in smoke [3]. As a result, the international community is putting significant effort into finding solutions. However, Batchelor et al. [4] note that whilst past attempts have focussed on improving the efficiency of biomass cooking, there is an emerging opportunity to leverage progress in electrification to drive forward access to clean cooking solutions. The United Nations set the optimistic goal of achieving universal access to both electricity and clean cooking by 2030 with SDG 7 (Sustainable Development Goal 7) [2]. Whilst the former may be within sight, at current rates of progress with predominantly

[1]This material has been funded by UK aid from the UK government; however the views expressed do not necessarily reflect the UK government's official policies.

biomass-based solutions, the world will fall far short of the latter for cooking [2, 4]. This paper highlights a key opportunity to address this globally important challenge.

Regionally, East Africa presents a strategic opportunity, as it contains many of the world's largest charcoal markets, whilst at the same time, electricity grids are becoming stronger and reaching more people than ever before. This paper focusses in on four politically stable East African countries, Tanzania, Kenya, Uganda and Ethiopia, where the uptake of energy-efficient products has already revolutionised other sectors such as lighting [5]. To date, electric cooking has seen limited uptake in East Africa, due to the intertwined challenges on both the supply (reliability, access, poor quality wiring) and demand (perception of cost, taste, behavioural change). However, the supply side barriers are decreasing rapidly and as will be shown in this paper, energy efficient appliances offer a new opportunity to overcome many of the demand side challenges.

Southern Africa presents a different, but equally important opportunity, as electric cooking is already the aspirational solution for many, however the legacy of old and inefficient equipment makes cooking with electricity unnecessarily expensive. Many Southern African grids are dominated by hydropower (excluding South Africa, which is predominantly coal based), leading to frequent seasonal power shortfalls and load shedding. This paper gives the example of Zambia, where over 10% of the population already cook on electricity. However, recent load shedding caused a significant number of these users to revert back to charcoal, rapidly accelerating deforestation. As a result, the national utility, ZESCO, is looking for ways to reduce demand for electricity, so finding a more efficient alternative to inefficient hotplates is vitally important.

The paper presents insights from empirical data collected in Kenya, Tanzania and Zambia[2] where the cooking practices and associated energy consumption of households were recorded. Households then switched to electric cooking with a range of conventional (hotplate) and energy efficient appliances, including EPCs. The data presented includes perception of cost, taste and experiences of behavioural change. (This material has been funded by UK Aid from the UK Government, however the views expressed do not necessarily reflect the UK government's official policies.)

1.1 Electricity in East and Southern Africa

1.1.1 Generating Capacity

On the supply side, driven by long term economic growth ambitions, East African grids have been growing stronger and broader, presenting an opportunity to expand electrical demand into new sectors such as cooking. Whilst utilities in the region have historically shied away from stimulating demand due to shortfalls in supply, Batchelor et al. [4] note that several countries in the region now have surplus

[2] At the time of writing, the collection of comparable data is underway in Uganda and Ethiopia.

electricity. Kenya's national utility, KPLC (Kenya Power and Lighting Company) has already begun to promote electric cooking through a television series, Pika na Power (Cook with Electricity) [6]. Recent installations in Uganda have increased generating capacity to 950 MW, creating a generating surplus, for the moment. Power Africa has identified a further 1900 MW of projects for completion by 2030. The World Bank estimate that generating capacity in Kenya double from 2300 MW in 2015 to 5000 MW in 2020 [7]. Generating capacity in Tanzania was roughly 1500 MW in 2017 [8], and with a further 1600 MW planned, this capacity is projected to double imminently [9]. More recently, the Stiegler's Gorge hydropower project has been given the go-ahead, which will bring an additional 2100 MW online [9], so the government's aim to reach 5000 MW by 2020 [10] appears feasible. The government's Better Results Now initiative (2013) contains a longer term ambition to reach 10,000 MW by 2025 [11].

1.1.2 Reliability and Security of Supply

Table 1 shows that with the notable exception of Tanzania, the majority of electricity in the region is generated from renewable sources and that reliability in urban areas is now relatively high. Batchelor et al. [4] note that there are still many outstanding transmission, infrastructure and management issues within the private and public sector including utilities. However, Table 1 suggests that in major cities, reliability is already sufficient to consider cooking—the SAIDI and SAIFI from each country's economic centre indicate that in all four countries, power outages average less than 5 h per month.

Table 1 Electricity supply factors in selected East and Southern African nations [12–18]

	Electricity access—total, urban	Blackout frequency and duration (SAIFI,[a] SAIDI[b])	Electricity tariff (USD/kWh)	Lifeline tariff and allowance (USD/kWh, kWh/month)	Generation mix (% renewable)
Ethiopia	44% (urban 97%)	n/a	0.09	0.01 $/kWh 50 kWh/month (0.03, 100 0.06, 200 0.07, 300 0.08, 400)	100
Kenya	64% (urban 81%)	13/year, 60 h/year	0.23	0.17 $/kWh, 100 kWh/month	87
Tanzania	33% (urban 65%)	47/year, 21 h/year	0.15	0.04 $/kWh 75 kWh/month	34
Uganda	22% (urban 57%)	42/year, 59 h/year	0.20	0.06 $/kWh 15 kWh/month	93
Zambia	40% (urban 75%)	5/year, 50 h/year	0.09	0.02 $/kWh 200 kWh/month	97

[a]SAIFI (System average interruption frequency index) is the average number of service interruptions experienced by a customer in a year

[b]SAIDI (System average interruption duration index) is the average total duration of outages over the course of a year for each customer served, measured in hours

1.1.3 Urbanisation and Biomass Cooking

Africa is rapidly urbanising and many areas that were previously rural are becoming peri-urban, meaning that many people who used to collect firewood are now forced to purchase charcoal. As nearby forests are exhausted, charcoal has to be brought from further and further away, pushing up the price in urban centres [19]. Another Nigeria will be added to the continent's total urban population by 2025 and urban centres are set to double in size over the next 25 years, reaching one billion people by 2040 [20]. Ironically, despite having a surplus, while 22% of Ugandans and 64% of Kenyans are covered by electricity, electric cooking doesn't even register on national surveys (0%). Even in urban areas, where access rates are 57% and 81% respectively, electricity is only used as a primary cooking fuel by 1% in both nations [21]. Figure 1 shows that cooking in urban areas of East and Southern Africa is still dominated by charcoal, with all its associated problems of respiratory illness, deforestation, general air quality, climate change contribution, and ever rising monetary cost. Only in Zambia (27%) and Ethiopia (18%) do significant fractions of the population cook with electricity, most likely due to the low unit cost, which at consumption levels below 200 kWh/month (which as will be shown later in this paper, is more than enough to cook with), averages approximately 0.02 USD/kWh in both nations. There is therefore a considerable latent opportunity for a relatively easy switch to clean cooking that has to date been held back by reliability and security concerns.

Collectively, Kenya, Uganda, Tanzania, Zambia and Ethiopia are home to 38 million people who have a grid connection, yet choose to cook with commercialised polluting fuels (charcoal and kerosene). Figure 2 shows that Kenya, Uganda and

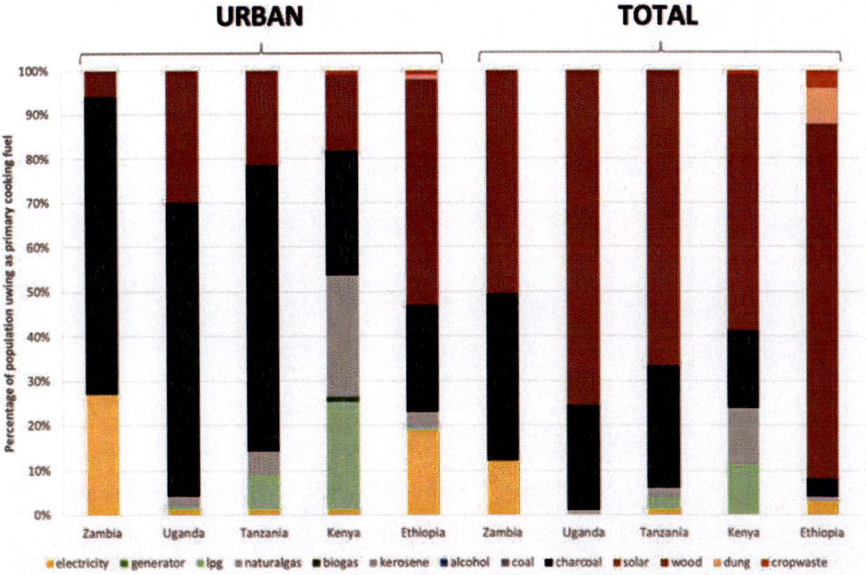

Fig. 1 Breakdown of fuel users for selected East and Southern African nations. (Adapted from eCook Global Market Assessment [22] with data from WHO Household Energy Database [21])

Tanzania each have approximately 10 million people who are paying to cook with charcoal and kerosene, whilst approximately the same number have a grid connection, but do not use it for cooking. In Zambia, electrification rates are modest (40%), but uptake of electric cooking amongst those that are connected is relatively high (12% of total population). As a result, roughly half the number of people cooking with commercialised polluting fuels (6 million) have a grid connected, but do not yet use it for cooking (3 million). Ethiopia is a much bigger country than Zambia and electrification rates are similar (44%), however, Fig. 1 shows that the majority of the population (even in urban areas) cook with firewood, which is often collected for free. As a result, relatively few people (5 million) cook with commercialised polluting fuels compared to those who are grid connected but not cooking with electricity (25 million).

1.1.4 Affordability of Electric Cooking

Whilst charcoal prices have risen significantly in most East and Southern African nations, electricity tariffs have remained relatively affordable. Table 1 shows that despite hiking prices by 75% in 2017, electricity prices in Zambia are still below

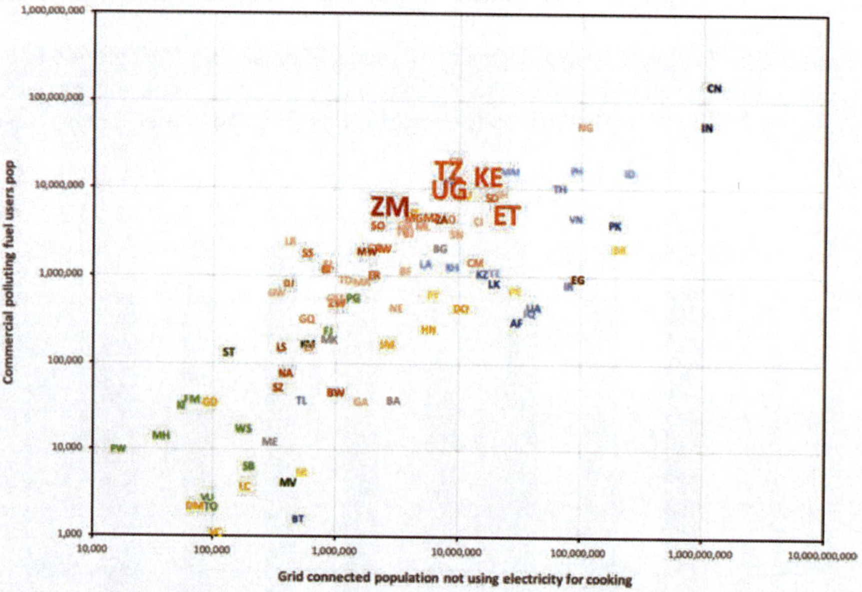

Fig. 2 Size of key market segments by country: commercialized polluting fuel (kerosene, charcoal, coal) users and grid connected population that are not using electricity as their primary cooking fuel (Adapted from eCook Global Market Assessment [22] with data from World Development Indicators [12] and WHO Household Energy Database [21]). Countries coloured by region: AIMS, Central Africa, Central America & Caribbean, Central Asia & North Korea, East Africa, Europe, India & China, Middle East, North Africa, Pacific Islands & PNG, South America & Mexico, South Asia (excl. India), Southeast Asia, Southern Africa and West Africa.

. ET Ethiopia, ZM Zambia, TZ Tanzania, KE Kenya, UG Uganda. All two-letter country codes listed in Appendix 2

0.10 USD/kWh. In fact, the national utility, ZESCO, offer a generous lifeline tariff of just 1.5 cents (USD) per unit for the first 200 kWh per month, which is actually more than enough for most households to cook with. Researchers undertaking a global review of price data found that even when utilising an inefficient hotplate, there are a number of African countries (including Ethiopia, Tanzania and Zambia) where it is already affordable for grid-connected households cooking with charcoal to switch to cooking with electricity [22]. Of course, when energy-efficient appliances are considered, many more countries rise above the price parity line between electricity and charcoal. Figure 1 shows that there is a clear correlation between the cost of electricity and uptake of electric cooking. Several East and Southern African nations with low to moderate tariffs already have reasonably high levels of uptake of electric cooking (amongst those that have access). Zambia, Ethiopia, Zimbabwe and South Africa all have over 1 million people using electricity as their primary cooking fuel. Of particular note is South Africa, which appears second only to China,[3] offering a market of over 40 million people already cooking with electricity and a moderate tariff of 0.12 USD/kWh (Fig. 3).

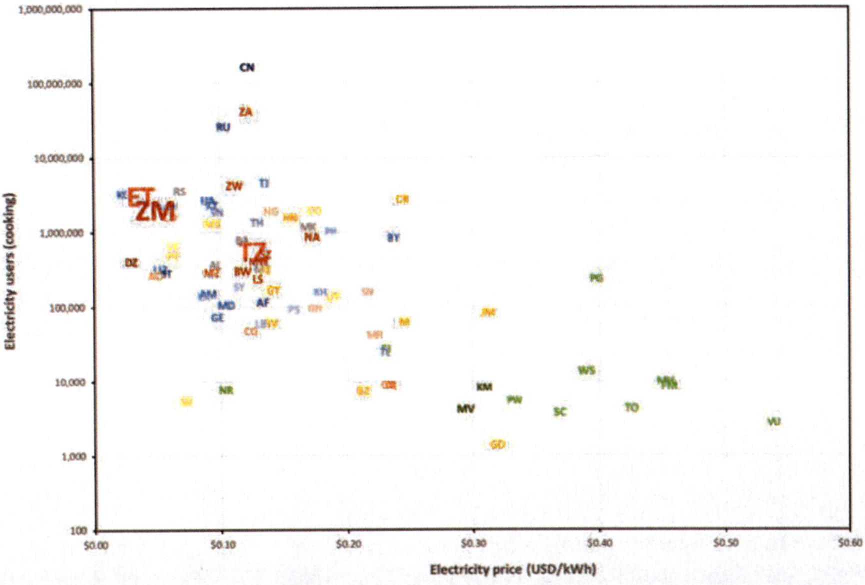

Fig. 3 Size of existing electric cooking markets by country (Adapted from eCook Global Market Assessment [22] with data from World Development Indicators [12] and WHO Household Energy Database [21]). Countries coloured by region: **AIMS, Central Africa, Central America & Caribbean, Central Asia & North Korea, East Africa,** Europe, **India & China,** Middle East, **North Africa, Pacific Islands & PNG,** South America & Mexico, **South Asia (excl. India), Southeast Asia, Southern Africa and** West Africa.

. ET Ethiopia, ZM Zambia, TZ Tanzania. Kenya & Uganda not shown, as household surveys show 0% electric cooking [21]. All two-letter country codes listed in Appendix 2

[3] High Income Countries (HICs) were excluded from the analysis.

1.1.5 The Impact of Energy-Efficient Products/Services in East/Southern Africa

The paper argues that the EPC has the potential to revolutionise cooking markets across the region in the same way that the Light Emitting Diode (LED) has transformed lighting markets. The LED has enabled access to lighting across the region in ways that simply were not possible before. The LED reduces energy demand for lighting by an order of magnitude compared to incandescent light bulbs, whilst simultaneously increasing reliability and product size. Notably, East Africa in particular has embraced this new technology through the development of solar home systems designed to replace inefficient, expensive and polluting technologies such as kerosene lanterns, candles and torches with disposable batteries. Pay as you go (PAYGO) business models have enabled the high upfront costs of such systems to be broken down into manageable repayments in tune with how kerosene, candles and dry cell batteries are purchased. GOGLA [23] state that East Africa represents about 80% of total sales volumes of solar lighting products in Sub-Saharan Africa, with 1.49 million products sold, generating US$ 44.07 million revenue in the second half of 2016 alone.

While this example clearly demonstrates that significant improvements in quality of life for poor households can result from by strategic use of an energy efficient appliance, it is important to note the differences between lighting and cooking. Cooking is deeply cultural, and it is not enough to ensure a supply chain of an energy efficient appliance; people will need to know that it can cook their food and that the food will be just as tasty. In the data below, we approach the potential of EPCs as an energy efficient device for East and Southern Africa by showing how it fits existing cooking practices. However, before getting to the specifics of cooking processes and acceptability of EPCs in East and Southern African culture, we consider why the EPC is so efficient in its energy use.

1.2 Why the EPC?

In developed economies, EPCs are attractive not so much for their energy saving capability, but for their convenience and speed [1]. The first Electric Pressure Cooker (EPC) patent was filed by a Chinese scientist in 1991 [24], but the appliance has recently gained popularity in other parts of the world, with North America now dominating the market [1]. In fact, newer models can even be controlled remotely via a smartphone app, allowing food to be loaded in the morning and the cooking process triggered when the user leaves work in the evening. The accelerated speed of cooking can also support the busy lifestyle of many modern households [1].

However, the features of automatic control, insulation and pressurisation also enable the EPC to save a lot of energy, and therefore money. What is more, the insulation allows it to continue cooking during short blackouts and also keeps food warm after cooking has finished. These may not be important parts of the value

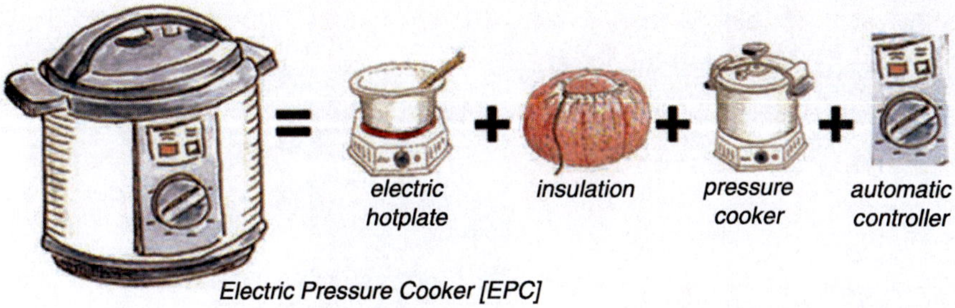

Fig. 4 The fundamental components of an EPC

proposition in North America and Europe, but as we will show in this paper with data from East and Southern Africa, they become all the more important when relating this to impoverished societies with more unreliable electricity supplies.

The EPC (or multicooker) simply combines an electric hotplate, a pressure cooker, an insulated box and a fully automated control system (Fig. 4). Batchelor et al. [25] explain that unlike other cooking fuels that rely on combustion, electricity does not need air flow to create heat. It therefore opens up the possibility of the food being cooked in a highly insulated environment (Fig. 5). This principle is used in many popular electric cooking appliance, such as rice cookers, slow cookers and thermo pots. Having raised the temperature of the device to the cooking temperature, the insulation drastically reduces heat loss, meaning that little to no extra energy is required to continue to cook the food (see Fig. 6). Indeed this is the basis of the 'fireless' cooker, sometimes called Wonderbag or Lindamoto. A pot of beans, for instance, is cooked for some minutes to remove toxins, and then taken off (any) stove and placed in the fireless cooker. With the highly insulated bag keeping the temperature high, the beans continue to cook—thus saving fuel.

In addition to minimising heat losses through insulation, the EPC adds the option to pressurise. This raises the boiling point of water and enables the food to be cooked faster. Figure 4 shows that after the initial pressurisation, the hotplate in an EPC only comes on periodically to maintain the temperature in the sealed environment inside and resulting in considerable energy savings. As Prof. R. Khan states: "*it is temperature that cooks food, not energy per se*" [25].

In contrast, whilst rice cookers are also insulated and automated, they are not sealed and their control system is much simpler, merely dumping full power into the pot until all the water has been vaporised. However, they are much more useful than their name suggests, as one participant noted: "I have learnt that rice cookers are badly named—they can cook so much more than rice!" It should also be noted that because of the insulation, 'full power' on a rice cooker is generally much lower than on a hotplate, which has important implications for systems where peak power is a constraint, such as battery-supported cookers or mini-grids.

As stated above, the EPC goes further by pressurising the system; during this stage the boiling point of water is raised up from 100 °C to around 120 °C. The increased temperature enables the food to cook faster, resulting in shorter cooking

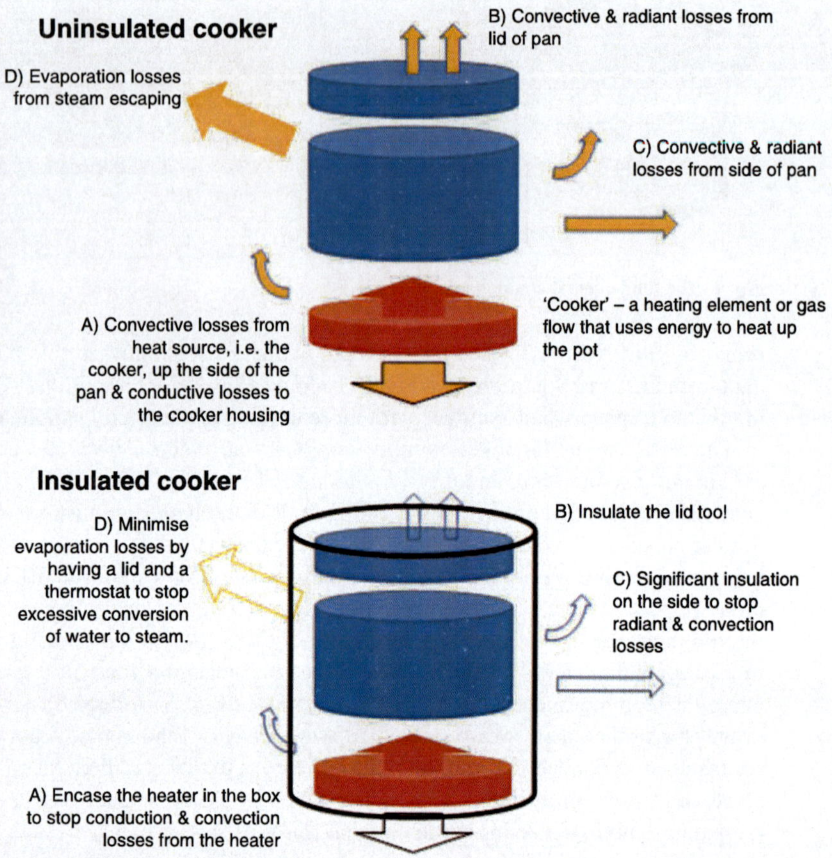

Fig. 5 Heat loss mechanisms mitigated by insulating the cooking pot and heating device. (Adapted from Batchelor et al. [25])

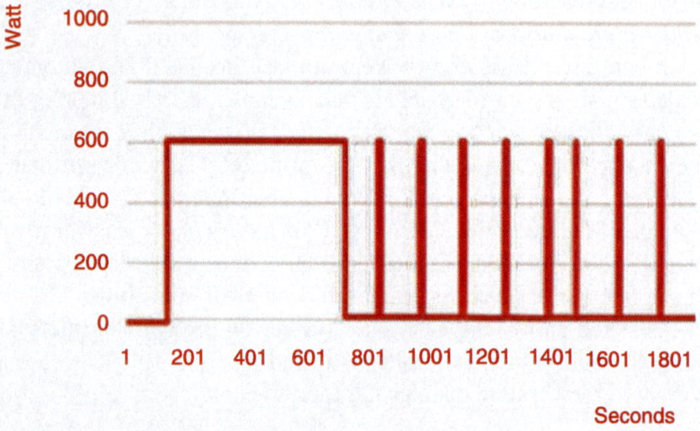

Fig. 6 Typical load profile for a 700 W rated EPC on a half hour cooking cycle [26]

times and therefore reduced energy consumption. 'Manual' stove-top pressure cookers (heated by charcoal and gas) are common in East Africa, although their safety is of concern to many users. EPCs integrate an array of safety and control features, offering multiple redundancies if any one were to fail (see Fig. 7). It controls the energy input into the device, such that the cook can walk away and leave the device cooking autonomously.

While the sealed environment has a positive effect on energy consumption, the sealed, blind, nature of pressure cooking can make inexperienced cooks nervous. They believe that more stirring is required, or they need to see the food to make sure it is cooking, or has not overcooked. Such responses hold back many cooks from

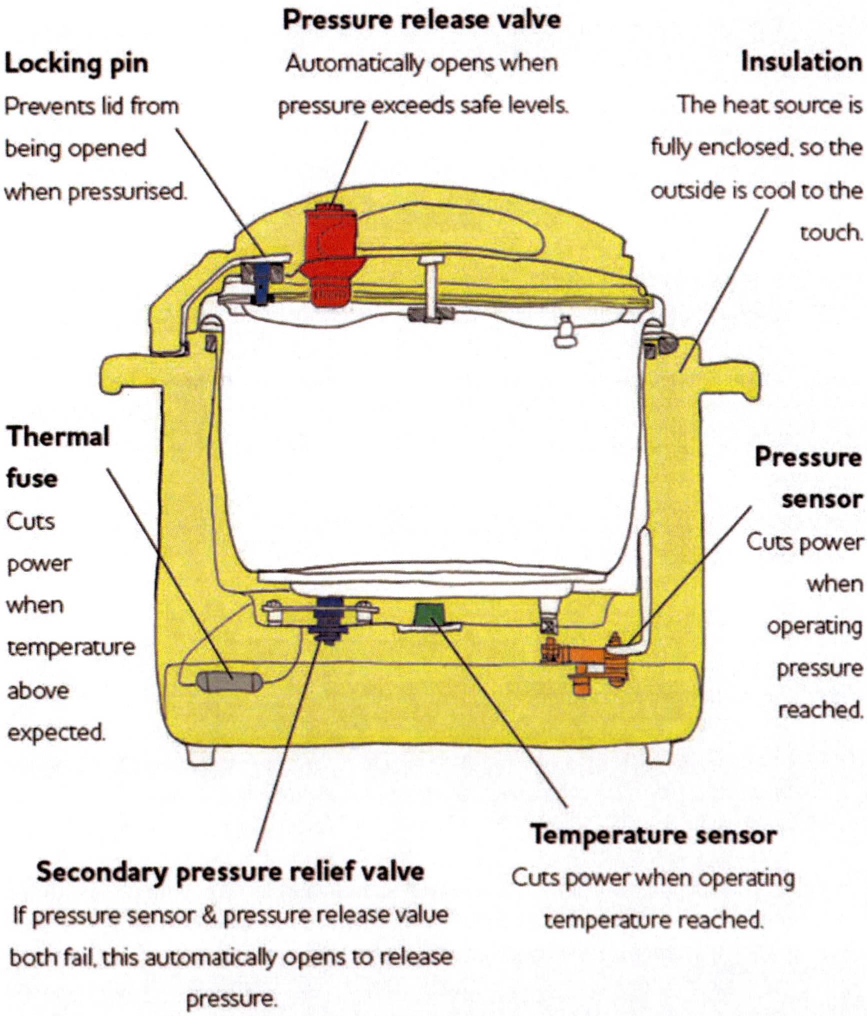

Fig. 7 Automatic control and safety features of a typical EPC [27]

utilising the EPC. In fact, much less stirring is needed, as no water escapes from the sealed environment during pressure cooking and the temperature is automatically limited to 120 °C, so it is almost impossible to burn the food. In the data below we identify whether these beliefs are an insurmountable barrier to using EPCs in East and Southern Africa or whether the other benefits might outweigh this particular challenge.

2 Methodology

This paper seeks to demonstrate the compatibility of East and Southern African cooking practices with EPCs by answering the following research questions:

- How much energy is really needed to cook popular East/Southern African dishes with an EPC?
- How does the EPC fit into the kitchen routines of East/Southern African cooks?
- What is the user experience for everyday cooks in East/Southern Africa with EPCs?

A range of multidisciplinary techniques including cooking diaries, focus groups and kitchen laboratories were employed to understand how households in East and Southern Africa currently cook and how they aspire to cook. To date, these techniques have been applied in Kenya, Tanzania and Zambia. At the time of writing, similar studies are underway in Ethiopia and Uganda. This paper focusses on the data from Kenya, which is the most detailed dataset currently available.

2.1 Cooking Diaries

Despite decades of work on improving the efficiencies of biomass stoves, there seems to be little available data on 'how' people cook. Modern fuels such as gas and electricity are more controllable and can be turned on/off in an instant. There are also a huge range of electric cooking appliances, each designed for specific processes (e.g. kettles for heating water). Therefore, it is important to know how often people are frying, boiling, reheating or something else entirely.

To date international improved cookstove tests have focused on the Water Boiling Test and the Kitchen Performance Test [28, 29]. Neither of these tests were designed to give insight into 'how' a cook cooks, and whether, when they transition to a different fuel or appliance, their cooking practices change. Cooking is a deeply cultural experience, as the foods people cook and the practices they use to prepare them vary widely. To date studies of 'how' people cook have largely been based on observational qualitative data.

The cooking diaries study was designed to offer a deeper exploration into the unique cooking practices of individual households, paired with quantitative measurements of energy consumption. In each country, 20 households were selected to participate in the study, based upon the fuels they cooked with and their willingness/ability to record high quality data for the duration of the study. This mixed methods approach gathers data from various sources:

- *Cooking diary forms*: foods cooked, cooking processes/times, appliances used.
- *Energy measurements*: manual measurements of fuel use and electricity consumption taken by participants.
- *Registration surveys*: simple demographic data on participants.
- *Exit surveys*: qualitative user experience feedback from participants (Fig. 8).

Data was recorded in two stages:

- *Baseline (2 weeks)*: cooking as normal, simply recording data.
- *Transition (4 weeks)*: cooking with electric appliances only.

Energy measurements were taken before and after each heating event to give '*meal-level resolution*' data (Table 3). Solid, liquid and gaseous fuels were measured using a hanging balance and calculating the difference in weight between before and after cooking measurements. Electricity consumption was measured using a plug-in electricity meter (Fig. 7). Paper records kept by participants were transcribed into digital form by the enumerators. Subsequent analysis of the complete database was performed in both SPSS and Excel.

In the second part of the experiment, the households were asked to transition to using solely electricity for cooking. In Kenya,[4] each household was given a hotplate,

Fig. 8 An enumerator training a participant to record data during the cooking diaries in Nairobi, Kenya. The electric cooking appliances are plugged into an energy meter in the top right of the photo

[4]In Zambia, participants used hotplates and EPCs and in Tanzania, participants were given free choice from a range of seven different electric cooking appliances.

Table 3 Measured and modelled energy consumption for 100% electric cooking on a mixture of inefficient and efficient appliances

	No. complete days of data	Median daily energy consumption (kWh/household/day)	Household size (no. ppl)	Median per capita daily energy consumption (kWh/person/day)
Zambia				
100% electricity measured, median	99	1.63	7.9	0.21
Proportion of energy consumed by EPC cooking 50% of meals		0.55		
Total consumption if EPC at 90% of menu		*1.1*		*0.14*
Tanzania				
100% electricity measured, (with EPC proportion modelled)	423	2.06	4.2	0.49
Proportion of energy consumed by EPC cooking 50% of meals		0.69		
Total consumption if EPC at 90% of menu		*1.44*		*0.34*
Kenya				
100% electricity measured, (with EPC proportion modelled)	431	1.4	3.1	0.46
Proportion of energy consumed by EPC cooking 50% of meals		0.47		
Total consumption if EPC at 90% of menu		*0.96*		*0.30*

a rice cooker and an EPC, and received basic training on how to use each appliance. The three appliances were plugged into an extension cable, which fed into a plug-in energy meter (Fig. 7). Participants were also able to continue using any electrical appliances that they already owned, as long as they were plugged into the meter, so that energy consumption data could be captured. Data was recorded for a further 4 weeks, allowing participants time to adapt their cooking practices around the new appliances.

The study finished with an exit survey, asking participants about their experience with cooking with different electric appliances. Participants were also invited to share their energy-efficient cooking practices by participating in the Githeri eCooking Challenge. A prize was offered to the participant who could cook half kg of githeri using the least energy possible, whilst the enumerators observed and recorded their cooking practices to understand exactly where energy was being saved/wasted.

The cooking diaries protocols offer a more complete guide to this methodology for those looking to replicate the cooking diaries study: https://elstove.com/forward-looking-guidance/

2.2 *'Kitchen Laboratory'*

A mixture of 'ethno-engineering' techniques were employed to explore the compatibility of East and Southern African cooking practices with a range of electric cooking appliances in a 'kitchen laboratory' setting (Fig. 9). 'Ethno-engineering' blends anthropological and engineering approaches to create more holistic and culturally-informed development solutions. Initially, this focussed on simply observing the cooking practices of everyday cooks. Evidence was recorded as recipes, tips and reflective notes in a field diary and supplemented by photography. The plug-in energy meters from the cooking diaries study enabled a more quantitative dimension. In Kenya, this methodology evolved to be much more prescriptive, delving deeper into the findings from the cooking diaries study by exploring where energy is saved/wasted within a specific dish. This resulted in the production of an eCookBook [27], designed to inform everyday cooks of how to save time and money in the kitchen with smarter cooking practices, in particular adopting EPCs.

Fig. 9 Experiments in the Nairobi 'kitchen laboratory' during the production of the first eCookBook

2.3 Focus Groups

Focus groups were carried out in each country to gain further insight into current and aspirational cooking practices in a range of different contexts. A series of questions were designed to guide the discussion, however open dialogue was encouraged when unforeseen issues were brought up by the participants. An EPC was demonstrated during each session, inviting comments from the audience on how compatible the device was with the current and aspirational cooking practices (see Fig. 10).

3 Results

3.1 How Energy Efficient Are EPCs at Cooking East/Southern African Foods Under Controlled Conditions?

Controlled tests in the 'kitchen laboratory' for the eCookBook in Kenya revealed that EPCs can save up to 85% of the cost of cooking 'heavy foods' on charcoal [27]. 'Heavy foods' typically involve boiling for an hour or more on conventional stoves.

Fig. 10 Participants interacting with a range of energy-efficient electric cooking appliances during a focus group in Kibindu, Tanzania

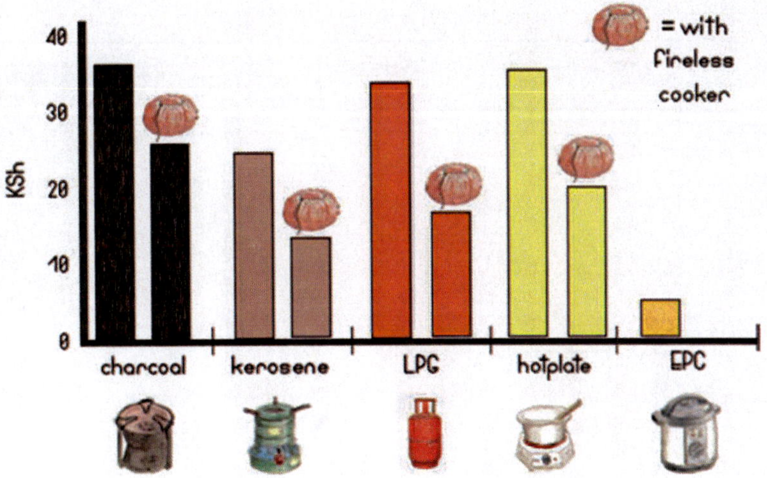

Fig. 11 Cost comparison for ½ kg dried yellow beans on the most popular fuels in urban Kenya (Nairobi costs, July 2018) [27]

They include beans, tripe, githeri (beans and maize stew) and stews with tougher cuts of meat.

A fireless cooker utilises the principles of insulation (but not pressurisation) as a means to save fuel on any conventional cooking device during the simmering section of a recipe. For beans, the pot is heated until they are partially cooked (there is a need to cook until the toxins are removed) and then the pot is transferred into the fireless cooker and sealed in an insulated environment. Because the temperature is maintained with minimal heat losses, the food continues to cook with no further input of energy. Figure 5 shows that judicious use of the fireless cooker can save between 10 to 15 KSh (0.10–0.15 USD) on fuel for charcoal, kerosene, LPG or an electric hotplate.

As it is an insulated appliance, a fireless cooker is effectively inbuilt into every EPC, allowing it to prevent heat from escaping from the pot throughout the entire recipe (not just the simmering stage). As a result, Fig. 9 shows that whilst cooking on LPG or an electric hotplate works out roughly the same cost as charcoal, the pressurisation and automatic control features of the EPC make it an order of magnitude cheaper. Kerosene is slightly cheaper than charcoal, LPG or an electric hotplate, however still several times more than the EPC (Fig. 11).

3.2 How Energy Efficient Are EPCs When Used by Everyday East/Southern African Cooks Under Semi-controlled Conditions?

The results of the Githeri eCooking Challenge show that almost all households were capable of cooking very efficiently (80–90% savings) when they want to (Fig. 12). On a hotplate, cooking ½ kg githeri usually exceeds 2 kWh and can even reach

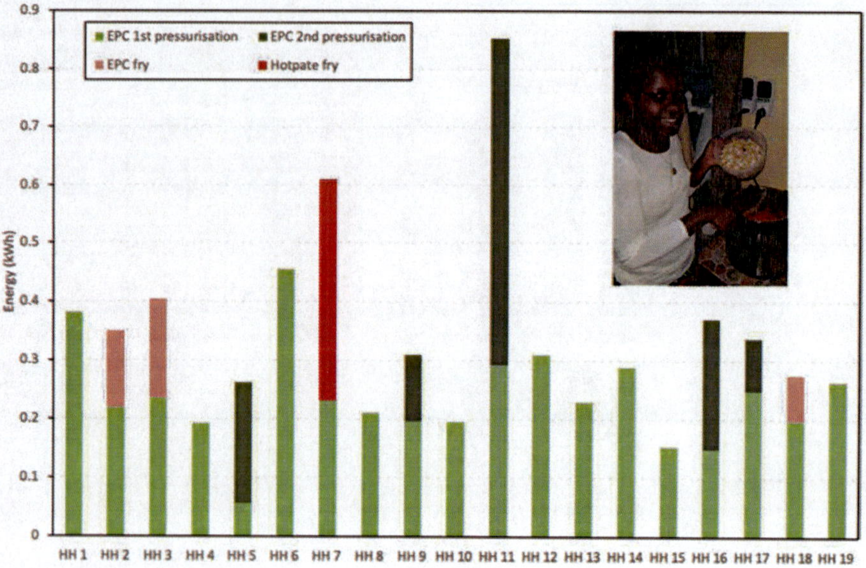

Fig. 12 Energy consumption during the Githeri eCooking Challenge by participant, appliance and process

4 kWh if no efficiency measures are in place (using the slowest cooking beans, leaving the lid off, etc.). Figure 6 shows that almost all households (16/19) were able to complete the challenge with under 0.4 kWh—an 80% saving over hotplates.

3.3 *How Useful Are EPCs and How Energy Efficient Are They When Used by Everyday Cooks in Real Kitchens?*

In the Kenya cooking diaries, households were able to cook all their food on electricity as they had all three key devices available: a hotplate, rice cooker and EPC. In these circumstances, the menu did not vary significantly from the baseline data obtained during the preceding weeks with their existing stoves and fuels. The analysis below shows that it is possible to cook over 90% of this typical Kenyan menu in an EPC. However, after limited training, with the free choice of three appliances, participants chose to cook approximately half their menu in efficient appliances (EPC or rice cooker), and that for these dishes, they used about half the energy of a hotplate.

3.3.1 Beyond 'Heavy Foods'

Energy savings on 'heavy foods' are clearly substantial in controlled and semi-controlled conditions; however, it is important to understand how they fit into the kitchen routines of everyday cooks. The evidence from the cooking diaries shows that 'heavy foods' comprise approximately one third of all dishes on a typical urban East African household's menu (see Table 3). In fact, many other dishes can also be cooked on an EPC: some are intuitive (e.g. rice), whilst others require some behaviour change (e.g. using a heatproof glove to hold the pot still whilst stirring ugali), however there are several that are extremely challenging on most models of EPC available on the market today (e.g. chapati).

A typical East/Southern African menu can be understood as composing of a set of categories of dishes, each with varying degrees of compatibility with EPCs. An overview of typical preparation techniques for popular Kenyan foods is given in Appendix 1. Table 2 proposes the following categories:

- 'Heavy' foods—usually require boiling the main ingredient (e.g. beans) for over an hour on a conventional stove and may also contain a frying stage with extra ingredients to add flavour (e.g. a tomato and onion sauce).
- Staples—normally boiled for approximately half an hour. Some require stirring (e.g. ugali, porridge), but others are simply left to boil (e.g. rice).
- Quick fryers—usually fried for 5–15 min, a shallow pan and high heat is often preferred, but not essential. Access to the pan is usually required to stir the food and prevent burning.
- Deep fryers—food is completely submerged in oil at 175–190 °C.
- Flat breads—medium heat, evenly distributed across a shallow pan is required to cook the whole of the flat bread at the same rate. Access to the pan is required to turn the bread frequently.

Analysis of the Kenya cooking diaries data allows us to deduce that EPCs use roughly half the energy of electric hotplates across the full range of dishes that they are able to cook. On average, rice cookers used 39% (median of 0.09 kWh/person/event, n = 46) and EPCs used 76% (0.18 MJ/person/event, n = 49) of the energy of a hotplate (0.23 MJ/person/event, n = 119). However, Fig. 13 reveals that EPCs were chosen to cook 'heavier' (and therefore more energy intensive) dishes, when in fact they can also be used for the lighter staples (e.g. rice), which had been cooked in the rice cooker. As all participants in the Kenya cooking diaries had an electric hotplate, a rice cooker and an EPC, it can be assumed that all the dishes that were cooked in a rice cooker could also have been cooked in the EPC with the same energy consumption. Averaging the per capita, per heating event energy consumption figures for rice cookers and EPCs comes to just under half (45%) that of the electric hotplate.

Further analysis of the Kenya cooking diaries dataset suggests that with minimal training, households would choose to use an EPC to cook half their menu if it were the only electric appliance available. Figure 13 shows that a total of 645 dishes were cooked on EPCs and rice cookers. Ignoring all other appliances (which totalled only 150 dishes and were mainly microwaves) and comparing directly to the 739 dishes cooked on a hotplate, roughly half (47%), of a total of 1387 dishes were cooked by choice on an EPC or rice cooker. We can therefore conclude that without additional training or design modifications, households with an EPC as their efficient appliance are likely to choose to cook roughly half their menu with it.

Table 2 Categorisation of typical Kenyan foods by their compatibility with EPCs

Food category	Frequency on urban Kenyan menu (%)	Typical dishes	Compatibility with EPCs	Energy savings with EPCs	Enablers
'Heavy foods'	32	Beans, matumbo (tripe), meat stews	Users instinctively use EPCs	High (50–90%)	Cooking times and water quantities for popular local foods
Staples	39	Ugali (maize meal), rice	Users use EPCs if encouraged	Moderate (20–50%)	Demonstrations, extra EPC
Quick fry	20	Sukuma wiki (kales), eggs	Users use EPCs if encouraged	Low (5–20%)	Demonstrations, manual heat control, extra EPC, shallow pan
Deep fry	2	Mandazi (donut), fried chicken, chips	Users cannot currently use EPCs	Low (5–20%)	Manual heat control or deep fry settings (175–190 °C)
Flat breads	4	Chapati (flat bread)	Users cannot currently use EPCs	Low (5–20%)	Manual heat control and shallow pan
Other	3	Unknown			

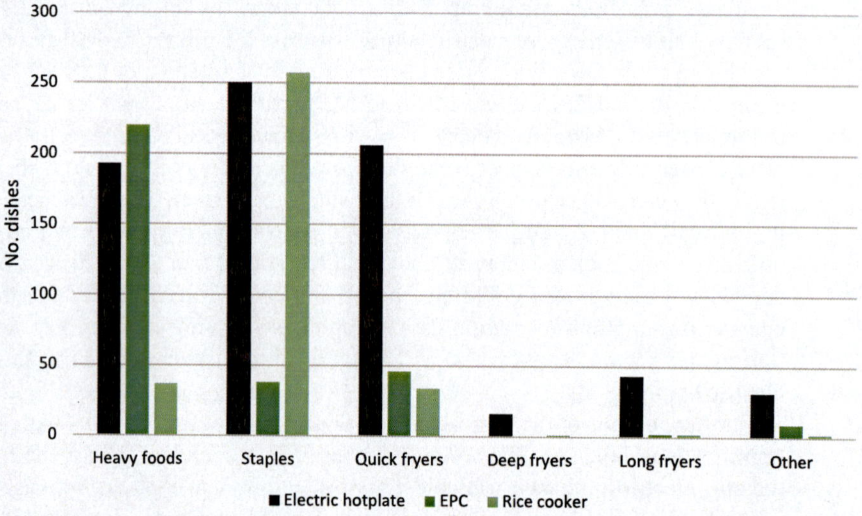

Fig. 13 Number of each category of dish cooked on inefficient (hotplate) and efficient (rice cooker and EPC) appliances during the Kenya cooking diaries (127 records for dishes cooked on microwaves and kettles already owned by some participants have been omitted). A full dish by dish breakdown of this data is available in Appendix 1

However, the data also suggests that it is actually possible for urban Kenyan households to cook over 90% of their menu on an EPC. Referring to Table 3 and Fig. 13, 'heavy foods', staples and quick fryers can all be cooked on an EPC, which together make up 91% of the urban Kenyan menu. With the exception of sausages, every dish in these three categories was cooked in an EPC at least once. For instance, there are 102 meal events for ugali with a hotplate, but there are also 105 events with a rice cooker and 11 with an EPC. As the cooking diary study only looked at the first month that participants used these appliances, it is likely that experimentation with cooking a broader range of dishes in the EPC didn't occur until the end of that period. What is more, as many participants were used to cooking on a 4-plate gas stove, the hotplate may well have often been chosen simply to allow more dishes to be cooked simultaneously. In contrast, poorer households with only a single burner cooking device tend to cook each dish one after the other.

Cooking solely on electricity across the three appliances was found to have a median of approximately 2 kWh per household per day, whilst the 50% of the menu cooked on efficient appliances was had a median consumption of roughly 0.6 kWh per household per day. Table 3 shows the median daily energy consumption figures from the 100% electric cooking stage of the cooking diaries in each country. Using the rules described above (50% of the menu cooked on EPCs, using 50% of the energy of the hotplate), estimations are made for the median daily energy consumptions of these households of the EPC only. However, if the EPC were to cook 90% of the menu (with training and experience), total consumptions would drop further.

3.4 *User Experience of EPCs*

Whilst cost, driven by energy efficiency, may be a strong driver, if the cooker is not easy to use and the food is not as tasty as usual, households will be unlikely to adopt it. This section presents insights from the exit survey from the Kenya cooking diaries, which asked the households who had been using EPCs (plus rice cookers and hotplates) for a month, about their experience with this new cooking device.

'Heavy foods' such as beans or matumbo (tripe) that usually require boiling for an hour or more to soften are unsurprisingly rated as much easier to cook on the EPC than the hotplate (Fig. 14). In contrast, foods that require manual heat control and/or a shallow pan, such as chapati or mandazi, are rated much easier on the hotplate.

Perhaps surprisingly to some, food cooked on electricity was rated as the tastiest, just ahead of LPG and charcoal (Fig. 15). Wood and kerosene lag far behind. Figure 16 shows that whist some respondents missed the smokey flavour in specific foods, many did not miss it at all.

In Figs. 16, 17 and 18 the results of interviews with the users of the electrical appliances have been summarised in word clouds. The words are sized according to the number of times they were mentioned. As one can see, while the mythology of the development sector suggests that people really like the smokey flavour of food,

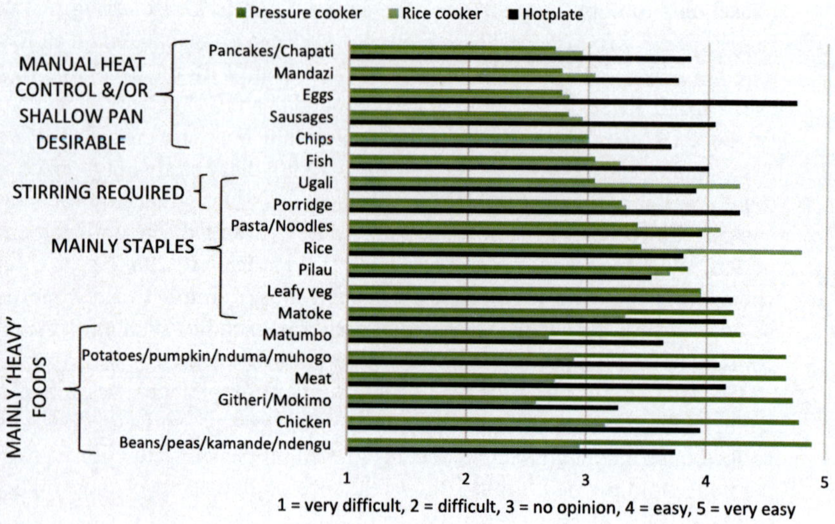

Fig. 14 Average responses to the question from 20 trial households in Kenya: "how easy is it to cook each food on the eCookers?" Ranked by ease of cooking on an EPC

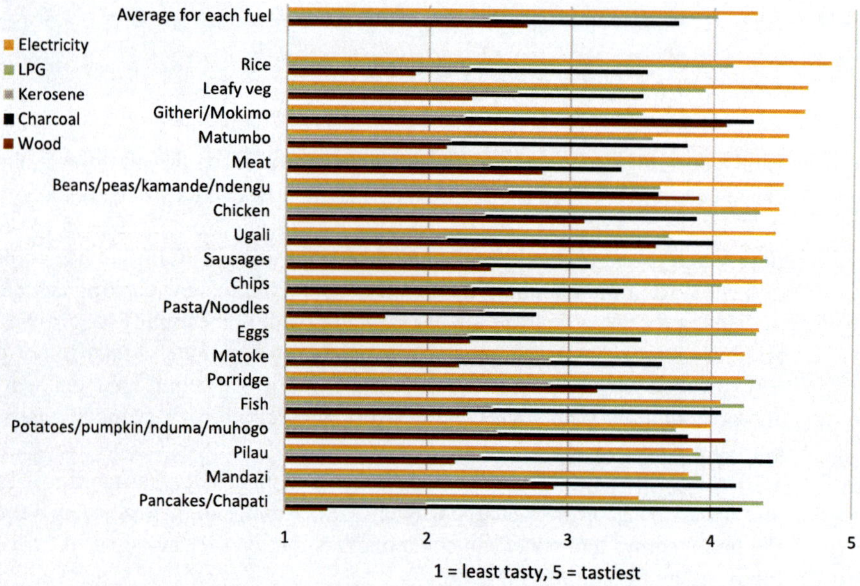

Fig. 15 Average responses to the question from 20 trial households in Kenya: "Do foods taste different when cooked on different fuels? If so, please rank each fuel for each food." Foods ranked by tastiness when cooked with electricity

Fig. 16 Responses to the question from 20 trial households in Kenya: "Do you miss the smokey flavour of food? If so, for which dishes in particular?". Words sized according to the number of responses

Fig. 17 Responses to the question from 20 trial households in Kenya: "What were the best/worst things about cooking with electricity?"

and want to retain it the majority of the participants state they do not miss the smokeyness (Fig. 16).

The automated control systems of the EPC and rice cooker makes cooking easier, enabling multi-tasking and preventing food from burning (Fig. 17). Being able to cook faster and keep the kitchen clean are also both highly valued by the urban participants of the Kenya cooking diaries study, however, priorities may well be different in rural areas. Figure 18 shows that the rice cooker and EPC have clearly found a place in almost every participant's home.

3.5 *Automatic vs. Manual Control*

EPCs' automatic control systems make cooking certain dishes much easier, however the lack of manual control makes them undesirable for others. The EPC is the appliance of choice for long boiling dishes and many of the stapes, as it enables the cook to multitask, avoid burning the food and cut cooking times in half. Figure 15 suggests that existing models of EPC are already well suited to user needs, but there are clearly still minor tweaks, such as manual heat control, that could make them even more attractive. The heating element of an EPC is fundamentally the same as an electric hotplate, but controlled by a thermostat. Although most electric hotplates have the ability to control heat manually, they generally work on two principles, neither of which are satisfactory for frying:

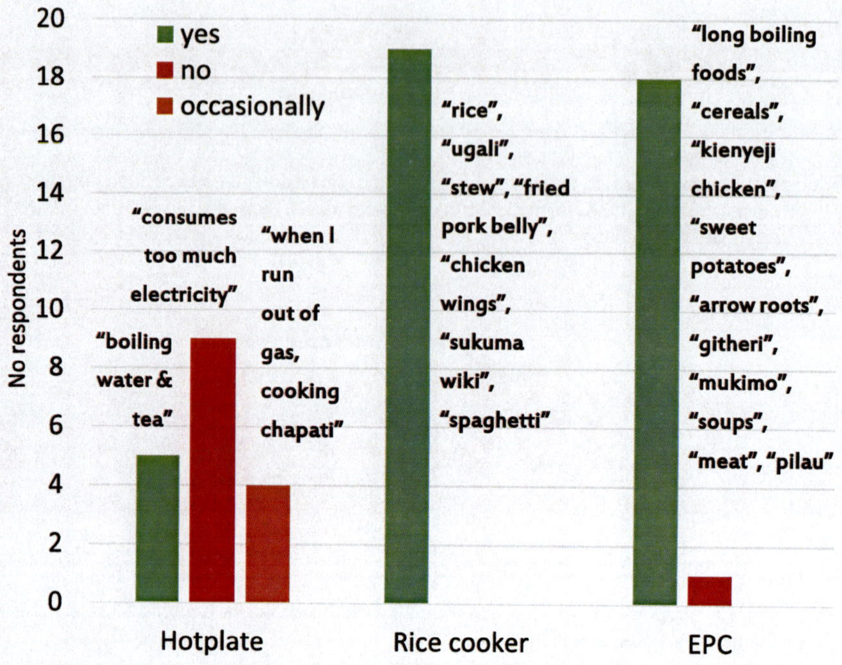

Fig. 18 Responses to the question from 20 trial households in Kenya: "We are done with our survey and are leaving the cookers with you. Will you continue using the e-cookers or will you switch back to your old stove?"

1. Turning the entire hotplate on and off using a thermostat, which when frying usually results in oscillation between burning and getting cold as the time delay between on/off cycles is usually several seconds.
2. Turning on parts of the heating coil independently, which results in uneven heating of the pan and means that only a few discreet power levels are available.

In contrast, LPG can be turned up and down almost instantly across a continuous range, from just keeping warm to full power. It is this manual controllability that makes LPG the fuel of choice for dishes that require fine control of the heating process, notably flat breads (e.g. chapati), quick fry dishes (e.g. sukuma wiki, kales) or deep fry dishes (e.g. mandazi).

Again in Fig. 19 we present the respondents comments to the question—If you could design your own completely new eCooker, what would it be like?

4 Discussion

Importantly, the EPC seems to fit into East/Southern African kitchen routines where charcoal is typically most favoured, offering an attractive pathway to achieving 100% modern cooking when combined with LPG or electric hotplates. To date,

Fig. 19 If you could design your own completely new eCooker, what would it be like?

Fig. 20 Cooking diary participants in Nairobi, Kenya who used to fuel stack kerosene with charcoal and now use an EPC for the 'heavy foods' that were previously cooked on charcoal

many households who gain access to modern energy for cooking chose to continue purchasing charcoal for 'heavy foods', as they believe it is cheaper. Figure 11 shows that in highly deforested urban contexts such as Nairobi, this is no longer the case, as the price of charcoal has risen considerably. In such contexts. Even cooking 'heavy foods' with a hotplate is now cost comparable with charcoal. Doing so with an EPC is an order of magnitude cheaper and can cut the cooking time in half, both of which provide strong drivers to households who have partially transitioned to modern energy to stop buying charcoal and transition to a 100% modern cooking solution (see Fig. 20).

4.1 EPC Design Modifications That Could Open Up New Opportunities in East/Southern African Markets

The evidence in this paper has shown that EPCs are clearly highly compatible with East/Southern African cooking practices, however, their potential is even greater. In European, North American and many Asian markets, EPCs are designed to be used alongside other modern cooking devices and are therefore optimised for a small range of dishes. In East and Southern African markets, it would be beneficial for EPCs to be able to cook a much broader range of dishes, so that households transitioning from charcoal are forced to light their stoves as little as possible. Table 2 shows that 91% of a typical urban Kenyan menu can be cooked on EPCs already available on the market. However, Fig. 13 shows that in practice, difficulties with accessing the pot to stir and the lack of manual heat control mean that everyday cooks typically only choose to use EPCs (or rice cookers) to cook half of their menu.

Several design modifications to the standard EPC designs available on the market today could open up new opportunities:

- *Deep frying*—most EPCs reach 120 °C and simply cut the power, as the control system believes it is pressurised and adding more heat to the pot could be dangerous. Enabling deep frying simply requires adding an alternative set point for the control system at a higher temperature (175–190 °C). This feature is already available on several models available on the market today already.
- *Manual heat control*—by enabling the user to control the amount of heat going into the pot when the lid is unlocked, the EPC may become the preferred choice for dishes that require fine control of heating, but can be cooked in a deep pot (e.g. sukuma wiki, scrambled eggs). This could be achieved very simply by having a variable thermostat that can be controlled by a knob on the front of the device, just like many hotplates. However, it is important that the time delay between on/off cycles of the hotplate is as small as possible to keep the temperature in the pot as constant as possible and avoid the food cycling between burning and getting cold. This could be achieved by using power electronics to vary input power continuously, such as PWM (Pulse Width Modulation), voltage transformers or thyristors. However, care would need to be taken to ensure that any distortion of the waveform is not detrimental to the grid that supplies it.
- *Highlighting how to cook local foods*—simply communicating to the user which foods can be cooked with the EPC and how can greatly increase the utilisation rate, likely even well beyond 50%, as over 90% of the urban Kenyan menu can be cooked in an EPC. This can be achieved in a number of ways, including clearly labelling the buttons or timer set points on the appliance itself with popular local foods, packaging EPCs with recipe books for local foods, or partnering with local distributors to engage local food bloggers and TV shows, or run live cooking demonstrations. The North American brand, Instant Pot, offers a great example of how developing supplementary resources and building a community of users can drive forward sales.

- *Battery-integration*—integrating a battery into an EPC that has been modified to run on DC power could unlock two completely new market segments, weak-grid and off-grid [22, 30–34]. Several DC EPCs have now appeared on the Chinese market; however, the authors are unaware of any that integrate a battery into the device. A 0.4 kWh battery could enable efficient cooks to cook almost any dish, whilst a 1 kWh battery would likely enable a full day's worth of EPC cooking (assuming the EPC is used for 50% of dishes, 80% DoD,[5] 90% charge/discharge efficiency). A lead acid battery discharged to 80% DoD at a rate of 1C would likely degrade, making lithium iron phosphate batteries a more appropriate choice. Due to the higher cost of battery-integrated devices, a financing mechanism that allows users to pay back the cost of the device over time is likely to be necessary. Fortunately, pay as you go and micro-credit business models have spread rapidly across East Africa in recent years, meaning that a range of established solutions are now on offer, in particular in the solar lighting sector.

5 Conclusion

This paper has shown that there is a huge and largely untapped opportunity to market EPCs in East and Southern Africa. In particular, Uganda, Kenya, Tanzania, Ethiopia and Zambia are politically stable countries, where new energy-efficient products have already revolutionised other sectors such as lighting, and together are home to 38 million people who are grid connected yet still pay for polluting cooking fuels (charcoal and kerosene). The focus groups and cooking diaries carried out in East and Southern Africa have shown that EPCs are highly desirable to everyday cooks because they cook faster, allow multi-tasking, save money and keep the kitchen clean.

The evidence in this paper shows that EPCs are significantly more energy efficient than electric hotplates in both laboratory and real kitchen environments. The empirical data from the kitchen laboratory shows that EPCs can cook the most energy intensive dishes with just one fifth of the energy of electric hotplates. This is complimented by the results of the cooking diaries studies, which show that everyday cooks choose EPCs for about half of their cooking and that across the full range of dishes they were used for, they use approximately half the energy of electric hotplates. Cooking with both hotplates and EPCs was found to use approximately 2 kWh per household per day, with the cook choosing to cook 50% of the menu on an EPC, which was estimated to use roughly 0.6 kWh per household per day. However, analysis of the range of dishes that make up a typical menu and experimentation in the kitchen laboratory has shown that EPCs are capable of cooking over 90% of the typical urban Kenyan menu. Training and experience are likely to move the proportion of EPC use from 50% nearer to 90%. In poorer households

[5] Depth of Discharge.

which are used to only having one 'device' for cooking, the EPC is likely to be used for a greater proportion of the menu. To increase the utilisation factor even further, the design of EPCs could be modified to include deep frying, allow manual heat control and most importantly, clearly indicating how to cook local foods.

Developing battery-integrated DC EPCs could unlock two huge market segments: weak-grid and off-grid. Currently just a few factories (mainly in China) are manufacturing DC EPCs and the authors are unaware of any that are manufacturing battery-integrated DC devices. In the same way that the mobile phone unlocked a much larger market than land line phones, in particular on the African continent, DC battery-integrated cooking devices have the potential to open up huge new markets for electric cooking. As one of the most energy efficient electric cooking devices available that is culturally well matched to East and Southern African cooking practices, there is no doubt that the EPC will be the first of many DC battery-integrated cooking devices to reach scale across the region.

Forthcoming activities under the Modern Energy Cooking Services (MECS) programme (www.MECS.org.uk) that will support the development of EPCs for the East and Southern African markets include:

- Global LEAP Awards for EPCs—in collaboration with the Efficiency for Access programme, the Global LEAP Awards aim to identify the 'best in class' efficient appliances. Manufacturers are invited to submit their products for testing and appliances are rated by energy-efficiency, usability, durability and cost and to produce a consumer's guide. Past winners of Global LEAP awards have been supported to disseminate their products with cash prizes, access to grant funding and results-based financing schemes.
- Challenge funds—a series of challenge funds enable organisations with innovative ideas to take the next step towards developing commercially viable modern energy cooking products and services. This can include initial feasibility studies, design, prototyping, field testing, social marketing and more.
- New research methodologies and new contexts—further cooking diaries, focus groups, the kitchen laboratory and other innovative methodologies will be employed in more countries and across a broader range of society throughout Sub-Saharan Africa and South/Southeast Asia. This aims to offer a greater understanding how different cultures cook and what the key opportunities are for transitioning to modern energy for cooking are in each context.

Appendix 1: Typical Kenyan Foods

'***Heavy***' foods like beans, meat stew or makande/githeri generally require boiling for 60 min or more. They are easy to cook on an EPC, which can offer significant energy and time savings over electric hotplates, or a rice cooker with moderate energy savings.

- **Githeri/mokimo**—beans and maize stew, usually wet fried/mashed potatoes with maize/beans/peas/pumpkin leaves. Many people will pre-cook (boil) githeri in bulk and wet fry portions throughout the week.
- **Beans/peas/kamande/ndengu**—beans/peas/lentils/green grams, usually stewed. Typically dried, so require rehydrating as well as cooking—some people soak before cooking, others just cook for longer. Many people will pre-cook (boil) in bulk and wet fry portions throughout the week.
- **Chicken, meat**—Usually wet fry (stew) or dry fry. Many people will pre-cook (boil) meat in bulk and wet fry portions throughout the week.
- **Matumbo**—Tripe, usually wet fried

'Staple' foods and water that require boiling for 15 min or more can also be cooked on an EPC, with moderate energy and time savings or rice cooker with moderate energy savings.

- **Heating water**—for tea/coffee, bathing, drinking etc.
- **Pasta/noodles**—Boiled and then often wet fried.
- **Porridge**—Requires regular stirring, but perhaps not in the electric pressure cooker. Need to do more experimentation on this.
- **Potatoes/pumpkin/nduma/muhogo**—Nduma = arrow roots, muhogo = cassava. Usually boiled, sometimes wet fried. Will need to check process to differentiate boiled and stewed.
- **Matoke**—Bananas. Usually wet fried, sometimes boiled. Will need to check process to differentiate boiled and stewed.
- **Rice**—Just boiled.
- **Pilau**—A combination of meat stew and rice. May use meat stew/stock pre-cooked on a previous occasion, or may cook the meat especially for this dish. May involve some frying of onions too. Sometimes potato is even thrown in!
- **Ugali**—Kenyans usually bring water to the boil, turn down the heat, add maize flour, stir, repeating a few times, then leaving to simmer until the mixture has reached the desired consistency.

'Quick fry' foods can also be cooked on an EPC or rice cooker, but some households may be reluctant to try and/or there are limited energy savings.

- **Eggs**—Could be boiled, fried or omelette. If omelette, can often be combined with potatoes (chips mayai), which may need deep frying first.
- **Fish**—Typically wet or dry fried whole or in fillets.
- **Leafy veg**—Sukuma wiki, spinach, etc. Typically dry fried, sometimes with onions.
- **Sausages**—Typically shallow fried.

'Long fry and deep fry' foods are very difficult to cook on an EPC or rice cooker, as they require precise temperature control.

Table 4 Number of each dish cooked on inefficient (hotplate) and efficient (rice cooker and EPC) appliances during the Kenya cooking diaries[a]

	Electric hotplate	EPC	Rice cooker	Totals	% Efficient appliance (EPC + rice cooker)
Heavy foods	*193*	*220*	*37*	*450*	*57*
Beans/peas	80	87	6	173	54
Matumbo	3	8	0	11	73
Githeri/mokimo	24	13	8	45	47
Meat	50	80	9	139	64
Pilau	4	2	4	10	60
Fish	23	11	8	42	45
Chicken	9	19	2	30	70
Staples	*250*	*38*	*257*	*545*	*54*
Ugali	102	11	105	218	53
Potatoes/pumpkin	30	14	6	50	40
Pasta/noodles	25	1	17	43	42
Porridge	57	1	2	60	5
Rice	36	11	127	174	79
Quick fryers	*206*	*46*	*34*	*286*	*28*
Sausages	10	0	0	10	0
Eggs	86	1	1	88	2
Leafy veg	110	45	33	188	41
Deep fryers	*16*	*0*	*1*	*17*	*6*
Chips	7	0	1	8	13
Mandazi	9	0	0	9	0
Long fryers	*43*	2	2	*47*	*9*
Chapati/pancake	43	2	2	47	9
Other	*31*	*9*	2	*42*	*26*
Other	31	9	2	42	26
Totals	739	315	333	1387	47

[a]One hundred and twenty-seven records for dishes cooked on microwaves and kettles already owned by some participants have been omitted

- ***Pancakes/Chapati***—Shallow fried one by one in a shallow pan, as they must be flipped and swapped over many times. Requires low heat evenly distributed throughout the pan.
- ***Chips***—Deep fried. If oil too hot, they burn, if too cold, they go soggy.
- ***Mandazi***—Donuts. As above (Table 4).

Appendix 2: Regional Colour Coding and Two-Letter Country Codes

See Table 5.

Table 5 Regional colour coding and two-letter country codes used throughout this report

AIMS		*Central Africa*		*Central America & Carribbean*	
CV	Cabo Verde	CM	Cameroon	BZ	Belize
KM	Comoros	CF	Central African Republic	CR	Costa Rica
MV	Maldives	TD	Chad	CU	Cuba
MU	Mauritius	CG	Congo	DM	Dominica
ST	Sao Tome & Principe	CD	DRC	DO	Dominican Republic
SC	Seychelles	GQ	Equatorial Guinea	SV	El Salvador
		GA	Gabon	GD	Grenada
				GT	Guatemala
				HT	Haiti
				HN	Honduras
				JM	Jamaica
				NI	Nicaragua
				PA	Panama
				LC	Saint Lucia
				VC	Saint Vincent & the Grenadines
Central Asia & North Korea		*East Africa*		*Europe*	
AM	Armenia	BI	Burundi	AL	Albania
AZ	Azerbaijan	DJ	Djibouti	BA	Bosnia & Herzegovina
BY	Belarus	ER	Eritrea	BG	Bulgaria
GE	Georgia	ET	Ethiopia	MK	Macedonia
KZ	Kazakhstan	KE	Kenya	ME	Montenegro
KG	Kyrgyzstan	RW	Rwanda	RO	Romania
MD	Moldova	SO	Somalia	RS	Serbia
MN	Mongolia	SS	South Sudan		
KP	North Korea	SD	Sudan		
RU	Russia	TZ	Tanzania		
TJ	Tajikistan	UG	Uganda		
TM	Turkmenistan				
UA	Ukraine				
UZ	Uzbekistan				
India & China		*Middle East*		*North Africa*	
CN	China	IR	Iran	DZ	Algeria

(continued)

Table 5 (continued)

IN	India	IQ	Iraq	EG	Egypt
		JO	Jordan	LY	Libya
		LB	Lebanon	MA	Morocco
		PS	Palestine	TN	Tunisia
		SY	Syria	EH	Western Sahara
		TR	Turkey		
		YE	Yemen		
Pacific Islands & PNG		*South America & Mexico*		*South Asia (Excl. India)*	
AS	American Samoa	AR	Argentina	AF	Afghanistan
CK	Cook Islands	BO	Bolivia	BD	Bangladesh
FJ	Fiji	BR	Brazil	BT	Bhutan
KI	Kiribati	CL	Chile	NP	Nepal
MH	Marshall Islands	CO	Colombia	PK	Pakistan
FM	Micronesia	EC	Ecuador	LK	Sri Lanka
NR	Nauru	GY	Guyana		
NU	Niue	MX	Mexico		
PW	Palau	PY	Paraguay		
PG	Papua New Guinea	PE	Peru		
WS	Samoa	SR	Suriname		
SB	Solomon Islands	UY	Uruguay		
TO	Tonga	VE	Venezuela		
TV	Tuvalu				
VU	Vanuatu				
Southeast Asia		*Southern Africa*		*West Africa*	
KH	Cambodia	AO	Angola	BJ	Benin
ID	Indonesia	BW	Botswana	BF	Burkina Faso
LA	Laos	LS	Lesotho	CI	Côte d'Ivoire
MY	Malaysia	MG	Madagascar	GM	Gambia
MM	Myanmar	MW	Malawi	GH	Ghana
PH	Philippines	MZ	Mozambique	GN	Guinea
TH	Thailand	NA	Namibia	GW	Guinea-Bissau
TL	Timor-Leste	ZA	South Africa	LR	Liberia
VN	Vietnam	SZ	Swaziland	ML	Mali
		ZM	Zambia	MR	Mauritania
		ZW	Zimbabwe	NE	Niger
				NG	Nigeria
				SN	Senegal
				SL	Sierra Leone
				TG	Togo

References

1. IMARC, "Multi Cooker Market: Global Industry Trends, Share, Size, Growth, Opportunity and Forecast 2019-2024," 2019.
2. Sustainable Energy for All, "SEforAll.org," 2019. [Online]. Available: https://www.seforall.org/. [Accessed: 12-Jul-2019].
3. WHO, "Household air pollution and health," 2018. [Online]. Available: https://www.who.int/en/news-room/fact-sheets/detail/household-air-pollution-and-health. [Accessed: 25-Mar-2019].
4. S. Batchelor, E. Brown, N. Scott, and J. Leary, "Two Birds, One Stone—Reframing Cooking Energy Policies in Africa and Asia," *Energies*, vol. 12, no. 9, p. 1591, 2019.
5. D. Ockwell *et al.*, "Can Pay-As-You-Go, Digitally Enabled Business Models Support Sustainability Transformations in Developing Countries? Outstanding Questions and a Theoretical Basis for Future Research," *Sustainability*, vol. 11, no. 7, p. 2105, 2019.
6. KPLC, "Pika na Power," 2019. [Online]. Available: https://www.facebook.com/PikaNaPower/. [Accessed: 12-Jul-2019].
7. Power Africa, "Development of Kenya's power sector 2015-2020," Nairobi, Kenya, 2015.
8. EWURA, "Annual Report for the Year Ended 30th June 2016," Dar es Salaam, Tanzania, 2017.
9. A. Eberhard, K. Gratwick, and L. Kariuki, "A review of private investment in Tanzania's power generation sector," *J. Energy South. Africa*, vol. 29, no. 2, pp. 1–11, 2018.
10. Export.gov, "Tanzania—Energy," 2019. [Online]. Available: https://www.export.gov/article?id=Tanzania-Energy. [Accessed: 25-Mar-2019].
11. EEG, "Tanzania Energy Sector Overview," Energy for Economic Growth Research Programme, 2016.
12. World Bank, "World Development Indicators (WDI) | Data Catalog," 2019. [Online]. Available: https://datacatalog.worldbank.org/dataset/world-development-indicators. [Accessed: 12-Jul-2019].
13. World Bank, "Getting Electricity—Doing Business," 2019. [Online]. Available: https://www.doingbusiness.org/en/data/exploretopics/getting-electricity. [Accessed: 12-Jul-2019].
14. ZESCO, "Current ZESCO Tariffs," 2019. [Online]. Available: http://www.zesco.co.zm/customerCare/tariffs. [Accessed: 12-Jul-2019].
15. KPLC, "Implementation of New Tariffs," 2019. [Online]. Available: https://kplc.co.ke/img/full/rZr6LLCjVqVw_1st November 2018—Tarrifs.jpg. [Accessed: 12-Jul-2019].
16. TANESCO, "Bei Za Umeme Zilizoidhinishwa," 2019. [Online]. Available: http://www.tanesco.co.tz/index.php/customer-service/tariffs/7-bei-za-umeme-zilizoidhinishwa/file. [Accessed: 12-Jul-2019].
17. Umeme, "Power Tariffs," 2019. [Online]. Available: https://www.umeme.co.ug/tariffs. [Accessed: 12-Jul-2019].
18. Ethiopian Electric Utility, "New Tariff," 2019.
19. Adam Smith International, "Black gold—the real cost of charcoal in Africa," 2016. [Online]. Available: https://medium.com/@adamsmithinternational92/black-gold-the-real-cost-of-charcoal-in-africa-7d241a2f3084. [Accessed: 25-Mar-2019].
20. S. V. Lall, J. V. Henderson, and A. J. Venables, *Africa's Cities—Opening Doors to the World*, vol. 15, no. 3. World Bank, UK AID, 2017.
21. WHO, "Household energy database," *WHO*. World Health Organization, 2017.
22. J. Leary, S. Batchelor, M. Leach, E. Brown, and A. Alsop, "eCook Global Market Assessment Where will the transition take place first?," Implemented by Gamos, Loughborough University, University of Surrey. Funded by DfID, Innovate UK, Gamos, 2018.
23. GOGLA, Lighting Global, and Berenschot, "Global Off-Grid Solar Market Report Semi-Annual Sales and Impact Data—JULY-DECEMBER 2016, PUBLIC REPORT," 2016.
24. Instant Pot, "History of the Pressure Cooker," 2019. [Online]. Available: https://www.instantpot.co.uk/pressure-cooker-history/. [Accessed: 13-Jul-2019].

25. S. Batchelor, A Talukder, R Uddin, S K Mondal, S Islam, R Redoy, R Hanlin and M. Rezwan Khan; "Solar e-Cooking: A Proposition for Solar Home System Integrated Clean Cooking," *Energies*, vol. 11, no. 11, p. 2933, Oct. 2018.
26. T. Couture and D. Jacobs, "Beyond Fire: How to Achieve Electric Cooking," HIVOS & World Future Council, 2019.
27. J. Leary *et al.*, *The Kenya eCookBook: Beans & Cereals Edition*, no. September. Nairobi, Kenya: Available from: www.MECS.org.uk, 2015.
28. R. Bailis and R. Edwards, "Kitchen Performance Test (KPT) Version 3.0," no. January, pp. 1–32, 2007.
29. R. Bailis, D. Ogle, N. MacCarty, and D. Still, "The Water Boiling Test (WBT) Version 3.0," *Heal. (San Fr.)*, no. January, pp. 1–38, 2007.
30. G. Zubi, F. Spertino, M. Carvalho, R. S. Adhikari, and T. Khatib, "Development and assessment of a solar home system to cover cooking and lighting needs in developing regions as a better alternative for existing practices," *Sol. Energy*, vol. 155, pp. 7–17, 2017.
31. S. Batchelor, E. Brown, J. Leary, N. Scott, A. Alsop, and M. Leach, "Solar electric cooking in Africa: Where will the transition happen first?," *Energy Res. Soc. Sci.*, vol. 40, 2018.
32. E. Avila, G. Lawrence, J. Lin, N. Howard, B. Mohamed, and H. Ally, "The Desirability of Clean Cooking in Off-Grid Households," A2E1, Arusha, Tanzania, 2019.
33. S. Batchelor *et al.*, "eCook Tanzania Country Report—Opportunities and Challenges in Tanzania," Dar es Salaam, Tanzania. Available from: https://elstove.com/innovate-reports/, 2018.
34. J. Leary *et al.*, "eCook Zambia Prototyping Report," Available from: https://elstove.com/innovate-reports/, 2019.

Project "Intermittence Plus", on the Path of Real-Time DSM

Jerome Gilbert

Abstract "Intermittence Plus" is a collaborative R&D project, which has been done in the frame of the French national Investments for the Future program, which has been selected, supervised and evaluated by the French Environment and Energy Management Agency in charge of Climate change—ecological, energy transition.

This project aimed to propose new solutions for increasing the renewable share in the generation mix, possibly up to 100%, safely, by controlling appropriate massive amounts of power loads, simultaneously and in real time for adjusting the balance dynamically by shedding, forcing consumption or storing energy at the right moment at demand side.

The solutions, which have been explored in this project make households a key part of the smart grids, with storage and demand response/flexibility, and valuate flexibility as a Distributed Energy Resource.

These new solutions allow balancing unpredictable variations of production and unexpected network failures and therefore reduce the risks of blackout; it also represents valuable virtual primary and secondary reserves.

Moreover, the service of aggregation of the right amount of diffuse power loads to be deactivated or to be activated, is automatically made for free by the control system itself.

The choice of the right mass of loads to be aggregated can be done in open loop according to deterministic criteria. This choice can also be made in closed loop through a continuous monitoring of the activatable shedding, consumption and storage capacities of the electrical network.

J. Gilbert (✉)
Degetel, Boulogne-Billancourt, France
e-mail: jgilbert@degetel.com

P. Bertoldi (ed.), *Energy Efficiency in Domestic Appliances and Lighting*, Springer Proceedings in Energy, https://doi.org/10.1007/978-3-030-79124-7_26

1 The Project Data

A project supervised by the French Environment and Energy Management Agency in the frame of the research program Smart Electric Networks, co-financed by the national Program of Investments for the Future.

A consortium of nine complementary partners, a duration of 3 years (2016–2019), a budget of 5.1 M€.

1.1 Project Overarching Objectives

Provide innovative solutions to increase the renewable share in the generation mix, possibly up to 100%, safely, by controlling appropriate massive amounts of power loads, simultaneously, within 1 min max, for adjusting the balance dynamically by shedding, forcing consumption or storing energy at the right moment at demand side.

Provides innovative solutions aiming to be:

- Suitable for balancing unpredictable variations of production and network failures
- Virtual primary and secondary reserve
- Durable, reliable and energy efficient
- Easy and inexpensive to deploy and operate
- Acceptable by users
- Exploitable in the context of numerous present and future business models
- Low requirements for interoperability
- Compliant with European regulations

2 The Consortium of the Project in Its National Ecosystem

The project team has been built for covering all the aspects of the project through the selection the national leaders of each complementary specialty (please, see Fig. 1).

3 The Genesis of the Project

At the beginning, the ambition was to find smarter solutions, and more suitable to Europe, than those of the American company Sequentric, who was aiming in 2015 to control 100 million water heaters in North America and planned to come in Europe.

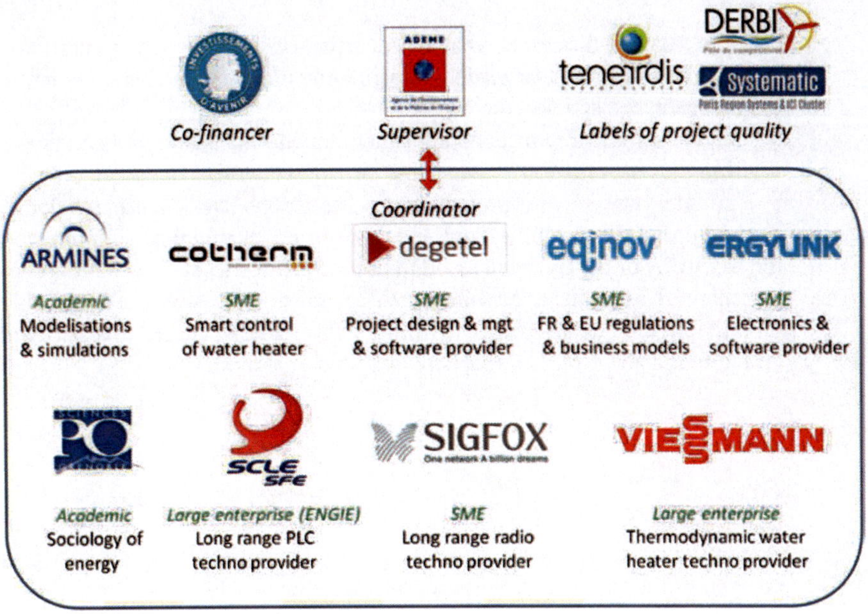

Fig. 1 The consortium of the project in its national ecosystem

France was the best place to start this project because of its track record in the domain, the presence of industrial leaders and the size of its park of electric water heaters (15 million having an average capacity of 200 L).

4 Issues, Context and Objectives of the Project

4.1 The Context

In its report "The Power of Transformation: Wind, Sun and the Economics of Flexible Power System" published in February 2014 [1], the International Energy Agency cites as the four essential flexibility levers for the implementation of the energy transition: flexible power plants, power infrastructure, storage and demand management.

In terms of energy storage, this reference report ranks diffuse thermal storage at the top of the ranking for its cost performance and benefits relative to other electrical energy storage solutions. However, the report indicates that there is some uncertainty about the possibility of massive deployment of such storage solutions in real-life situations.

In its report "Towards a 100% renewable electricity mix in 2050" [2], Ademe demonstrates the feasibility of increasing the share of renewable energies beyond

80% by 2050 by assuming that the distribution of a second generation of smart meters by 2050 will have made it possible to generalize the deployment of devices to manage usage at a very low cost.

In addition, this report concludes that demand management is a key element in limiting the cost of a 100% renewable energy scenario.

The Intermittence Plus project, whose main objective is a research demonstrator, aims to prove the feasibility and relevance of novel technical solutions to increase the flexibility of the electricity grid to unprecedented levels in order to prepare it for the energy transition by enabling it to accept a larger share of renewable energy production, the random nature of which will require solutions to adjust demand, in both directions, erasure and storage, and this in real time.

4.2 *The Issues at Stake*

One of the main technological challenges of this project is to find reliable and economical solutions to remotely control simultaneously and in real time massive quantities of remote objects such as power loads located in consumers' downstream installations. Indeed, neither the infrastructures that have been designed to meet the needs of the smart meters, nor the traditional telecom networks allow this.

Another challenge is to remove the technological barriers that limit the massive deployment of diffuse thermal storage solutions and prevent access to this potentially considerable storage resource at any time, depending on the network's needs (for the record, in France, the 15 million electric water heaters represent an installed capacity four times greater than that of WWTPs and three times greater than the current wind power).

For example, on the North American plate, Sequentric is demonstrating the value of the massive use of electric water heaters as a means of choice for balancing intermittent power generation (Sequentric aims to connect 100 million water heaters).

The "Intermittence Plus" project has a higher ambition in several respects. This will include, for example, extending the field beyond the means of domestic hot water production by adding controllable thermal storage capacities to electric heating appliances and effectively combining energy storage, consumption reduction and energy efficiency.

5 Technical and Scientific, Environmental and Societal Objectives

The objective of this project is to provide innovative technical solutions in the following areas:

Remote control of power loads from upstream decision-making systems. The project solutions must be able to control massive quantities of power devices simultaneously and in real time, in an approach compatible with existing electrical systems, without requiring heavy investment or high operating costs.

Power loads and their control in low-voltage installations downstream of the meters, which we will call "Smart Loads", as well as new solutions for their remote controls upstream of the meters. These innovative solutions will be simple and cost-effective to deploy, to provide the necessary flexibility for electricity networks in both directions (erasure and storage), to offer new system-oriented services, new customer-oriented services and to improve the quality of service perceived by electricity users, while continuing to take into account in a transparent manner the information from the meter (e.g. tariff signals, planned erasures if necessary etc.).

The project solutions also indirectly aim to contribute to the solution of the problems of congestion or point deficit that may appear on the low voltage network with the multiplication of diffuse renewable energy sources and with the expected increase in the park of electric vehicles. Indirect benefits are also expected to limit the need to strengthen the infrastructure of the distribution network to meet the ever-increasing demand for power from end-users, which poses economic challenges, particularly in rural facilities and low user density networks.

In addition, the project will extend the scope of its field tests by simulating, on the scale of relevant territories, the impact on the electricity grid of massive deployments of the "Smart Loads" resulting from the project.

Economic and regulatory aspects, which are structuring in the energy sector, and in particular in terms of flexibility, will also be the subject of research aimed at making the most of the project's results.

Finally, the sociological dimension will also be explored through research aimed at enabling the adoption, facilitating the use and increasing the dissemination of the project's solutions.

This Project Also Has the Following Objectives

- Providing the necessary flexibility to the electricity grid in both directions (load shedding and storage) in a multioperator context.
- Enabling dynamic management of peak consumption in addition to more massive and/or longer cycle storage solutions (STEP, Power-to-gas…).
- Proposing solutions that can be easily appropriated by the industry (low cost, easy interoperability).
- Proposing solutions adapted to the new build and renovation markets.
- Providing non-technocentric ergonomics targeted at the general population.
- Bringing out new "B to B" services (between operators).
- Bringing out new "B to C" services (tariff offers, DHW per m3…).
- Improving the perception of service quality in the electricity grid (zero blackouts).
- Improving the quality of interactions between the power system and its users.
- In-depth study of the underlying regulatory issues that open and structure markets as well as business models whose profitability will determine the intensity of future deployments of project solutions.

- In-depth study of socio-energetic issues, the modalities of appropriation of an energy management tool at home by users and their monetary and non-monetary motivations during the different phases of their relationship with the tool.
- Moving the regulatory guidelines to make demand management attractive.
- Making the electricity system environmentally sound (no more thermal production in support of wind power, preferential local consumption of local renewable energy production…).

5.1 Industrial Objectives

The industrial objectives aim to fill the gap between the state-of-the-art upstream flexibility control information systems and the power devices that can potentially be controlled in the downstream installations of users/customers. Given the particular importance of the electric water heater in this project, it is also planned to offer an innovative water heater in many aspects, in particular for its exceptional energy efficiency. The demonstrator will thus make it possible to evaluate the impact of the water heater's energy efficiency compared to the use of traditional water heaters to modulate demand on the electricity grid. The demonstrator will also validate the compatibility of thermodynamic heat production means with the low latency reactivity needs of the project management solutions.

The prototype devices that will be produced as part of the project and implemented in the demonstrator are:

- The demand modulation manager for controlling "Smart Loads".
- The "electric water heater of the future" (which is natively a "Smart Load").
- The adapter to transform a conventional power load into a Smart Load.
- The demand modulation manager for controlling conventional loads.
- A long-range PLC remote control transmitter for equipping power distribution dispatching.
- A prototype radio network implementing an innovative "targeted broadcasting" mode.

The services made possible by the existence of a "Smart Loads" park:

- The aggregation of erasure of consumption and/or diffuse energy storage.
- The contribution to frequency adjustment.
- Optimization and valuation of primary and secondary reserves.
- The sale of "real-time" remote control transmission capacities.
- The sale of services guaranteeing the effectiveness of actions on Smart Loads.
- The supply of electricity as part of innovative tariff offers.
- The sale of domestic hot water on the residential and professional diffuse markets.
- The sale of heat on the residential and/or professional diffuse markets.
- The sale of cold on the residential and/or professional markets.

- Improving the quality of the electricity supply service (i.e. reducing the risk of voltage jig, reducing the risk of blackout).

6 Project Solutions

The project plans to innovate on several levels in relation to the state of the art, intensely on the technological level of course, but also on the level of business models taking into account the highly regulated context that is characteristic of the energy sector and in terms of energy sociology.

6.1 The Technological Aspect

On the technical level, the five main innovations of the project are:

- Near real-time remote controls in two variants, a first variant based on low-power long-range radio, completely independent of the infrastructure of electricity distributors and grid operators, and a second variant based on long-range PLC (ripple control). This second variant is for demonstrating that the legacy infrastructure of the French electricity grid could be advantageously reused for doing rapid DSM by just updating the software in the power transmitter equipments.
- A complete paradigm shift in load management methods with the adoption of a new approach to energy objective management (transmission of simple commands assigning power loads an objective to be achieved in terms of the quantity of energy to be stored or deleted). This is to enable the use of unidirectional transmission solutions with a low amount of data transported and to reduce interoperability constraints between equipment to a level never before achieved, which is conducive to widespread adoption by industry.
- An innovative technical approach to accounting for stored or deleted energy quantities as part of flexibility management to enable the operators concerned to implement compensation or invoicing arrangements between themselves and to make the best use of project solutions at the appropriate points in the value chain.
- New intrinsically safe solutions to give power loads a capacity for early local detection of catastrophic incidents that could lead to a blackout and to provide them with an autonomous decision-making capacity for withdrawal from the network.
- New solutions for managing multidimensional storage spaces for electrical energy in thermal form with the ability to self-adapt according to users' hot water consumption. This is to allow standard size electric water heaters to continue to provide the storage capacity necessary to meet recurring storage needs at off-peak nuclear times, while providing additional storage capacity that can meet random storage needs for storing excess REN production at any time.

6.2 *The Project Demonstrator*

This demonstrator aims to provide proof of concept relating to the control chain from upstream systems to power devices (by both long-range radio and long-range PLC) in terminal installations.

A modulation manager is installed in the electrical panel in the form of a module in the standard DIN rail format that does not require interaction with the end user. The modulation manager provides the interface between the upstream real-time remote control systems and the appropriate fusion of the data coming from the "client information output" of the smart meter, from the internal power and energy meter, from the local measurement of network voltage, frequency and derivate of frequency and from the controlled "Smart Loads" inside the premise. The Smart Loads communicate bidirectionally and reliably with the modulation manager by means of the "Digital Pilot Wire", the future standard for bidirectional "1-wire bus" dedicated to communications between electrical heating and air conditioning equipment attached to the building. Interactions with the user, if necessary, are carried out in the most natural way possible in terms of ergonomics: on Smart Loads devices themselves. As part of an ergonomics that must remain as simple as possible to be understood by everyone without having to read a user manual. A push button and clearly presented status indicators are proposed on the front panel to allow an immediate and unconditional return to a transparent operating mode as if the device were not installed. An automatic return delay to the normal operating mode in which the system is operational will be provided to prevent the user from forgetting to restore it manually.

6.3 *The Structuring Orders of Magnitude that Guide the Project's Researches*

The controls system must drive a massive quantity of connected "modulation manager" objects (i.e. at least a potential park of 15 million in France, as many as electric water heaters, half the total number of residential installations).

The latency time in the sense of the project expresses the time that elapses between a need expressed by an upstream system to automatically aggregate a storage or erasure capacity of at least 1 MW in a part of the low-voltage grid, and the actual observation of the corresponding effect on the electricity grid.

The objectives of the project are to achieve the following latency times:

- ≤1 min for the SIGFOX radio solution.
- ≤1 min for the SCLE-SFE long-range PLC solution.

The acceptable unit cost for the manufacturers consulted for transforming an already intelligent power device in a "Smart Load" according to the project is around € 2 (for large industrial quantities). Beyond this value, this cost is likely to have an impact on sales prices that could be a handicap to deployments.

The objectives of the project are thus to limit the additional bill of materials cost of the existing electronics of a HVAC apparatus to 2 €.

The estimated costs of the project's radio modem and long-range PLC receiver/decoder meet this objective, as well as those of the "Digital Pilot Wire" interfaces, which are intended to provide two-way links between the modulation manager and the power equipment of the installations.

7 The Positioning of the Project's Solutions Relatively to the State of the Art

See Table 1.

8 Achievements of the Project at Technical Side

8.1 *An Innovative Load Controller Tailored for Residential Market*

- Allows simultaneous aggregation from hundreds of kW to hundreds of MW of loads with a short latency for flexibility actions.
- Manages all communication channels used for remote operations (radio broadcast RX + radio unicast RX/TX and PLC broadcast RX).
- Monitors mains frequency, derivate of mains frequency and voltage.
- Controls the loads according to the algorithm through two outputs for external power relays and one bidirectional bus (1 wire 230 V).
- Implements its own smart meter dedicated to the energy management of the controlled load(s) (up to 60 A).
- Can be connected to an external smart meter (e.g. Linky in France).
- Allows "subscribe, install and forget" type of services.
- A compact design in a standard 3 U DIN module, which achieves a MTBF of 20 years, a self-consumption <1 W in operation, and which can be manufactured in volume at a unit price <40 €.

(Please, see a photo of this load controller in Fig. 2)

Table 1 The positioning of the project's solutions relatively to the State Of The Art

Project solutions	Project objectives	Strengths/SOTA	Weaknesses/SOTA
Load management solutions as part of a second remote control plan that is separated from the one implemented for the management of recurring "nuclear off-peak hours" as well as in the context of new tariff offers.	To allow a control of the diffuse power loads, which are in the terminal installations of the low voltage network according to the random needs of the network in "intraday" mode. This is done by maintaining the current off-peak management that is still needed and by being transparent to any load control related to pricing (compatibility with possible future tariffs including planned deletions).	Addition of a direct and fast action capacity on the mass of diffuse charges. A capacity for two-way action required by the need for flexibility: – Absorption of short peaks of excess power in all loads. – Absorption of large quantities of excess energy in loads with storage capacity. – Reduction of one-off deficits by stealth and selective load shedding according to the uses of the power subassemblies on board the equipment concerned.	Need solutions to deliver new remote controls to terminal installations in a very short time.
Solutions for load management on energy objectives by transmitting a single command in "fire-and-forget" mode.	Allow a complete decoupling between the energy manager of the installation and the loads he has to control. As soon as the instruction setting the energy objective to be achieved is received, the charge is managed independently.	A significant reduction in bandwidth requirements and transmission time control for the telecommunications systems used. The manager no longer has to know the characteristics of the loads, which frees manufacturers from the need for complex interoperability standards to bring new "Smart Load" compatible products to market.	No disadvantages have been identified. (the partners have technical solutions for embedding energy undercounting means in the loads without additional costs).

(continued)

Table 1 (continued)

Project solutions	Project objectives	Strengths/SOTA	Weaknesses/SOTA
Solutions for autonomous control of power loads by continuous monitoring of physical quantities such as the voltage and frequency of the electrical network at the point of consumption and by innovative signal processing algorithms.	Embedding effective means of protecting the integrity of the electrical network in devices with power loads at zero marginal cost by acting in a certain, safe and appropriate manner in considerably shorter times than if remote measuring means, remote information system and telecommunications means were involved. Provide an ultimate protection solution for electrical distribution networks when upstream adjustment and protection mechanisms have failed to maintain the network balance. Participation in the local voltage adjustment. Participation in frequency adjustment.	Embedding grid integrity protection capabilities in next-generation power devices at zero marginal cost (only a little more embedded software). Only offloads or regulates the power subassemblies embedded in the devices concerned (which generally offer an inertia that is used to mask short-term actions on loads for the end users' senses). Low-energy-consuming features and low and medium-power uses in installations remain powered. Makes load shedding stealthy in that it does not impact uses that have an immediate effect on users' senses (lighting, computers, audiovisual, etc.). Improves the perceived quality of electricity supply service provided to customers (which justifies their contribution to the "system effort" by involving their power devices connected to the grid). A significant reduction in the risk of general blackout in high-risk areas of the interconnected continental grid.	No disadvantages have been identified (provided that all associated locks are removed).

(continued)

Table 1 (continued)

Project solutions	Project objectives	Strengths/SOTA	Weaknesses/SOTA
Solutions for accounting and managing electricity consumption by use in diffuse terminal installations of the low-voltage network.	Provide service operators concerned with data that can be used in their customer invoices and in compensation mechanisms between operators in the event of an implementation involving several operators (for example, to "divert" and compensate for consumption specific to water heating or space heating/air conditioning in hot water and heat/cooling service offers).	Simple and cost-effective accounting and management solutions to be deployed by design (these solutions will be directly integrated into the design of Smart Loads, which will make it possible to minimize implementation costs by using sub-metering mechanisms based mainly on software solutions).	No disadvantages have been identified (provided that all associated locks are removed).
Solutions to manage storage space dedicated to random needs in storage devices, in addition to the storage space reserved for recurring storage needs.	Manage several "logical" storage spaces assigned to distinct uses within a shared "physical" storage capacity of energy in thermal form (heat or cold).	Innovative solutions that eliminate the need to modify manufacturing processes, which is conducive to their adoption and deployment. Several solutions are emerging from the consortium to address this need at several levels of cost and functionality. A first solution allows an external implementation in the manager which controls a standard water heater to satisfy the needs of the legacy Joule effect water heater market. A second solution aims to be embedded in the thermostats of the devices allows to increase the energy storage capacity at constant tank volume by offering automatic adjustment capacities to the users' needs and to the electric system. These solutions will be easily transposable to the storage of cold.	No disadvantages have been identified (provided that all associated locks are removed).

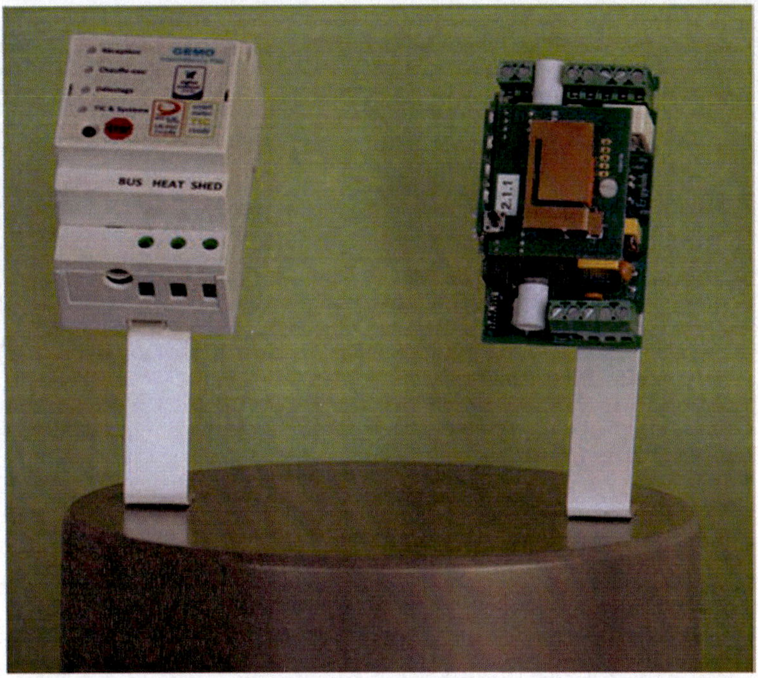

Fig. 2 The "GEMO", the ultimate smart load controller (research prototype) (DEGETEL)

8.2 A New Type of Communicating Regulation to Be Integrated in Highly Efficient Water Heaters

- Manage the thermodynamic and the Joule heating sources.
- Implement an adaptive algorithm, which predicts user needs in order to always satisfy them first (The algorithm observes continuously the user habits during every weekdays and adapts forward the water heating cycles for having the right temperature at the right moment, just before the user needs hot water. This algorithm reduces significantly thermal losses and improves water heater efficiency).
- Virtual additional storage capacity for being able to respond to electrical network needs at any time (all system services).
- Continuous monitoring of the remaining storage capacity (e.g. remaining storage capacity in kWh, time needed to reach 100% capacity, water temperature…).
- Fully compliant with all relevant EU safety standards.

(Please, see a photo of this communicating regulation in Fig. 3)

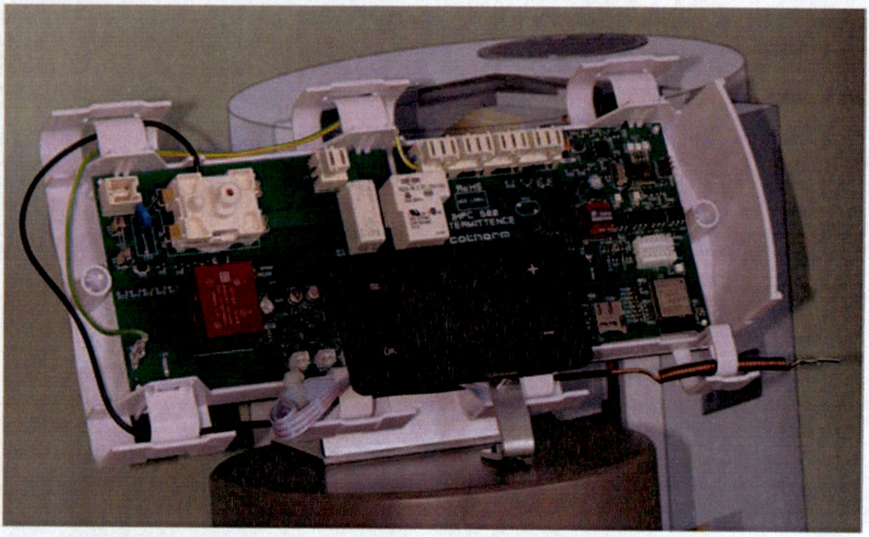

Fig. 3 The smart thermodynamic water heater controller (COTHERM)

9 The Sociological Aspect in the Intermittence Plus Project

By courtesy of M. Stephane La Branche, researcher affiliated to Grenoble's Institute of Political Science and to Pacte (a social science research center linked to the National Research Council).

He was nominated to the IPCC by the French government, contributing to the fifth Assessment Report as well as to several national and local strategies and laws on energy and climate, among which the National Adaptation strategy and the National law on Energy and Ecological Transition.

Stephane La Branche is now involved in the final review of the IPCC sixth Assessment Report.

The sociology of energy is a recent discipline [4–25]—especially in its link to climate change—but it offers results and data that can contribute to both the understanding and efficiency of energy systems. We have already conducted studies on the Positive Energy Family Challenge (FAEP, a device that focuses on behaviors); GreenLys (a smartgrid project) and Empowering (focusing on the ability of information to change energy behavior in the home). These studies confirm the importance of sociology in efforts aimed at energy objectives.

Indeed, the process of appropriation (or not) of techno-energy offers by users is an important factor (1) in the time it takes for investments to become profitable; (2) in the growth of demand management potential, which is still very suboptimal; (3) in better integrated management of the proposed scheme; (4) in energy efficiency and thus; (5) in greenhouse gases reduction. Our sociological component will therefore contribute to a detailed and empirical understanding of this problem.

Social representations are mental realities that guide us in the way we name and define together the different aspects of our everyday reality, in the way we interpret them, decide on them and, if necessary, take a position towards them and defend it—or ignore an issue as unimportant. Our studies show that social representations play a more important role in behavior than information.

New information even tends to be reinterpreted by an individual's cognitive system. If the information is not consistent with the individual's dominant social representation, he or she will reinterpret it to integrate it and in the event that this is impossible, the information may be rejected as invalid, false or uninteresting. Developing support and a meaningful catchy language for the different profiles is therefore key in efforts to achieve energy objectives.

They play an important role in:

- The perception of energy or its habitat as an energy medium (important or not, interesting or not, useful or not and why?)...
- The legitimization of innovation (technical or behavioral): the user integrates it or not, to what degree... according to his perception of its utility, to his understanding, to his energy profile...
- The integration or not in the organization of the individual's life, according to his habits, his daily routines, his ability and propensity to act in a new way...
- The contribution, even, the reinforcement of new practices, which may or may not be part of strategies of sobriety or control of one's habitat, of one's intimacy.

9.1 What About the Water Heater?

In different energy projects analyzed in the last 10 years [3], we have observed that the water heater is not an invisible object for the experimenters since it is thought of as a major energy consumption device. But beyond this observation, participants were concerned about several aspects and our secondary analysis for Intermittence Plus project allowed us to identify differences, particularly the perception of the possibility of controlling it—using it as an energy ***management*** tool—that can be attributed in large part to energy profiles. For example, many participants mentioned the age of their water heater, wondering if it was worthwhile to do a detailed water heater management with an older appliance. A new device seemed more promising to them but, as an energetic enthusiast reminds us, "only if the difference in final consumption is worth the (financial) effort between the old and the new!"

Seven categories or energy profiles exist. Here they are presented in their relationships to our high performance connected water-heaters. Here are some examples of the questions explored through our sociological analyses

1. **Modalities of accession and adoption of techno-energy offers**

   (a) How do users address global and individual energy issues?

(b) How do they or do they not adhere to the new equipment and services, and why?
(c) What are the modalities of variable appropriation of the equipment homes that can be controlled by third parties?
(d) What type of information and support, and in what form and for what profiles, before, during and after the deployment of new equipment and systems?

2. **Terms of use of offers**
   The ergonomics will be tested but also:
   (a) Both the HMI (Human Machine Interfaces) and,
   (b) What we call the IHE the Human-Energy Interface, with the forms and types of information for the energy profiles developed in our studies,
   (c) What are the modalities of sustainability (or not) of behavioral changes, differentiating between behavioral change and energy technical management.

To improve our understanding of HEIs and make accompanying recommendations, we will use the notion of energy profiles (developed in the studies cited above—FAEP, GreenLys and Empowering) applied to a multidisciplinary and multifactorial approach of the MDE system of this project, associated with water heaters.

9.2 Socio-Energy Profiles in Interactions with the Water Heater

In their relations with energy, individuals interact with the water heater on the basis of a seven types of behavioral logics and representations identified in previous studies, referred to as "energy profiles". It must be understood that these profiles explain the motivations behind the practices. This is why the notion has a real heuristic power: profiles go beyond the simple description of a gesture ('A' puts a lid on the pot when cooking) and explains its roots and the framework within which this gesture is embedded. This simple gesture can take numerous meanings for the individuals. User may want to: reduce greenhouse gases; reduce the bill in euros; reduce the number of kW/h consumed on the bill; do it as an exercise on oneself; maintain or increase their comfort; enjoy new technologies associated to energy-control. But on the whole, individuals tend to be more inertial and resistant to the adoption of technical or behavioral innovation than proactive.

In this discussion on the posture of each profile relative to the water heater, none of the previous studies had been conducted specifically on the water heater. We therefore make assumptions here for each profile based on our understanding of them. The sociological component of Intermittence Plus makes it possible to understand better these relations. The aim is to produce recommendations to improve their use and overall effectiveness. We have identified four 'active' energy profiles and three present in the general population.

- ***Technophiles***: Neither energy per se nor kW/h interest them and, moreover, they do not understand it or only very little. What attracts them are new technologies and getting them before others. They like to appropriate it and to play with it. This logic is very present in high technology intensive energy projects and much less on those based on behavioral changes alone—considered to be too much of an effort. For Intermittence Plus, the new technology component and the connected management capacity represent the main interest of the water heater for this profile.
- ***Energyphiles***: they are mainly interested in energy itself, before technology or behavioral changes. Mostly found in technology intensive projects, this profile also exists in behavior-change based ones but always related to the ecological issue. This energy enthusiast understands kW/h, network stability issues and peak shavings—this is the only profile who understands this. Reducing kW/h is more important than reducing euros or the ecological footprint. Powergrid stability at the national level is important, in order to guarantee access to energy for all, according to a principle of energy ethics. This issue is basically never mentioned by others. While management and technical efficiency are the most important factors for this profile, user does not entirely ignore behavioral changes—as long as it does not impact physical comfort or does not require too much effort. In Intermittence Plus, the water heater was evaluated to be so energy-efficient that none of the testers modified the default settings. Interfaces were used mostly to follow energy consumption, which, in all cases, decreased compared to the older heater.
- ***Thrifties***: they want to reduce their energy budget, in euros. kW/h do not mean much to them and they do not care much about it. This profile is present in all the projects analyzed but in an unequal way, depending on the experimentation. If the thrifty is a more representative profile of the disadvantaged classes, it should be noted that it is among this group that we find most a double logic: economic gains are for some, closely associated to ecological issues. For the thrifty, energy management takes precedence, and secondarily, technical efficiency, with the aim of reducing the bill in euros. Behavioral change does not attract them (except for those who are also eco-friendly) and they are the most inclined to think that energy savings go first through renovation work and lastly, to behavioral changes. The high-performance water heater would help them to better manage their energy and pay less. They will want to automate everything very quickly, so that they do not have to worry about it while being economically efficient.
- ***Ecophiles***: for them, energy is an ecological problem. Sociologically more representative of the middle class interested in environmental issues, they are most attracted to energy programs aimed at behavioral change. They are interested in self-control practices first, with the environment and energy appearing as specific fields allowing practices of this deeper trend. More generally, energy sobriety is part of a family culture of frugality. This may even be the key factor in their involvement in this system. Bills and consumption tracking are used to determine whether or not their practices are effective, they are not goals as such. In this case, controlling the water heater is more a matter of evaluating behavioral

efforts (heating, washing, bathing, showering…). They raise questions that others do not or much less: what is the balance sheet between a dishwasher and doing it by hand? Is doing cold water laundry as effective as hot water? Most will abandon having a bath and opt for the shower instead. In Intermittence Plus, this type of audience would likely be very interested in the possibility of improving their behavioral efforts, adapted to the new knowledge they will have acquired through the interfaces. In this case, the proposed technique and the high-performance water heater would help behavior control.

Energy saving programs, as they are practiced today, almost all involve volunteers who are not necessarily representative of the general population. The four energy profiles presented so far do not represent the majority of the population at its current level of energy knowledge and sensitivity—although recent developments are to be noted. So, what about the general population?

9.3 The General Population

At this stage of knowledge, we can identify three main categories in the average population.

- For the ***helpless***, energy is a concern but they don't think they can act or do otherwise—they offer significant potential for reduction. They have insufficient knowledge to improve their consumption, but they are interested. The smart water heater could then be a way to improve their knowledge and perception of being able to control part of their energy consumption. We could see a positive "spillover" effect, i.e. that the learning taking place through interactions with the water heater will induce effects on other areas of energy and water consumption.
- For the ***indifferents***, energy is invisible and uninteresting, they do not care—the reduction potential is less important or will require more effort on the part of the energy provider to convince this profile to enter into an energy management or efficiency process—sobriety will not appear here. It is likely that the financial argument will be the most attractive but perhaps not sufficient.
- For the ***resistants***, behavioral change requires too much cognitive and physical efforts. They will be interested in reducing energy consumption under conditions of equal comfort and through management and efficiency. They will only be interested in the water heater if they believe that the entry costs (cognitive and behavioral efforts required) are considered minimal and beneficial. If the interfaces and management tools are easy to control and allow a reduction in consumption at equal comfort, then they will be adapted.

Without this understanding of the motivations and social representations of energy, the modes and language to support behavioral changes can be meaningless; while the message may be conveyed, it is not received. This is one of the major problems of energy saving campaigns today: most mention the fight against climate change or environmental protection when it is not a factor of change—nor a source of motivation to make efforts—for the majority of people. The second major problem is that most campaigns assume that if their users are informed, they will wish to reduce—and actually do it—their energy consumption on their own, which is not the case. This can have significant negative effects on the expected reduction targets. In all cases, trade-offs between different factors and different representations and motivations, including comfort and habits, are made, but the final choice, the weighting between the factors—is guided by the logic of the profile (Figs. 4, 5 and 6).

10 Conclusion

All the research objectives of the project have been achieved to date.

This project has validated new concepts in the field of flexibility management in electrical networks by an appropriate management of large amounts of diffuse loads.

The understanding of user behavior has also increased, thanks to the project's work on sociology of energy.

With these encouraging results, the main partners of this project are now planning to develop and test new innovative solutions to go further in their quest of DSM in real time for bringing "hyperflexibility" to the electricity networks.

Their objective is now obtaining transmission times of the order of the second, for transmitting simultaneously relevant remote controls to massive quantities of diffuse loads or renewable production sources, for then achieving robust and efficient dynamic production-consumption balancing.

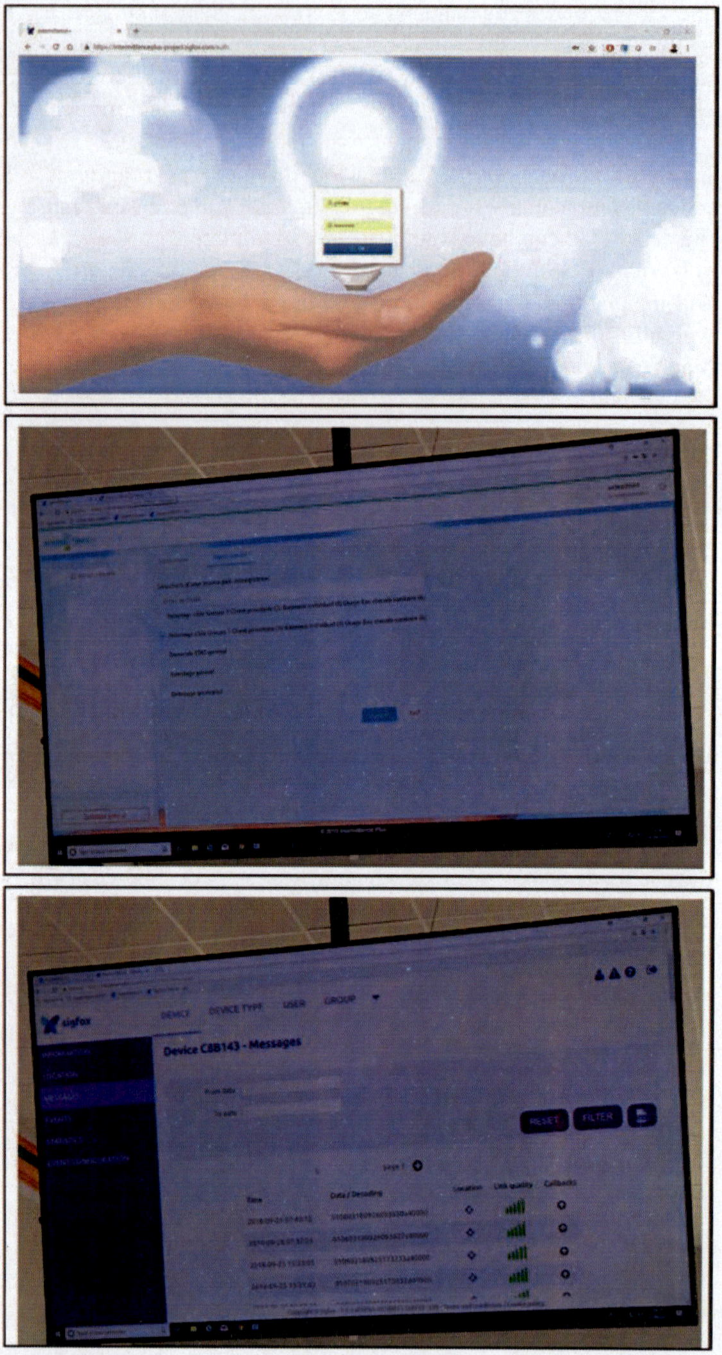

Fig. 4 User interfaces for driving and supervising the field test of the project (SIGFOX)

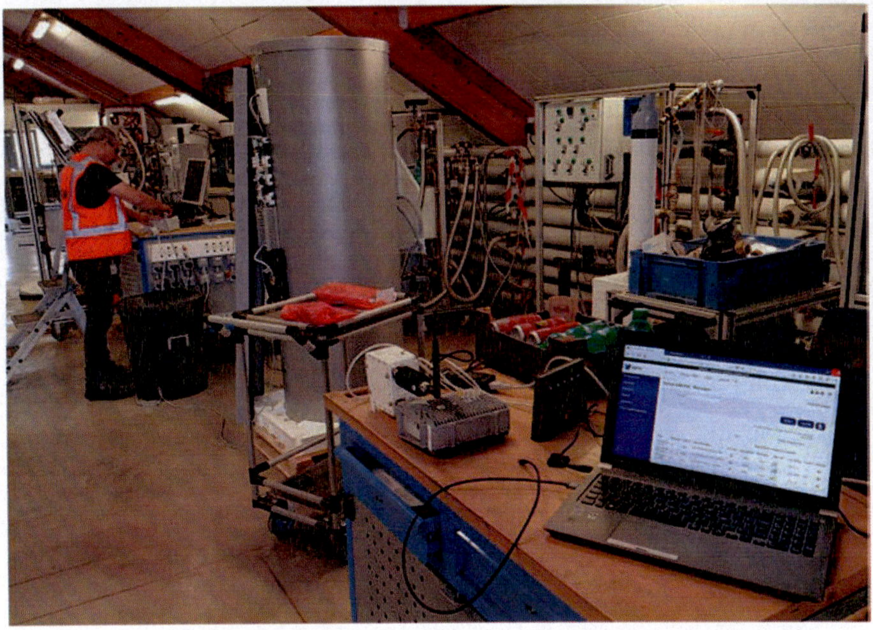

Fig. 5 Integration of project's super-efficient thermodynamic water heater (VIESSMANN)

Fig. 6 The showroom of the project, with a booth for demos, in Toulouse, France (ENGIE)

References

1. "The Power of Transformation: Wind, Sun and the Economics of Flexible Power System" published in February, the International Energy Agency. 2014.
2. Report "Towards a 100% renewable electricity mix in 2050", Ademe. 2015.
3. C. Gauquelin, C. Lutoff and S. La Branche. "Which status and place for social science data in local climate services?", *Vertigo*, Vol. 17 no 3. 2018.
4. G. Debizet and La Branche. "Energy transition scenarios: the critical virtues of an actor-based analysis". *Approches critiques de la transition énergétique en SHS* (ACTESHS). 2018.
5. S. La Branche and P. Bosboeuf. "Institutional Barriers in Local Adaptation Policies". In W. Leal, *Handbook of Climate Change Resilience*, Springer. 2018.
6. S. La Branche "Designing an Adaptation Strategy in a Complex Socioecosystem: the Case of Territorial Climate and Energy Plans in France". In *Handbook of Climate Change Adaptation*, edited by Walter Leal, Springer. 4 2015.
7. S. La Branche "Energy Transition and local authorities: a delicate appropriation". In A. Artis (Dir.) *Cooperation, Governance, Territories and energy Transition*, Éditions Campus Ouvert. 2017.
8. S. La Branche and P. Bosboeuf. "The appropriation of energy by local authorities: obstacles and motors of change". *Revue Environnement Urbain*, vol 11. 2017.
9. S. La Branche and A. Tabourdeau. "The operational impacts of a large scale heat storage apparatus on a heating network." *Revue de l'Energie*, N° 631. 2016.
10. S. La Branche "Sociological factors in woodheating-related pollution in Grenoble: main results". *Revue Pollution atmosphérique* - N°228 Janvier – Mars. 2016.
11. S. La Branche "Governance and actors interactions in ecodistricts". *Revue InnovatiO*, la revue pluridisciplinaire en innovation. http://innovacs-innovatio.upmf-grenoble.fr/index.php?id=113 2015.
12. Wanneau K. and La Branche S., "Challenges to local climate adaptation measures; between science and society", in Beringuier P., Blot F., Desailly B. et Saqalli M., *Environment, public policies and local practices*, L'Harmattan, pp. 259-286. 2015.
13. S. La Branche "Political and sociological elements of energy transition in France". *Revue Vie Économique*, vo 6, no 2. 2015.
14. N. Buclet, G. Debizet, C. Gauthier, F. Forest, S. La Branche, P. Menanteau, P. Schneuwly, A. Tanourdeau. "Four scenarios for energy coordination in urban settings at 2040". In (Dir. G. Debizet) *Scénarios de transition énergétiques en ville. Acteurs, régulations, technologies*. La documentation Française. 2015.
15. S. La Branche "Elements for a sociale science of energy", *Encyclopedia of energy*. 2014.
16. S. La Branche "Innovations in ecodistricts: a few lessons for energy transition governance". *Vertigo*, vol.14, No.3. 2014.
17. S. La Branche Review author for the IPCC 5e Assessment Report on the following chapters: 7, "Energy Systems"; 8, "Transports"; 12 "Human Settlements, Infrastructure And Spatial Planning"; 15, "National and Sub-national Policies and Institutions". 2014.
18. S. La Branche "Paradoxes and Harmony in the Energy-Climate Governance Nexus". In H. Dyer and J. Trombetta, *The International Handbook of Energy*. Edward Elgar Publ. 2013.
19. S. La Branche and C. Bethoux. "The internalization of participation norms by international organizations". *International Journal*, Winter 2011-12, 195-217. 2012.
20. S. La Branche and E. Schmid, B. Knopf, M. Fink. "Social Acceptance in Quantitative Low Carbon Scenarios". In O. Renn, A. Reichel, J. Bauer (Eds.), *Civil Society for Sustainability: A Guidebook for Connecting Science and Society*. EHV. 2012.
21. S. La Branche "Daily mobility and ecological schizophrenia. The case if Lyon". No. Spécial de *Vertigo*. Hors-série 11, mai 2012. 2012.
22. S. La Branche and L. Charles. "The Social acceptability of a Low Emission Zone in Grenoble: sociological analysis and results", *Revue Pollution Atmosphérique*, no. Spécial, nov. Pp.226-230. 2012.
23. W. Chou, A. Bigano, A. Hunt, S. La Branche, A. Markandya, R. Pierfederici. "Households' WTP for the reliability of gas supply". In *Nota di Lavoro*, 58. 2011.
24. S. La Branche "Climate governance versus daily mobility". *Revue Environnement Urbain/Urban environment*. No. 5 dossier thématique *ville durable et changement climatique*. 10-23. 2011.
25. S. La Branche (Dir.). *Climate Change: from meta-risk to meta-governance*. Lavoisier. 2011.

Appliance Efficiency in Pacific Island Countries

George Wilkenfeld

Abstract The developing countries of Melanesia and Polynesia have a combined population of over ten million people. The rate of household electrification varies, from nearly 100% in some of the smaller countries to 12–15% in Papua New Guinea. The main demand for electricity comes from lighting, refrigeration and air conditioning, which is still rare in households but common in the commercial and hotel sectors. All appliances are imported, as is nearly all of the fuel for electricity generation.

Until 2011 none of the countries had minimum energy performance standards (MEPS) for appliance or lighting energy efficiency, or rules regarding energy labelling. Many products carried the energy labels of the countries of manufacture, but this confusion of rating systems did not help consumers. In 2011 Fiji became the first Pacific country to introduce mandatory energy labelling and MEPS, when it adopted the Australian and New Zealand (ANZ) test standards, MEPS and energy labels.

In 2012 the Secretariat of the Pacific Community (SPC), which services the Pacific Forum, received an Australian Aid grant to develop the Pacific Appliance Labelling and Standards (PALS) program. By the time of its formal conclusion in June 2019, the PALS program had assisted the governments of Samoa, Solomon Islands, Vanuatu and Tuvalu to introduce MEPS and labelling for refrigerators, freezers, air conditioners and lighting products. These joined Fiji in adopting the ANZ system, allowing all the countries to share an online product registration system and simplifying compliance for regional product distributors.

This paper sets out the objectives and scope of the PALS program, estimates its impacts and draws conclusions about the difficulties and opportunities facing developing countries that wish to improve the energy efficiency of the appliances and lighting products they import. Although many Pacific countries have ambitious renewable energy targets, diesel will remain the marginal generation fuel for decades to come. This means that every kWh saved is valuable for reducing petroleum fuel

G. Wilkenfeld (✉)
George Wilkenfeld and Associates, Energy Policy and Planning Consultants, Sydney, NSW, Australia

P. Bertoldi (ed.), *Energy Efficiency in Domestic Appliances and Lighting*, Springer Proceedings in Energy, https://doi.org/10.1007/978-3-030-79124-7_27

imports, which represent the greatest balance of payment issue for almost all Pacific countries.

As most of the appliances and lighting products supplied in the Pacific are sourced from China, the program has worked with product suppliers as well as importers, to ensure that they are aware of the regulations and their legal obligations.

The program has also helped to build a baseline of information from which future changes in product energy efficiency and consumer behaviour can be evaluated.

1 Electricity Use in Pacific Countries

Figure 1 is a partial map of Pacific Island Countries (PICs). All the countries shown and others further west are members of the Pacific Community, as are Australia, New Zealand and the USA.[1] The Secretariat of the Pacific Community (SPC) is the executive agency of the Pacific Community. Most PICs are sovereign independent nations, but some are "overseas collectives" of France (New Caledonia, French Polynesia) or in various types of association with the USA (American Samoa, Federated States of Micronesia).

For historical reasons the PICs north of the equator have adopted the US standards for electricity supply: 60 Hz and 110 V. The southern PICs have 230–240 V 50 Hz supply systems. Electrification rates in the southern PICs vary considerably, from 12% to 13% in the least developed and most geographically dispersed countries (i.e. where the population is dispersed across many islands) to near 100% in the PICS that are more developed and where the population is more concentrated (Table 1).

Papua New Guinea (PNG), Fiji and Vanuatu have significant electricity use for industry, agriculture or mining, but most electricity use in the PICs is residential and commercial (public facilities, public lighting, retail, offices and tourist accommodation). Refrigerators and freezers account for about 33% of residential electric use, lighting for 20% and televisions for about 14%. Residential air conditioning accounts for only 4%, since most household cooling is still done with fans. All other appliances, including fans, cooking and electric water heating account for about 30%. Water heating is not a major electricity load because unboosted solar water heating is common. Commercial sector electricity use is dominated by air conditioning, refrigeration and lighting [2].

There is no appliance manufacturing in the PICs. All appliances and lighting products are imported, mainly from China, Thailand and south-east Asia. Although higher quality and more energy-efficient products are available, most consumers are low-income and will purchase the cheapest products irrespective of how much they

[1]The membership of the Pacific Community (formerly the Pacific Islands Forum) is at https://www.spc.int/our-members

Fig. 1 Pacific Island Countries. Notes: (1) Wallis & Futuna, (2) Samoa, (3) Tokelau, (4) American Samoa, (5) Niue

cost to run. Electricity tariffs in PICs are relatively high since the majority of generation relies on imported diesel fuel (despite a rising share of renewables), so purchasing more efficient products has a short payback time. For efficient, medium-quality lighting products the payback is often less than a year. For major appliances it is generally 3–5 years.

2 Regional Appliance Energy Efficiency Programs

In 1996, the Forum Secretariat Energy Division commissioned a study on the establishment of a regional energy labelling program [3]. The study concluded that the implementation of energy labelling and Minimum Energy Performance Standard

Table 1 Estimated population and electrification rates, PALS PICs, 2017

Pacific Island Country	Population ('000)	Persons per household	Households (HH)	Electrification rate (%)	HH with 230 V electricity
Cook Islands	19,500	4.0	4875	97	4729
Fiji	865,611	4.5	191,589	70	134,112
Kiribati	112,850	7.5	15,087	51	7695
Niue	1611	5.8	280	100	280
PNG	7,300,000	7.7	948,052	12	113,766
Samoa	219,998	7.0	31,428	96	30,171
Solomon Islands	595,613	5.5	108,293	13	14,078
Tonga	120,898	5.7	21,210	87	18,453
Tuvalu	12,373	6.0	2062	98	2021
Vanuatu	243,304	5.0	48,661	22	10,705
PALS PICs totals	9,491,758	6.9	1,371,538	24	336,010

Source: [1]. Households relying solely on photovoltaic systems are not considered "electrified" for the purpose of the PALS program since they are not able to operate the mains-voltage appliances and lights covered by the MEPSL standards. Efficiency standards for low-voltage products were outside the scope of PALS

(MEPS) for selected appliances was feasible. Given the different levels of economic development, it recommended that standards and labelling (S&L) be implemented first in Fiji before expanding to other PICs.

A voluntary pilot S&L program for refrigerators and freezers commenced in Fiji in 2002. Most products at the time were imported from Australia or New Zealand (ANZ) suppliers even if they were manufactured elsewhere, and already had the ANZ energy labels on them. Given the high level of recognition among consumers, it was logical to adopt the ANZ energy labelling scheme. After a further review [4] the government of Fiji decided to make S&L mandatory for refrigerators and freezers in 2007, but it was not actually enforced until the end of 2011. In the meantime, work continued on a sub-regional S&L program, with an IIEC study covering Samoa, Tonga and Vanuatu [5].

The Pacific Appliance Labelling and Standards (PALS) program originated at the end of 2011, with the endorsement of the PIC leaders and with a A$3 million grant to the SPC from Australian Aid, with oversight by the Australian Department of Industry, Innovation and Science and Resources (DIISR), which manages the Australian appliance S&L program. The aim was to assist PICs to enact legislation for S&L, so they could realise the economic benefits estimated in an initial cost-benefit analysis conducted in 2011 [6].

Rather than progressive payments, all the funds were transferred at the beginning, which enabled SPC to take responsibility for program budgeting and administration, including the engagement of a program manager and a technical advisor, both of whom stayed with PALS from its formal start in April 2012 until it wound up at the end of June 2019.[2]

In the first phase of the program the SPC invited Forum member PICs to sign PALS agreements, covering the shared objectives and, in some cases, financial support for staff to work on PALS in that PIC. Initially 13 PICs signed on to participate in PALS. A key part of the agreement was a Cabinet-level commitment to S&L legislation. In the end, North Pacific countries with 110 V electricity systems were not able to adopt Australian and New Zealand S&L, unlike the 10 South Pacific PICs in Table 1, which all have 230 V systems and were able to use the ANZ labels (Fig. 2). Nevertheless, SPC funded the participation of the North Pacific PICS in regional PALS meetings, and with the assistance of DIISR tried to engage the US authorities in helping their appliance energy efficiency efforts.

Fig. 2 ANZ energy label, also used in Pacific Island Countries

[2] The PALS Manager was Ms Makereta Lomaloma of SPC, and the technical adviser was the present author. PALS was originally intended to end in June 2015, but the Australian government approved its extension, without additional funding, until June 2019.

3 Pacific Appliance Labelling and Standards (PALS) Program Outcomes

By the conclusion of the program, five PICs had drafted, passed and implemented legislation to enforce MEPS and energy labelling for the products shown in Table 2. This means that importers must first demonstrate that the products comply with the same MEPS and energy labelling requirements as apply in Australia and New Zealand: AS/NZS 4474 for household refrigerators and freezers, and AS/NZS 3823 for air conditioners. The regulations cover all imports, including new and used products and products imported for "own use" (e.g. by hotels) as well as for resale.

In most cases S&L was phased in over years, with household refrigerators and freezers first, air conditioners next and then lighting products. While commercial refrigeration is not (yet) covered in any PIC, residential size refrigerators and especially freezers, are used in many small food businesses. The scope of AS/NZS 3823 is air conditioners up to 65 kW cooling capacity, which covers the great majority of models used in the PICs.

Some PICs wanted to include other products as well, but investigations indicated that the additional administrative costs were greater than the likely benefits [7]. The only other product with sufficient energy use was commercial refrigeration. The energy savings potential from televisions was much lower, and there is already a high level of voluntary use of the ANZ label. There was no case for including clothes washers, despite their relatively high level of ownership, because virtually all clothes washing in PICs is done with cold water and there is very little energy

Table 2 Situation in PALS PICs, 2019

Pacific Island Country	Status of S&L	Date of effect	Refrig and freezers	Air conditioners	Lighting products	Other products
Fiji	Enacted	1/2012	✓	Proposed	Proposed	Televisions proposed
Tuvalu	Enacted	4/2016	✓	✓	✓	
Vanuatu	Enacted	3/2017	✓	✓	✓	
Solomon Is	Enacted	4/2017	✓	✓	✓	
Samoa	Enacted	5/2018	✓	✓	✓	
Kiribati	Draft 2018	NA	Proposed	Proposed	Proposed	
Tonga	Draft 2017	NA	Proposed	Proposed	Proposed	
Cook Is	Draft 2014	NA	Proposed	Proposed	Proposed	Washers proposed
Niue	Draft 2018	NA	Proposed	Proposed	Proposed	
PNG	Draft 2017	NA	Proposed	Proposed	Proposed	

difference between models if water heating energy is excluded.[3] There was no case for including dishwashers, clothes dryers or electric water heaters because sales in the PICs are negligible.

Fans are widely used for used for cooling in the residential sector, but as there are no Australian or New Zealand test standards or MEPS for them, they did not fit easily into the PALS model, which relied heavily on the ANZ database of MEPS-compliant products. However, there is nothing to prevent the PICs from adopting MEPS standards from other countries or regions in the future, now that the legislative framework is in place,

Five of the 10 PALS PICs are yet to enact the necessary legislation, despite having prepared drafts (with legal assistance funded by PALS). Some of the delays were due to wider legislative changes, such as late decisions to include S&L regulations in overarching Energy Acts rather than more limited S&L Acts.

4 Establishing Consistency

4.1 Labels

When a regional S&L program was first considered in 1996 there were relatively few brands in PIC appliance markets and the ANZ energy label, introduced in Australia ten years earlier, was the only one with a presence in the south Pacific. By 2011 however, appliances carrying Singapore, Thai, Chinese, Korean, Japanese, European and US energy labels were common. Even so, consumer research showed that over 80% of PIC residents recognized the ANZ label, and over 70% of label-aware consumers could only recall the ANZ label—the other 30% were aware of the other label types as well [1].

There was some debate early in the life of PALS whether other energy labels should also be legally permitted. It was pointed out that the different energy tests, scales and languages made it impossible for consumers to compare the efficiency or running costs of models carrying different types of labels. Furthermore, some labels had annual running costs calculated according to the currency and tariffs of their country of origins, whereas local tariffs (expressed in the local currency) were often much higher in the Solomon Islands, for example, annual running costs were seven times the dollar value on the Singapore label. Therefore permitting display of labels with running costs would be misleading.

Once it was decided that the PALS countries should permit only one type of energy label, it was logical to adopt the ANZ label, which had already been legislated in Fiji. There was no commercial advantage to Australia or New Zealand from

[3] Given the shortage of fresh water in the outlying islands of some PICs, there is a case for adopting *water*-efficiency labelling and standards for clothes washers, as in the Australian WELS program—see https://www.waterrating.gov.au/

this decision, since neither country manufactures refrigerators, freezers or lighting products any more. They are in the same position as the PICs—all the product types subject to S&L (with the exception of some locally made air conditioners) are imported.

All of the PICs require the display of ANZ type labels, and forbid the display of any other label types for the products covered by their legislation (see Table 2). However, this has not prevented the continuing display of other label types on products *not* covered by the S&L legislation, such as televisions or fans. This somewhat weakens the consistency and impact of energy labelling in the PICs.

4.2 Product Registration

All of the products subject to S&L regulations in PALS PICs must be registered before they are imported (unless the same model has already been registered previously in that country). This provides a point of verification, where the importer must demonstrate that the product complies with the regulated S&L rules. It also establishes a list of complying products available in each PIC, which helps both the regulators and those consumers who wish to compare the energy-efficiency of different models, not all of which would be on display in every store.

Each PALS PIC has the same three categories of registration:

- Category A products are already registered in Australia or New Zealand, so compliance has already been verified by another regulator. The PIC regulator only requires proof of the ANZ registration number.[4] About 90% of PIC registrations are in this category.
- Category B products are registered in Australia, New Zealand or another PALS PIC, but under a different brand name or model number. This is fairly common, as many PIC appliance importers will use their own brand names, due to geographical brand copyright arrangements. The PIC regulator requires a statement from the manufacturer that the product is technically identical to an ANZ-registered model.
- Category C models do not fit Category A or B. The PIC regulator requires a complete test report conducted to the specific AS/NZS standards (AS/NZS 4474, AS/NZS 3823 etc.).

Some PALS PICs charge registration fees, while others do not. The methods for assuring Customs that imports are compliant also vary between PICs. Some require new permits for each import consignment, while in others the listing of that model on the electronic register is enough.

[4] Products registered as compliant in Fiji would also be Category A, but Fiji has not yet implemented a full registration system.

4.3 *Pacific Appliance Database (PAD)*

To help PICs with limited resources to administer the registration system, the PALS project funded the development of a Pacific Appliance Database (http://pad.spc.int/). Anyone can check the list of models registered in the PICs linked to the PAD (at present Samoa, Solomon Islands, Vanuatu and Fiji). Intending importers can create their own accounts and register models on the PAD. For Category A and B applications, the PAD will transfer the technical data from the ANZ register, and for Category C the test reports and other information must be uploaded.

The PIC regulator can then review all the documentation electronically and if approved, issue a computer-generated printable registration certificate. (The PICs have also retained the option of paper-based applications for those without internet access). Applications for import permits can be made and approved in the same way. After about a year in development, the PAD became active in July 2018.

At the time of writing there were 290 refrigerator and freezer models registered in Samoa, 73 in Solomon Islands and 81 in Vanuatu (in some cases the same model was registered in more than one country). Air conditioner registrations were 46, 59 and 23 models respectively. Due to budget constraints and other reasons (see below) the PAD was not designed to handle registration of lighting products.

The PAD represents a joint investment by the PICs, and a visible manifestation of the regional nature of PALS. The PAD home page is linked to the information page of each PIC energy agency, and its internal structure embodies, in a durable form, both regulatory rules and administrative processes. This will extend the value of PALS beyond its funding period, and will help with the training of new energy officials in due course. However, while the PAD is hosted on the SPC's servers, there will inevitably be a need for ongoing software maintenance, and this is at present not securely funded.

4.4 *Training*

Training and capacity building were central elements of PALS. Training workshops for officials from all PALS countries were held at least annually, in Fiji, which is the focal point for Pacific air routes. In addition, the PALS program manager and technical adviser visited each PALS country at least once, and for most PICs several times.

The training program for each PIC was organized in three phases:

1. While the legislation was being drafted: clarification of the technical and legal requirements, engagement of other responsible agencies (usually Customs, in some cases Consumer Affairs) and preliminary discussions with private stakeholders (appliance retailers, and others such as hotel groups, who import products for use in their own facilities).

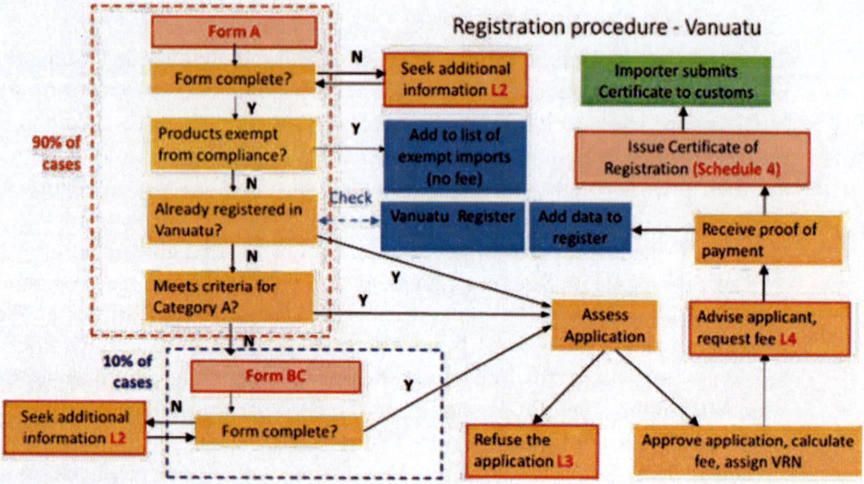

Fig. 3 Example of process flowchart form a Standard Operating Procedures manual

2. After the legislation was passed by that PIC's parliament: ensuring that officials understand the legislation, developing registration and record-keeping procedures, practicing model identification in the field (through site visits to retailers).
3. Once all systems were in place, holding joint workshops for officials and private stakeholders, including live training with the PAD.

The training material for each PIC was summarised in a Manual of Standard Operating Procedures, designed as both a reference for the initial PALS administrator/s (often only a single person in the smaller PICs) and as an aid for training future staff. Figure 3 shows an example from the Vanuatu manual. This was prepared for the paper form-based system which Vanuatu implemented before the PAD was developed. The same processes are now embedded in the Vanuatu section of the PAD.

The public documents produced by PALS meetings and workshops are available on SPC's Pacific Regional Data Repository website [8] and at the website of the Tonga-based Pacific Centre for Renewable Energy and Energy Efficiency [9].

5 Lighting

Most PALS PICs have implemented MEPS and mercury content limits for a range of lighting products (see Table 2):

- Incandescent lamps for general lighting service (GLS) within the scope of AS 4934. These effectively prohibit the traditional tungsten filament lamps, an objective endorsed by PIC energy ministers in 2012.

- Double-capped fluorescent lamps within the scope of AS/NZS 4782.
- Self-ballasted lamps (CFLs) within the scope of AS/NZS 4847.
- Ballasts for fluorescent lamps within the scope of AS/NZ 4783.

These standards cover efficacy MEPS only, not consumer energy labelling on packaging, although some cover the permanent marking of model number and energy efficiency class on the products themselves to preserve a verification and audit trail.

Despite the share of PIC energy allocated to lighting, this was always seen as a more complex area than refrigeration or air conditioning due to the very large number of product categories and individual models, the difficulty of detecting small import consignments and the large number of retail outlets (supermarkets, hardware and convenience stores). Perhaps the greatest challenge, however, was the rapid introduction of LED lighting products in the Pacific, which coincided with the PALS period.

In 2014 the SPC arranged with the UNEP enlighten program to develop a Pacific Efficient Lighting Strategy (PELS) with the aim of "transitioning to efficient lighting" [10]. A regional assessment found that in the PICs with reliable data, inherently efficient lamps types (CFLs and linear fluorescent lamps, LFLs) already represented the great majority of indoor lamp types and lighting energy, and incandescent lamps accounted for less than 20% of lighting energy [11]. The main problem was the poor quality and short service life of CFLs and the LEDs that were beginning to displace them. However, there were no performance or quality standards for LEDs in Australia or New Zealand that could be used as a model for the PICs.

This may change. In April 2018 the Australian and New Zealand Energy Ministers agreed to introduce minimum standards for LED lamps in line with European Union (EU) standards. The timing of the new regulation will align with revised EU minimum standards that will apply to LED light bulbs (planned for September 2020) [12]. It would then be open to the PALS PICs to include reference to the ANZ standards (and so indirectly the EU standards) in their regulations.

The Pacific Efficient Lighting Strategy published in 2015 [13] covered not only S&L but also financing and other measures for upgrading of street, outdoor and commercial lighting in PICs. The total budget for these programs was estimated at US$ 5.83 million (AU$8.3 million), of which AU$1.1 million was to be allocated for MEPS. However, no funding was secured from aid donors, so it was left to PICs to implement lighting MEPS through the S&L framework developed for PALS.

Lighting remains a lower priority in PALS, because:

- Resources are limited, and need to be concentrated on the products with the highest lifetime energy use per unit.
- The model registration approach that suits refrigerators, freezers and air conditioners is too complex and costly to apply to lighting products (hence the PAD does not have a lighting registration capability).
- The lighting market is changing rapidly, and there is little point in allocating resources to regulating essentially obsolete technologies (e.g. CFLs) or encouraging transitional technologies (e.g. mains voltage halogens). Once LED

performance and quality standards are developed these should become the focus of regulation.

Therefore, on the advice of SPC, the PALS countries have adopted simplified "deemed to comply" rules for lighting products, which in most cases can be verified by visual inspection. The import of GLS incandescent lamps is prohibited, all LFL imports are permitted so long as the diameter does not exceed 25.4 mm and all electronic ballasts are permitted. CFLs are permitted as long as the model is already registered in Australia or New Zealand. Anyone who wishes to import a regulated lighting product that is not "deemed to comply" can still do so, but must provide documentation proving compliance with the relevant standard.

6 PALS Evaluations

The only PIC where the post-implementation impacts of S&L have been estimated is Fiji, and only for refrigerators and freezers. The other PALS PICS have implemented S&L too recently for a quantified evaluation (see Table 2). After examining import and sales data covering the period before S&L and for the first three years after (2012–2014), Energy Efficient Strategies (EES) estimated that the energy consumption of new products in Fiji was about 20% lower than it would otherwise have been [14]. While product prices had increased slightly, the additional purchase price would have been repaid within 3 years of operation.

EES reported that all the stakeholders interviewed supported the program and its expansion. The vast majority of respondents were satisfied with the level of government support and the quality of training programs and materials received. All respondents were of the view that the administration of the program was well managed. Consumer information programs were also generally seen as adequate.

The terms of the Australian Aid grant required the Australian Department of Industry and Science to undertake an external "end of project" evaluation of the PALS program in 2015, when it was originally scheduled to conclude, even though it was by then already extended to 2017 (and subsequently to 2019). The evaluation concluded that "the structure of the programme, working through SPC and bringing PICT countries together for workshops, was very effective in developing skills and capacity in the region. PALS has been managed efficiently" but "in hindsight, the approach taken was ambitious; more effort could have been made to assess the extent of the risk that it would be harder than expected to secure legislative change and to identify mitigation options" [15].

Energy labelling works by influencing purchasers of labelled products to take energy running costs into account and to select more energy-efficient models. In 2016, the PALS program commissioned market research on consumer awareness and appliance purchase behavior in all the PALS PICs, as a baseline for tracking the future impacts of energy labelling [16]. The opportunity was also taken to survey and analyse the ownership of household appliances in a consistent way [1].

An end-of-project evaluation was carried out in mid-2019 by an independent external assessor, working with the SPC's internal evaluation team. The report concluded:

> "Based upon a qualitative assessment of key success indicators, PALS made high progress toward three of its four desired outcomes: political commitment, support of MEPSL adoption/operation, and regional capacity building. PALS' progress varied from low to high on the fourth desired outcome, establishment of enabling environments ...
>
> PALS is a success story with a recognized "brand" in regional governmental agencies. It has built a regionwide network of MEPSL supporters and PALS' staff and contractors are sought out as trusted advisors. The four PICs that have enacted MEPSL are concerned about a future without PALS, both within their countries and regionwide, since MEPSL is a new endeavor for most of them. While they are committed to MEPSL, they also hope that PALS can continue to provide services until their efforts are further established. Finally, they hope more PICs will adopt legislation so that MEPSL becomes a regional standard" [17].

7 Conclusions

The first serious investigation of the potential for appliance energy labelling and MEPS programs in the pacific region took place in 1996. It then took until 2011 for Fiji to become the first Pacific Island Country to implement S&L legislation. By the time the PALS program ended in 2019 a further four PICs had implemented S&L legislation and another five were on the way. While this process took twice as long as envisaged when PALS was set up, the initial expectations were never realistic.

It takes far longer to engage with governments, draft legislation, set up administrative structures and train both officials and private stakeholders than is normally expected by aid donors' project cycles and funding horizons. Hence the preference of aid donors for shorter-lived programs where the main drivers are financial incentives, and which can be delivered relatively quickly by external contractors.

Once an appliance S&L measure becomes law, however, it has a much better chance of persisting, with or without further financial support. It also acts as a foundation for further financial incentive programs, in the event that these can be funded, and makes them far more efficient, since the technical standards are in place, the market participants are known and the information on the comparative energy efficiency of products is already available.

That said, the mandatory S&L programs set up under PALS do need modest but constant funding, whether from the resources of the PICs themselves, the SPC or from new development partners and aid donors. PALS represented the only regional scale energy efficiency program in the Pacific, and arguably the most successful, but efforts to secure modest ongoing funding have not so far been successful. This highlights systemic defects in sustainable energy support for developing countries, and not just in the Pacific. It is more difficult to get aid donor funding for a six year than a three year program, even if the total amount is the same. There is also a bias towards highly visible renewable energy projects over less visible but usually more cost-effective energy efficiency measures.

Developing countries benefit greatly from regional programs, which avoid both duplication of effort and the risk of conflicting standards and labels in neighbouring countries with small markets. In the case of the south Pacific, PICs accepted that there was a strong case for adopting the ANZ S&L system, since most of the products came from the same sources (China and other Asian countries) and were technically identical to products supplied to ANZ. When all the PALS PICs in Table 2 implement this S&L program, the "footprint" of the ANZ energy label will have increased from about 30 million people (the population of Australia and New Zealand) to about 40 million.

Applying consistent S&L rules across the entire region offers economies of scale for manufacturers and importers, keeping down the price of energy-efficient products to all PIC households and businesses. The United Nations Environment Programme (UNEP) is aware of PALS, and the experience is being shared with other regional energy efficiency groupings, including South-East Asia, the Caribbean and Africa [18].

References

1. George Wilkenfeld and Associates (GWA), *Pacific Appliance Labelling and Standards (PALS) Programme: Consumer Awareness Baselines, Strategies and Targets*, for Secretariat of the Pacific Community, October 2017
2. George Wilkenfeld and Associates (GWA), *Energy Labelling and Minimum Energy Performance Standards for Appliances and Lighting: Impacts in Cook Islands, Fiji, Kiribati, Samoa, Tonga and Vanuatu,* for Pacific Forum Secretariat 2015. https://edd.spc.int/en/document-download/finish/11-reports/2498-energylabellingandminimumenergyperformancestandardsforappliancesandlighting-impactsincifjkisatova
3. George Wilkenfeld and Associates (GWA) with Energy Efficient Strategies (EES), *Electrical Appliance Energy Labelling and Minimum Energy Performance Standards for the Pacific Island Nations: Baseline Study, December 1996*
4. George Wilkenfeld and Associates (GWA), *The costs and benefits of energy labelling and minimum energy performance standards for refrigerators and freezers in Fiji,* for Australian Greenhouse Office, February 2006
5. International Institute for Energy Conservation (IIEC) *Situation Analysis and Feasibility Study on the Impacts of Introducing an appliance labelling programme in Samoa, Tonga and Vanuatu,* for Pacific Forum Secretariat October 2010. https://edd.spc.int/en/document-download/finish/11-reports/26-final-study-report-appliance-labeling
6. George Wilkenfeld and Associates (GWA) *The costs and benefits of introducing standards and labels for electrical appliances in Pacific island countries,* for the Secretariat of the Pacific Community, supported by Department of Climate Change and Energy Efficiency and AusAID, September 2011. http://prdrse4all.spc.int/node/4/content/cost-and-benefits-introducing-standards-and-labels-electrical-appliances-pacific
7. George Wilkenfeld and Associates (GWA), *Energy Labelling and Minimum Energy Performance Standards for Appliances and Lighting in Fiji: Expanding the Coverage of the Program to Additional Products, for the Department of Energy, Fiji and the Secretariat*

of the Pacific Community, November 2014 https://www.reinfofiji.com.fj/wp-content/uploads/2017/11/06-15-MEPSL-Expansion-Final-Report.pdf

8. http://prdrse4all.spc.int/node/4/content/pacific-appliance-labelling-and-standards-project-pals
9. https://www.pcreee.org/search/node/PALS
10. *Development of Pacific regional efficient lighting strategy under way*, http://prdrse4all.spc.int/system/files/mr_development_of_pacific_regional_efficient_lighting_strategy_under_way.pdf
11. George Wilkenfeld and Associates (GWA), *Regional Status Report on Efficient Lighting in Pacific Island Countries and Territories,* prepared for UNEP enlighten, SPC and Australian Aid, 2015 http://prdrse4all.spc.int/node/4/content/regional-status-report-efficient-lighting-pacific-island-countries-and-territories
12. https://www.energyrating.gov.au/products/light-emitting-diodes-leds
13. International Institute for Energy Conservation (IIEC) *Pacific Efficient Lighting Strategy (PELS) 2016-2020,* prepared for UNEP enlighten, SPC and Australian Aid, 2015
14. Energy Efficient Strategies (EES) *Evaluation of Fiji's Minimum Energy Performance Standards and Labelling Program (MEPSL),* for Department of Energy, Fiji, December 2014
15. Michaelis, Charles, *Evaluation of PALS Programme Report*, Databuild for the Australian Department of Industry, Innovation and Science, October 2015. http://prdrse4all.spc.int/system/files/pals_-_databuild_evaluation_presentation_-_6_aug_2015.pdf
16. Tebbutt Research, *Survey of consumer awareness and use of energy rating labels in PICs: REGIONAL REPORT Prepared for The Pacific Community* (SPC), 2017 http://prdrse4all.spc.int/content/survey-consumer-awareness-and-use-energy-rating-labels-pics-regional-report
17. Dethman, Linda et al. *Pacific Appliance Labelling and Standards Programme Final Evaluation,* Prepared for the Pacific Community Geosciences, Energy & Maritime Division, Final Report July 4, 2019 http://prdrse4all.spc.int/data/pacific-appliance-labelling-and-standards-programme-pals-final-evaluation-final-report
18. https://www.unenvironment.org/news-and-stories/press-release/pacific-efficient-light-strategy-will-reduce-greenhouse-gas

Creating a Business Model Framework for New Small and Medium-Sized Enterprise Energy Service Provision Companies in China

Luke Allan, Michael Otreba, and Qinhua Wang

Abstract The energy consumption during the operational phase of buildings and structures one of the major contributions to overall energy use in the building sector both in developed and developing countries. It comprises of space conditioning (heating and cooling), lighting, heating of water, and running home and office appliances. Increasing the efficiency of energy use will make a huge impact on the worlds energy demand, and thereby reduce the carbon dioxide production, reduce the dependency of imported energy and improve the environmental conditions. New business models need to be developed to cater for the changes to energy markets and the creation of energy service providers. Energy Service Companies (ESCO) have the potential to provide new business models for facility management companies, building managers, energy providers, maintenance providers and many other stakeholders currently working in the area of building design, construction, building operation, energy management, maintenance, etc. since customers may wish to get access to all services related to energy supply, energy consumption and building services maintenance through an ESCO. This paper describes how the concept of a Business Modelling Framework could be applied in the area of Energy Service Companies for new Small and Medium-Sized Enterprise (SME) start-up companies in China. Although the improvement of energy efficiency in new build and refur-

L. Allan (✉)
School of Architecture and Civil Engineering, Huizhou University,
Huizhou, Guangdong Province, China

M. Otreba
School of Engineering, University College Cork, Cork, Ireland
e-mail: Michal.Otreba@cit.ie

Q. Wang
Department of Civil and Environmental Engineering, Shantou University,
Shantou, Guangdong Province, China
e-mail: qinhuawang@stu.edu.cn

P. Bertoldi (ed.), *Energy Efficiency in Domestic Appliances and Lighting*, Springer Proceedings in Energy, https://doi.org/10.1007/978-3-030-79124-7_28

bished buildings is slow considering the ambitious goal of China to improve energy efficiency there is great potential for ESCO's in the domestic and commercial area—an example being through neighbourhood management systems (NMS)—and there are opportunities for large-scale cost-effective investments. This paper will examine a Business Model Framework which may help SME's enter the aforementioned arena of ESCO development in China.

Keywords Energy Service Companies (ESCO) · Business Model Framework (BMF) · Small and Medium-Sized Enterprise (SME) · Neighbourhood Management System (NMS)

1 Introduction

From 1978 China began to move from a centrally-planned economy to a more market-oriented economy. This led to the enormous growth that has seen China become a major world economy. With this growth came massive industrial and social change, which has necessitated a rethink in the use of renewable and sustainable energy to offset the effects of fossil fuels and deleterious emissions such as greenhouse gases [1]. As China adapts to its relatively new market economy many changes have occurred particularly in energy consumption and efficiency and this has helped the standard of living of people becoming more affluent as consumer companies have also helped change the landscape as they try to tempt a richer population to buy consumer items as more and more rural residents move to towns and cities. Another reason is that from late 2016, provincial governments embarked on a massive wave of investment in infrastructure and real estate, which prompted a recovery in energy-intensive industries such as steel and cement. All these changes require energy in the form of electrical power of one sort or another.

To feed this phenomenal growth China has traditionally relied on coal of which there is abundance. Figure 1 shows that although coal is the biggest source of energy in China other forms such as renewable energy are gaining in use e.g. the $36 billion Three Gorges dam, was completed in 2012, becoming the largest hydroelectric plant in the world at 22.5 GW thus adding to the production of renewable energy.

Ten years ago China overtook the U.S. as the world's largest energy consumer. This meant that inexpensive and dirty coal plants grew, but pollution from these plants and other industries meant that many cities were choking by the early 1990s. Within the past decade however, China has been working on a plan to curb its fossil-fuel pollution [2]. Although demand for energy has increased as a result of market changes Chinese citizens are demanding cleaner air, and they want immediate improvements. Air quality is now a political priority for the Chinese Communist

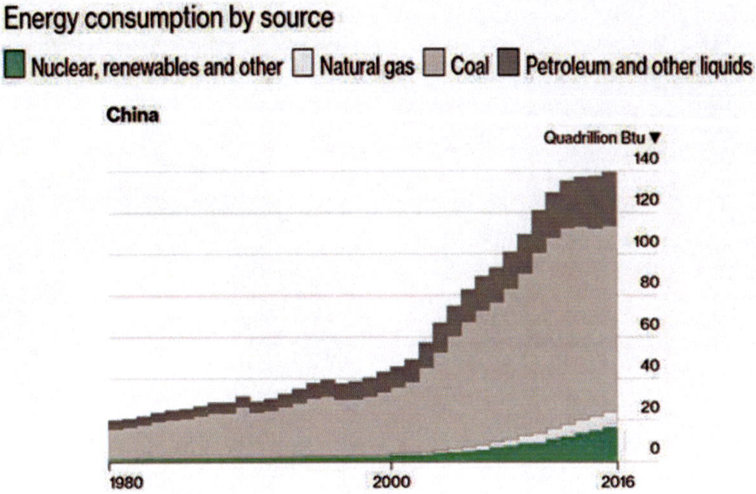

Fig. 1 Source: U.S. Energy Information Agency

Party on par with economic growth and transparency. This means that China cannot continue to run the same high-pollution coal plants that were considered acceptable decades ago. Beijing's solution is to move towards renewable energy while simultaneously investing in what may become the most efficient, least polluting coal fleet the world has ever seen [3].

Drivers such as pollution reduction, consumer demand and reducing climate change have allowed private Chinese SME's enter a more liberalised market as shown in Fig. 2. According to the IEA 2018 marked the 40th anniversary of China's Economic Reform and 20th anniversary of China's ESCO industry. China's ESCO market now represents 59% of ESCO global revenue. Since the beginning of China's reform there has been much progress, and the industry is gearing towards greater focus on quality, productivity and deeper structural changes. According to the EMCA's Annual Report 2018 the ESCO market grew by 15.1%, project investment by 5.2%, employment grew by 44,000 people. Total number of ESCOs is now 6439 (+302 compared to 2017). Industry still takes up majority share of ESCO market (67%), but the number of ESCO projects in buildings sector is growing as well as size of projects [4].

This paper will examine a Business Model Framework which may help SME's enter the aforementioned arena of ESCO development in China.

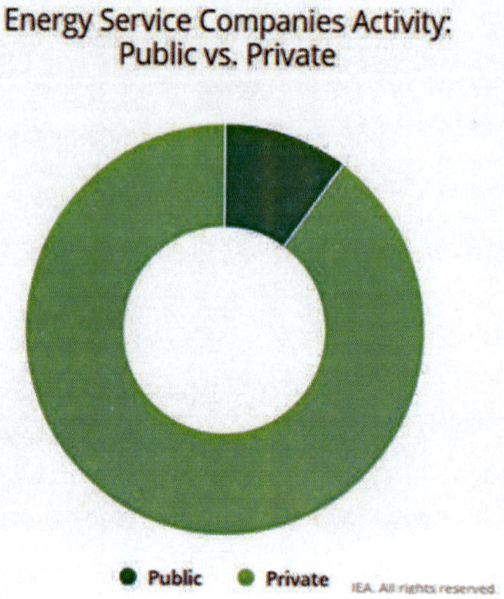

Fig. 2 Source: International Energy Agency

2 Energy Service Provision Business Models

Although there are many forms of business model available this paper will apply a well known business model that will be used for this paper based on a template created by and currently used by the Business Model Generation Group [5]. Using a generic business model framework will enable the creation of business model suitable for an energy service provider company [6]. The framework is suggested to assist companies that may be existing or new start-ups hoping to enter the ESCO market.

2.1 Energy Service Provision Business Models Description

For a successful enterprise or business the business model purpose is to take all required factors, consider and analyse them and ensure that these factors are valid and viable and will produce the required outcome [7]. There are four areas to be considered for the energy service provider as shown in Table 1:

The business model template is made up of nine building blocks each of which will be considered individually in the context of creating the business model for energy service providers.

These nine building blocks and their relationships are shown in Fig. 3 [8]:

Table 1 Areas to be considered for the energy service provider

Customer	(a) Consumers\purchases—those who buy create the business revenue
Offer	(b) Offer—the product that sells
Infrastructure	(c) Organisation—the entity that is required to exploit the business opportunity
Financial viability	(d) Money matters—this determines viability, financial performance and profitability

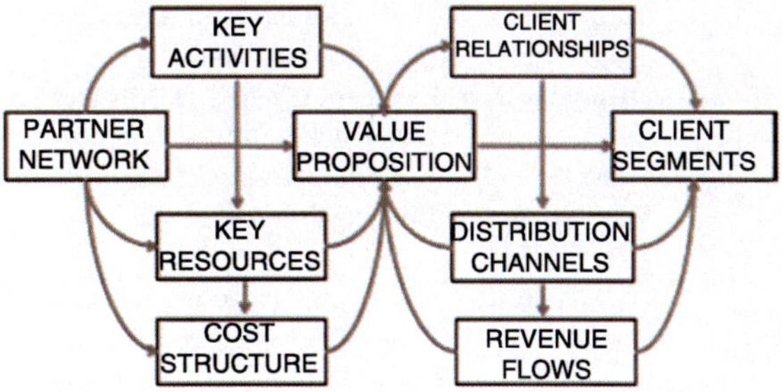

Fig. 3 Source: http://business-model-design.blog-spot.com/2005/11/what-is-business-model.html

The business model canvas takes the nine building blocks and plots onto a canvas which is displayed as a poster.

2.1.1 Main Areas of a Business

The nine basic building blocks that show the logic of how an energy service provider intends to make money cover four main areas of business: customers, offer, infrastructure and financial viability. For the energy service provider the business model is the blueprint for a strategy to be implemented through organisational structures, process and systems. These are shown in Table 2:

ESCO Customers

The customers or end users of the energy managed by the energy service provider (or ESCO as it is now known) are the ones who generate the revenue. The offer or value proposition will describe the demographics purchasing patterns and power and the location of the potential customers or purchasers of the product or energy. The customer part of the business will also dictate the distribution channels that the ESCO will pursue in order to manage the provision of the energy. The ESCO may provide services in relation to energy efficiency solutions for energy provided from the main grid managed by Distribution or from a standalone energy source i.e. wind

Table 2 Main areas of business

	Business model Building block	Description
Product	Value Proposition	Gives an overall view of a company's bundle of services and products
Customer Interface	Target Customer	Describes the segment of customers the company wants to offer value to
	Distribution Channel	Describes the various means of the company to get in touch with its customers
	Relationship	Explains the kind of links a company establishes between itself and its different customer segments
Infrastructure Management	Core Competency	Outlines the competencies and the key resources required to execute the company's business model
	Value Configuration	The key activities necessary to implement the business model and the arrangement of activities and resources
	Partner Network	Describes the key partners, their motivations to participate in the business model, and the network of cooperative agreements with other companies necessary to efficiently offer and commercialise value
Financial Aspects	Cost Structure	The cost structure resulting from the business model; Sums up the monetary consequences of the means employed in the business model
	Revenue Model	The revenue streams generated by the business model (constituting the revenue model), and the way a company makes money through a variety of revenue flows

Source: Adapted from Osterwalder [5]

turbine, Combined Heat and Power fuelled by biomass or by geothermal means. Distribution will involve how the ESCO markets and what their strategy for distribution supply and energy efficiency is. The ESCO will have to develop relationships with the end-user be they owner\operation, facility managers, apartment or house-owners. As part of the customer relationship or Customer Relationship Management (CRM) the ESCO will have to communicate what is on offer, the various price options depending on the source of fuel—fossil or renewable customer support and assistance in purchasing particularly where alternatives are offered.

ESCO Offer

The offer from the energy service provider is what they sell—and produce under certain circumstances i.e. as part of an neighbourhood management system NMS. This is the ESCO value proposition which describes what forms of energy are being provided and why end-users or consumers should purchase from this provider. The ESCO will analyse and audit the problems experienced by the end-users and will advise on how the provider can provide offers in terms of products and services that will help solve the problem. The offer could also include trading with other providers. An example could be selling excess energy power generation back to main stream providers.

ESCO Infrastructure

Another area of business the ESCO will have to consider will be infrastructure which includes core capabilities, partner networks and value configuration. The ESCO entrepreneur will have to examine the core competencies and capabilities required to operate the business. Apart from standard administrative and service skills will specialist personnel be needed if say wind energy is provided? Are special wind energy engineers required or can this be outsourced? If part of an NMS or community scheme is there land available to develop, are the facilities in place to connect and provide and is the equipment readily available and if not what are the lead in times from order to delivery?

The ESCO will form a partner network to create alliances that will be needed to run the business. There will also be various agreements and licenses to be obtained and there will be the need for specialist service providers of insurance, legal matters, security etc.

The process by which the ESCO provides their service will entail value configuration which describes what materials, supplies and other resources will be obtained in order to be transformed into the delivered end product for the customers. It describes the process used to supply the energy supply as described in the value proposition or offer.

ESCO Financial Viability

The ESCO provider will need to consider what they need to invest, what the cost structure will consist of, how revenue will be earned and what profit will be made. This will also be dependent on which financing scheme the ESCO will operate i.e. whether the ESCO provides the finance and thereby takes the financial risk or whether the client provides the finance and who in turn takes the risk. The ESCO at start-up will need capital to buy the equipment needed to monitor the energy usage and also working capital to start and sustain the distribution of energy. As with any business the initial outlay and expenses are usually large, with losses often made during the first 2 or 3 years. Very often the price for obtaining capital involves forfeiting a share of the business to others.

The cost structure described in the ESCO's value proposition includes the expenses involved in supplying energy to the end user if this service is provided. These can include all costs both direct and indirect. These expenses need to be calculated for future operations to allow budgets be created or forecast for working capital future requirements.

The income the ESCO receives from providing the energy service provision and other related services will provide revenue. Revenue will include sales volume forecasts and revenue projections and assumptions that will also help form part of budget requirements for the future. When all has been considered the ESCO will need to see if the main objective is achievable profit. This also however takes into consideration the return on Investment (ROI) and cash flows for potential future investment.

3 ESCO Business Model Building Blocks

Market changes moved the energy market in many cases from a traditional monopolistic business model towards a more service oriented business model or networked business models. In traditional business models relating to distribution most of the tasks were carried out "in house". Figure 4 [9] shows a traditional business model that would have applied to energy suppliers.

In many energy markets in the EU (where massive liberalisation has occurred [10, 11]), China and the US however this traditional form no longer exists as market changes, laws and regulations have changed. Another driver for change from in-house operations to outsourcing has been the entry of non-indigenous energy companies into Chinese markets resulting in a call for better efficiency. The overall result is a shift from capital intensive to information intensive current business models which itself is being increasingly affected by service provision. Figure 5 shows a networked business model with outsourced services.

Further developing from this is the creation of an ESCO that is part of an NMS and that enters the Chinese market. The scenario is one whereby the ESCO provided alternative forms of energy as well as other services such as auditing or existing energy services and recommendations on energy efficiency and cost reductions and

Fig. 4 Source: Trygg [9]

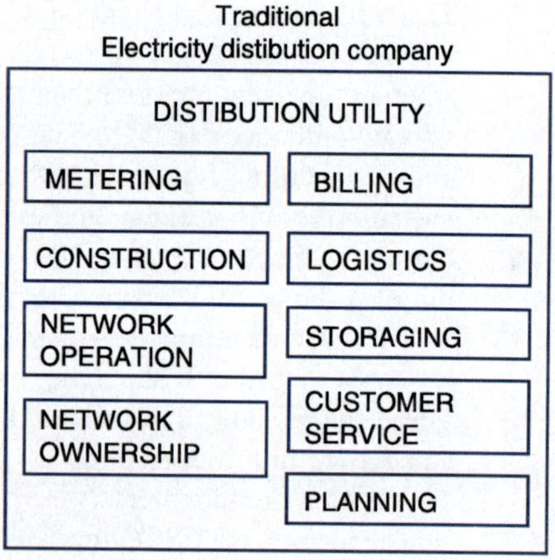

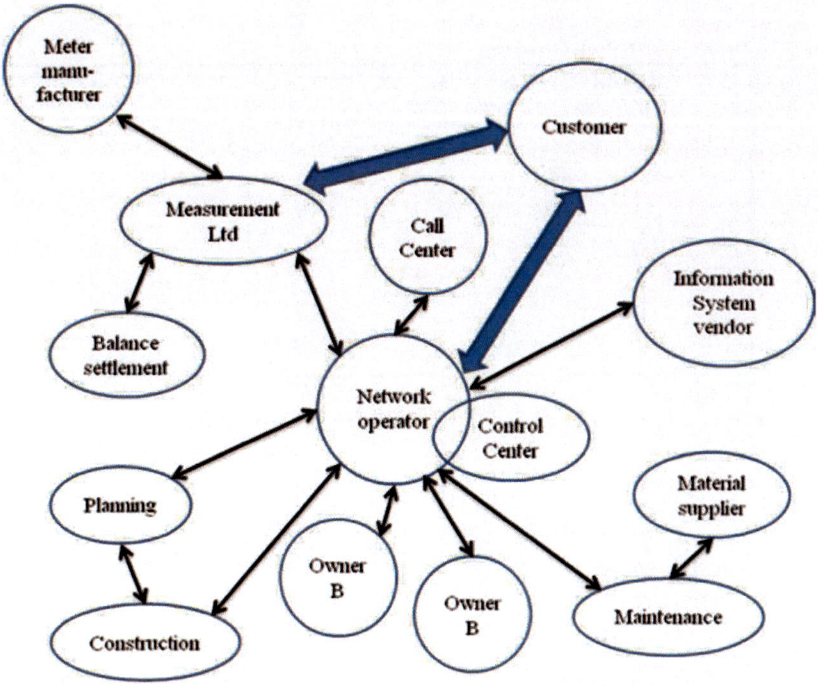

Fig. 5 Source: Partanen [10]

savings—providing energy efficiency and guaranteed energy savings is a key role. The provision of energy in different forms will allow customers make choices not only in terms of cost but also in terms of the source of the energy fossil or renewable. The objective is to a build distribution business model using the following template. This distribution business model can best be described using the nine building blocks shown in Fig. 6 below.

For the business plan and in order to create a good picture of the energy service provider business model the ESCO management should describe the following [12]:

- Customer segments: ESCO groups of customers with distinct characteristics
- Value proposition: The ESCO bundles of products and services that satisfy the customer segments needs
- Distribution channels: the channels through which ESCO communicate with its customers and through which the ESCO offer its value propositions
- Customer relationship: the types of relationships ESCO entertains with each customer segment
- Revenue streams: the streams through which the ESCO will earn revenue from customers for value creating and customer facing activities
- Key resources: the key resources on which the ESCO business model is built

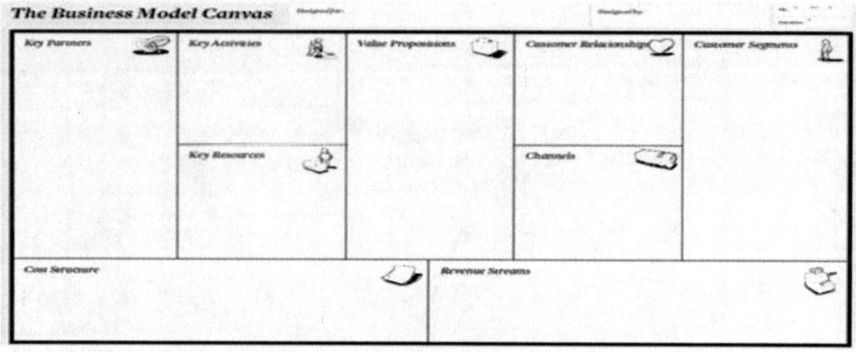

Fig. 6 Source: www.businessmodelgeneration.com, 26-3-2010

- Key activities: the most important activities performed to implement the ESCO business model
- Partner network: the partners and suppliers the ESCO works with
- Cost structure: the costs the ESCO will incur in order to run the business model

The ESCO consists of a group or organisation set up to provide products and services relating mainly to energy efficiency. In this context assume that a group of Chinese entrepreneurs have come together to create such an organisation. The group will form an ad-hoc task force working as a team to make a whole examination in the business logic of the newly formed organisation. This would mean looking at the contributions of the different entities of the organisation. The multi disciplinary team would establish the following requirements applicable to the business model project:

- Manage the business model project as a business venture to support the ESCO strategy and to create shareholder value
- Develop the basics of the project business model
- Communicate effectively through senior management about the business model by relating it to financial and strategic goals
- Influence the ESCO business model toward an appropriate number of products and services which be viably offered
- Manage the business model according to the overall ESCO business plan
- Analyse the business impact of decisions brought about by implementing the business model
- Transition areas to those who are responsible for the business model lifecycle which maybe relatively short as models are not static
- Contribute lessons learned should the business model have to be changed once or a number of times before gaining the highest levels of commercial and financial efficiency

3.1 ESCO Customer Segments

Since the ESCO is going to become a service provider they are going to rely on customers to provide revenue and profit. Customers will be part of the core of the company as without them the company will not last long. The business model will help define the Customer Segments. Management will need to identify their most valuable customers, acquire them, keep them and hopefully get them to increase their purchases. The likely customers for an NMS based ESCO will include householders, apartment dwellers, SME's, and possibly hospitals if in the area.

There is a general rule in relation to the segmentation of customers called the 80/20 rule [13]. The rule states that 20% of customers account for 80% of profit, whilst the other 80% only contribute to the remaining 20%. An integral part of the model entails having a clear customer description and understanding. This means that the ESCO will need to ask a number of key questions in order to clarify and identify who their customers are. Since the ESCO, in this case may have a number of services, customers may be grouped into different categories or groups. These questions are standard [14] and can be applied to any industry in the right context.

Having identified the different customer segments and defining the most valuable (potential) customers, the ESCO should undertake the following steps:

- Gather as much relevant information about (potential) customers as possible, in particular the big and top customers.
- Analyse this information and if necessary, redesign the information requirements.
- Set goals for how the ESCO wants customers to perceive them as a provider of energy and other services and/or experiences.
- Choose media, systems and content for communicating and interacting with customers.
- Develop rules of engagement and 'packages' for each customer segment i.e. gas and electricity bundles.
- Embed a customer-driven culture in the company.
- Develop customer management systems as the company grows and learns.

Developing and maintaining CRM in targeting, getting and keeping customers is not the be all and end all for the ESCO. Other factors influence success, such as the value proposition of the ESCO service and the prices offered for the alternative forms of energy supply. Other generic problems also exist, such as an incoherent marketing strategy or the support of a customer-centred approach.

3.2 Energy Provider Value Propositions

The Value Proposition building block for the ESCO describes the bundle of products and services that create value for a specific Customer Segment. The objective for the ESCO is to examine each Customer Segment to identify the value proposition building blocks for the ESCO as shown in Fig. 7 [15]:

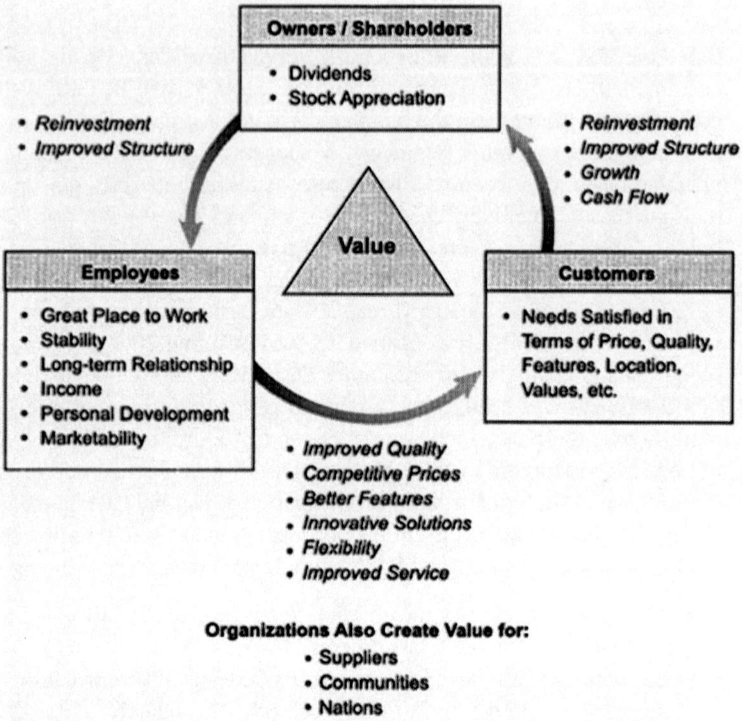

Fig. 7 Source: SMG [15]

The ESCO needs to ascertain what offer attracts the customers and what is that offer to the market. Since this ESCO will be created as a single firm that manages and coordinates energy on offer to each of the customer segments. In this case, some of the services offered could include energy audits of commercial, industrial or residential complexes. In the case where the ESCO is hired to provide an NMS scheme consisting of alternative forms of energy through Combined Heat and Power or wind farms the ESCO could provide construction management services that would include preparing tender documentation and performance specifications, the design and commissioning of the plant. Other services could include project financing, monitoring and energy saving guarantees. Equipment maintenance and operations may be offered as a service to facility managers. Other services might include administrative services, consulting, meter monitoring etc. This would suggest that the ESCO would require the expertise and knowledge of many types of technologies in order to carry out such a broad range of activities and historically ESCOs have tended to provide a complete package of services. The services mentioned above are broken down into greater detail:

(a) Energy Audit—The ESCO could be employed by an SME, an owner/operator, a hospital etc. to provide an energy audit to analyse the operation of major energy using systems and to determine if improvements can be made to increase efficiency and save costs or reduce fuel bills.

    Alternatively, a company may request an investment grade audit which is a very comprehensive audit that examines the likes of HVAC equipment, heating, AC, hot water, lighting, controls and energy generation systems to produce a cost-effective report with recommendations for improvement.

(b) Project Guarantees—ESCOs will guarantee project performance in terms of energy efficiency and potential future savings. To do this a baseline has to be agreed and acceptable to all parties. The information can be from monitored metered equipment. One problem is where a meter cannot be placed, which makes providing information difficult and possibly subjective. Since some of the energy saving calculations may be based on assumptions it might be difficult for clients to argue against the ESCO findings.

(c) Equipment Maintenance and Servicing—If providing a guarantee an ESCO might insist on a maintenance contract. It may also be one of the standard services that an ESCO provides. The maintenance service could include monitoring equipment, repairs and replacement, reports on equipment operating, reviewing strategies of operation. The level of service will determine the cost.

(d) Project and Construction Management Services—an ESCO may take responsibility for designing and managing the construction of an energy project. For example, as part of an NMS the ESCO might design, manage construction of and commission a Combined Heat and Power plant or a wind turbine. As part of commissioning the ESCO may contract to run the plant or may train client staff to overtake that responsibility.

(e) Provision of Energy Service Provision—the ESCO as an energy service provider will source and provide alternative forms of energy from which the customer can choose. This means that the ESCO will buy wholesale fossil and renewable energy and offer same to its clients. The customer will be able to choose from what form the energy was created and at what cost. It may be that some customers may prefer to buy from renewable sources at a higher (or lower) cost. If the ESCO represents an NMS owner generated energy through Combined Heat and Power or a wind turbine may reduce energy costs and possibly create earnings from sale to the national grid.

(f) Other ESCO services—As mentioned above, training may be given by the ESCO either as part of commissioning or as part of a consultancy service to clients. Should a client decide to opt out of an ESCO maintenance contract the ESCO could provide the client's in-house staff with training to operate and maintain whatever the ESCO was responsible for installing. The ESCO may, on an ongoing basis, offer specialised technical staff by monitoring electricity and gas usage, CO_2 levels, room temperatures and air quality.

As part of this building block, the ESCO has described the value propositions it will offer to clients in each customer segment. Each customer segment may have many value propositions whilst a single value proposition may have several customer segments with a distinct mix of elements.

3.3 ESCO Communication and Distribution Channels

The objective of this building block is to identify the channels through which the ESCO will offer value propositions to the client. In other words, the channels through which the ESCO will communicate with its customers and through which it will offer its value propositions.

The communication distribution channel represents the interface between the ESCO and its customers and can be done by many means. Retail outlets, sales teams and conferences are some of the more traditional methods. Advertising through the media is another form and given the prevalence of websites as a means of communication, many companies now choose to place their names on Chinese social websites such as Renren and Weibo as well as having their logos on various trade related websites.

The ESCO will determine through which communication and distribution channels they will reach their markets. Further, each channel will be analysed as to how well it works. Residential customers may be attracted by leaflets and advertisements posted through the letterbox. Reaching larger scale customers may be done through trade fairs and specialist conferences. Of particular importance will be the expense and cost efficiency of each channel. With profitable and regular customers cost intensive channels may be justified—low profit clients require cost efficient channels. The Business Model Group contend that Channels have five distinct phases. These are shown in Fig. 8:

<table>
<tr><th colspan="3">Channel Types</th><th colspan="5">Channel Phases</th></tr>
<tr><td rowspan="3">Own</td><td rowspan="2">Direct</td><td>Sales Force</td><td rowspan="5">1. Awareness
How do we raise awareness about our company's products and services?</td><td rowspan="5">2. Evaluation
How do we help customers evaluate our orgainisations Value proposition?</td><td rowspan="5">3. Purchase
How do we allow customers to purchase specificproducts and services?</td><td rowspan="5">4. Delivery
How do we deliver a Value Proposition to customers?</td><td rowspan="5">5. After sales
How do we provide post-purchase customer support?</td></tr>
<tr><td>Web Sales</td></tr>
<tr><td rowspan="3">Own</td><td>Own Stores</td></tr>
<tr><td rowspan="2">Partner</td><td>Partner stores</td></tr>
<tr><td>Wholesaler</td></tr>
</table>

Fig. 8 Source: Osterwalder [14]

3.4 ESCO Customer Relationships

In Channels, the ESCO is prepared to invest in cost intensive channels for profitable clients. It follows therefore, that these clients will require a high touch relationship. Using this block will help determine the types of relationship the ESCO will form with each of its customer segments. This will entail identifying relationships and which will be maintained with each customer segment. The ESCO will formulate a clear strategy for CRM for each customer segment, thereby helping build a robust business model.

To help create a strategy, the ESCO will utilise the value disciplines of key management model states and there are three generic value disciplines that will help the ESCO provide value to its customers [5]. Since the ESCO will have a number of Customer Segments the value disciplines model will help the ESCO determine the key issues in delivering unique value by fulfilling customer needs. The three value disciplines are:

- Operational excellence—in pursuit of optimal running costs.
- Product leadership—to offer the best product (technically and using the latest technology); and above all to be the first to do so.
- Customer intimacy: to offer the best total solution by being most dependable and responsive to customers' needs.

It is argued by that trying to pursue all three disciplines at the one time will give rise to conflict, confusion and inefficiencies and therefore the ESCO should determine which discipline applies and act upon it vigorously in order to determine a significant product value.

Users of operational excellence offer high-quality products at low prices. The focus for such companies is efficient streamlined processes, supply chain integration, low stocks of goods, no frills and the ability to manage large volumes. Product and process standardisation is the norm.

Product leaders are such because they are inventors and brand leaders. High risks mean margins can be high due to dynamic or unknown markets. In order to create a once-off hit they concentrate on Research and Development, design and short-time to market. Very often their products are copied by others and the market becomes flooded. The keys for such companies are product life cycle management and technological innovation.

The discipline the ESCO should follow to help build their Customer Relationship strategy is the customer intimacy discipline. This is because this discipline is about leaders who will do their utmost to satisfy a relatively small set of customers as long as they are worth it. The ESCO will have a portfolio of customers, some of which may be on a once-off (energy audit) or short term contract (once off construction project). There will also be a core of preferential customers whom the ESCO will want to keep. These will be those who are supplied with energy on an on-going basis or those with whom the ESCO has maintenance and operation contracts, particularly in relation to a facility management role. Value disciplines are shown in Fig. 9 [16].

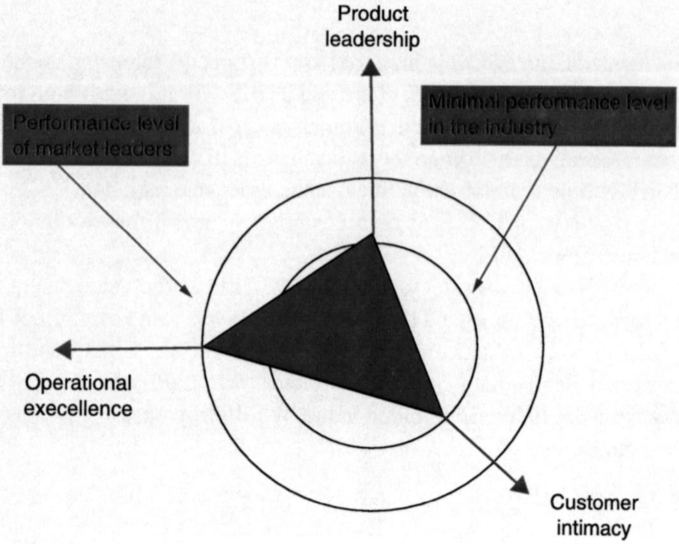

Fig. 9 Source: Treacy and Wiersema [16]

Along with the value proposition and operating model, the inclusion of the value discipline will offer the ESCO a very good business strategy. The value discipline will focus on value for the customer and will allow the ESCO to be pro-active to changing customer needs and new market developments. This should help avoid customers leaving. The ESCO will be aware however, that there is a fine line between neglecting its own company obligations and responsibilities by over focusing on the customer relationships.

3.5 *ESCO Revenue Streams*

By creating value for its customers, the ESCO will earn revenue in return. The main objective of this building block is to determine which types of revenue streams the ESCO will earn from the customer segments and value propositions. In Sect. 3.2 the products/services offered by the ESCO and generating revenue are:

(a) Energy Audits
(b) Project guarantees
(c) Equipment maintenance and service
(d) Project and Construction management
(e) Provision of energy
(f) Other services

These are the revenue streams on which the ESCO is dependent for sustainability and profit. The ESCO will determine how much each revenue stream's contribution is to overall revenues in terms of percentages. Asset sales in the form of selling energy will be the largest revenue stream for the ESCO. Project guarantees and maintenance and service contracts will also be a major earner of revenue as these will be on an on-going basis or as long as the contract runs for. Energy audits and Project/Construction Management will occur on a once-off basis most likely at irregular intervals. The two types of Revenue Streams that the ESCO will have will therefore be transaction revenues resulting from on-going payments.

To determine these revenue streams the ESCO will create a revenue model within its business model. In creating the revenue model the ESCO will determine who will buy, how often they will buy, how soon they will buy and at what cost. The ESCO will also determine the revenue model will essentially be price multiplied by quantity where quantity can be thought of in terms of market share, purchase frequency, ancillary sales etc. The revenue model will form part of the profit formula which will also include the cost structure, target unit margin and resource velocity. Cost structure will include direct costs, overheads and economies of scale. Target unit margin will dictate how much each value proposition should net in order to cover direct and indirect costs and achieve the desired profit levels. Finally, resource velocity will examine how quickly resources need to be used to support target volume and includes lead times (transformer manufacture 9–12 months) throughout, stock turnover, asset utilisation (wind turbines rarely operate 100%) etc.

The pricing mechanisms employed by the ESCO will be a mixture that is dependent on the service purchased. The following Fig. 10 shows typical Pricing Mechanisms:

Pricing Mechanisms

Fixed "Menu" Pricing		Dynamic Pricing	
Predefined prices are based on static variables		Prices change based on market conditions	
List Price	Fixed prices for individual products, services or other Value Propositions	Negotiation (bargaining)	Price negotiated between two or more partners depending on negotiation power and/or negotiation skills
Product feature dependent	Price depends on the number or quality of Value Proposition features	Yield management	Price depends on inventory and time of purchase (normally used for perishable resources such as hotel rooms or airline seats)
Customer segment dependent	Price depends on the type and characteristic of a Customer Segments	Real -time-market	Price is established dynamically based on supply and demand
Volume dependent		Auctions	Price determined by outcome of competitive bidding

Fig. 10 Pricing mechanisms

3.6 ESCO Key Resources

The ESCO will make its business model function through the use of key resources. In order to do this it is necessary to identify the key tangible and intangible resources by asking key questions. The first is what are the key resources the ESCO relies on to run its business model and secondly how do these resources relate to the ESCO value propositions and corresponding customer segments, channels and relationships. The ESCO key resources required will be physical, intellectual, human and financial.

Physical resources: The ESCO will provide headquarter office space with Information and communications technology (ICT) capabilities of connecting to remote generation stations through SCADA. Vehicles will be provided for the sales team and technicians. Storage space will be made available for monitoring equipment although it is likely that much equipment will be provided by outsourced contractors employed by the ESCO. Partners will also provide some physical resources not provided for by the ESCO.

Intellectual resources: When monitoring for energy audits, project guarantees and maintenance the ESCO will create customer databases. The data ownership will remain with the client organisation. All other proprietary knowledge, designs, copyright and partnership material will remain the property of the ESCO.

Human Resources: The ESCO will rely heavily on human resources as its business model is dependent on project managers, experienced engineers, technicians and administrative staff such as customer support and a skilled sales team. As part of its business model the ESCO will implement a structured approach to develop a HR strategy and produce an action plan. By utilising a strategic Human Resources (HR) management model the ESCO will follow a systematic approach to develop a structured HR plan for its organisational purposes. The Strategic Human Resources Management (HRM) model is an ideal approach which will lead to staff employee commitment who will work efficiently and effectively together [17].

Preparation: The ESCO will create a team that will be responsible for developing the strategic HR plan. Following the next three steps will format the HR strategy.

Analysis of the future ESCO organisational profile: Since the ESCO is being formed as a new company there is no current profile to analyse or change. The ESCO will therefore start from scratch and analyse the desired organisational profile.

Determining key issues for ESCO HRM: The critical success factors mentioned in the previous step will be weighted and clustered into key issues for the ESCO HRM. The HR action plan will be linked with the organisational strategy through these issues. These key issues are normally associated with cultural change, human capital reduction, outsourcing and quality of leadership.

Organisation of the ESCO HRM: Having determined the key issues the ESCO HRM strategy will be put in place. The strategy will be created by examining and distinguishing between people, processes, and structure and technology perspectives. The people perspective will include employees such as the Project Manager's, engineers, technicians and administrative staff as well as ESCO management. The processes will include recruiting training (Continuing Professional Development will be the norm), salaries and other benefits, such as company vehicle, pension fund, out of hours working payment, holidays etc.

Creating an ESCO action plan: In order for action 4 above to be put in place the ESCO will produce an action plan which will track completion of the objectives; and also review the commitment of staff to prioritise the required actions [5].

Implementation: During implementation, the ESCO will make additions (not previously included), revaluate the plan and make any necessary changes.

The execution and value of the model will be dependent on the ESCO HR manager who is responsible for its implementation. To improve on the organisational structure and management system in order to build an excellent organisation the ESCO will then implement the recognised European Foundation for Quality Management (EFQM) excellence model [18] as part of its business model and to further strengthen its human resource policy and structure. The EFQM model will throw light on performance gaps and highlight areas for improvement. This model is used also for assessing and further designing the ESCO company architecture in terms of best practices. It will help the ESCO management implement their strategy (designed in the Strategic HRM model) and redesign and further develop the company structure and processes. Figure 11 shows the models enablers and results:

The model distinguishes the enablers—five organisational areas and the results—four performance areas. The key elements for effectively managing the ESCO are the organisational areas made up of leadership; policy and strategy; people; partnerships and resources; and processes. To assess how well the ESCO performs and the health of the company generally the key results can be examined. The ESCO customer, people and society results will be measured in perception measures and performance indicators and there will be a feedback loop to help the ESCO establish a co-ordinated learning curve.

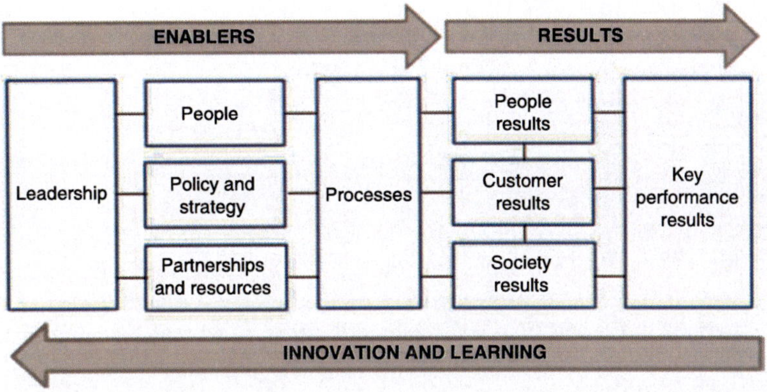

Fig. 11 Source: EFQM, Belgium 2010

Financial Resources: The ESCO financial plan will form part of the business model. As part of the financial plan the ESCO recognises that due to initial set up and establishment outgoings it is unlikely that the ESCO will be cash positive in the first or second year. Loans and credit lines will be set up. The ESCO will consider stock options to entice and hire key employees.

3.7 ESCO Key Activities

Key activities are the most important activities that the ESCO needs to perform in order to run the business model. The ESCO objective in this building block is to identify these key activities. Apart from the question of what are the key resources needed to run the business model the other question of how does each resource relate to the value proposition and corresponding customer segments, channels and relationships will also be examined.

The main activities for the ESCO will centre around provision of energy/district heating as part of an NMS to residential and apartment owners, multi-building owner/operators, SME's and public authority, hospitals, schools etc. Associated with this will be energy audits to reduce energy and operational costs for clients.

It is envisaged that project/construction management activities will only apply where a community or NMS scheme is being put into place whereby the construction and maintenance and operating will be of a Combined Heat and Power or wind turbine attached to the NMS to provided additional energy or district heating. Guarantees and other services will more than likely occur on a once off basis and at regular intervals. The ESCO will therefore begin with its most important value proposition and its related channels and relationships.

The ESCO will then list the key activities necessary to offer the value proposition to the customer. This action will be repeated for the aforementioned key activities (many of the ESCO activities will more than likely serve several value propositions).

3.8 ESCO Key Partnerships

The partners and suppliers that the ESCO works with will be part of this building block. The ESCO will therefore outline which partners and suppliers will be worked with as part the business model. Another key question will be what key resources these partners and suppliers relate to. Further, to what value propositions, channels or relationships do the partners and suppliers contribute.

The ESCO may form a partnership with Distribution which is responsible for transporting electricity from the transmission grid through the distribution systems. This is because, as a supplier of excess energy, the ESCO will feed into the electricity pool which will then feed to the final end-user. Lessons learnt from ESCOs in the

EU would be an advantage in helping firms in China [19].There may be a partnership with Distribution if the generation of excess energy is large. The first step the ESCO will take in the connection process is to establish whether a transmission connection or a distribution connection is needed.

In the provision of gas to NMS customers the ESCO will form a partnership with the supplier of gas.

The ESCO will see its main supply to industrial and commercial and residential customers as part of an NMS. The ESCO will form partnerships with these clients. Other partnerships will be formed with clients availing of other services provided by the ESCO and may be long term or once-off partnerships depending on the service contract.

For services other than the provision of energy which will have its own customer contracts, the ESCO will make provision for new contracts. There are a number of ways to contract depending on the risk the ESCO is willing to accept. The ESCO costs for its sources will be higher if the ESCO assumes and guarantees project performance.

The ESCO will propose the following types of contracts—(special negotiated contracts will also be considered):

3.8.1 Shared Savings Contracts

The ESCO will create a mutually agreed method for measuring cost savings which will be shared with the client. These cost savings will be either fixed or variable. There are a number of areas that the ESCO will take in consideration. Since the ESCO will provide initial funding on shared savings contracts, the customer does not have to show the financing in their balance sheets. However, since the ESCO agrees to carry financing, credit and performance risks fees and costs will be higher than in other forms of contract. With government and public bodies where tax exempt financing is used and sharing shaving is not permitted the ESCO will charge commercial interest rates. The ESCO will be vigilant with respect to the price of energy, irrespective of type, as increases in prices will increase the share of savings achieved by the ESCO. The ESCO will accept that it will not be paid if there no cost savings.

3.8.2 Typical Shared Contracts

- Scaled fee: As the ESCO recovers investment the fee will decrease over time.
- Variable fee: The monthly savings will have a fixed percentage fee which will fluctuate in line with monthly savings fluctuations.
- Fixed fee—Any excess savings—after the fixed fees will be shared between the ESCO and the customer
- Fixed savings total: The ESCO will receive all savings amounts to a point or agreed specified amount. Once this limit has been reached the ESCO will share savings with the customer.

3.8.3 Guaranteed Savings Contracts

The benefit offered to the customer under this form of contract is that the ESCO guaranteed savings will equal or be greater than the equipment payments or debt service carried by the customer unless the price of energy falls below a specified floor price. Another benefit offered to the customer is that the customer's financier does not have to rely on the savings guaranteed as the customer will make provision for the project through a loan capital lease or operating lease. Since tax—exempt financing offers the best interest rates the customer can benefit as this form of contract allows tax exempt financing and allows the customer keep more of the savings than under the shared savings contract. The ESCO contract will be arranged to guarantee that energy savings will meet or exceed an agreed minimal amount i.e. the full amount or an agreed percentage of the savings that the energy audit for example might specify.

3.8.4 Non-guaranteed Savings Contracts

Under this form of contract the customer accepts project responsibility and all financial risk but in return gets all the savings from the project. The ESCO will provide the energy audit the project design, construction management and commissioning fee. The ESCO will not however, provide any guarantee in relation to energy savings or performance.

Through partnerships and contracts the ESCO accepts some risks but given uncertainty in both a competitive market and changing technology aims to reduce risk exposure. This will be evident where the ESCO outsources due to not having its own in-house resources to carry out the tasks. The risks in this case will be transferred to the out-sourced company. Optimising the allocation of resources and activities and utilising economy of scale partnerships will help reduce costs.

3.8.5 Alternative Forms of Contract

Two other types of contract options are the Build-Own-Operate-Transfer (BOOT) contract and the frequently used contract in Europe—the 'chauffage' contract. The BOOT contract operates with the contractor owning and running a facility for a number of years before transferring it back to the client. Perhaps the best known example would be toll roads whereby the toll is built and run for a number of years before being handed back to a government department such as the transport authority.

Chauffage is where an ESCO takes over complete responsibility for the provision to the client of an agreed set of energy services and provides in effect an extreme form of energy management outsourcing [20].

3.9 *ESCO Cost Structure*

This important building block describes the important costs sustained whilst operating the ESCO under this business model and plan. The ESCO will identify the most important costs applicable to the business model and also which cost positions can be easily connected to a business model building block. Further the ESCO will have to determine what costs can be calculated for each customer segment. The ESCO Company cash cycle is shown in Fig. 12.

Important costs include fixed and variable costs and this particular model will be value driven whilst still minimising costs. This will mean the ESCO will focus on premium value propositions and a high-level personalised service. The fixed costs for the will include salaries, rent or lease costs for the building and the cost of fixed assets.

4 Conclusion

There is a global need to find alternative forms of energy to replace traditional forms and in efforts to reduce green house emissions many countries such as China are reviewing what their future energy needs will be and how will they be supplied. Renewable alternative forms of energy are seen as the way forward. Technology allows ESCOs or other energy generators play their part in providing alternative

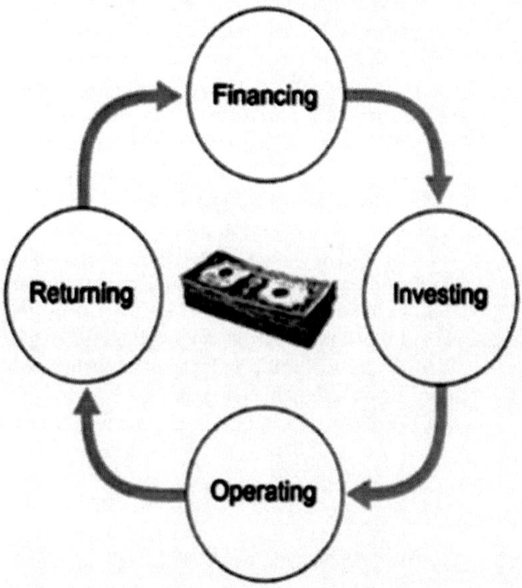

Fig. 12 ESCO Company cash cycle. Source: SMG [15]

forms of energy. Other energy generators may not be ESCOs but could be Energy Efficient Services providers [21]. Indeed individual households can now help combat the demand on the system and even help supply excess energy through for example the introduction of wind generators or micro-generation. This paper has set out the key actions the ESCO will carry out in setting up a company in China. This includes all actions relating to personnel, structure and assets that will be required in order to operate a viable organisation. These building blocks will therefore form a company charter that will be referenced to ensure the ESCO company is meeting its obligations and to help create the business model framework that will become a core part of the organisation. China has proved that it embraces the need for change with respect to changes in energy consumption and the opening up of markets. Like the US and EU is doing this both in industry and academia [22, 23]. Using the aforementioned building blocks will help provide a Chinese ESCO company with the necessary requirements on which to build its operations.

References

1. Allan, L., Wang, Q. "Using BIM to enhance the ESCO services used in the setting up of Neighbourhood Management Systems in China", 9th International Conference on Energy Efficiency in Domestic Appliances and Lighting (EEDAL'17) 13-15 September 2017 Irvine, USA.
2. Leatherby, L., Martin, C. (2019), available at https://www.bloomberg.com/graphics/2019-international-energy-use-renewables-coal-oil/
3. Hart, M., Bassett, L., and Johnson, B. (2017) available at https://www.americanprogress.org/issues/green/reports/2017/05/15/432141/everything-think-know-coal-china-wrong/
4. Energy Service Companies—At the heart of innovative financing models for efficiency (2019), available at https://www.iea.org/topics/energyefficiency/escos/China/
5. Osterwalder, A. (2004). The Business Model Onthology A Proposition in a Design Science Approach. Lausanne: Licencie en Sciences Politiques de l'Universite de Lausanne.
6. Allan, L., (2012), "A New Business Model Framework for Energy Service Provision", PhD Dissertation University College Cork (UCC).
7. Bertoldi, Paolo; Berrutto, Vincent; De Renzio, Mario; Adnot, Jerome; Vine, Edward; How are EU ESCOs behaving and how to create a real ESCO market Proceedings of the Summer Study Conference, ECEEE, Saint Raphaël, France, 2-7 June 2003.
8. Osterwalder. (2005, Novemeber). Business model design blog spot. Retrieved April 2011, from What is business model: http://business-model-design.blog-spot.com
9. Trygg, P. T. (2005). Changes of Business Models in Electricity Distribution. Tampere: Tampere University of Technology.
10. Partanen, J. T. (2010). Cost and benefit analysis for a distribution management system in electricity distribution networks. *International Journal of Energy Sector Management, Vol. 4 No. 2*, pp. 256–272.
11. Bertoldi, Paolo; Hinnells, Mark; Rezessy, Silvia (2006); Liberating the power of energy services and ESCOs in a liberalised energy market. Proceeding of the International Energy Efficient Domestic Appliances and Lighting Conference (EEDAL06 London, 21-23 June), Citeseer.
12. Osterwalder, & Pigneur. (2009). Business Model Innovation. Geneva: ULURU.
13. Van Assen, Van Den Berb, & Pietersma. (2009). Key Management Models. Harlow: Pearson Education Limited.

14. Osterwalder, A. P. (2010). Business Model Generation. Self Published.
15. SMG. (2000). Business Skills for Project Managers. Project Management. Philadelphia, PA, USA: Strategic Management Group.
16. Treacy, M., & Wiersema, F. (1995). The Dicipline of Market Leaders: Choose your customers, narrow your focus, dominate your market. London: Harper Collins.
17. Curry, J., & Curry, A. (2000). The Customer Marketing Method: How to implement and profit from customer relationship management. New York: Free Press.
18. European Foundation for Quality Management (2013). EFQM Excellence Model.
19. Bertoldi, Paolo; Boza-Kiss, Benigna (2017); Analysis of barriers and drivers for the development of the ESCO markets in Europe. Energy Policy 107, 345-355, Elsevier.
20. Bertoldi, Paolo; Rezessy, Silvia (2005); Energy service companies in Europe. Status report. European Commission ISBN: 92-894-9424-7.
21. Hansen, Shirley J; Langlois, Pierre; Bertoldi, Paolo (2009); ESCOs around the world: lessons learned in 49 countries. The Fairmont Press, Inc.
22. Nicola Labanca, Felix Suerkemper, Paolo Bertoldi, Wolfgang Irrek, Bruno Duplessis (2015), Energy efficiency services for residential buildings: market situation and existing potentials in the European Union, Journal of Cleaner Production, Volume 109, 2015, Pages 284–295, ISSN 0959-6526, https://doi.org/10.1016/j.jclepro.2015.02.077
23. Dandridge, Cyane; Bertoldi, Paolo; Walton, Jennifer (2005); Engaging the small commercial sector: new approaches for solving market barriers. Energy Savings: What Works and Who Delivers? eceee Summer Study Proceedings. V. 1-3.

Transforming Air Conditioner Markets Through Effective Labeling Programs: Case Studies from India, Europe, Brazil, Thailand, and Vietnam

Colin Taylor, Marie Baton, Ana María Carreño, and Eric Gibbs

Abstract Energy labeling is a critical component of effective air conditioner (AC) energy efficiency policy. While minimum energy performance standards (MEPS) remove the least efficient ACs from the market, energy labels are key to drive AC markets to higher efficiency. Energy labels allow consumers to make informed purchasing decisions by differentiating high efficiency products from average and low efficiency products. They incentivize manufacturers to produce more efficient products by helping them to market their high efficiency products, as the label provides unbiased evidence that their products are more efficient. In addition, they provide the foundation for market transformation programs by allowing policymakers to easily identify high efficiency products to target for bulk purchasing, financing, and incentives. Ensuring that energy labeling programs are effective in transforming markets requires that the label clearly and credibly differentiates high efficiency products for consumers and encourages manufacturers to innovate and upgrade their production lines.

This paper examines the AC labeling programs in India, the European Union, Brazil, Thailand, and Vietnam, as well as the evolution of the AC markets in these countries over the past 5 to 10 years, in order to identify crucial elements in labeling program design for influencing market evolution. These elements include effective label design, the frequency and magnitude of labeling tier revisions, public awareness of the label, and the efficiency metric used for the label. The comparison of these case studies provides empirical evidence on which approaches to labeling program design are most effective at moving markets to high efficiency ACs.

C. Taylor (✉) · M. Baton · A. M. Carreño · E. Gibbs
CLASP, Washington, DC, USA
e-mail: ctaylor@clasp.ngo

P. Bertoldi (ed.), *Energy Efficiency in Domestic Appliances and Lighting*, Springer Proceedings in Energy, https://doi.org/10.1007/978-3-030-79124-7_29

1 Introduction

Labeling programs that promote highly efficient products are a critical element of energy efficiency policy. They are a valuable tool for different stakeholders, including consumers, industry, government energy agencies, and other institutions interested in market transformation. While minimum energy performance standards (MEPS) remove the least-efficient products from the market, energy labels drive product markets to higher efficiency in three ways by:

- **Informing consumers as to which products have a higher or lower level of efficiency**, allowing them to make an informed decision about the trade-off between up-front and operating cost, as well as the environmental benefits of more efficient products. They can inform consumers by categorizing products into different efficiency tiers, identifying a product's efficiency relative to a continuous spectrum representing the range of efficiencies available in the market, or by endorsing the highest efficiency products.
- **Incentivizing manufacturers to produce more efficient products** by helping them to market their high efficiency products, as the label provides unbiased evidence that their products are more efficient. An official label can be viewed as an impartial signal that the product is highly efficient, and can therefore help to justify what may be a higher up-front cost. In this way, energy efficiency labeling can help appliance manufacturers increase their revenues by marketing products with higher upfront costs. The labels also encourage manufacturers to improve efficiency of their products, as these highly efficient products will be differentiated in the market; manufacturers use the labels from the top rated products as a marketing tool.
- **Providing the foundation for market transformation programs** by allowing policymakers to easily identify high efficiency products to target for bulk purchasing, financing, and incentives.

Ensuring that AC energy efficiency labels move markets in these three ways requires broad public awareness of the label, effective label design, periodic labeling tier revisions, and an appropriate efficiency metric for the label. This paper sheds light on what are viable and tested practices for maintaining an effective AC labeling program.

2 Case Studies

As of 2019, approximately 90 economies around the world have some form of energy labeling policy for ACs, showcasing a wide array of experiences and best practices for AC energy labeling policy design and revision. The case studies below reflect the experiences of some of the largest AC markets with significant local manufacturing.

2.1 *India*

The Indian Bureau of Energy Efficiency (BEE) launched the labeling program for fixed-speed ACs in 2006 as a voluntary initiative, and the program became mandatory in 2009. BEE revised the energy performance thresholds for ACs covered under the program on a biennial basis from 2010 to 2016. In 2015, BEE launched a voluntary labeling program for inverter ACs, and made the program mandatory in January 2018. The labeling program for ACs now covers both fixed and inverter units under a common rating plan. The increases in stringency have resulted in a substantial efficiency improvement of 35% to the MEPS for split units, which are the most popular type of AC. Since the inception of the AC labeling program, 46 TWh of electricity have been saved and 38 million tons of carbon emissions have been avoided.[1] The Indian energy efficiency label for ACs appears in Fig. 1.

Fig. 1 Indian energy efficiency label for split ACs

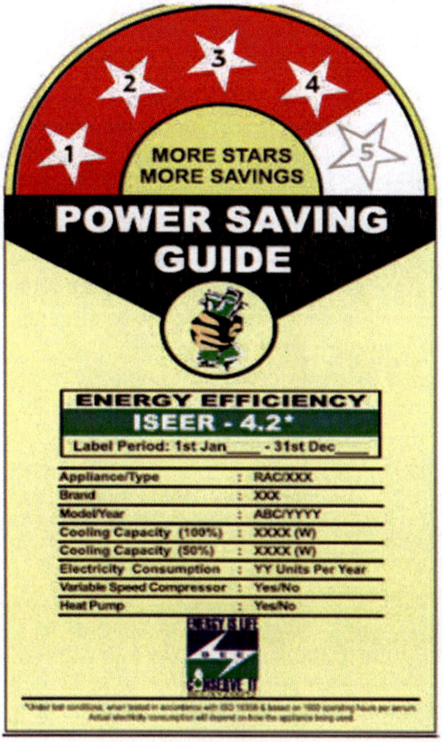

[1] The data and qualitative information contained in this case study has been gathered by the CLASP India office with support from BEE.

2.1.1 The Evolution of the Indian AC Labeling Program

BEE developed distinct star rating plans for split[2] and window/unitary type ACs and has been revising the star rating levels since launching the program, increasing the stringency of the energy performance thresholds. Table 1 shows the star rating plan for split AC units, which make up the majority of the Indian market. These revisions have been conducted based on analysis of the registered labeled products in BEE's database, with a view to ensuring that each star rating contains a meaningful share of the products available on the market. Such a distribution of products across all rating tiers allows consumers to clearly distinguish between the efficiency levels of the various available products.

While the revisions for window ACs have been limited due to the technological and size constraints inherent in window ACs, the split type ACs have periodically seen more substantial revisions. For example, the existing 5-Star level (more efficient) in 2009 became 3-Star in 2015 and 1-Star (least efficient) in 2018 as per new star levels and the new rating methodology.

As per Table 1, efficiency was originally measured in terms of the energy efficiency ratio (EER). Starting on a voluntary basis in 2016, BEE adopted an improved rating methodology that factors in variance in temperature across the various climatic zones in India and operating hours, and therefore captures the efficiency benefits of inverter ACs. Using a metric that captures the benefits of inverter technology is vital for improving energy efficiency, as inverter AC units are up to 51% more efficient than fixed speed units [1]. The new metric is called the Indian Seasonal Energy Efficiency Ratio (ISEER), which is the ratio of the total annual amount of heat that the equipment may remove from the indoor air when operated for cooling in active mode to the total annual amount of energy consumed by the equipment during the same period.

Table 1 Revisions in star rating levels for split ACs

Star level	1st January 2009 to 31st December 2011	1st January 2012 to 31st December 2013	1st January 2014 to 31st December 2015	1st January 2016 to 31st December 2017	1st January 2018 to 31st December 2019
	EER	EER	EER	EER	ISEER
1 Star	2.3	2.5	2.7	2.7	**3.1**
2 Star	2.5	2.7	2.9	2.9	3.3
3 Star	2.7	2.9	**3.1**	3.1	3.5
4 Star	2.9	3.1	3.3	3.3	4.0
5 Star	**3.1**	3.3	3.5	3.5	4.5

Source: BEE

[2] The split AC rating plan covers wall, ceiling, and floor-mounted ACs.

2.1.2 Impacts of the Indian Labeling Program

The Indian labeling program has driven a dramatic transformation of the Indian AC market over the last decade, as can be seen in Fig. 2. The production-weighted average EER/ISEER of ACs has increased from 2.8 W/W in 2011–2012 to 3.6 W/W in 2017–2018, which represents a 29% increase in efficiency due to the tightening of standards and the introduction of a labeling program for variable speed ACs in 2015.

The effect of the move to the ISEER in the Indian market has been tremendous. While inverter units only made up 1.9% of the market in 2012 and were only projected to make up 5.7% of the market by 2018, they came to make up 11% of the market in 2016 and around 30% of the market in 2017. This rapid growth in inverter market share was built on the foundation of the new test metric and supported by government and bulk procurements that specified high ISEER values only attainable by inverter units (Fig. 3) [2].

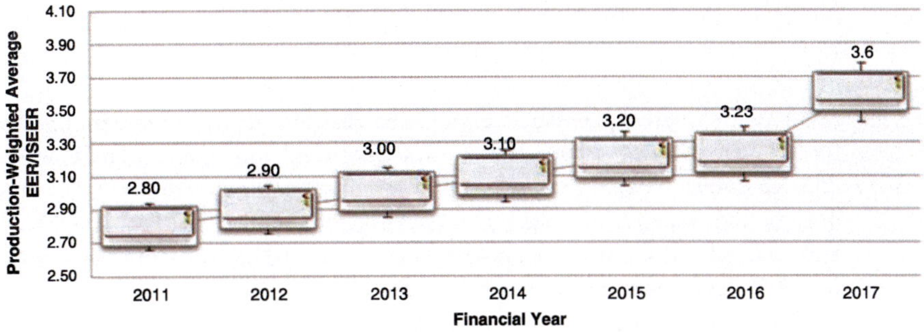

Fig. 2 Production-weighted average EER/ISEER of ACs, 2011–2017

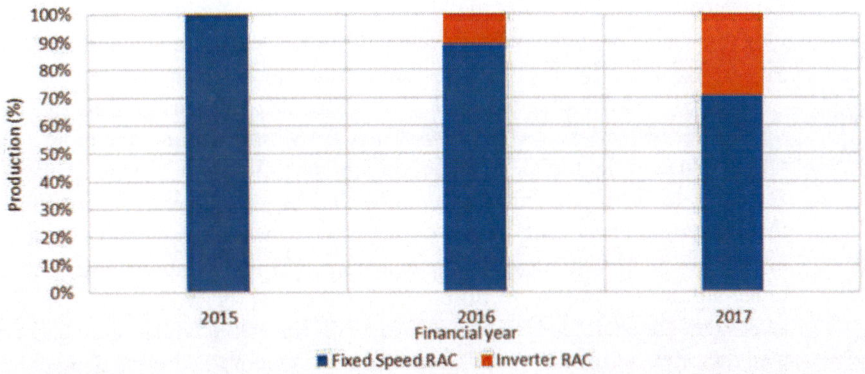

Fig. 3 Market share of fixed speed and inverter ACs

2.2 European Union

The European Union introduced categorical labeling for household appliances in 1992 with the *Council Directive 92/75/EEC of 22 September 1992 on the indication by labelling and standard product information of the consumption of energy and other resources by household appliances*. This directive established the A to G scale and the general design of the EU Energy Label as is still used today. After a few years of enforcement of the label, it became obvious that the highest efficiency classes for some products were already overpopulated whereas the lower classes were empty. The clustering of models in the top classes meant that the label no longer allowed for visible differentiation between products on the market. The scale of the label therefore needed to be revised in order to restore its ability to help consumers make well-informed purchasing choices. The discussion preceding the adoption of a new Directive in 2010 centred on the question of the rescaling. However, in view of the industry resistance to proceed to a complete rescale of the label, EU Member States decided to maintain the existing classes and add higher efficiency classes (A+, A++, and A+++). The 2010 Directive therefore did not trigger a rescale of the energy labels but rather an extension of the scale to those three additional classes.

In 2013, CLASP published a consumer research study [3] to examine the effectiveness of the new label design in supporting consumers in making informed choices about the energy efficiency of appliances during purchase. The study was designed to assess how consumers use, understand, and are motivated by the new and revised labels. This was achieved by holding 10 consumer focus groups and 30 in-depth interviews across ten cities in the EU.

Evidence from this study demonstrated that consumers understand both versions of the label and that both positively affect purchase decisions. It also showed some differences in appeal and understanding between the two versions. The new design was deemed more attractive and clear but the appeal of the best class compared to the rest of the scale appeared higher in the A to G scale than in the A+++ to D scale. Consumers were less likely to choose an A+++ model over an A model under the new regulation than they were to choose an A model over a D model under the previous regulation. See Fig. 4 for a comparison of the two labels.

The 2010 label was considered 'clearer' by 50% of focus groups, and 'less cluttered' and 'better designed' by 60%. The CLASP study found that more consumers would consider the middle class acceptable in an A+++ to D label scale than in an A to G scale. When selecting products from an A+++ to D scale, consumers declared that they would be willing to pay 44% more for the highest energy efficiency class as opposed to middle-range products, compared to 50% more for an A to G class. Other studies found a larger difference between the motivational power of the two scales (see for example Heinzle and Wüstenhagen [4], in which researchers dissociated the effect of the A+++ to D scale from the rest of the design changes).

The study also investigated other parts of the label, such as energy consumption per year, water consumption, icons and others. Detailed feedback from consumers

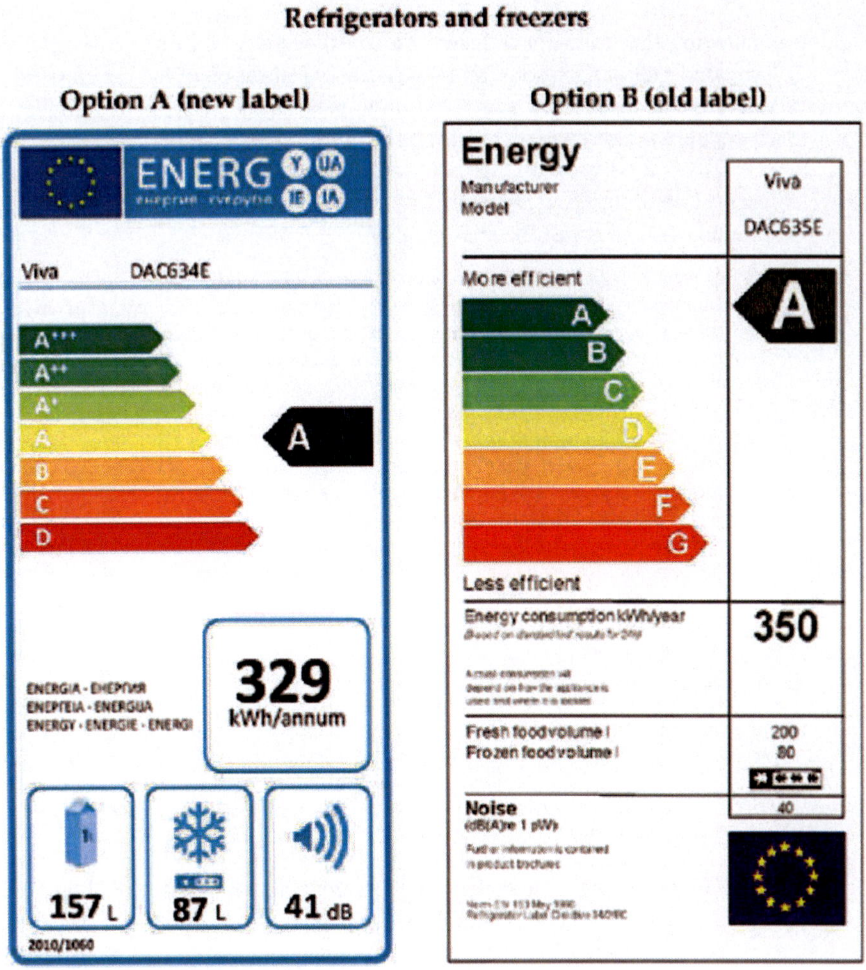

Fig. 4 2010 (left) and 1992 (right) versions of the EU energy label for refrigerators

was extremely useful to justify shifting back to the original A to G scale in the 2017 revision of the legal framework, as well as to improve the presentation of certain elements and identify what may have to be further investigated for each product group. The survey also highlighted the importance of improving communication around the energy label.

In addition, just 3 years after the revisions to the EU Energy Label in 2010, there was already a consensus that a new rescaling was needed and that adding plus-signs was not a long term solution. The European Environmental Bureau (EEB), the European Environmental Citizens Organisation for Standardisation (ECOS) and the European Committee of Domestic Equipment Manufacturers (CECED) co-authored a paper on this issue in 2013: *Revising EU energy label: evolution or revolution?*, confirming that even for industry, the 2010 revision with its scale extension only "provided a short term solution to the issue of saturation of the top classes" [5]. This paper also lays out a few suggested principles for the revision of the label, some of which were accepted by the European Commission.

In 2014, the European Commission published a report on the evaluation of the Energy Labelling Directive (initiated in 2012) [6]. The first priority identified by this study was to revise the energy label:

> *A key priority is the revision of the present energy label so that higher efficiency levels can be communicated in the future. This will help to ensure future relevance and effectiveness of the energy label. While a new label design will inevitably require a rebasing of the efficiency classes currently applied, consumer understanding should be the chief concern for future label revisions (…). It is also becoming increasingly clear that the A+ categories are less effective at attracting consumers to the higher classes than the A class on an A–G scale. The evolution of energy labels to the A+++ categories is one that has little support among stakeholders, and where there is an overwhelming recognition of the need for change. In addition, labels should also not show empty classes at the lower end of the scale without in some way indicating that they are no longer active. The possibility to display environmental information on the label should be maintained. Future options to explore in greater depth are the opportunities offered by ICT to convey additional information or provide electronic labels, display of lifecycle cost information, the development of guidelines for how to revise existing labels, an in-depth assessment of transition issues, as well as a number of advanced label design options.*

In 2017 the European Commission adopted a revised legal framework for the energy efficiency label [7]. Not only does this new framework restore the original A to G scale, but it also institutes rules about how efficiency classes shall be defined and revised in the future:

- Leaving top of the scale empty:
  - "Where a label is introduced or rescaled, the commission shall ensure that no products are expected to fall into energy class A at the moment of the introduction of the label and the estimated time within which a majority of models falls into that class is at least 10 years later.
  - By way of derogation […], where technology is expected to develop more rapidly, requirements shall be laid down so that no products are expected to fall into energy classes A and B at the moment of the introduction of the label."
- Rescale trigger: "the commission shall review the label with a view to rescaling if it estimates that:
  - 30% of the units of models belonging to a product group sold within the union market fall into the top energy efficiency class A and further technological development can be expected, or
  - 50% of the units of models belonging to a product group sold within the union market fall into the top two energy efficiency classes A and B and further technological development can be expected."

For ACs, the revisions to the labeling levels will be dramatic. According to the labeling tiers proposed by the EU Commission in May of 2018, though not yet adopted, the current 'A' class will become an 'F' class. The full revisions can be seen below, in Table 2 [8].

2.3 *Brazil*

There are two energy labels for electricity-consuming products in Brazil: (1) the mandatory Brazilian Labeling Program (PBE, for its initials in Portuguese) comparative label with categories from 'A' to 'C' or 'G,' depending on the product and (2) the voluntary Selo PROCEL endorsement label. The Brazilian National Metrology, Quality, and Technology Institute (INMETRO) manages the PBE. The Selo PROCEL, which was first launched for window air conditioners (ACs) in 1996, is managed by the Electricity Conservation Program (PROCEL) of the state-owned electricity generation and transmission company, Eletrobras. These two labels are closely interconnected; for example, any room AC that achieves the 'A' class on the PBE can also receive the Selo PROCEL (Fig. 5).

Consumers in Brazil respond to the labels, and manufacturers, importers, and retailers all recognize that products that do not attain an 'A' rating and the Selo PROCEL do not sell well. A 2015 study conducted by INMETRO found that 91% of consumers recognized the comparative label, 79.9% said they understood the label, and 68.3% said that they would pay 10% more for a product bearing the Selo PROCEL [9]. Because of the preference for 'A' rated products that bear the Selo PROCEL, many manufacturers seek to primarily or exclusively produce 'A' rated products, and some retailers only carry 'A' rated products.[3] The influence of the labeling program on manufacturers' production decisions is evident in the products available on the market; the most common efficiency level for a split AC is an energy efficiency ratio (EER) of 3.24 W/W, which is just above the 'A' class and

Table 2 Existing and proposed EU AC label tiers

Existing label tiers		Proposed label tiers	
Tier	SEER	Tier	SEER
A+++	SEER ≥ 8.50	A	SEER ≥ 11.5
A++	6.10 ≤ SEER < 8.50	B	9.7 ≤ SEER < 11.5
A+	5.60 ≤ SEER < 6.10	C	8.1 ≤ SEER < 9.7
A	**5.10 ≤ SEER < 5.60**	B	6.8 ≤ SEER < 8.1
B	4.60 ≤ SEER < 5.10	E	5.7 ≤ SEER < 6.8
C	4.10 ≤ SEER < 4.60	**F**	**4.8 ≤ SEER < 5.7**
D	3.60 ≤ SEER < 4.10	G	SEER < 4.8
E	3.10 ≤ SEER < 3.60		
F	2.60 ≤ SEER < 3.10		
G	SEER < 2.60		

[3] Based on interviews with manufacturers and retailers, conducted in August 2018.

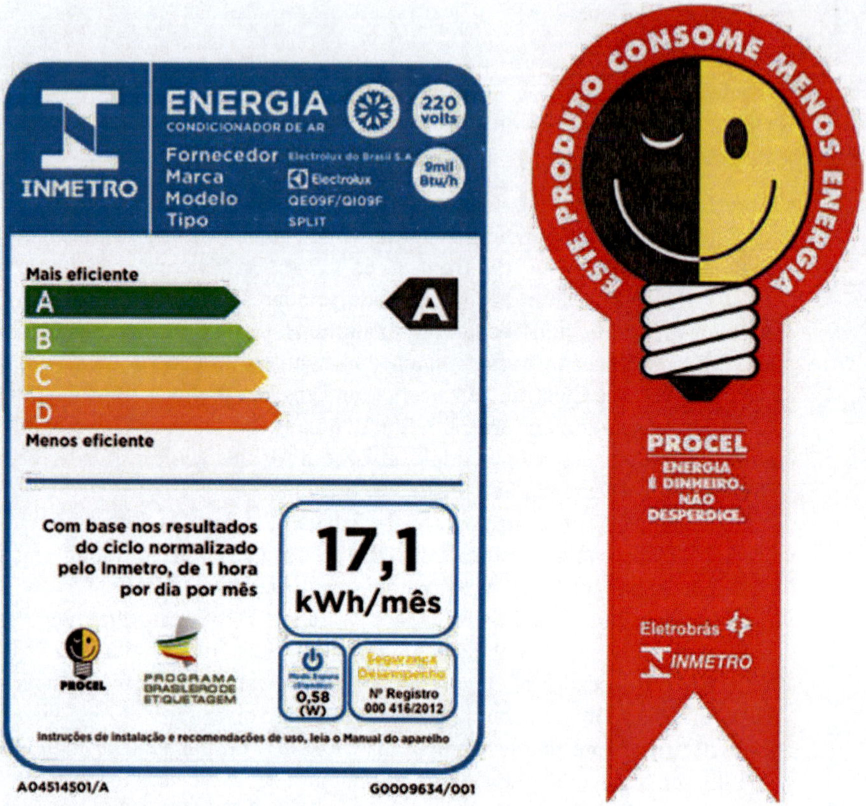

Fig. 5 Brazilian comparative label for ACs (left) and Selo PROCEL (right)

Selo PROCEL threshold of 3.23 W/W.[4] The two AC labels combined have had a significant impact in reducing energy demand in Brazil, saving an estimated 2 TWh of electricity in 2009 alone [10].

Despite the advantageous position of the PBE, resulting from its widespread recognition, the program has not, in recent years, been capitalizing on this position to drive the AC market to more efficient products. The criteria for the 'A' class label and Selo PROCEL for split ACs have been virtually unchanged over the past 10 years. As of 2019, only categories 'A' and 'B' may be sold in the market, as 'C' and below do not meet the current MEPS [11]. In addition, all 'A' class products are eligible for the voluntary Selo PROCEL endorsement label. This means that 77% of split ACs being sold in the market are now 'A' class and eligible for the Selo PROCEL. This has greatly reduced the value of both the PBE and Selo PROCEL, because neither clearly differentiates highly efficient products from average efficiency or even below average efficiency products. This lack of differentiation

[4] Based on data from the PBE product database (from 2004 to 2018).

has slowed the improvement of energy efficiency for split ACs, with the median efficiency only increasing 10.2% over 8 years, since 2010.[5]

2.4 *Southeast Asia*

In Southeast Asia, national labeling programs have been significantly influenced by regional initiatives. In particular, members of the Association of Southeast Asian Nations (ASEAN)[6] agreed in 2015 to harmonize their standards for ACs to a seasonal rating metric. This agreement has led to ASEAN countries moving to seasonal metrics that capture the efficiency benefits from inverter ACs. Two countries that have already made the shift to seasonal metrics for their labeling programs are Vietnam and Thailand. The experiences of these two countries show how shifting to a seasonal metric promotes a market shift to inverter technology. However, the differences in the two experiences show that the metric must be applied to all products in the same way in order to have maximum effect on the market.

2.4.1 Vietnam

Energy performance labeling is mandatory in Vietnam. Vietnam's Ministry of Industry and Trade (MOIT) oversees the energy labeling program. The Vietnamese energy label is a comparative label that provides star ratings from 1 to 5. The more stars an AC receives, the more efficient the model is. A certified energy label provides the following information: manufacturer's name, product origin, model number, rated power, energy efficiency, the relevant regulation, and certification number (Fig. 6).

In 2013, Vietnam began the process of moving to a seasonal metric by publishing a test standard based on international standards requiring the use of seasonal performance factors [12]. The use of this standard was voluntary for the first 2 years. However, the 2015 MEPS and labeling revision mandated the use of the cooling seasonal performance factor (CSPF) energy performance metric and extended the regulation to cover inverter ACs. Notably, under the previous 2012 standard, MEPS and labels applied to all ACs with capacities under 48,000 Btu/h; however, the 2015 standard only covers ACs up to 41,000 Btu/h [12].

The effect of moving to a seasonal metric is clear. Compared to market data from 2013, inverter market share has increased by approximately 31%, from 34% of the market to 65% of the market in 2018. Despite inverter technology now having a larger overall market share than fixed speed technology, fixed speed ACs are still

[5]Based on data from the PBE product database (from 2004 to 2018).

[6]ASEAN member states are Indonesia, Thailand, Malaysia, Singapore, Philippines, Vietnam, Cambodia, Myanmar (Burma), Brunei, and Laos.

Fig. 6 Vietnam energy label

more prevalent at cooling capacities above 36,000 Btu/h. The higher market share of fixed speed units at these higher capacities is likely because units over 41,000 Btu/h are not labeled and consumers, therefore, cannot readily identify the efficiency benefits of the inverter models. Without a label, manufacturers have no incentive to develop high-capacity inverter models for the Vietnamese market. The consumer preference for ACs with energy labels indicating higher efficiency is also clear, as high efficiency 5-star labeled products account for 54% of the market. This preference can also be seen in the fact that inverter ACs are more popular despite their slightly higher average prices [12].

2.4.2 Thailand

Thailand has maintained an energy efficiency labeling program for ACs since 1995. The label is voluntary, implemented by the Electricity Generating Authority of Thailand (EGAT), with five levels. Because the label is voluntary, manufacturers only choose to label products achieving the fifth labeling level (EGAT No. 5). The label is well recognized by Thai consumers and the vast majority of AC units sold on the Thai market are labeled EGAT No. 5 [13]. Notably, government procurement often requires that products have the EGAT No. 5 label (Fig. 7).

Fig. 7 2018 EGAT No. 5 labels for inverter (left) and fixed speed (right) ACs

The label levels have been revised several times since the program was launched. Until 2015, all AC units had their efficiency measured by EER. However, in 2015, Thailand began the move to harmonize its labeling tiers to a seasonal metric by introducing new label levels for inverter AC units, based on the seasonal energy efficiency ratio (SEER). This was followed by the 2017 revision of the label levels for fixed speed units, also based on SEER.

This movement to a single metric, SEER, has accompanied a significant increase in the market share of inverter ACs in Thailand. In 2013, inverter ACs accounted for 16% of the Thai AC market—this figure had increased to 32% by 2018. However, this increase is less dramatic than in Vietnam. A likely reason for the different results in the two countries is that Thailand has maintained different labeling tiers for fixed speed and inverter ACs, while Vietnam has moved to one set of labels for all AC technologies [13].

Maintaining different labeling requirements for fixed speed and inverter ACs has likely slowed the market transformation towards higher efficiency, inverter AC units. Thai consumers have a strong preference for EGAT No. 5 labeled products and the EGAT No. 5 label is often required for government procurement and bulk purchases by real estate developers. However, maintaining different labeling requirements levels for different technologies allows less efficient fixed speed ACs to continue to receive a No. 5 label. Eliminating this difference would result in few, if any, fixed speed ACs meeting the No. 5 label and would therefore lead the Thai market to move rapidly to inverter ACs if the strong preference for EGAT No. 5 labeled products continues [13].

3 Conclusion and Lessons Learned

The case studies above show important lessons about label public awareness, design, frequency of revisions, and the use of an efficiency metric that can drive markets to high efficiency ACs and achieve maximum program impacts.

3.1 *Public Awareness*

Label recognition and understanding are key for impactful AC labeling programs. Labels that communicate in a clear and transparent way to consumers the relative performance of equipment can effectively inform and promote the purchase of high efficiency ACs. In Brazil, the substantial impact of the labeling program on the AC market is due to widespread consumer awareness and response to the label, as consumers exhibit a strong preference for 'A' class products. As a result, manufacturers in Brazil seek to primarily sell such products. Similarly, the widespread public preference for the EGAT No. 5 label has led to most products on the Thai market achieving this label. However, the label is also confusing to consumers, as they use the presence of the label (as intended) as the criteria for purchasing a product, without widespread understanding of the different labeling criteria for the two main technologies (fixed and inverter), which leads them to purchase lower efficiency fixed speed units over higher efficiency inverter units. This problem could likely be resolved by eliminating the separate criteria for the two technologies.

3.2 *Label Design*

Label tiers must be designed in a way that encourages consumers to purchase high efficiency products. The challenge of ensuring effective label design can be addressed by consumer research to identify which labeling designs best motivate consumers to purchase highly efficient products. As can be seen from the European case, adding plus signs to the top tier did not encourage consumers to purchase the most efficient products. Consumers were willing to pay a larger premium for an (top-tier) 'A' product over a (lower-tier) D product than for an (top-tier) 'A+++' product over an (lower-tier) 'A,' product, even though the difference in efficiency was similar.

3.3 *Label Revisions and Rescaling*

In order for the labels to clearly differentiate high efficiency products, their efficiency criteria must be periodically revised as technologies improve and the market shifts. The approaches followed by the European Union or India are good practices for other countries to replicate. The European Union has chosen to allow more time between rescaling by leaving the top classes empty, and triggering rescaling when a percentage of the units of models sold within the union market fall into the top two energy efficiency classes A and B. India has chosen to make smaller revisions every 2 years, publishing a multi-year schedule in advance that allows manufactures time to adjust production lines.

Any revision should ensure that there are products falling into multiple categories so that consumers can identify a variety of different efficiencies in the market. This allows consumers who may not be able to afford the highest efficiency products to identify more affordable products that are more efficient than the lowest efficiency products in the market. This has proven important in India, where middle-rated products (3-Star ACs) dominate the market.

3.4 Efficiency Metric

With inverter technology becoming increasingly mainstream, it is essential that the metric used by the label capture the efficiency benefits of this technology. In countries such as India, Thailand, and Vietnam, the transition to seasonal performance metrics has led to a doubling or more of the market share of inverter ACs [14]. In India, this increase in inverter market share accounts for nearly half of the total improvement in the average efficiency of AC units sold.[7]

However, it is critically important that the label use a common rating plan for both inverter and fixed speed units. As can be seen from the case in Thailand, the impact of shifting to a seasonal metric has been decreased by the application of different requirements for the two main technologies (fixed and inverter ACs). A technology-neutral approach, as has been applied in Vietnam and India, would more quickly move markets to high efficiency ACs.

References

1. Yoon, M.S., J. H. Lim, T. S. M. Al Qahtani, Y.J. Nam. "Experimental Study on Comparison of Energy Consumption between Constant and Variable Speed Air-Conditioners in Two Different Climates." Proceedings of the 9th Asian Conference on Refrigeration and Air-conditioning. June 2018.
2. Mukherjee, P.K. Personal Interview. December 2018.
3. CLASP. "Assessing Consumer Comprehension of the EU Energy Label." 2013. Available online at: https://clasp.ngo/publications/assessing-consumer-comprehension-of-the-eu-energy-label
4. Heinzle S. and Wüstenhagen R., 2010. "Disimproving the European energy label's value for consumers?" Results from a consumer survey, University of St. Gallen, February 2010.

[7] Based on data collected by CLASP from the Bureau of Energy Efficiency product database.

5. Arditi, S., Toulouse, E., and Meli, L., "Revising EU energy label: evolution or revolution?" 2013. Available online at: http://www.ecostandard.org/wp-content/uploads/Revising-EU-energy-label-evolution-or-revolution.pdf
6. ECOFYS. "Final Technical Report: Evaluation of the Energy Labelling Directive and specific aspects of the Ecodesign Directive" 2014. Available online at: http://www.energylabelevaluation.eu/tmce/Final_technical_report-Evaluation_ELD_ED_June_2014.pdf
7. Regulation (EU) 2017/1369 of the European Parliament and of the Council of 4 July 2017 setting a framework for energy labelling and repealing Directive 2010/30/EU. Available online at: https://eur-lex.europa.eu/eli/reg/2017/1369/oj
8. Huang Baijia, Philippe Riviere, Peter Martin Skov Hansen, Jan Viegand, Hassane Asloune, Florian Dittmann. Air conditioners and comfort fans, Review of Regulation 206/2012 and 626/2011 Final report. [Research Report] European Commission, DG Energy. 2018.
9. BRACIER. "USO DE ETIQUETAS DE CONSUMO DE ENERGIA GERÁ ECONOMIA DE R$ 2,9 BI EM DEZ ANOS." 2015. Available online at: http://bracier.org.br/noticias/brasil/5288-uso-de-etiquetas-de-consumo-de-energia-gera-economia-de-r-2-9-bi-em-dez-anos
10. Balbino Cardoso, Rafael. "Estudo dos impactos energéticos dos Programas Brasileiros de Etiquetagem Energética: Estudo de caso em refrigeradores de uma porta, condicionadores de ar e motores elétricos." UNIVERSIDADE FEDERAL DE ITAJUBÁ. 2012.
11. Diário Oficial da União. "PORTARIA INTERMINISTERIAL N° 2, DE 31 DE JULHO DE 2018."
12. CLASP. Vietnam Room Air Conditioner Market Assessment and Policy Options Analysis. June 2019.
13. CLASP. Thailand Room Air Conditioner Market Assessment and Policy Options Analysis. June 2019.
14. CLASP. "Cooling in a Warmer World." January 2019. Available online here: https://issuu.com/claspngo/docs/clasp_-_cooling_in_warming_world?e=0